HANDBOOK OF DEFEASIBLE REASONING AND UNCERTAINTY MANAGEMENT SYSTEMS

VOLUME 4

HANDBOOK OF DEFEASIBLE REASONING AND UNCERTAINTY MANAGEMENT SYSTEMS

EDITORS:

DOV M. GABBAY
King's College, London, U.K.

PHILIPPE SMETS
IRIDIA - Université Libre de Bruxelles, Belgium

Volume 1: Quantified Representation of Uncertainty and Imprecision
Edited by P. Smets
Volume 2: Reasoning with Actual and Potential Contradictions
Edited by P. Besnard and A. Hunter
Volume 3: Belief Change
Edited by D. Dubois and H. Prade
Volume 4: Abductive Reasoning and Learning
Edited by Dov M. Gabbay and Rudolf Kruse
Volume 5: Algorithms for Uncertainty and Defeasible Reasoning
Edited by Jürg Kohlas and Serafín Moral

HANDBOOK OF DEFEASIBLE REASONING AND UNCERTAINTY MANAGEMENT SYSTEMS

VOLUME 4

ABDUCTIVE REASONING AND LEARNING

Volume Editors:

DOV M. GABBAY
King's College, London, U.K.

and

RUDOLF KRUSE
Otto-von-Guericke-University, Magdeburg, Germany

SPRINGER-SCIENCE+BUSINESS MEDIA, B.V.

A C.I.P. Catalogue record for this book is available from the Library of Congress.

ISBN 978-90-481-5560-6 ISBN 978-94-017-1733-5 (eBook)
DOI 10.1007/978-94-017-1733-5

Printed on acid-free paper

CONTENTS

PETER A. FLACH AND ANTONIS C. KAKAS

ON THE RELATION BETWEEN ABDUCTION AND INDUCTIVE LEARNING

1 INTRODUCTION

This Handbook volume is devoted to abduction and learning as they appear in various subfields of artificial intelligence. Broadly speaking, abduction aims at finding explanations for, or causes of, observed phenomena or facts. Learning occurs when an agent adapts its state or behaviour on the basis of experience. Both processes have a number of common characteristics. They both aim at improving their picture or model of the universe of discourse; they are both hypothetical, in the sense that the results may be wrong; and they can both be seen as reasoning processes, with the observations and the current knowledge of the world as input statements and the learned or abduced hypotheses as output statements.[1] In the case of learning from examples, which is the most common form of learning studied in artificial intelligence, this form of reasoning is called induction, and that is the term we will be mostly using in this chapter.

Given these common characteristics of abduction and induction, it makes sense to study them together. Once we have a clear picture of their similarities as well as their differences, understanding one contributes to the understanding of the other. Such an integrated study is the subject of this introductory chapter. As abduction and induction have been studied in philosophy and logic as well as artificial intelligence, we review selected approaches from each of these disciplines before attempting to come up with an integrated perspective.

Even a casual investigation of the literature on abduction and induction will reveal that there is a significant amount of controversy and debate. For instance, Josephson writes that 'it is possible to treat every good (...) inductive generalisation as an instance of abduction' [Josephson, 1994, p.19], while Michalski has it that 'inductive inference was defined as a process of generating descriptions that imply original facts in the context of background knowledge. Such a general definition includes inductive generalisation and abduction as special cases' [Michalski, 1987, p.188].

One can argue that such incompatible viewpoints indicate that abduction and induction themselves are not well-defined. Once their definitions have been fixed, studying their relation becomes a technical rather than a conceptual matter. However, it is not self-evident why there should exist absolute, Platonic ideals of abduction and induction, waiting to be discovered and captured once and for all by an appropriate definition. As with most theoretical notions, it is more a matter

[1] We exclude non-declarative learning systems such as neural networks, which learn without explicit symbolic representations.

D. M. Gabbay and R. Kruse (eds.), Abductive Reasoning and Learning, 1–33.

of pragmatics, of how useful a particular definition is going to be in a particular context. Sometimes we want to stress that abduction and induction spring from a common root (say hypothetical or non-deductive reasoning), and sometimes we want to take a finer grained perspective by looking at what distinguishes them (e.g. the way in which the hypothesis extends our knowledge).

As already remarked, the relation between abduction and induction is subject of an ongoing debate, and one of our aims in this introductory chapter is to chart the terrain where the debate is taking place. We will retrace some of the main issues in this debate to their historical background. We will also attempt a synthesis of some of these issues, primarily motivated by work in artificial intelligence.

The outline of this chapter is as follows. In Section 2 we discuss the philosophical and logical origins of abduction and induction. In Section 3 we analyse previous work on abduction and induction in the context of logic programming and artificial intelligence, and attempt a (partial) synthesis of this work. Section 4 considers the integration of abduction and induction in artificial intelligence. In Section 5 we briefly survey the other chapters in this volume, indicating how they perceive abduction and induction. Finally, Section 6 concludes.

2 ABDUCTION AND INDUCTION IN PHILOSOPHY AND LOGIC

In this section we discuss various possible viewpoints on abduction and induction that can be found in the philosophical and logical literature.The philosophical issue is mainly one of categorisation (which forms of reasoning exist?), while the logical issue is one of formalisation.

As far as categorisation is concerned, it seems uncontroversial that deduction should be singled out as a separate reasoning form which is fundamentally different from any other form of reasoning by virtue of its truth-preserving nature. The question, then, is how non-deductive reasoning should be mapped out. One school of thought holds that no further sub-categorisation is needed: all non-deductive logic is of the same category, which is called induction. Another school of thought argues for a further division of non-deductive reasoning into abduction and induction. We will discuss these two viewpoints in the next two sections.

2.1 Induction as non-deductive reasoning

Let us start by taking a look at a textbook definition of induction.

> Arguments can be classified in terms of whether their premisses provide (1) conclusive support, (2) partial support, or (3) only the appearance of support (that is, no real support at all.) When we say that the premisses provide conclusive support for the conclusion, we mean that if the premisses of the argument were all true, it would be impossible for the conclusion of the argument to be false. Arguments that

> have this characteristic are called deductive arguments. When we say that the premisses of an argument provide partial support for the conclusion, we mean that if the premisses were true, they would give us good reasons—but not conclusive reasons—to accept the conclusion. That is to say, although the premisses, if true, provide some evidence to support the conclusion, the conclusion may still be false. Arguments of this type are called inductive arguments. [Salmon, 1984, p.32]

This establishes a dichotomy of the set of non-fallacious arguments into either deductive or inductive arguments, the distinction being based on the way they are supported or justified: while deductive support is an absolute notion, inductive support must be expressed in relative (e.g. quantitative) terms.

Salmon further classifies inductive arguments into arguments based on samples, arguments from analogy, and statistical syllogisms. Arguments based on samples or *inductive generalisations* have the following general form:

> X percent of observed Fs are Gs;
> therefore, (approximately) X percent of all Fs are Gs.

Arguments from analogy look as follows:

> Objects of type X have properties F, G, H, ...;
> objects of type Y have properties F, G, H, ..., and also property Z;
> therefore, objects of type X have property Z as well.

Finally, *statistical syllogisms* have the following abstract form:

> X percent of all Fs are Gs;
> a is an F;
> therefore, a is a G.

Here X is understood to be a high percentage (i.e. if X is close to zero, the conclusion must be changed to 'a is not a G').

There are several important things to note. One is that some premisses and conclusions are statistical, talking about relative frequencies ('X percent of'), while others are categorical. In general, we can obtain a categorical special case from arguments involving a relative frequency X by putting $X = 100\%$. Obviously, the categorical variant of statistical syllogism is purely deductive. More importantly, *categorical inductive generalisation* has the following form:

> All observed Fs are Gs;
> therefore, all Fs are Gs.

As argued in Section 2.3, most inductive arguments in artificial intelligence are categorical, as this facilitates further reasoning with the inductive conclusion.

Regardless of whether inductive arguments are statistical or categorical, we must have a way to assess their strength or inductive support, and this is the second way in which statistics comes into play. Given evidence E collected in the premisses of an inductive argument, we want to know the degree of belief we should attach to the hypothetical conclusion H. It is widely believed that degrees of belief should be quantified as (subjective) probabilities – in particular, the degree of belief in H given E is usually identified with the conditional probability $P(H|E)$. The probabilistic formalisation of inductive support is known as *confirmation theory*.

It is tempting to consider the degree of confirmation of hypothesis H by evidence E as the degree of validity of the inductive argument 'E, therefore H', and treat this 'inductive validity' as analogous to deductive validity. Following this line of thought, several authors speak of confirmation theory as establishing an 'inductive logic':

> 'What we call inductive logic is often called the theory of nondemonstrative or nondeductive inference. Since we use the term 'inductive' in the wide sense of 'nondeductive', we might call it the theory of inductive inference... However, it should be noticed that the term 'inference' must here, in inductive logic, not be understood in the same sense as in deductive logic. Deductive and inductive logic are analogous in one respect: both investigate logical relations between sentences; the first studies the relation of [entailment], the second that of degree of confirmation which may be regarded as a numerical measure for a partial [entailment]... The term 'inference' in its customary use implies a transition from given sentences to new sentences or an acquisition of a new sentence on the basis of sentences already possessed. However, only deductive inference is inference in this sense.' [Carnap, 1950, §44B, pp.205–6]

In other words, confirmation theory by itself does not establish a consequence relation (a subset of $L \times L$, where L is the logical language), since any evidence will confirm any hypothesis to a certain degree. Inductive logic based on confirmation theory does not have a proof theory in the traditional sense, and therefore does not guide us in generating possible inductive hypotheses from evidence, but rather evaluates a given hypothesis against given evidence. The inductive logic arising from confirmation theory is a logic of hypothesis *evaluation* rather than hypothesis *generation*. This distinction between hypothesis generation and hypothesis evaluation is an important one in the present context, and we will have more to say about the issue in Sections 2.3 and 3.

To summarise, one way to categorise arguments is by dividing them into non-defeasible (i.e. deductive) and defeasible but supported (i.e. inductive) arguments. A further sub-categorisation can be obtained by looking at the syntactic form of the argument. Confirmation theory quantifies inductive support in probabilistic terms, and deals primarily with hypothesis evaluation.

2.2 Deduction, induction and abduction

After having discussed the view that identifies induction with all non-deductive reasoning, we next turn to the trichotomy of deductive, inductive and abductive reasoning proposed by the American philosopher Charles Sanders Peirce (1839–1914).

Peirce was a very prolific thinker and writer, but only a fraction of his work was published during his life. His collected works [Peirce, 1958][2] therefore reflect, first and foremost, the evolution of his thinking, and should be approached with some care. With respect to abduction and induction Peirce went through a substantial change of mind during the decade 1890 – 1900 [Fann, 1970]. It is perhaps fair to say that many of the current controversies surrounding abduction seem to be attributable to Peirce's mindchange. Below we will briefly discuss both his early, *syllogistic* theory, which can be seen as a precursor to the current use of abduction in logic programming and artificial intelligence, and his later, *inferential* theory, in which abduction represents the hypothesis generation part of explanatory reasoning.

Peirce's syllogistic theory

In Peirce's days logic was not nearly as well-developed as it is today, and his first attempt to classify arguments (which he considers 'the chief business of the logician' (2.619), follows Aristotle in employing syllogisms. The following syllogism is known as Barbara:

> All the beans from this bag are white;
> these beans are from this bag;
> therefore, these beans are white.

The idea is that this valid argument represents a particular instantiation of a reasoning scheme, and that any alternative instantiation represents another argument that is likewise valid. Syllogisms should thus be interpreted as argument schemas.

Two other syllogisms are obtained from Barbara if we exchange the conclusion (or Result, as Peirce calls it) with either the major premiss (the Rule) or the minor premiss (the Case):

> *Case.* — These beans are from this bag.
> *Result.* — These beans are white.
> *Rule.* — All the beans from this bag are white.

> *Rule.* — All the beans from this bag are white.
> *Result.* — These beans are white.
> *Case.* — These beans are from this bag.

[2]References to Peirce's collected papers take the form $X.Y$, where X denotes the volume number and Y the paragraph within the volume.

The first of these two syllogisms (inference of the rule from the case and the result) can be recognised as what we called previously a categorical inductive generalisation, generalising from a sample of beans to the population of beans in the bag. The sort of inference exemplified by the second syllogism (inference of the case from the rule and the result) Peirce calls making a hypothesis or, briefly, hypothesis – the term 'abduction' is introduced only in his later theory.[3]

Peirce thus arrives at the following classification of inference (2.623):

$$\text{Inference}\begin{cases}\text{Deductive or Analytic}\\ \text{Synthetic}\begin{cases}\text{Induction}\\ \text{Hypothesis}\end{cases}\end{cases}$$

Comparing this classification with the one obtained in Section 2.1, we can point out the following similarities. That what was called induction previously corresponds to what Peirce calls synthetic inference (another term he uses is ampliative reasoning, since it amplifies, or goes beyond, the information contained in the premisses). Furthermore, what Peirce calls induction corresponds to what we called inductive generalisation in Section 2.1.[4]

On the other hand, the motivations for these classifications are quite different in each case. In Section 2.1 we were concentrating on the different kinds of support or confirmation that arguments provide, and we noticed that this is essentially the same for all non-deductive reasoning. When we concentrate instead on the syllogistic form of arguments, we find this to correspond more naturally to a trichotomy, separating non-deductive reasoning into two subcategories. As Horn clause logic is in some sense a modern upgrade of syllogistic logic, it is perhaps not surprising that the distinction between abduction and induction in logic programming follows Peirce's syllogistic classification to a large extent. This will be further taken up in Section 3.

Peirce's inferential theory

In his later theory of reasoning Peirce abandoned the idea of a syllogistic classification of reasoning:

> '(...) I was too much taken up in considering syllogistic forms and the doctrine of logical extension and comprehension, both of which I made more fundamental than they really are. As long as I held that opinion, my conceptions of Abduction necessarily confused two different kinds of reasoning.' [Peirce, 1958, 2.102, written in 1902]

[3]Peirce also uses the term 'retroduction', a translation of the Greek word απαγωγή used by Aristotle (translated by others as 'reduction').

[4]It should be noted that, although the above syllogistic arguments are all categorical, Peirce also considered statistical versions.

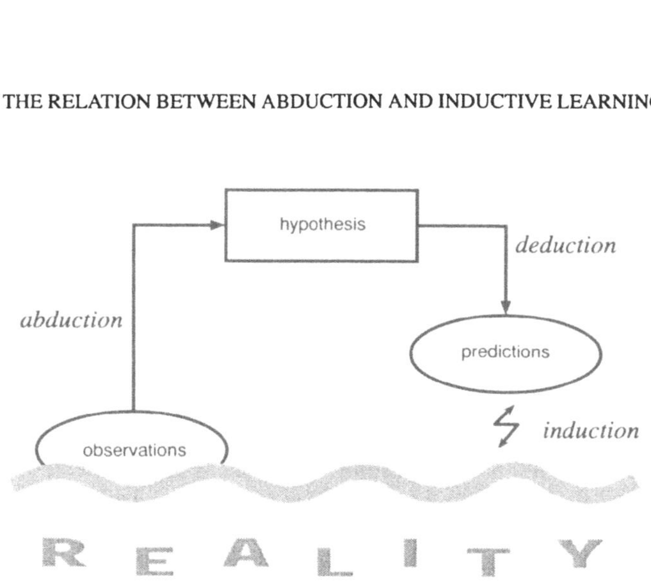

Figure 1. The three stages of scientific inquiry.

Instead, he identified the three reasoning forms – abduction, deduction and induction – with the three stages of scientific inquiry: hypothesis generation, prediction, and evaluation (Figure 1). The underlying model of scientific inquiry runs as follows. When confronted with a number of observations she seeks to explain, the scientist comes up with an initial hypothesis; then she investigates what other consequences this theory, were it true, would have; and finally she evaluates the extent to which these predicted consequences agree with reality. Peirce calls the first stage, coming up with a hypothesis to explain the initial observations, abduction; predictions are derived from a suggested hypothesis by deduction; and the credibility of that hypothesis is estimated through its predictions by induction. We will now take a closer look at these stages.

Abduction is defined by Peirce as the process of forming an explanatory hypothesis from an observation requiring explanation. This process is not algorithmic: 'the abductive suggestion comes to us like a flash. It is an act of insight, although of extremely fallible insight' [Peirce, 1958, 5.181]. Elsewhere Peirce describes abduction as 'a capacity for 'guessing' right', a 'mysterious guessing power' underlying all scientific research [Peirce, 1958, 6.530]. Its non-algorithmic character notwithstanding, abduction

> 'is logical inference (...) having a perfectly definite logical form. (...) Namely, the hypothesis cannot be admitted, even as a hypothesis, unless it be supposed that it would account for the facts or some of them. The form of inference, therefore, is this:
>
> The surprising fact, C, is observed;
> But if A were true, C would be a matter of course,

> Hence, there is reason to suspect that A is true.' [Peirce, 1958, 5.188–9]

Let us investigate the logical form of abduction given by Peirce a little closer. About C we know two things: that it is true in the actual world, and that it is surprising. The latter thing can be modelled in many ways, one of the simplest being the requirement that C does not follow from our other knowledge about the world.

Then, 'if A were true, C would be a matter of course' is usually interpreted as 'A logically entails C'.[5] Peirce calls A an explanation of C, or an 'explanatory hypothesis'. Whether or not this is an appropriate notion of explanation remains an issue of debate. See [Console and Saitta, 2000] for arguments supporting explanation as entailment, and [Josephson, 2000] for counterarguments.

Besides being explanatory, Peirce mentions two more conditions to be fulfilled by abductive hypotheses: they should be capable of experimental verification, and they should be 'economic'. A hypothesis should be experimentally verifiable, since otherwise it cannot be evaluated inductively. Economic factors include the cost of verifying the hypothesis, its intrinsic value, and its effect upon other projects [Peirce, 1958, 7.220]. In other words, economic factors are taken into account when choosing the best explanation among the logically possible ones. For this reason, abduction is often termed 'inference to the best explanation' [Lipton, 1991].

Induction is identified by Peirce as the process of testing a hypothesis against reality through selected predictions. 'Induction consists in starting from a theory, deducing from it predictions of phenomena, and observing those phenomena in order to see how nearly they agree with the theory' [Peirce, 1958, 5.170]. Such predictions can be seen as experiments:

> 'When I say that by inductive reasoning I mean a course of experimental investigation, I do not understand experiment in the narrow sense of an operation by which one varies the conditions of a phenomenon almost as one pleases. (...) An experiment (...) is a question put to nature. (...) The question is, Will this be the result? If Nature replies 'No!' the experimenter has gained an important piece of knowledge. If Nature says 'Yes,' the experimenter's ideas remain just as they were, only somewhat more deeply engrained.' [Peirce, 1958, 5.168]

This view of hypothesis testing is essentially what is called the 'hypothetico-deductive method' in philosophy of science [Hempel, 1966]. The idea that a verified prediction provides further support for the hypothesis is very similar to the notion of confirmation as discussed in Section 2.1, and also refutation of hypotheses through falsified predictions can be brought in line with confirmation theory,

[5] Note that interpreting the second premiss as a material implication, as is sometimes done in the literature, renders it superfluous, since the truth of $A \rightarrow C$ follows from the truth of the observation C.

with a limiting degree of support of zero.[6] The main difference from confirmation theory is that in the Peircean view of induction the hypothesis is, through the predictions, tested against *selected* pieces of evidence only. This leads to a restricted form of hypothesis evaluation, for which we will use the term *hypothesis testing.*

Peirce's inferential theory makes two main points. It posits a separation between hypothesis generation and hypothesis evaluation; and it focuses attention on hypotheses that can explain and predict. Combining the two points, abduction is the process of generating explanatory hypotheses (be they general 'rules' or specific 'cases', as in the syllogistic account), and induction corresponds to the hypothetico-deductive method of hypothesis testing. However, the two points are relatively independent: e.g., we can perceive generation of non-explanatory hypotheses. We will come back to this point in the discussion below.

2.3 Discussion

In the previous two sections we have considered three philosophical and logical perspectives on how non-deductive reasoning may be categorised: the inductivist view, which holds that no further categorisation is needed since all non-deductive reasoning must be justified in the same way by means of confirmation theory; the syllogistic view, which distinguishes between inductive generalisation on the one hand and hypothesis or abduction as inference of specific 'cases' on the other; and the inferential view, which holds that abduction and induction represent the hypothesis generation and evaluation phases in explanatory reasoning. As we think that none of these viewpoints provides a complete picture, there is opportunity to come to a partial synthesis.

Hypothesis generation and hypothesis evaluation

The most salient point of Peirce's later, inferential theory is the distinction between hypothesis generation and hypothesis evaluation. In most other accounts of non-deductive reasoning the actual hypothesis is already present in the argument under consideration, as can be seen clearly from the argument forms discussed in Section 2.1. For instance, when constructing an inductive generalisation

> X percent of observed Fs are Gs;
> therefore, (approximately) X percent of all Fs are Gs.

our job is first to conjecture possible instantiations of F and G (hypothesis generation), and then to see whether the resulting argument has sufficient support (hypothesis evaluation).

One may argue that a too rigid distinction between generation and evaluation of hypotheses is counter-productive, since it would lead to the generation of many, ultimately useless hypotheses. Indeed, Peirce's 'economic factors', to be considered

[6] From a Bayesian perspective $P(H|E)$ is proportional to $P(E|H)P(H)$, where $P(H)$ is the prior probability of the hypothesis; if E is contrary to a prediction $P(E|H) = 0$.

when constructing possible abductive hypotheses, already blur the distinction to a certain extent. However, even if a too categorical distinction may have practical disadvantages, on the conceptual level the dangers of confusing the two processes are much larger. Furthermore, the distinction will arguably be sharper drawn in artificial reasoning systems than it is in humans, just as chess playing computers still have no real alternative to finding useful moves than to consider all possible ones.

In any case, whether tightly integrated or clearly separated, hypothesis generation and hypothesis evaluation have quite distinct characteristics. Here we would argue that it is hypothesis generation, being concerned with possibilities rather than choices, that is most inherently 'logical' in the traditional sense. Deductive logic does not help the mathematician in selecting theorems, only in distinguishing potential theorems from fallacious ones. Also, as [Hanson, 1958] notes, if hypothesis evaluation establishes a logic at all, then this would be a 'Logic of the Finished Research Report' rather than a 'Logic of Discovery'. An axiomatic formalisation of the logic of hypothesis generation is suggested by Flach in his chapter in this volume.

We also stress the distinction between generation and evaluation because it provides a useful heuristic for understanding the various positions of participants in the debate on abduction and induction. This rule of thumb states that *those concentrating on generating hypotheses tend to* distinguish *between non-deductive forms of reasoning; those concentrating on evaluating hypotheses tend* not *to distinguish between them.*

Inductive generalisation

Turning next to the question 'What is induction?', we expect that any form of consensus will centre around the argument form we called inductive generalisation (see above). In the inductivist approach such sample-to-population arguments were separated out on syntactic grounds. They also figured in Peirce's syllogistic theory as one of the two possible reversals of Barbara.

As we remarked above, hypothesis generation here amounts to instantiating F and G. In general the number of possibilities is large, but it can be reduced by constraining the proportion X. Many artificial intelligence approaches to induction actually choose F and G such that X is (close to) 100%, thereby effectively switching to categorical inductive generalisations:

> All observed Fs are Gs;
> therefore, all Fs are Gs.

For instance, instead of observing that 53% of observed humans are female, such approaches will continue to refine F until all observed Fs are female (for instance, F could be 'humans wearing a dress').

The point here is not so much that in artificial intelligence we are only interested in infallible truths. Often, we have to deal with uncertainties in the form of noisy data, exceptions to rules, etc. Instead of representing these uncertainties explicitly in the form of relative frequencies, one deals with them semantically, e.g. by attaching a degree of confirmation to the inductive conclusion, or by interpreting rules as defaults.

The above formulation of categorical inductive generalisation is still somewhat limiting. The essential step in any inductive generalisation is the extension of the universal quantifier's scope from the sample to the population. Although the universally quantified sentence is frequently a material implication, this need not be. A more general form for categorical inductive generalisation would therefore be:

All objects in the sample satisfy $P(x)$;
therefore, all objects in the population satisfy $P(x)$.

where $P(x)$ denotes a formula with free variable x. Possible instantiations of $P(x)$ can be found by pretending that there exist no other objects than those in the sample, and looking for true universal sentences. For instance, we might note that every object in the sample is either female or male.

Confirmatory and explanatory induction

This more comprehensive formulation of categorical inductive generalisation also indicates a shortcoming of Peirce's inferential theory: not all hypotheses are explanatory. For instance, take the inductive generalisation 'every object in the population is female or male'. This generalisation does not, by itself, explain that Maria is female, since it requires the additional knowledge that Maria is not male. Likewise, an explanation of John being male is only obtained by adding that John is not female. This phenomenon is not restricted to disjunctive generalisations: the rule 'every parent of John is a parent of John's brother' does not explain parenthood.

In line with recent developments in inductive logic programming, we would like to suggest that inductive generalisations like these are not explanatory at all. They simply are generalisations that are confirmed by the sample. The process of finding such generalisations has been called *confirmatory* induction (also *descriptive* induction). The difference between the two forms of induction can be understood as follows. A typical form of explanatory induction is concept learning, where we want to learn a definition of a given concept C in terms of other concepts. This means that our inductive hypotheses are required to explain (logically entail) why particular individuals are Cs, in terms of the properties they have.

However, in the more general case of confirmatory induction we are not given a fixed concept to be learned. The aim is to learn relationships between any of the concepts, with no particular concept singled out. The formalisation of confirmatory hypothesis formation thus cannot be based on logical entailment, as in Peirce's abduction. Rather, it is a qualitative form of degree of confirmation, which explains its name. We will have more to say about the issue in Section 3.2.

Abduction

Turning next to abduction, it may seem at first that Peirce's syllogistic and inferential definitions are not easily reconcilable. However, it is possible to perceive a similarity between the two when we notice that the early syllogistic view of abduction or hypothesis (p. 5) provides a special form of explanation. The Result (taking the role of the observation) is explained by the Case in the light of the Rule as a given theory. The syllogistic form of abduction can thus be seen to meet the explanatory requirement of the later inferential view of abduction. Hence we can consider explanation as a characterising feature of abduction. This will be further discussed in Section 3.2.

Even if the syllogistic and inferential view of abduction can thus be reconciled, it is still possible to distinguish between approaches which are primarily motivated by one of the two views. The syllogistic account of abduction has been taken up, by and large, in logic programming and other work in artificial intelligence addressing tasks such as that of diagnosis and planning.

3 ABDUCTION AND INDUCTION IN LOGIC PROGRAMMING AND ARTIFICIAL INTELLIGENCE

In this section, we will examine how abduction and induction appear in the field of artificial intelligence (AI) and its specific subfield of logic programming. In Section 3.1 we will argue that in these fields abduction and induction are generally perceived as distinct reasoning forms, mainly because they are used to solve different tasks. Consequently, most of what follows should interpreted from the viewpoint of Peirce's earlier, syllogistic theory. In Section 3.2 we argue that abductive hypotheses primarily provide explanations, while inductive hypotheses provide generalisations. We then further investigate abduction and induction from a logical perspective in Section 3.3, pointing out differences in the way in which they extend incomplete theories. In Section 3.4 we investigate how more complex reasoning patterns can be viewed as being built up from simple abductive and inductive inferences. Finally, in Section 3.5 we address the computational characteristics of abduction and induction.

3.1 A task-oriented view

In AI the two different terms of abduction and induction exist separately and are used by different communities of researchers. This gives the impression that two distinct and irreducible forms of non-deductive reasoning exist. We believe this separation to be caused by the fact that in AI, irrespective of the level at which we are examining the problem, we are eventually interested in tackling particular tasks such as planning, diagnosis, learning, and language understanding. For instance, a prototypical AI application of abductive reasoning is the problem of diagnosis. Here abduction is used to produce a reason, according to some known theory of a system, for the observed (often faulty) behaviour of the system. A typical inductive task, on the other hand, is the problem of concept learning from examples. From a collection of observations which are judged according to some background information to be similar or related we draw hypotheses that generalise this observed behaviour to other as yet unseen cases.

What distinguishes this AI view from the philosophical and logical analyses discussed in the previous section is the more practical perspective required to tackle these tasks. Hence in AI it is necessary to study not only the issue of hypothesis evaluation but also the problem of hypothesis generation, taking into account the specific characteristics of each different task. These tasks require different effects from the non-deductive reasoning used to address them, resulting in different kinds of hypotheses, generated by different computational methods. As we will argue in Section 3.2, abductive hypotheses are primarily intended to provide explanations and inductive hypotheses aim at providing generalisations of the observations.

The point we want to stress here is that in AI hypothesis generation is a real issue, while in philosophy and logic it often seems to be side-stepped since the analysis usually assumes a given hypothesis. Since abduction and induction produce different kinds of hypotheses, with different relations to the observations and the background theory, it seems natural that this increased emphasis on hypothesis generation reinforces the distinguishing characteristics of the two reasoning forms. However, despite this emphasis on hypothesis generation in AI it is not possible to avoid the problem of hypothesis evaluation and selection amongst several possible alternatives. Returning to this problem we see that work in AI where the emphasis lies on hypotheses selection tends to conclude that the two forms of reasoning are not that different after all – they use the same kind of mechanism to arrive at the conclusion.

Peirce revisited

AI's emphasis on solving practical tasks notwithstanding, most research is still aimed at providing general solutions in the form of abductive and inductive engines that can be applied to specific problems by providing the right domain knowledge and setting the right parameters. In order to understand what these systems are doing, it is still necessary to use abstract (logical) specifications. Let us examine this more closely, using the case of logic programming and its two extensions of abductive and inductive logic programming.

Logic programming assumes a normal form of logical formulae, and therefore has a strong syllogistic flavour. Consequently, the logic programming perception of abduction and induction essentially follows Peirce's earlier, syllogistic characterisation. Here are Peirce's two reversals of the syllogism Barbara, recast in logic programming terms:

Case. — `from_this_bag(b).`
Result. — `white(b).`

Rule. — `white(X):-from_this_bag(X).`

Rule. — `white(X):-from_this_bag(X).`
Result — `white(b).`
Case — `from_this_bag(b).`

The first pattern, inference of a general rule from a case (description) and a result (observation) of a particular individual, exemplifies the kind of reasoning performed by *inductive logic programming* (ILP) systems. The second pattern, inferring a more complete description of an individual from an observation and a general theory valid for all such individuals, is the kind of reasoning studied in *abductive logic programming* (ALP).

The above account describes ILP and ALP by example, and does not provide a general definition. Interestingly, attempts to provide such a general definition of abduction and induction in logic programming typically correspond to Peirce's later, inferential characterisation of explanatory hypotheses generation. Thus, in ALP abductive inference is typically specified as follows:

> 'Given a set of sentences T (a theory presentation), and a sentence G (observation), to a first approximation, the abductive task can be characterised as the problem of finding a set of sentences Δ (abductive explanation for G) such that:
> (1) $T \cup \Delta \models G$,
> (2) $T \cup \Delta$ is consistent. ' [Kakas *et al.*, 1992, p.720]

The following is a specification of induction in ILP:

> 'Given a consistent set of examples or observations O and consistent background knowledge B find an hypothesis H such that: $B \cup H \models O$' [Muggleton and De Raedt, 1994]

In spite of small terminological differences the two specifications are virtually identical: they both invert a deductive consequence relation in order to complete an incomplete given theory, prompted by some new observations that cannot be deductively accounted for by the theory alone.[7] If our assessment of the distinction between abduction and induction that is usually drawn in AI is correct, we must conclude that the above specifications are unable to account for this distinction. In the remainder of Section 3 we will try to understand the differences between abduction and induction as used in AI in modern, non-syllogistic terms. For an account which stays closer to syllogisms, the reader is referred to [Wang, 2000].

[7]Extra elements that are often added to the above definitions are the satisfaction of integrity constraints for the case of abduction, and the avoidance of negative examples for the case of induction; these can again be viewed under the same heading, namely as being aimed at exclusion of certain hypotheses.

3.2 Explanation and generalisation

Let us further analyse the logical processes of abduction and induction from the utility perspective of AI, and examine to what extent it is possible to distinguish two such processes on the basis of the function they are intended to perform. We will argue that such a distinction is indeed possible, since the function of abduction is to provide explanations, and the function of induction is to provide generalisations. Some of our views on this matter have been influenced directly by the paper by Console and Saitta [2000] where more discussion on this possibility of distinction between abduction and induction can be found.

First, it will be convenient to introduce some further terminology.

Observables and abducibles

We will assume a common first-order language for all knowledge (known, observed, or hypothetical). We assume that the predicates of this language are separated into *observables* and *non-observables* or *background* predicates. *Domain knowledge* or *background knowledge* is a general theory concerning non-observable predicates only. *Foreground knowledge* is a general theory relating observable predicates to background predicates and each other. *Instance knowledge* (sometimes called scenario knowledge) consists of formulae containing non-observable predicates only, possibly drawn from a restricted subset of such predicates. Known instance knowledge can be part of the background knowledge. *Observations* are formulae containing observable predicates, known to hold; *predictions* are similar to observations, but their truthvalue is not given.

It will often be useful to employ the notion of an *individual* to refer to a particular object or situation in the domain of discourse. For example, instance knowledge will usually contain descriptions of individuals in terms of non-observable predicates (thence the name). An *unobserved* or *new* individual is one of which the description becomes known only after the abductive or inductive hypothesis has been formed. As a consequence, the hypothesis cannot refer to this particular individual; however, the hypothesis may still be able to provide a prediction for it when its description becomes available.

Given this terminology, we can specify the aim of induction as *inference of foreground knowledge* from observations and other known information. Typically, this information consists of background and instance knowledge, although other known foreground knowledge may also be used. In some cases it may be empty, for instance when we are learning the definition of a recursive predicate, when we are learning the definitions of several mutually dependent predicates, or when we are doing data mining. The observations specify incomplete (usually extensional) knowledge about the observables, which we try to generalise into new foreground knowledge.

On the other hand, in abduction we are *inferring instance knowledge* from observations and other known information. The latter necessarily contains foreground information pertaining to the observations at hand. Possible abductive hypotheses are built from specific non-observable predicates called *abducibles* in ALP. The intuition is that these are the predicates of which the extensions are not completely known as instance knowledge. Thus, an abductive hypothesis is one which completes the instance knowledge about an observed individual.

Explanation

Non-deductive reasoning as used in AI provides two basic functions that are generally useful in addressing different problems. These two functions are (a) finding how a piece of information came about to be true according to a general theory describing the domain of interest, and (b) constructing theories that can describe the present and future behaviour of a system. Purely from this utility point of view, non-deductive reasoning is required to provide these two basic effects of explanation and generalisation. Informally, for the purposes of this chapter it is sufficient for *explanation* to mean that the hypothesis reasoned to (or generated) by the non-deductive reasoning does not refer to observables (i.e. consists of instance knowledge) and entails a certain formula (an observation), and for *generalisation* to mean that the hypothesis can entail additional observable information on unobserved individuals (i.e. predictions).

As we have seen before, both abduction and induction can be seen as a form of reversed deduction in the presence of a background theory, and thus formally qualify as providing explanations of some sort. The claim that abduction is explanatory inference indeed seems undisputed, and we do not find a need to say more about the issue here. We only point out that if an abductive explanation Δ is required to consist of instance knowledge only, then clearly abduction needs a given theory T of foreground knowledge, connecting observables to background predicates, in order to be able to account for the observation with Δ. An abductive explanation thus makes sense only relative to this theory T from which it was generated: it explains the observation according to this particular theory.

However, if induction provides explanations at all, these explanations are of a different kind. For instance, we can say that 'all the beans from this bag are white' is an explanation for why the observed beans from the bag are white. Notice however that this kind of explanation is universal: 'observed Xs are Y' is explained by the hypothesis that 'all Xs are Y'. This explanation does not depend on a particular theory: it is not according to a particular model of the 'world of beans'. It is a general, meta-level explanation that does not provide any insight to why things are so. As Josephson puts it, inductive hypotheses do not explain particular observations, but they explain the frequencies with which the observations occur (*viz.* that non-white beans from this bag are never observed).

Generalisation

We thus find that inductive hypotheses are not explanatory in the same way as abductive hypotheses are. But we would argue that being explanatory is not the primary aim of inductive hypotheses in the first place. Rather, the main goal of induction is to provide generalisations. In this respect, we find that the ILP definition of induction (p. 14) is too much focused on the problem of learning classification rules, without stressing the aspect of generalisation. An explanatory hypothesis would only be inductive if it generalises. The essential aspect of induction as applied in AI seems to be the kind of sample-to-population inference exemplified by categorical inductive generalisation, reproduced here in its more general form from Section 2.3:

> All objects in the sample satisfy $P(x)$;
> therefore, all objects in the population satisfy $P(x)$.

As with Peirce's syllogisms, the problem here is that $P(x)$ is already assumed to be given, while in AI a major problem is to generate such hypotheses. The specification of confirmatory or descriptive induction follows this pattern, but leaves the hypothesis unspecified:

> Given a consistent set of observations O and a consistent background knowledge B, find a hypothesis H such that: $M(B \cup O) \models H$
> [Helft, 1989; De Raedt and Bruynooghe, 1993; Flach, 1995]

Hence the formal requirement now is that any generated hypothesis should be true in a certain model constructed from the given knowledge and observations (e.g. the truth-minimal model).

This specification can be seen as sample-to-population inference. For example, in Peirce's bean example (p. 5), B is 'these beans are from this bag' (instance knowledge), O is 'these beans are white' (observation), and H – 'all the beans from this bag are white' – is satisfied by the model containing 'these beans' as the only beans in the universe. Under the assumption that the population is similar to the sample, we achieve generalisation by restricting attention to formulae true in the sample. Note that the induced hypothesis is not restricted to one explaining the whiteness of these beans: we might equally well have induced that 'all white beans are from this bag'.

Above we defined a hypothesis as generalising if it makes a prediction involving an observable. We have to qualify this statement somewhat, as the following example shows (taken from [Console and Saitta, 2000]). Let our background theory contain the following clauses:

```
measles(X):-brother(X,Y),measles(Y).
red_spots(X):-measles(X).
brother(john,dan).
```

The observation is `red_spots(john)`. A possible explanation for this observation is `measles(john)`. While this explanation is clearly completing instance knowledge and thus abductive, adding it to our theory will lead to the prediction `red_spots(dan)`. Thus, the hypothesis that John has measles also seems to qualify as a generalisation. We would argue however that this generalisation effect is already present in the background theory. On the other hand, an inductive hypothesis produces a *genuinely new* generalisation effect, in the sense that we can find new individuals for which the addition of the hypothesis to our knowledge is necessary to derive some observable property for these individuals (usually this property is that of the observations on which the induction was based). With an abductive hypothesis this kind of extension of the observable property to other new individuals does not necessarily require the *a priori* addition of the abductive hypothesis to the theory but depends only on the properties of this individual and the given background theory: the generalisation, if any, already exists in the background theory.

We conclude that abductive and inductive hypotheses differ in the degree of generalisation that each of them produces. With the given background theory T we implicitly restrict the generalising power of abduction as we require that the basic model of our domain remains that of T. The existence of this theory separates two levels of generalisation: (a) that contained in the theory and (b) new generalisations that are not given by the theory. In abduction we can only have the first level with no interest in genuinely new generalisations, while in induction we do produce such new generalisations.

3.3 Extending incomplete theories

We will now further examine the general logical process that each of abduction and induction takes. The overall process that sets the two forms of reasoning of abduction and induction in context is that of theory formation and *theory development*. In this we start with a theory T (that may be empty) which describes at a certain level the problem domain we are interested in. This theory is incomplete in its representation of the problem domain, as otherwise there is no need for non-deductive ampliative reasoning. New information given to us by the observations is to be used to complete this description. As we argue below, abduction and induction each deal with a different kind of incompleteness of the theory T.

Abductive extensions

In a typical use of abduction, the description of the problem domain by the theory T is further assumed to be sufficient, in the sense that it has reached a stage where we can reason with it. Typically this means that the incompleteness of the theory can be isolated in some of its non-observable predicates, which are called abducible (or open) predicates. We can then view the theory T as a representation of all of its possible abductive extensions $T \cup \Delta$, usually denoted $T(\Delta)$, for each abducible hypothesis Δ. An enumeration of all such formulae (consistent with T) gives the set of all possible abductive extensions of T. Abductive entailment with T is then defined by deductive entailment in each of its abductive extensions.

Alternatively, we can view each abductive formula Δ as supplying the missing instance knowledge for a different possible situation or individual in our domain, which is then completely described by $T(\Delta)$. For example, an unobserved individual and its background properties can be understood via a corresponding abductive formula Δ. Once we have these background properties, we can derive – using T – other properties for this new individual.[8] Given an abductive theory T as above, *the process of abduction is to select one of the abductive extensions* $T(\Delta)$ *of* T in which the given observation to be explained holds, by selecting the corresponding formula Δ. We can then reason deductively in $T(\Delta)$ to arrive at other conclusions. By selecting Δ we are essentially enabling one of the possible associations between Δ and the observation among those supplied by the theory T.

It is important here to emphasise that the restriction of the hypothesis of abduction to abducible predicates is not incidental or computational, but has a deeper representational reason. It reflects the relative comprehensiveness of knowledge of the problem domain contained in T. The abducible predicates and the allowed abductive formulae take the role of 'answer-holders' for the problem goals that we want to set to our theory. In this respect they take the place of the logical variable as the answer-holder when deductive reasoning is used for problem solving. As a result this means that the form of the abductive hypothesis depends heavily on the particular theory T at hand, and the way that we have chosen to represent in this our problem domain.

Typically, the allowed abducible formulae are further restricted to simple logical forms such as ground or existentially quantified conjunctions of abducible literals. Although these further restrictions may be partly motivated by computational considerations, it is again important to point out that they are only made possible by the relative comprehensiveness of the particular representation of our problem domain in the theory T. Thus, the case of simple abduction – where the abducible hypothesis are ground facts – occurs exactly because the representation of the problem domain in T is sufficiently complete to allow this. Furthermore, this restriction is not significant for the purposes of comparison of abduction and induction: our analysis here is independent of the particular form of abducible formulae. The important elements are the existence of an enumeration of the abductive formulae, and the fact that these do not involve observable predicates.

Inductive extensions

Let us now turn to the case of induction and analyse this process to facilitate comparison with the process of abduction as described above. Again, we have a collection of possible inductive hypotheses from which one must be selected. The

[8] Note that this type of abductive (or open) reasoning with a theory T collapses to deduction, when and if the theory becomes fully complete.

main difference now is the fact that these hypotheses are not limited to a particular subset of predicates that are incompletely specified in the representation of our problem domain by the theory T, but are restricted only by the language of T. In practice, there may be a restriction on the form of the hypothesis, called *language bias*, but this is usually motivated either by computational considerations, or by other information external to the theory T that guides us to an inductive solution.

Another essential characteristic of the process of induction concerns the role of the selected inductive hypothesis H. The role of H is to extend the existing theory T to a new theory $T' = T \cup H$, rather than reason with T under the set of assumptions H as is the case for abduction. Hence T is replaced by T' to become a new theory with which we can subsequently reason, either deductively of abductively, to extract information from it. The hypothesis H changes T by requiring extra conditions on the observable predicates that drive the induction, unlike abduction where the extra conditions do not involve the observable predicates. In effect, H provides the link between observables and non-observables that was missing or incomplete in the original theory T.

Analogously to the concept of abductive extension, we can define inductive extensions as follows. Consider a common given theory T with which we are able to perform abduction and induction. That is, T has a number of abductive extensions $T(\Delta)$. Choosing an inductive hypothesis H as a new part of the theory T has the effect of further conditioning each of the abductive extensions $T(\Delta)$. Hence, while in abduction we select an abductive extension of T, with induction we extend each of the abductive extensions with H. The effect of induction is thus 'universal' on all the abductive extensions.

If we now consider the new abductive theory $T' = T \cup H$, constructed by induction, we can view induction as a process of selecting a collection of abductive extensions, namely those of the new theory T'. Hence *an inductive extension can be viewed as a set of abductive extensions of the original theory* T that are further (uniformly) conditioned by the common statement of the inductive hypothesis H. This idea of an inductive extension consisting of a set of abductive extensions was used in [Denecker *et al.*, 1996] to obtain a formalisation of abduction and induction as selection processes in a space of possible world models over the given theory in each case. In this way the process of induction can be seen to have a more general form than abduction, able to select a set of extensions rather than a single one. Note that this does not necessarily mean that induction will yield a more general syntactic form of hypotheses than abduction.

Analysis

Comparing the possible inductive and abductive extensions of a given theory T we have an essential difference. In the case of abduction some of the predicates in the theory, namely the observables, cannot be arbitrarily defined in an extension. The freedom of choice of abduction is restricted to constrain directly (via Δ) only the abducibles of the theory. The observable predicates cannot be affected except through the theory: the observables must be grounded in the existing theory T by the choice of the abductive conditions on the abducible part of the extension. Hence in an abductive extension the extent to which the observables can become true is limited by the theory T and the particular conditions Δ on the rest of the predicates.

In induction this restriction is lifted, and indeed we can have inductive extensions of the given theory T, the truthvalue of which on the observable predicates need not be attributed via T to a choice on the abducibles. The inductive extensions 'induce' a more general change (from the point of view of the observables) on the existing theory T, and – as we will see below – this will allow induction to genuinely generalise the given observations to other cases not derivable from the original theory T. The generalising effect of abduction, if at all present, is much more limited. The selected abductive hypothesis Δ may produce in $T(\Delta)$ further information on abducible or other predicates, as in the measles example from the previous section. Assuming that abducibles and observables are disjoint, any information on an observable derived in $T(\Delta)$ is a generalisation already contained in T.

What cannot happen is that the chosen abductive hypothesis Δ alone (without T) predicts a new observation, as Δ does not affect directly the value of the observable predicates. Every prediction on an observable derived in $T(\Delta)$, not previously true in T (including the observation that drives the abductive process), corresponds to some further instance knowledge Δ', which is a consequence of $T(\Delta)$, and describes the new situation (or individual) at hand. Such consequences are already known to be possible in the theory T, as we know that one of its possible extensions is $T(\Delta')$. In the measles example (p. 17), the observation `red_spots(john)` gives rise to the hypothesis Δ = `measles(john)`. Adopting this hypothesis leads to a new prediction `red_spots(dan)`, corresponding to the instance knowledge Δ' = `measles(dan)`, which is a consequence of $T(\Delta)$. This new prediction could be obtained directly from T(`measles(dan)`) without the need of Δ = `measles(john)`.

Similarly, if we consider a previously unobserved situation (not derivable from $T(\Delta)$) described by Δ_{new} with $T(\Delta) \cup \Delta_{new}$ deriving a new observation, this is also already known to be possible as $T(\Delta \cup \Delta_{new})$ is one of the possible extensions of T. For example, if Δ_{new} = `measles(mary)`, then $T(\Delta) \cup \Delta_{new}$, and in fact $T \cup \Delta_{new}$ derives `red_spots(mary)`, which is again not a genuine generalisation.

In short, abduction is meant to select some further conditions Δ under which we should reason with T. It concerns only this particular situation described by Δ and hence, if Δ cannot impose directly any conditions on the observable predicates, the only generalisations that we can get on the observables are those contained in T under the particular restrictions Δ. In this sense we say that the generalisation is not genuine but already contained in T. Hence, abduction increases the intension of known individuals (abducible properties are now made true for these individuals), but does not have a genuine generalisation effect on the observables (it does not increase the extension of the observables with previously unobserved individuals for which the theory T alone could not produce this extension when it is given the instance knowledge that describes these individuals).

On the other hand, the universal conditioning of the theory T by the inductive hypothesis H produces a genuine generalisation on the observables of induction. The extra conditions in H on the observables introduce new information on the relation of these predicates to non-observable predicates in the theory T, and from this we get new observable consequences. We can now find cases where from H alone together with a (non-observable) part of T, describing this case, we can derive a prediction not previously derivable in T.

The new generalisation effect of induction shows up more when we consider as above the case where the given theory for induction has some of its predicates as abducible (different from the observables). It is now possible to have a new individual described by the extra abducible information Δ_{new}, such that in the new theory $T' = T \cup H$ produced by induction a new observation holds which was not known to be possible in the old theory T (i.e. it is not a consequence of $T \cup \Delta_{new}$). Note that we cannot (as in the case of abduction) combine H with Δ_{new} to a set Δ'_{new} of instance knowledge, under which the observation would hold from the old theory T. We can also have that a new observation holds alone from the hypothesis H and Δ_{new} for such previously unobserved situations not described in the given theory T. These are cases of genuine generalisation not previously known to be possible from the initial theory T.

Summarising this subsection, induction – seen as a selection of a set of extensions defined by the new theory $T \cup H$ – has a stronger and genuinely new generalising effect on the observable predicates than abduction. The purpose of abduction is to select an extension and reason with it, thus enabling the generalising potential of the given theory T. In induction the purpose is to extend the given theory to a new theory, the abductive extensions of which can provide new possible observable consequences.

Finally, we point out a duality between abduction and induction (first studied in [Dimopoulos and Kakas, 1996]) as a result of this analysis. In abduction the theory T is fixed and we vary the instance knowledge to capture (via T) the observable knowledge. On the other hand, in induction the instance knowledge is fixed as part of the background knowledge B, and we vary the general theory so that if the selected theory T is taken as our abductive theory then the instance knowledge in B will form an abductive solution for the observations that drove the induction. Conversely, if we perform abduction with T and we consider the abductive hypothesis Δ explaining the observations as instance knowledge, the original theory T forms a valid inductive hypothesis.

3.4 Interaction between abduction and induction

In the preceding sections we analysed basic patterns of abduction and induction. In practice hybrid forms of ampliative reasoning occur, requiring an interaction between these basic patterns. Such interaction is the subject of this section.

Let us consider a simple example originating from [Michalski, 1993]. We have the observation that:

O: all bananas in this shop are yellow,

and we want to explain this given a theory T containing the statement:

T: all bananas from Barbados are yellow.

An explanation for this is given by the hypothesis:

H: all bananas in this shop are from Barbados.

Is this a form of abduction or a form of induction, or perhaps a hybrid form? As we will show, this strongly depends on the choice of observables and abducibles.

Suppose, first, that we choose 'yellow' as observable and the other predicates as abducibles.[9] The hypothesis H selects amongst all the possible abductive extensions of the theory T (corresponding to the different abducible statements of instance knowledge consistent with T) a particular one. In this selected extension the observation is entailed and therefore the hypothesis explains according to the abductive theory T the observation. Note that this hypothesis H does not generalise the given observations: it does not enlarge the extension of the observable predicate 'yellow' over that provided by the statement of the observation O. In fact, we can replace the universal quantification in 'all bananas from this shop' by a typical representative through skolemisation. More importantly, the link of the observation O with the extra information of H is known *a priori* as one of the possible ways of reasoning with the theory T to derive new observable information.

There is a second way in which to view this reasoning and the hypothesis H above. We can consider the predicate 'from Barbados' as the observable predicate with a set of observations that each of the observed bananas in the shop is from Barbados. We then have a prototypical inductive problem (like the white bean example of Peirce) where we generate the same statement H as above, but now as an inductive hypothesis. From this point of view the hypothesis now has a genuine generalising effect over the observations on the predicate 'from Barbados'. But where did the observations on Barbados come from? These can be obtained from the theory T as separate abductive explanations for each of the original observations (or a typical one) on the predicate 'yellow'. We can thus understand this example as a hybrid process of first using (simple) abduction to translate separately each given observation as an observation on the abducibles, and then use induction to generalise the latter set of observations, thus arriving at a general statement on the abducibles.

Essentially, in this latter view we are identifying, by changing within the same problem the observable and abducible predicates, simple basic forms of abduction

[9] We can if we wish consider only the predicate 'from Barbados' as abducible.

and induction on which we can build more complex forms of non-deductive reasoning. Referring back to our earlier discussion in Section 3, these basic forms are: pure abduction for explanation with no generalisation effect (over what already exists in the theory T); and pure induction of simple generalisations from sample to population. This identification of basic distinct forms of reasoning has important computational consequences. It means that we can consider two basic computational models for the separate tasks of abduction and induction. The emphasis then shifts to the question of how these basic forms of reasoning and computation can be integrated together to solve more complex problems by suitably breaking down these problems into simpler ones.

It is interesting to note here that in the recent framework of inverse entailment as used by the ILP system Progol [Muggleton, 1995] where we can learn from general clauses as observations, an analysis of its computation as done in [Yamamoto, 2000] reveals that this can be understood as a mixture of abduction and induction. As described in the above example, the Progol computation can be separated into first abductively explaining according to the background theory a skolemised, typical observation, and then inductively generalising over this abductive explanation. The usefulness of explicitly separating out abduction and induction is also evident in several works of theory formation or revision. Basic computational forms of abduction and induction are used together to address these complex problems. This will be described further in Section 4 on the integration of abduction and induction in AI.

3.5 Computational characteristics

We will close this section by discussing further the computational distinction that the basic forms of abduction and induction have in their practice in AI and logic programming. Indeed, when we examine the computational models used for abduction and induction in AI, we notice that they are very different. Their difference is so wide that it is difficult, if not impossible, to use the computational framework of one form of reasoning in order to compute the other form of reasoning. Systems developed in AI for abduction cannot be used for induction (and learning), and *vice versa*, inductive AI systems cannot be used to solve abductive problems.[10]

We will describe here the main characteristics of the computational models of the basic forms of abduction and induction, discussed above, as they are found in practical AI approaches. According to these basic forms, abduction extracts an explanation for an observation from a given theory T, and induction generalises a set of atomic observations. For abduction the computation has the following basic form: extract from the given theory T a hypothesis Δ and check this for consistency. The search for a hypothesis is done via some form of enhanced deduction method e.g. resolution with residues [Cox and Pietrzykowski, 1986; Eshghi and

[10] With the possible exception of Cigol [Muggleton and Buntine, 1988], a system designed for doing unrestricted reversed deduction.

Kowalski, 1989; Kakas and Mancarella, 1990; Denecker and de Schreye, 1992; Inoue, 1992; Kakas and Michael, 1995], or unfolding of the theory T [Console *et al.*, 1991; Fung and Kowalski, 1997].

The important thing to note is that the abductive computation is primarily based on the computation of deductive consequences from the theory T. The proofs are now generalised so that they can be successfully terminated 'early' with an abductive formula. To check consistency of the found hypothesis, abductive systems employ standard deductive methods (these may sometimes be specially simplified and adapted to the particular form that the abductive formulae are restricted to take). If a hypothesis (or part of a hypothesis) is found inconsistent then it is rejected and another one is sought. Note that systems that compute constructive abduction (e.g. SLDNFA [Denecker and de Schreye, 1998], IFF [Fung and Kowalski, 1997], ACLP [Kakas and Michael, 1995]), where the hypothesis may not be ground but can be an existentially quantified conjunction (with arithmetic constraints on these variables) or even a universally quantified formula, have the same computational characteristics. They arrive at these more complex hypotheses by extending the proof methods for entailment to account for the (isolated) incompleteness on the abducible predicates.

On the other hand, the computational model for the basic form of induction in AI takes a rather different form. It constructs a hypothesis and then refines this under consistency and other criteria. The construction of the hypothesis is based on methods for inverting entailment proofs (or satisfaction proofs in the case of confirmatory induction) so that we can obtain a new theory that would then entail (or be satisfied by) the observations. Thus, unlike the abductive case, the computation cannot be based on proof methods for entailment, and new methods such as inverse resolution, clause generalisation and specialisation are used. In induction the hypothesis is generated from the language of the problem domain (rather than a given theory of the domain), in a process of iteratively improving a hypothesis to meet the various requirements posed by the problem. Furthermore, in induction the comparison of the different possible hypotheses plays a prominent and dynamic role in the actual process of hypothesis generation, whereas in abduction evaluation of the different alternative hypothesis may be done after these have been generated.

It should be noted, however, that the observed computational differences between generating abductive hypotheses and generating inductive hypotheses are likely to become smaller once more complex abductive hypotheses are allowed. Much of the computational effort of ILP systems is spent on efficiently searching and pruning the space of possible hypotheses, while ALP systems typically enumerate all possible abductive explanations. The latter approach becomes clearly infeasible when the abductive hypothesis space grows. In this respect, we again mention the system Cigol which seems to be the only system employing a unified computational method (inverse resolution) to generate both abductive and inductive hypotheses.

Computational distinctions of the two forms of reasoning are amplified when we consider the different works of trying to integrate abduction and induction in a common system. In most of these works, each of the two forms of reasoning is computed separately, and their results are transferred to the other form of reasoning as input. The integration clearly recognises two different computational processes (for each reasoning) which are then suitably linked together. For example, in LAB [Thompson and Mooney, 1994] or ACL [Kakas and Riguzzi, 1997; Kakas and Riguzzi, 1999] the overall computation is that of induction as described above, but where now – at the point of evaluation and improvement of the hypothesis – a specific abductive problem is computed that provides feedback regarding the suitability of the inductive hypothesis. In other cases, such as RUTH [Adé *et al.*, 1994] or Either [Ourston and Mooney, 1994] an abductive process generates new observable input for a subsidiary inductive process. In all these cases we have well-defined separate problems of simple forms of abduction and induction each of which is computed along the lines described above. In other words, the computational viability of the integrated systems depends significantly on this separation of the problem and computation into instances of the simple forms of abduction and induction.

4 INTEGRATION OF ABDUCTION AND INDUCTION

The complementarity between abduction and induction, as we have seen it in the previous section – abduction providing explanations from the theory while induction generalises to form new parts of the theory – suggests a basis for their integration. Cooperation between the two forms of reasoning would be useful within the context of theory development (construction or revision), where a current theory T is updated to a new theory T' in the light of new observations O so that T' captures O (i.e. $T' \models O$).

At the simplest level, abduction and induction simply co-exist and both function as revision mechanisms that can be used in developing the new theory [Michalski, 1993]. In a slightly more cooperative setting, induction provides new foreground knowledge in T for later use by abduction. At a deeper level of cooperation, abduction and induction can be integrated together within the process of constructing T. There are several ways in which this can happen within a cycle of development of T, as will be described below.

The cycle of abductive and inductive knowledge development

On the one hand, abduction can be used to extract from the given theory T and observations O abducible information that would then feed into induction as (additional) training data. One example of this is provided by [Ourston and Mooney, 1994], where abduction identifies points of repair of the original, faulty theory T, i.e. clauses that could be generalised so that positive observations in O become entailed, or clauses that may need to be specialised or retracted because they are

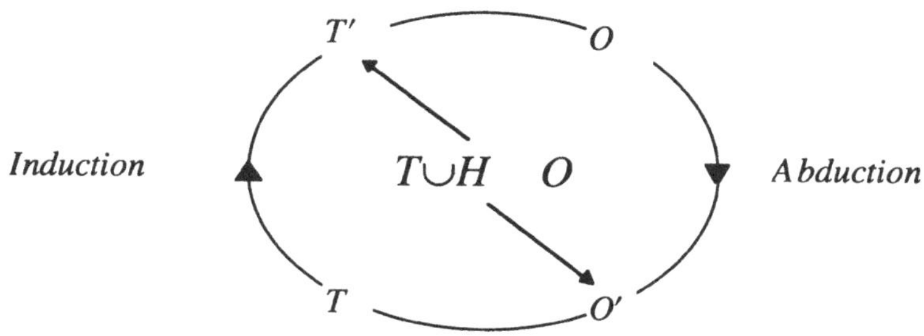

Figure 2. The cycle of abductive and inductive knowledge development.

inconsistent with negative observations.

A more active cooperation occurs when, first, through the use of basic abduction, the original observations are transformed to data on abducible background predicates in T, becoming training data for induction on these predicates. An example of this was discussed in Section 3.4; another example in [Dimopoulos and Kakas, 1996] shows that only if, before inductive generalisation takes place, we abductively transform the observations into other predicates in a uniform way, it is possible to solve the original inductive learning task.

In the framework of the system RUTH [Adé *et al.*, 1994], we see induction feeding into the original abductive task. An abductive explanation may lead to a set of required facts on 'inducible' predicates, which are inductively generalised to give a general rule in the abductive explanation for the original observations, similar to (one analysis of) the bananas example discussed previously.

These types of integration can be succinctly summarised as follows. Consider a cycle of knowledge development governed by the 'equation' $T \cup H \models O$, where T is the current theory, O the observation triggering theory development, and H the new knowledge generated. Then, as shown in Figure 2, on one side of this cycle we have induction, its output feeding into the theory T for later use by abduction, as shown in the other half of the cycle, where the abductive output in turn feeds into the observational data O for later use by induction, and so on.

Inducing abductive theories

Another way in which induction can feed into abduction is through the generation of confirmatory (or descriptive) inductive hypotheses that could act as integrity constraints for the new theory. Here we initially have some abductive hypotheses regarding the presence or absence of abducible assumptions. Based on these hypotheses and other data in T we generate, by means of confirmatory induction, new sentences I which, when interpreted as integrity constraints on the new theory T, would support the abducible assumptions (assumptions of presence would be consistent with I, assumptions of absence would now be inconsistent with I).

This type of cooperation between abductive and inductive reasoning is based on a deeper level of integration of the two forms of reasoning, where induction is perceived as hypothesising abductive (rather than deductive) theories. The deductive coverage relation for learning is replaced by abductive coverage, such that an inductive hypothesis H is a valid generalisation if the observations can be abductively explained by $T' = T \cup H$, rather than deductively entailed. A simple example of this is the extension of Explanation-Based Learning with abduction [Cohen, 1992; O'Rorke, 1994], such that deductive explanations are allowed to be completed by abductive assumptions before they are generalised.

Inducing abductive theories is particularly useful in cases where the domain theory is incomplete, and also when performing multiple predicate learning, as also in this case the background knowledge for one predicate includes the incomplete data for the other predicates to be learned. In these cases the given theory T is essentially an abductive theory, and hence it is appropriate to use an abductive coverage relation. On the other hand, it may be that the domain that we are trying to learn is itself inherently abductive or non-monotonic (e.g. containing nested hierarchies of exceptions), in which case the hypothesis space for learning is a space of abductive theories.

LAB [Thompson and Mooney, 1994] is one of the first learning systems adopting this point of view The class predicates to be learned are the abducible predicates, and the induced theory H describes the effects of these predicates on other predicates that we can observe directly with rules of the form *observation* $\leftarrow$ *class*. Then the training examples (each consisting of a set of properties and its classification) are captured by the induced hypothesis H when the correct classification of the examples forms a valid abductive explanation, given H, for their observed properties. Other frameworks for learning abductive theories are given in [Kakas and Riguzzi, 1997; Kakas and Riguzzi, 1999; Dimopoulos *et al.*, 1997] and [Lamma *et al.*, 2000]. Here, both explanatory and confirmatory induction are used to generate theories together with integrity constraints. Inoue and Haneda [2000] also study the problem of learning abductive logic programs for capturing non-monotonic theories.

With this type of integration we can perceive abduction as being used to evaluate the suitability or credibility of the inductive hypothesis. Similarly, abductive explanations that lead to induction can be evaluated by testing the induced generalisation. In this sense, the integration of abduction and induction can help to cross-evaluate the hypothesis that they generate.

5 ABDUCTION AND INDUCTION IN THE CONTRIBUTIONS TO THIS VOLUME

We now briefly survey the other eight chapters in this volume with regard to their approach to abduction and induction.

A first distinction can be made regarding the nature of the representation and reasoning formalism employed, which is predominantly logical in the first four chapters by Paul, Gabbay, Flach, and Bergadano *et al.*, and mainly probabilistic (or possibilistic) in the last four chapters by Dubois & Prade, Borgelt & Kruse, de Campos *et al.*, and Gebhardt. Of these, the last three employ graphical models to represent probabilistic assumptions regarding the domain.

A second distinction is that half of the chapters are concerned with abduction: Paul and Gabbay in the logical approach, and Dubois & Prade and Borgelt & Kruse in the probabilistic or possibilistic approach. Similarly, there are two 'logical' chapters dealing (mainly) with inductive learning, *viz.* Flach and Bergadano *et al.*, and two 'probabilistic' chapters, namely, Gebhardt and de Campos *et al.*.

Both Gabriele Paul and Dov Gabbay, in their respective chapters, adopt the explanatory view of abduction, in which abduction is the process of extending an incomplete theory with one or more abducibles such that given observations are entailed or explained. This corresponds most closely to Peirce's later, inferential theory, in which abduction plays the role of generating possible explanations for given observations, although there are similarities also with the syllogistic theory, in particular where they discuss the logic programming perspective. Both chapters tend to concentrate on generating possible abductive hypotheses rather than selecting the best one, although Paul describes a selection criterion based on a coherence metric.

The next two 'logical' chapters, by Flach, and Bergadano, Cutello & Gunetti, are written by authors with a machine learning background. Flach aims at logically characterising various forms of inductive learning, concentrating on the hypothesis generation part. With regard to explanatory induction, this (confusingly) corresponds to what Peirce called abduction in his inferential theory; the difference is that Flach is primarily interested in general hypotheses. Bergadano *et al.*, on the other hand, take the syllogistic perspective found in logic programming and distinguish between abduction and induction by means of the syntactic form of the hypotheses (specific facts in the case of abduction, general rules in the case of induction). A particular interest of their chapter is that it addresses ways of integrating abduction and inductive learning (see also Section 4 of the present chapter).

Next follow four 'probabilistic' chapters. Didier Dubois and Henri Prade consider causal diagnostic problem solving, i.e. inferences of the form: from "if disorder is present then symptom is observed" and "symptom is observed" infer "disorder is plausible". The main concern of their chapter is to assign degrees of plausibilities to possible disorders; thus, it concentrates on hypothesis evaluation rather than hypothesis generation. Evaluating abductive hypotheses is also the concern of the next chapter by Christian Borgelt and Rudolf Kruse, who propose probabilistic networks as a possible approach. A special interest of their chapter is that it considers possible connections between logical and probabilistic approaches to abduction.

The last two chapters study the problem of learning probabilistic or possibilistic graphical models. Gebhardt investigates one such alternative to probability theory, namely possibility theory. The chapter by de Campos, Huete & Moral concentrates on the underlying notion of stochastic independence, which allows decomposition of probability distributions and other uncertainty measures.Both chapters address the problem of finding a graphical model that best fits the data, which can be seen as an inductive learning problem.

6 CONCLUSIONS

The nature of abduction and induction is still hotly debated. In this introductory chapter we have tried to chart the terrain of possible positions in this debate. From a logico-philosophical perspective, there are broadly speaking two positions: either one holds that abduction provides explanations and induction provides generalisations; or one can hold that abduction is the logic of hypothesis generation and induction is the logic of hypothesis evaluation. AI approaches tend to adopt the first perspective (although there are exceptions) – abduction and induction each deal with a different kind of incompleteness of the given theory, extending it in different ways.

Abduction and induction can be usefully integrated when trying to solve complex theory development tasks. We have reviewed a number of AI approaches to such integration. Most of these frameworks of integration use relatively simple forms of abduction and induction, namely abduction of ground facts and basic inductive generalisations. Moreover, each of the two is computed separately and its results transferred to the other, thus clearly recognising two separate and basic computational problems. From these, they synthesise an integrated form of reasoning that can produce more complex solutions, following a cyclic pattern with each form of reasoning feeding into the other.

An important open question in artificial intelligence is the integration of logical and numerical uncertainty handling formalisms, and this is no less true in the areas of abduction and inductive learning. While such an integration has not been the focus of this introductory chapter, various other chapters in this volume do address it to some extent. We hope that the volume will be found useful as an attempt to further our understanding of abduction and learning, and contributes towards a fuller understanding of uncertainty handling in AI.

ACKNOWLEDGMENTS

This chapter is an adapted and extended version of [Flach and Kakas, 2000, Chapter 1]. We would like to express our thanks to all authors contributing to that volume, whose ideas have significantly contributed to this chapter. Part of this work was supported by Esprit IV Long Term Research Project 20237 (Inductive Logic Programming 2).

Peter A. Flach
University of Bristol, UK.

Antonis C. Kakas
University of Cyprus, Nicosia.

REFERENCES

[Adé *et al.*, 1994] Hilde Adé, Bart Malfait, and Marc Denecker. RUTH: an ILP theory revision system. In *Proceedings of the Eighth International Symposium on Methodologies for Intelligent Systems*, volume 869 of *Lecture Notes in Artificial Intelligence*, pages 336–345. Springer-Verlag, Berlin, 1994.

[Carnap, 1950] Rudolf Carnap. *Logical Foundations of Probability*. Routledge & Kegan Paul, London, 1950.

[Cohen, 1992] William W. Cohen. Abductive explanation-based learning: a solution to the multiple inconsistent explanation problem. *Machine Learning*, 8:167–219, 1992.

[Console and Saitta, 2000] Luca Console and Lorenza Saitta. On the relations between abductive and inductive explanation. In Flach and Kakas [2000].

[Console *et al.*, 1991] Luca Console, Daniele Theseider Dupré, and Pietro Torasso. On the relationship between abduction and deduction. *Journal of Logic and Computation*, 1(5):661–690, 1991.

[Cox and Pietrzykowski, 1986] P. T. Cox and T. Pietrzykowski. Causes for events: their computation and application. In *Proceedings of the Eighth International Conference on Automated Deduction*, volume 230 of *Lecture Notes in Computer Science*, pages 608–621. Springer-Verlag, 1986.

[De Raedt and Bruynooghe, 1993] Luc De Raedt and Maurice Bruynooghe. A theory of clausal discovery. In *Proceedings of the Thirteenth International Joint Conference on Artificial Intelligence*, pages 1058–1063, Chambéry, France, 1993.

[Denecker and de Schreye, 1992] Marc Denecker and Danny de Schreye. SLDNFA: An abductive procedure for normal abductive programs. In K. Apt, editor, *Proceedings of the 1992 Joint International Conference and Symposium on Logic Programming*, pages 686–700. MIT Press, 1992.

[Denecker and de Schreye, 1998] Marc Denecker and Danny de Schreye. SLDNFA: An abductive procedure for abductive logic programs. *Journal of Logic Programming*, 34(2):111–167, 1998.

[Denecker *et al.*, 1996] Marc Denecker, Bern Martens, and Luc De Raedt. On the difference between abduction and induction: a model theoretic perspective. In Peter A. Flach and Antonis C. Kakas, editors, *Proceedings of the ECAI'96 Workshop on Abductive and Inductive Reasoning*, pages 19–22, 1996. Available on-line at `http://www.cs.bris.ac.uk/~flach/ECAI96/`.

[Dimopoulos and Kakas, 1996] Yannis Dimopoulos and Antonis C. Kakas. Abduction and inductive learning. In Luc De Raedt, editor, *Advances in Inductive Logic Programming*, pages 144–171. IOS Press, Amsterdam, 1996.

[Dimopoulos *et al.*, 1997] Yannis Dimopoulos, Sašo Džeroski, and Antonis C. Kakas. Integrating explanatory and descriptive learning in ILP. In *Proceedings of the Fifteenth International Joint Conference on Artificial Intelligence*, pages 900–905. Morgan Kaufmann, 1997.

[Eshghi and Kowalski, 1989] K. Eshghi and Robert A. Kowalski. Abduction compared with negation by failure. In Giorgio Levi and Maurizio Martelli, editors, *Proceedings of the Sixth International Conference on Logic Programming*, pages 234–254. MIT Press, 1989.

[Fann, 1970] K. T. Fann. *Peirce's Theory of Abduction*. Martinus Nijhoff, The Hague, 1970.

[Flach and Kakas, 2000] Peter A. Flach and Antonis C. Kakas, editors. *Abductive and inductive reasoning: essays on their relation and integration.* Kluwer Academic Publishers, 2000.

[Flach, 1995] Peter A. Flach. *Conjectures — an inquiry concerning the logic of induction.* PhD thesis, Tilburg University, 1995.

[Fung and Kowalski, 1997] T.F. Fung and Robert A. Kowalski. The IFF proof procedure for Abductive Logic Programming. *Journal of Logic Programming*, 33(2):151–165, 1997.

[Hanson, 1958] N. R. Hanson. The logic of discovery. *Journal of Philosophy*, 55(25):1073–1089, 1958.

[Helft, 1989] Nicolas Helft. Induction as nonmonotonic inference. In *Proceedings of the First International Conference on Principles of Knowledge Representation and Reasoning*, pages 149–156, Toronto, Canada, 1989. Morgan Kaufmann.

[Hempel, 1966] Carl G. Hempel. *Philosophy of Natural Science.* Prentice Hall, Englewood Cliffs, NJ, 1966.

[Inoue and Haneda, 2000] Katsumi Inoue and Hiromasa Haneda. Learning abductive and nonmonotonic logic programs. In Flach and Kakas [2000].

[Inoue, 1992] Katsumi Inoue. Linear resolution for consequence finding. *Artificial Intelligence*, 56:301–353, 1992.

[Josephson, 1994] John R. Josephson. Conceptual analysis of abduction. In *Abductive Inference: Computation, Philosophy, Technology*, chapter 1, pages 5–30. Cambridge University Press, New York, 1994.

[Josephson, 2000] John R. Josephson. Smart inductive generalizations are abductions. In Flach and Kakas [2000].

[Kakas and Mancarella, 1990] Antonis C. Kakas and Paolo Mancarella. On the relation between truth maintenance and abduction. In *Proceedings of the Second Pacific Rim International Conference on Artificial Intelligence*, 1990.

[Kakas and Michael, 1995] Antonis C. Kakas and Antonia Michael. Integrating abductive and constraint logic programming. In *Proceedings of the Twelfth International Conference on Logic Programming ICLP-95*, pages 399–413, 1995.

[Kakas and Riguzzi, 1997] Antonis C. Kakas and Fabrizio Riguzzi. Learning with abduction. In Nada Lavrač and Sašo Džeroski, editors, *Proceedings of the Seventh International Workshop on Inductive Logic Programming*, volume 1297 of *Lecture Notes in Artificial Intelligence*, pages 181–188. Springer-Verlag, 1997.

[Kakas and Riguzzi, 1999] Antonis C. Kakas and Fabrizio Riguzzi. Abductive concept learning. *New Generation Computing*, 1999.

[Kakas *et al.*, 1992] Antonis C. Kakas, Robert A. Kowalski, and Francesca Toni. Abductive logic programming. *Journal of Logic and Computation*, 2(6):719–770, 1992.

[Lamma *et al.*, 2000] Evelina Lamma, Paola Mello, Fabrizio Riguzzi, Floriana Esposito, Stefano Ferilli, and Giovanni Semeraro. Cooperation of abduction and induction in logic programming. In Flach and Kakas [2000].

[Lipton, 1991] Peter Lipton. *Inference to the Best Explanation.* Routledge & Kegan Paul, London, 1991.

[Michalski, 1987] Ryszard S. Michalski. Concept learning. In S.C. Shapiro, editor, *Encyclopedia of Artificial Intelligence*, pages 185–194. John Wiley, Chicester, 1987.

[Michalski, 1993] Ryszard S. Michalski. Inferential theory of learning as a conceptual basis for multistrategy learning. *Machine Learning*, 11:111–151, 1993.

[Muggleton and Buntine, 1988] Stephen H. Muggleton and Wray L. Buntine. Machine invention of first-order predicates by inverting resolution. In *Proceedings of the Fifth International Conference on Machine Learning*, pages 339–352, San Mateo, CA, 1988. Morgan Kaufmann.

[Muggleton and De Raedt, 1994] Stephen H. Muggleton and Luc De Raedt. Inductive logic programming: theory and methods. *Journal of Logic Programming*, 19-20:629–679, 1994.

[Muggleton, 1995] Stephen H. Muggleton. Inverse entailment and Progol. *New Generation Computing*, 13(3-4):245–286, 1995.

[O'Rorke, 1994] Paul O'Rorke. Abduction and explanation-based learning: case studies in diverse domains. *Computational Intelligence*, 10:295–330, 1994.

[Ourston and Mooney, 1994] Dirk Ourston and Raymond J. Mooney. Theory refinement combining analytical and empirical methods. *Artificial Intelligence*, 66:311–344, 1994.

[Peirce, 1958] Charles Sanders Peirce. *Collected Papers of Charles Sanders Peirce.* Harvard University Press, Cambridge, Massachusetts, 1958. Edited by C. Harstshorne, P. Weiss & A. Burks.

[Salmon, 1984] M.H. Salmon. *Introduction to Logic and Critical Thinking*. Harcourt, Brace, Jovanovich, 1984.

[Thompson and Mooney, 1994] C. A. Thompson and Raymond J. Mooney. Inductive learning for abductive diagnosis. In *Proceedings of the Twelfth National Conference on Artificial Intelligence*, pages 664–669, Seattle, WA, August 1994.

[Wang, 2000] Pei Wang. Unified inference in extended syllogism. In Flach and Kakas [2000].

[Yamamoto, 2000] Akihiro Yamamoto. Using abduction for induction based on bottom generalization. In Flach and Kakas [2000].

GABRIELE PAUL

AI APPROACHES TO ABDUCTION

1 INTRODUCTION

Abductive reasoning has gained increasing interest in many fields of AI research. Its utility was first observed for diagnostic tasks (cf. [Pople, 1973] or, e.g., [Console and Torasso, 1991; Console *et al.*, 1991b]), but as many researchers have shown it is not limited to this use. Currently under investigation or suggested are such different applications as plan recognition (e.g., [Dragoni and Puliti, 1994; Helft and Konolige, 1990; Bauer and Paul, 1993; Bauer *et al.*, 1993]), text understanding and generation (e.g., [Stickel, 1990]), program debugging (cf. [Charniak and McDermott, 1985]), test generation (see [McIlraith, 1994]), planning (e.g., [Eshghi, 1991; Stone, 1998]), user modeling (cf. [Poole, 1988]), database updates (e.g., [Kakas and Mancarella, 1990a]), case-based reasoning (cf. [Leake, 1993; Satoh, 1998]), learning (cf. [Kakas *et al.*, 1998; Lamma *et al.*, 1997] or [Thompson and Mooney, 1994]), temporal reasoning (e.g., [Li and Pereira, 1996]), constraint handling (e.g., [Bürckert and Nutt, 1992; Wetzel and Toni, 1998]) or vision (cf. [Charniak and McDermott, 1985]).

The aim of this paper is to summarize the underlying formal models of abduction and investigate different methods for arriving at abductive hypotheses. Only in the last section will we regard some possible applications from the field of plan recognition and plan generation.

Abduction is used to generate explanations, i.e., abducible sentences, whereas deduction is normally used for testing derivability. In a deduction step, each consequence is obtained by applying a logically correct inference rule, so the deduced formulae are logical consequences of the theory under consideration. Reasoning by abduction can be roughly described by the following inference rule:

$$\frac{\varphi \rightarrow \omega,\ \omega}{\varphi}$$

This rule corresponds to some kind of inversion of Modus Ponens. Nevertheless, concerning the implication there is an important difference. Modus Ponens requires material implication as it is used in classical logic, but abduction allows one to characterize the relationship between φ and ω with a much wider degree of freedom. Usually, it is thought of as some kind of causal relationship, as e.g., *φ is the reason for ω being true*, but as Levesque (cf. [Levesque, 1989]) shows, not all abduction has to be concerned with cause and effect in the strict sense. He suggests the extension of the notion of explanation in order to grasp the case that φ is sufficient, not only necessary, to sanction the belief in ω, i.e., there does not have to be a direct causal relationship between both formulae but, in connection

D. M. Gabbay and R. Kruse (eds.), Abductive Reasoning and Learning, 35–98.

with what is known, φ is enough for ω to be true (see also [Boutilier and Becher, 1995]).

Most of the current approaches do not consider these representation problems. Normally, the rules used for abduction are written down using material implication that is implicitly, i.e., by its use, interpreted as some kind of causal relationship.

In this paper, we will first examine the origin of the notion of abduction in philosophical logic, where it is considered in the context of the two other important inference rules, namely deduction and induction. To give a more extensive description of abductive reasoning we will divide the reasoning process into three parts that are examined independently. First, we will give a *formal description* of the abductive task that determines the underlying model. We discuss the set-cover-based approach, the logic-based and knowledge-level approaches as well as semantic accounts for abduction. Methods for *hypothesis generation* are presented in section 3. In particular, we treat generation by resolution and generation by an Assumption-based Truth Maintenance System. As these methods produce *sets* of feasible hypotheses, it is convenient to further restrict these sets by applying heuristics. Approaches to *finding the "best" hypotheses* are investigated in section 4. In section 5.1, we will briefly look at the relationship between default reasoning and abduction. By determining a common subtask some remarks on the complexity of both abductive and default reasoning can be made. We conclude by presenting in section 6 some applications of abductive reasoning in the field of planning and plan recognition, respectively: the approaches of Eshghi (see [Eshghi, 1991]), and Helft and Konolige (see [Helft and Konolige, 1990]).

1.1 Philosophical Background

The notion of abduction was first introduced by the philosopher Peirce (see [Peirce, 1931-1958]). In the field of AI it was taken up in the 1970s by Pople and Morgan (cf. [Pople, 1973; Morgan, 1971]) and the research in this area was revived by Charniak and McDermott. In [Charniak and McDermott, 1985], they stress the importance of abduction as a third form of inference besides induction and deduction.

In philosophical logic, abduction has always been seen in close connection with induction. In his earlier papers (see [Peirce, 1931-1958]), Peirce sees abduction as a reasoning process from effect to cause, thus yielding explanations. In contrast to this, induction conjectures general laws from particulars, i.e., instead of synthesizing explanations, induction classifies and thus does not add new knowledge to the theory. Later, Peirce emphasizes the connection between both inference methods in the following way: "Induction is an argument which sets out from a hypothesis, resulting from previous Abduction ... " (see p. 198 in [Goudge, 1950]). This means that abduction should be used to synthesize an explanation, the induction axiom, that thereafter is verified by induction. Only the use of both processes together yields an acceptable explanation. The integration of abductive and inductive reasoning is treated, e.g., in [Lamma *et al.*, 1997; Kakas and Riguzzi, 1997; Thompson and Mooney, 1994] or [Adé and Denecker, 1994].

An important feature of abduction and as well of induction is their reducibility to a valid deduction, if the abductive process is sound. If φ explains ω abductively in connection with theory $\mathcal{T}$, then ω must be derivable from $\varphi \cup \mathcal{T}$. This connection gave rise to one method of constructing abductive hypotheses, namely by resolving the negated observation ω with theory $\mathcal{T}$. The dead-ends reached are candidate hypotheses (see section 3). Finding an explanation in such a way by deductive means thus ensures the soundness of the process.

In general, there are several possible abductive hypotheses and the problem arises of how to choose among them. Peirce gives several criteria that a good explanation should fulfill (see [Goudge, 1950]). First and foremost, a hypothesis should account for the facts. But it should also follow Ockham's Razor, i.e., it should be the "simplest" hypothesis available. In general, "simplicity" is interpreted as logical simplicity, which means that those hypotheses are preferred that contain fewer different predicates. These are the hypotheses that require the fewest additional assumptions to what has been observed. Peirce stresses the importance of logical simplicity, but what he estimates higher is "psychological simplicity" in the sense that an adequate hypothesis should be the most natural, i.e., one that would be intuitively preferred.

Current research favours logical simplicity as selection criterion and one reason for this is certainly the fact that psychological simplicity is hard to define. However, Mooney and Ng (cf. [Ng and Mooney, 1990] and [Ng and Mooney, 1992]) have suggested a new measure for the quality of explanations which they call *explanatory coherence* (see also section 4.2). This metric controls the choice of hypotheses so that those are selected that best "tie together" the various observations that are to be explained, i.e., that yield the most coherent explanation. As the authors claim, these are usually also the intuitively preferred ones. Especially in the field of text understanding and plan recognition or planning, where reasoning processes of human agents have to be simulated, such a measure that prefers the most natural hypothesis seems inevitable.

2 FORMAL MODELS

It is possible to determine different models that are used to specify the abductive process:

- set-cover-based approaches (e.g., [Allemang *et al.*, 1987]),
- logic-based approaches (e.g., [Eshghi and Kowalski, 1988]),
- the knowledge-level approach ([Levesque, 1989; Inoue and Sakama, 1995]), and

- semantic accounts of abduction ([Boutilier and Becher, 1995; Mayer and Pirri, 1996]).

In the following, we will describe each of them.

2.1 A Set-cover-based Approach

In set-cover-based approaches a set of explanations is found by selecting a suitable subset from a given set of hypotheses. This subset should best account for the observations and is determined by coverings, parsimony, plausibility or another suitable selection criterion. Since hypotheses are constructed using a set of previously known candidates, this approach is also called *hypothesis assembly.*

To be able to treat this form of abductive reasoning more formally, we will first introduce the terminology used in [Allemang *et al.*, 1987].

DEFINITION 1. [Allemang *et al.*, 1987] (hypothesis assembly) A *domain for hypothesis assembly* is defined by the triple (Φ, Ω, e), where Φ is a finite set of hypotheses, Ω is a set of observations and e is a mapping from subsets of Φ to subsets of Ω. $e(\Phi)$ is called the *explanatory power* of the set of hypotheses Φ and determines the set of observations Φ accounts for. An assembly problem is given by a set $\Omega' \subseteq \Omega$ of observations that have to be explained.

Example: To illustrate the ideas presented in this and the following sections, we will always use slight variations of the scenario outlined below.

Let $\mathcal{T}$ be a theory consisting of the following propositions:

$$\forall x\ (bird(x) \land \neg ab(x) \supset flies(x)) \tag{1}$$
$$\forall x\ (ufo(x) \supset flies(x)) \tag{2}$$
$$\forall x\ (penguin(x) \lor ostrich(x) \supset ab(x)) \tag{3}$$
$$\forall x\ (songbird(x) \supset bird(x)) \tag{4}$$
$$\forall x\ (songbird(x) \supset eats\_insects(x)) \tag{5}$$
$$\forall x\ (frog(x) \supset eats\_insects(x)) \tag{6}$$
$$\forall x\ (frog(x) \supset green(x) \land croaks(x)) \tag{7}$$
$$\forall x\ (frog(x) \supset ab(x)) \tag{8}$$

In addition, we assume axioms stating the disjointness of the extensions of predicates denoting different animal types, i.e., frogs are not birds, and the implications for animal subclasses, i.e., ostriches are birds.

Taking this example scenario, but with explicit knowledge about things that do not fly, a domain for hypothesis assembly can be defined as follows:

$$
\begin{aligned}
\Phi &= \{frog(x), songbird(x), bird(x), ufo(x), no\_bird(x),\\
&\quad\ penguin(x), ostrich(x)\}\\
\Omega &= \{flies(x), croaks(x), \neg flies(x), eats\_insects(x)\},\\
e(\{frog(x)\} &= \{eats\_insects(x), \neg flies(x), croaks(x)\},\\
e(\{songbird(x)\} &= \{eats\_insects(x), flies(x)\},\\
e(\{ufo(x), bird(x)\} &= \{flies(x)\},\\
e(\{penguin(x)\} &= \{\neg flies(x)\},\\
e(\{ostrich(x)\} &= \{\neg flies(x)\},\\
e(\{no\_bird(x)\} &= \{\neg flies(x)\}.
\end{aligned}
$$

However, in order to ensure the practical use of this formal model some additional assumptions must be made which a feasible domain should fulfil. The *computability assumption* is essential for all set-cover-based models.

Computability Assumption

For any subset Φ' *of* Φ, $e(\Phi')$ *can be computed.*

This means that it is always exactly known which observations are explained by which set of hypotheses. Thus, following our example, e.g., $e(\{bird(x)\}$ must also be computable. How this can be done is determined by the *independence assumption.*

This states that the union of two hypotheses sets Φ_1 and Φ_2 accounts for the observations explained by Φ_1 as well as for those explained by Φ_2. Thus, it is sufficient to give for each single hypothesis the observations accounted for.

Independence Assumption

Let $\Phi_1, \Phi_2 \subseteq \Phi$. *Then* $e(\Phi_1 \cup \Phi_2) = e(\Phi_1) \cup e(\Phi_2)$.

The independence assumption is necessary for the stepwise computation of the mapping e. If $e(\Phi_1)$ is known, then $e(\Phi_2) = e(\Phi_1 \cup \{\varphi\})$ can easily be generated by conjoining $e(\{\varphi\})$ to the already known set $e(\Phi_1)$. Thus,

$$e(\{bird(x), songbird(x)\}) = \{flies(x), eats\_insects(x)\}.$$

Allemang et al. point out that this assumption is satisfied by most diagnostic domains that allow decomposition in independent parts. The independence assumption is also used in the set-covering model defined by Reggia (cf. [Reggia, 1988]). The algorithm of Allemang et al. allows one to weaken it by the following two assumptions:

Monotonicity Assumption

Let $\Phi_1, \Phi_2 \subseteq \Phi$ *and* $\Phi_1 \subseteq \Phi_2$. *Then* $e(\Phi_1) \subseteq e(\Phi_2)$.

The observations accounted for by a subset of Φ_2 are a subset of those accounted for by Φ_2. This assumption is weaker than the previous one. We only have $e(\Phi_1) \subseteq e(\Phi_1 \cup \Phi_2)$ and $e(\Phi_2) \subseteq e(\Phi_1 \cup \Phi_2)$, i.e., the union of both hypotheses sets accounts for all data explained by Φ_1 and Φ_2, but possibly also for more. With the following assumption, those data can be determined for which a hypothesis is essential.

Accountability Assumption

The function $\alpha : \Phi \rightarrow \Omega^*$ *with*

$$\alpha(\varphi) = \{\, \omega \in \Omega |\, \exists \Phi_1 \subseteq \Phi \text{ with } \varphi \in \Phi_1 \text{ such that } \omega \in e(\Phi_1) \text{ and } \omega \notin e(\Phi_1 \backslash \{\varphi\}) \,\}$$

is computable.

By $\alpha(\varphi)$ we get the observations that cannot be explained without φ. If the monotonicity assumption holds, we have $\alpha(\varphi) \subseteq e(\varphi)$. Our example domain yields $\alpha(\varphi) = e(\varphi)$, as it obeys the stronger independence assumption that states that each hypothesis φ accounts precisely for $e(\varphi)$.

Set-cover-based approaches rely heavily on the previously known mapping e that already determines a superset of the explanations searched for. As a consequence, there exists no actual hypothesis generation mechanism, but rather a collection procedure that selects the relevant hypotheses out of this superset. Therefore, we will now describe the complete abductive algorithm.

Let (Φ, Ω, e) be an assembly domain and $\Omega' \subseteq \Omega$ an assembly problem. The abductive algorithm of Allemang et al. consists of four parts:

- Screening phase,
- Collection phase,
- Parsimony phase, and
- Critique phase.

Screening phase:

In the first phase, the plausibility of the hypotheses in Φ is determined,e.g., by some kind of heuristics or by predefined probability values that can be evaluated with a hierarchical classification system. In the following phases, only those hypotheses with a sufficiently high plausibility are taken into account. Considering our example we can, e.g., classify owls as birds and assign a higher priority to the more general explanation "bird".

Collection phase:

In the collection phase, the hypotheses accounting for each $\omega \in \Omega'$ are collected iteratively. Note that Φ is now the result of the screening phase. In detail, we have:

$Hyp \leftarrow \emptyset$

until $\Omega' = \emptyset$

do let ω be the "most salient" element of Ω';

find the most plausible hypothesis $\varphi \in \Phi$ with $\omega \in \alpha(\varphi)$;

$Hyp \leftarrow Hyp \cup \{\varphi\}$;

$\Omega' \leftarrow \Omega' \setminus e(Hyp)$;

od

Remarks:

- The meaning of "salient" still has to be interpreted for a given application domain.
- A hypothesis can account for several observations. All of these are then taken out of the set Ω' of observations that still have to be explained.
- A conflict arises if incompatible hypotheses are collected in Hyp. In an actual implementation of Allemang et al. this is resolved by maintaining only the last hypothesis found. As a consequence, the data accounted for by the rejected hypothesis must be explained again and strong heuristics may have to be used in order to avoid infinite loops.

Example: Assume the previously defined example domain of hypothesis assembly is given.
Let $\Omega' = \{\neg flies(F), croax(F)\}$.
First, consider the observation $\neg flies(F)$. We have to find the most plausible $\varphi \in \{penguin(F), frog(F), ostrich(F), no\_bird(F)\}$. Assume that $\varphi = penguin(F)$ is chosen, i.e., $Hyp^1 = \{penguin(F)\}$. Thus,

$$\Omega' = \Omega' \setminus e(\{penguin(F)\} = \{croaks(F)\}$$

In the second loop iteration we find $\varphi = frog(F)$ with $croaks(F) \in \alpha(frog(F))$. Hence,

$$Hyp^2 = \{penguin(F), frog(F)\}.$$

Those hypotheses are incompatible (as we assumed disjointness of incompatible predicates denoting animals), so $penguin(F)$ is rejected and $\neg flies(F)$

has to be explained again. If $no\_bird(F)$ is taken as hypothesis, we get

$$Hyp = \{no\_bird(F), frog(F)\}$$

as a solution.

Parsimony phase:

The parsimony phase guarantees that there is no proper subset of Hyp that also explains Ω'. This case can occur because the elements of Ω' are considered successively. A hypothesis φ_1, added to Hyp as an explanation of the observation ω_1, can later become superfluous, because hypothesis φ_2 explains ω_2 and ω_1 as well. We get the following procedure:

for each $\varphi \in Hyp$
do if $\Omega' \subseteq e(Hyp \backslash \{\varphi\})$
  then $Hyp \leftarrow Hyp \setminus \{\varphi\}$;
  fi
od

Example (continuation):
We had the solution $Hyp = \{no\_bird(F), frog(F)\}$. As

$$\begin{aligned}\{\neg flies(F), croaks(F)\} &\subseteq e(\{frog(F)\}) \\ &= \{eats\_insects(F), croaks(F), \neg flies(F)\}\end{aligned}$$

we can find a subset of Hyp that also accounts for the observations, i.e.,

$$Hyp \leftarrow \{frog(F)\}.$$

Remark: This procedure does not ensure that the smallest solution with respect to set cardinality is found. An algorithm solving this problem in the context of a set-cover-based model was designed by Reggia (cf. [Reggia, 1988]).

Critique phase:

Essential hypotheses, i.e., hypotheses without which no explanation can be found at all, are marked as such. This supplies the user with additional information about important parts of *every possible* explanation. The collection phase is repeated $|Hyp|$ times with each hypothesis in Hyp marked as unusable in turn.

for each $\varphi \in Hyp$
do $G \leftarrow Hyp \backslash \{\varphi\}$;
  repeat collection phase with $Hyp \leftarrow G$
    and $\Phi \leftarrow \Phi \setminus \{\varphi\}$
  if no explanation can be found
  then mark φ as essential
od

Remarks:

- The collection phase is executed starting with the solution set diminished by exactly one hypothesis φ in each loop iteration. This hypothesis is marked as unusable. If no explanation can be found, then φ is essential.
- Obviously, $frog(F)$ is essential in our example.

This set-cover-based algorithm has been implemented in the system RED (cf. [Allemang *et al.*, 1987], [Smith, 1985]) for a blood bank antibody analysis. It yields one possible plausible explanation that is subset-minimal.

Nevertheless, there are some drawbacks that cannot be disregarded. First and foremost, the computability of the mapping e is crucial for the choice of possible explanations. All causal relationships that might be relevant must be encoded in form of relations before starting the abductive procedure. This seems practicable only in restricted areas, e.g., repair problems. Apart from that, the domain must satisfy further assumptions. The independence assumption is quite strong and even if it is relaxed by the monotonicity and accountability assumptions the application seems to be restricted to domains that are easy to manage, i.e., no domains where databases of commonsense knowledge are involved. As a last point, one should mention that, as Levesque notes (see [Levesque, 1989]), small changes in the underlying theory most probably lead to difficulties, as the corresponding changes in the explanations are hard to express, i.e., the addition of new facts can enforce a quite extensive respecification of the function e.

So finally we could state that the set-cover-based model appears to be adequate for diagnostic tasks or repair problems where all causal relationships are well known and can easily be represented by a function. In addition, the underlying theory should not undergo any changes. In [Bylander *et al.*, 1991], it is outlined how consistency-based diagnosis can be mapped into a set-cover-based model (see also [Eiter and Gottlob, 1995]).

2.2 *Logic-based Approaches*

The majority of research in abduction that is also the most widely accepted is based on a logical model. It allows that knowledge represented in some logical language for the purpose of deductive inferences can also be used for abduction.

Before going into the detail of several logic-based approaches, we will introduce the general idea. An abductive system consists of

- a logical theory $\mathcal{T}$ defined over the language $\mathcal{L}$, and
- a set of sentences A of $\mathcal{L}$ that are called *abducible.*

If a sentence φ is found as the result of an abductive process in searching for an explanation of ω, it must satisfy the following conditions:

- $\mathcal{T} \cup \varphi \vdash \omega$,
- $\mathcal{T} \cup \varphi$ is consistent,
- φ is abducible, i.e., $\varphi \in A$.

Remark: Sometimes predicates instead of sentences are declared as abducible. Then, sentences φ are called abducible if they contain only abducible predicates.

Thus, the observation ω must be derivable from the logical theory $\mathcal{T}$ augmented with the explanation φ under the additional condition that φ is consistent with $\mathcal{T}$. This ensures what Peirce called soundness of the abductive procedure. Furthermore, the explanation must be an element of the set of abducible predicates or sentences. This last condition is no restriction, as A can be chosen to cover all predicate symbols of $\mathcal{L}$.

Thus, abduction is defined over global logical properties such as consistency and derivability and, as Levesque points out in [Levesque, 1989], this seems to be a drawback of this approach. In addition, the knowledge about causal relationships that is used for finding abductive explanations is represented implicitly in theory $\mathcal{T}$. The implication in $\mathcal{T}$ has to be interpreted in two different ways corresponding to actual use, i.e., in sentences containing abducible predicates implication is seen as a kind of cause-effect relationship, whereas otherwise the material implication is meant. Abduction, modeling sufficient reason for belief in the observation, and deduction within the same logical theory require different notions of implication (cf. [Levesque, 1989]).

In the following we will sketch three different abductive frameworks based on the logical model. The approach of Poole (cf. [Poole, 1988]) is outlined in connection with complexity considerations in section 5.1.

Simple Causal Theories

Konolige analyses in [Konolige, 1990] the relationship between abduction and Reiter's consistency-based approach to explanation (cf. [Reiter, 1987]). The formal model for abduction that he defines corresponds in general to the standard model already outlined.

DEFINITION 2. [Konolige, 1990] (explanation) Let $(C, E, \mathcal{T})$ be a simple causal theory defined over the first-order language $\mathcal{L}$, i.e., C is a set of causes, E a set of effects and $\mathcal{T}$ is a logical theory defined over $\mathcal{L}$. An *explanation* of a set of observations $\Omega \subseteq E$ is a finite set of sentences Φ such that

- Φ is consistent with $\mathcal{T}$,
- $\mathcal{T} \cup \Phi \vdash \Omega$, where Ω denotes the conjunction of all $\omega \in \Omega$,
- Φ is subset-minimal.

Example: Let $(C, E, \mathcal{T})$ be defined as follows: Let $\mathcal{T}$ be given by our example specification in section 2.

$$C = \{frog(x), songbird(x), bird(x), ostrich(x)\}$$
$$E = \{flies(x), green(x), croaks(x), eats\_insects(x)\}$$

If we have the set of observations $\Omega = \{\neg flies(F), croaks(F)\}$ then $\Phi = \{frog(F)\}$ is an explanation, because

- $frog(F)$ is consistent with $\mathcal{T}$,
- $\mathcal{T} \cup frog(F) \vdash \neg flies(F) \wedge croaks(F)$, and
- $frog(F)$ is subset-minimal.

The last point of the definition for explanation embodies a selection criterion for "good" explanations that Allemang called parsimony (see section 2.1). It prevents the choice of a set of sentences as explanations that contains a proper subset which itself constitutes a valid explanation.

Another similarity to the set-cover-based approach can also be noted: the abductive process relates a set of observations that have to be elements in a predefined set E to a set of causes restricted by the set C. However, there is not an equivalent to the mapping e which determines for every feasible hypothesis the data it accounts for. The corresponding relationships are expressed in the logical theory by material implication going from cause to effects.

The results of Konolige show that this form of abductive reasoning can also be carried out with the consistency-based method of Reiter. In fact, a transformation is described by which abductive explanations can be generated or deduced respectively. For details see [Konolige, 1990].

Abduction in Logic Programming

Eshgi and Kowalski (cf. [Eshghi and Kowalski, 1988]) consider abduction in the context of logic programming with integrity constraints. They show that "negation by failure can be simulated by making negative conditions abducible and by imposing appropriate denials and disjunctions as integrity constraints." Thereby they get a semantics for negation as failure that generalizes the stable model semantics. An application of this method in the field of planning is discussed in section 6.1.

In this approach, logic programs working with negation as failure (NAF) are transformed into an abductive framework, where more general integrity constraints than denials can be defined. For this purpose, abduction is introduced as follows:

DEFINITION 3. [Eshghi and Kowalski, 1988] (abductive framework) $(\mathcal{T}, I, A)$ is an *abductive framework* iff

- $\mathcal{T}$ is a Horn clause theory without denials,
- I is a set of integrity constraints, and
- A is a set of predicate symbols defined as abducible.

DEFINITION 4. [Eshghi and Kowalski, 1988] (abductive solution) Let $(\mathcal{T}, I, A)$ be an abductive framework. A hypotheses set Φ is an *abductive solution* for the query q iff

- Φ consists of a set of variable free abducible atoms,
- $\mathcal{T} \cup \Phi \vdash q$,
- $\mathcal{T} \cup \Phi \cup I$ is satisfiable.

Remarks:

- Integrity constraints are used as selection criterion for explanations. Hypotheses not satisfying them are ruled out by the last condition.
- Φ is restricted to being variable free in order to avoid Skolemization. This limitation is not necessary and if more general formulae are to be allowed, the authors refer to the combination of Skolemization and reverse Skolemization as outlined in [Cox and Pietrzykowski, 1986].

Abduction instead of Negation as Failure

Programs working with negation as failure are now converted into ones using abduction by the following algorithm:

Conversion from NAF-formulation into an abductive framework

(1) Negative conditions g are replaced by a new symbol g^*.

(2) Add the integrity constraints $\forall x\ (\neg(g^*(x) \land g(x)))$ and $\forall x\ (g(x) \lor g^*(x))$.

(3) Declare g^* abducible.

Remarks:

- By (1) all negative conditions are eliminated by introducing an unambiguous new symbol for them.

- (2) and (3) ensure that if g is not valid then g^*, the new symbol introduced for $\neg g$, is abducible. The second integrity constraint enforces that either g or g^* is valid.

- The search space of the conversion is almost equal to the original one, but instead of testing the provability of negated conditions by negation as failure, the consistency of abducible predicates is checked.

Example: Let the Horn clause theory $\mathcal{T}$ be given by

$$\begin{array}{rcl} \mathcal{T} = \{flies(x) & \leftarrow & bird(x) \land \neg ab(x), \\ bird(x) & \leftarrow & songbird(x), \\ eats\_insects(x) & \leftarrow & songbird(x), \\ frog(x) & \leftarrow & green(x) \land croaks(x), \\ ab(x) & \leftarrow & frog(x)\} \end{array}$$

A conversion yields the framework $(\mathcal{T}', I, A)$ with

$$\begin{array}{rcl} \mathcal{T}' & = & \{flies(x) \leftarrow bird(x) \land ab(x)^*, \\ & & bird(x) \leftarrow songbird(x), \\ & & eats\_insects(x) \leftarrow songbird(x), \\ & & frog(x) \leftarrow green(x) \land croaks(x), \\ & & ab(x) \leftarrow frog(x)\} \end{array}$$

$$\begin{array}{rcl} I & = & \{\forall x\ (\neg(ab^*(x) \land ab(x))), \\ & & \forall x\ (ab^*(x) \lor ab(x))\} \end{array}$$

$$A = \{ab^*\}$$

Now, consider the query $flies(Sam) \land eats\_insects(Sam)$. $\mathcal{T}$ with NAF yields the answer $true$ as $\neg ab(Sam)$ can be proved by NAF.

The proof in the abductive framework is essentially equivalent till we reach the clause $\leftarrow ab^*(Sam)$. $ab^*(Sam)$ can be abduced, if

$$\begin{array}{lcl} \mathcal{T} \cup \{ab^*(Sam)\} & \vdash & flies(Sam) \wedge eats\_insects(Sam) \text{ and,} \\ \mathcal{T} \cup \{ab^*(Sam)\} & \cup & \{\forall x\,(\neg(ab^*(x) \wedge ab(x))), \forall x\,(ab^*(x) \vee ab(x))\} \end{array}$$

is satisfiable. By showing the consistency of $ab^*(Sam)$ this is proved.

Remark: The abductive proof consists of two phases. In a first step, the *abductive phase*, standard SLD-resolution is used to reach abducibles of A, in the example case $\leftarrow ab^*(x)$. The following *consistency phase* checks the integrity constraints by forward reasoning from the corresponding denial, here $\forall x\,(\neg(ab^*(x) \wedge ab(x)))$, to the goal $\neg ab(x)$ which then should finitely fail by backward resolution.

Note: In [Eshghi and Kowalski, 1989] the disjunctive integrity constraint is introduced with the aid of the metalogical statement $Demo(\mathcal{T}, q) \vee Demo(\mathcal{T}, q^*)$ that holds iff $\mathcal{T} \vdash q$ or $\mathcal{T} \vdash q^*$.

The authors have shown that abduction can be seen as an extension of logic programming by transforming programs working with negation as failure into an abductive framework that is at least as powerful as the original formulation. In order to drive the generation of abductive explanations Φ deduction is used, which ensures the soundness of the process, i.e., $\mathcal{T} \cup \Phi \vdash \Omega$, where Ω describes the facts to be explained.

As the proof procedure described above is only sound with respect to a certain class of logic programs, different extensions have been proposed. Among others, Dung (cf. [Dung, 1991]) developed an alternative sound semantics without the disjunctive integrity constraints using *preferred extensions*, and in [Satoh and Iwayama, 1992] the abductive proof procedure is extended to cover stable models. For details see [Kakas *et al.*, 1998].

Abductive Logic Programming (ALP)

In [Kakas and Mancarella, 1990b] the generalized stable model semantics for abductive frameworks is defined. It covers the more general case where positive literals can also be declared abducible. An abductive framework is given by the tuple $(\mathcal{T}', A \cup A', I \cup I')$ with positive abducibles in A as well as negative abducibles in the converted form in A'. Thus, it is possible to distinguish between default reasoning (by means of negation as failure) and non-default hypothetical reasoning. A transformation of such logic programs to normal logic programs with negation as failure is given in [Toni and Kowalski, 1995]. Details of this and several other reductions can be found in [Kakas *et al.*, 1998].

Generalizing the ideas presented above an argumentation framework (cf. [Toni and Kowalski, 1995], [Dung, 1993]) is proposed as a semantics for nonmonotonic reasoning.

DEFINITION 5. [Toni and Kowalski, 1995] (argumentation framework) An *argumentation framework* is a tuple $(\mathcal{T}, \vdash, A, I)$ where

- $\mathcal{T}$ is a theory in some formal language;
- $\vdash$ is monotonic derivability for the given language;
- A is a set of assumptions, which are sentences of the language, and
- I is a set of denial integrity constraints with retractibles, i.e. at least one literal in the constraint is marked as withdrawable in case of an integrity violation.

A sentence is said to be a nonmonotonic consequence, if it follows monotonically from the theory extended by an *acceptable* set of assumptions. In [Toni and Kowalski, 1995] different possibilities for acceptable sets are presented, depending on the notion of attacks between assumption sets. In simple terms, a set attacks an assumption if there exists an integrity constraint which is valid for the theory plus assumption, but violated if the set is added as well. In this way, various sets of assumptions can be defined, e.g., stable sets, partial stable sets or preferred sets which correspond to the stable models, partial stable models etc. for normal logic programs.

An alternative semantics which comprises logic programs, default logic and autoepistemic logic is given in [Brewka and Konolige, 1993]. For further details and a general overview of ALP see [Kakas *et al.*, 1998]. ALP has been used in many areas where abductive reasoning has been incorporated, e.g., in [Lamma *et al.*, 1997], [Kakas and Riguzzi, 1997], [Li and Pereira, 1996], [Jung, 1998], or [Satoh, 1998].

Abduction and Nonmonotonic Theory Change

Inoue and Sakama (see [Inoue and Sakama, 1995]) extend the notion of abduction to autoepistemic theories. Their aim is to characterize nonmonotonic update or revision of an autoepistemic theory with a formula ω with respect to the minimal explanation of ω.

To this end, they allow an autoepistemic background theory $\mathcal{T}$ and a set A of autoepistemic abducible formulae. Apart from defining the "classical" explanation of a formula ω (here called *positive explanation*) by $\mathcal{T} \cup \varphi \models \omega$ for $\varphi \in A$, it is also taken into account that the removal of a formula may explain the observation. These *negative explanations* of ω are characterized by $\mathcal{T} \setminus \varphi \models \omega$. In addition, positive (negative) *anti-explanations* are defined by $\mathcal{T} \cup \varphi \not\models \omega$ and $\mathcal{T} \setminus \varphi \not\models \omega$, respectively. An explanation is thus determined by a set which should be added to the theory and a set to be removed. We have

DEFINITION 6. [Inoue and Sakama, 1995] (explanation) An *abductive framework* is a pair $(\mathcal{T}, A)$ in which both $\mathcal{T}$ and A are autoepistemic theories. For a formula ω a pair (Φ_+, Φ_-) with Φ_+ and Φ_- consisting of instances of elements from A, is an *explanation* of ω with respect to the abductive framework, if

- $(\mathcal{T} \cup \Phi_+) \setminus \Phi_- \models \omega$;
- $(\mathcal{T} \cup \Phi_+) \setminus \Phi_-$ is consistent.

An *anti-explanation* is defined correspondingly with the first condition being $(\mathcal{T} \cup \Phi_+) \setminus \Phi_- \not\models \omega$.

Example: (taken from [Inoue and Sakama, 1995]) Let the theories $\mathcal{T}_1$ and A be given by

$$\begin{aligned} \mathcal{T}_1 &= \{bird(x) \wedge \neg ab(x) \supset flies(x), \\ &\quad\; bird(x) \wedge ab(x) \supset \neg flies(x), \\ &\quad\; bird(Tweety)\} \\ A &= \{\neg ab(x), ab(x)\} \end{aligned}$$

An explanation of $flies(Tweety)$ is $(\{\neg ab(Tweety)\}, \emptyset)$.
If $\neg ab(Tweety)$ is added to $\mathcal{T}_1$, i.e., $\mathcal{T}_2 = \mathcal{T}_1 \cup \{\neg ab(Tweety)\}$ and we later learn that Tweety has lost its ability to fly, an explanation of $\neg flies(Tweety)$ with respect to the new abductive framework $(\mathcal{T}_2, A)$ is given by

$$(\{ab(Tweety)\}, \{\neg ab(Tweety)\}).$$

An anti-explanation of $flies(Tweety)$ with respect to $(\mathcal{T}_2, A)$ is

$$(\emptyset, \{\neg ab(Tweety)\}).$$

Remarks:

- The definition of explanation comprises the "classical" first-order case, if $\Phi_- = \emptyset$ for every explanation.
- Minimality of an explanation is defined with respect to subset-minimality of Φ_- and Φ_+.
- Update of an autoepistemic theory $\mathcal{T}$ is regarded as the construction of $(\mathcal{T} \cup \Phi_+) \setminus \Phi_-$ for the minimal explanation (Φ_+, Φ_-) of a formula ω if ω has to be inserted into $\mathcal{T}$, and the minimal anti-explanation (Φ_+, Φ_-) if ω is deleted.

As also the authors state, this approach defines a more general framework than previous ones and better accounts for Peirce's theory of abduction (cf. section 1.1). Besides the generation of hypothetical explanations for some observation, also their revision is formalized. If experimental testing yields new information, i.e., new data can be observed at a later point in time, the verification of the previous explanations may lead to a revision of the hypotheses.

2.3 The Knowledge-level Approach

The knowledge-level approach was proposed by Levesque (cf. [Levesque, 1989]) and is based on a model of belief. Abduction is defined with respect to epistemic states and thus is independent of the symbolic representation. In the following we first outline abduction in Leveque's logic of implicit and explicit belief. Then we consider semantic accounts for abduction where the notions of explanation and preference between different explanations are defined in terms of a logic for abduction and the observation need not be consistent with the domain theory. We outline the ideas of Boutilier and Becher (see [Boutilier and Becher, 1995]) and sketch the logic of Mayer and Pirri (cf. [Mayer and Pirri, 1996]).

Abduction in a logic of implicit and explicit belief

The knowledge-level approach of Levesque (cf. [Levesque, 1989]) considers abduction in the context of implicit and explicit belief. For implicit belief, this approach coincides with the general ideas underlying the logical models as presented in the preceding section. But furthermore, it allows the exchange of the model of belief, thus yielding a very general definition of abduction that does not depend on the respective knowledge representation.

We suppose that a propositional language $\mathcal{L}$ is given. Beliefs are formulated in the language $\mathcal{L}^*$, whose atomic sentences are of the form $B\alpha$ with $\alpha \in \mathcal{L}$. A subscript is used to distinguish different types of belief.

DEFINITION 7. [Levesque, 1989] (epistemic state) An *epistemic state* e determines which sentences of $\mathcal{L}$ are believed, i.e., $e \models B_\lambda\alpha$ for beliefs of type λ.

Explanation is defined as follows:

DEFINITION 8. [Levesque, 1989] (explanation) φ expl$_\lambda$ ω with respect to e iff $e \models (B_\lambda(\varphi \supset \omega) \wedge \neg B_\lambda \neg\varphi)$.

Note: This means that φ is an explanation for ω if it is believed that φ implies ω, regarding material implication, and if the negation of φ is not believed. The last conjunct incorporates some kind of consistency check w.r.t. the belief set under consideration.

However, this definition allows multiple explanations. Therefore, a syntactic selection criterion is defined that will restrict the feasible explanations, thereby influencing the knowledge operator $EXPLAIN$ that is defined below. Following Ockham's Razor, explanations with fewer propositional letters are preferred, where p and $\neg p$ are considered as different. More formally we have:

DEFINITION 9. [Levesque, 1989] (literals of a formula) The *set of literals* $LITS(\alpha)$ of a formula α is defined by
$LITS(false) = \emptyset$;
$LITS(p) = \{p\}$ for an atom p;

$LITS(\neg\alpha) = \{\, \overline{m} \mid m \in LITS(\alpha) \,\}$;
$LITS(\alpha \wedge \beta) = LITS(\alpha \vee \beta) = LITS(\alpha) \cup LITS(\beta)$.

DEFINITION 10. [Levesque, 1989] (simplicity) A formula α is *simpler* than a formula β (written $\alpha \prec \beta$) iff $LITS(\alpha) \subset LITS(\beta)$. $\preceq$ is defined straightforwardly.

With this in mind we are able to clarify what the statement "φ explains ω minimally" means.

DEFINITION 11. [Levesque, 1989] (minimal explanation) φ *minimally explains* ω with respect to an epistemic state e and a belief type λ, i.e., $\varphi\ min\_expl_\lambda\ \omega$ with respect to e iff

$$\begin{cases} \varphi\ expl_\lambda\ \omega \text{ with respect to } e \text{ and} \\ \text{there is no } \varphi^* \prec \varphi \text{ with } \varphi^* expl_\lambda\ \omega \text{ with respect to } e. \end{cases}$$

An abductive process should return the disjunction of all minimal explanations, i.e, as a semantic characterization we get

DEFINITION 12. [Levesque, 1989] (explanation operator)

$$EXPLAIN_\lambda[\![e, \omega]\!] = \|\{\, \varphi \mid \varphi\ min\_expl_\lambda\ \omega \text{ with respect to } e \,\}\|,$$

where $\|\varphi\|$ is the set of all models in which φ is true.

An algorithm is called *correct* if it returns exactly all minimal explanations with respect to an epistemic state. Assume that the results of the abductive procedure are described by the function $explain(\mathcal{T}, \omega)$, i.e., $explain(\mathcal{T}, \omega)$ yields the explanations of ω with respect to the theory $\mathcal{T}$. Then, correctness can be characterized by the equation:

$$EXPLAIN_\lambda[\![\mathcal{R}_\lambda(\mathcal{T}), \omega]\!] = \|explain(\mathcal{T}, \omega)\|,$$

where the function $\mathcal{R}_\lambda$ maps sets of sentences into epistemic states with respect to the belief type λ.

The model described gives rise to different abductive procedures corresponding to the underlying notion of belief. In section 3.2, we will discuss explicit and implicit belief and the resulting connection between abduction and an Assumption-based Truth Maintenance System.

Abduction as Belief Revision

In [Boutilier and Becher, 1995], Boutilier and Becher give a logical specification of explanation based on the revision of an epistemic state of an agent. Thereby they use a qualitative notion of plausibility. As they argue, the definition of an abductive explanation in terms of deductive inference results in several shortcomings:

- The defeasibility of abductive explanations cannot be expressed by monotonic derivability.

- The preference of some explanations over others has to be expressed by metalogical conditions.
- Inconsistent observations cannot be explained.

As the authors point out, probabilistic approaches (cf. [Pearl, 1988], [Poole, 1991], see also section 4.1) or choosing the right representation of knowledge in terms of abnormal exceptions can cope with the first two points. Looking at the broader context of knowledge assimilation advocated in [Kakas *et al.*, 1998] or the revision model (cf. [Inoue and Sakama, 1995]) described in section 5, inconsistent observations can be explained if previous assumptions are responsible for the contradiction. In this case, the contradictory hypotheses are rejected.

In [Boutilier and Becher, 1995], a family of bimodal logics coping with all three points is proposed. A plausibility ordering over possible worlds, which captures preference in a qualitative manner, semantically describes the revision process and allows for the definition of different notions of explanation.

We assume a language $\mathcal{L}$ consisting of a denumerable set of propositional variables, the classical connectives, and the modal operators $\Box$ and $\overleftarrow{\Box}$. A semantics is given in terms of possible worlds:

DEFINITION 13. [Boutilier and Becher, 1995] (CT4O-model) A *CT4O-model* is a triple $M = (W, \leq, \mu)$, where W is a set of possible worlds, $\leq$ is a reflexive, transitive binary relation on W, and μ maps the propositional variables of $\mathcal{L}$ into 2^W, i.e., $\mu(P)$ is the set of worlds where P is true.
Sentences in $\mathcal{L}$ are interpreted as usual. The truth of a modal formula F at world $w \in W$, written $M \models_w F$ is given by

- $M \models_w \Box F$ iff for each v such that $v \leq w$, $M \models_v F$;
- $M \models_w \overleftarrow{\Box} F$ iff for each v such that $v \not\leq w$, $M \models_v F$.

Thus, $\Box F$ holds in w, if F holds in all equally or more plausible worlds, and $\overleftarrow{\Box} F$ is true, if F is true in all less plausible worlds.

If K is a belief set, then in a K*-revision-model* the most-plausible worlds are exactly those in which K holds. The revision of K by F, denoted by K^*_F, is defined by

$$\| K^*_F \| = \{w : w \models F, \text{and } v < w \text{ implies } v \not\models F\}$$

for $w \in W$ of a fixed CT4O-model. $\| S \|$ denotes the worlds satisfying each sentence $F \in S$. Thus, if K is revised by F, then F holds in the most-plausible worlds.

The intuition behind an explanation φ for ω is that belief in φ should be sufficient for belief in ω. This means in terms of revision-models that ω should be in

K^*_φ, i.e., ω is true in all most-plausible worlds of φ. This can be expressed in the logical language by defining the conditional connective $\varphi \Rightarrow \omega$ as follows:

$$\varphi \Rightarrow \omega \equiv_{def} \overleftrightarrow{\Box} (\varphi \supset \Diamond(\varphi \wedge \Box(\varphi \supset \omega)))$$

where $\overleftrightarrow{\Box}$ and $\Diamond$ are derived modal operators denoting truth in all worlds and truth in some equally or more plausible world, respectively.

Remark: A plausibility ordering should be determined by the application (agent or program) to reflect its epistemic state.

Using the presented ideas, the notion of predictive explanation is defined as follows.

DEFINITION 14. [Boutilier and Becher, 1995] (predictive explanation)

Let M be a K-revision-model reflecting the epistemic state of an agent with belief set K. A *predictive explanation* for an observation ω (relative to M) is any $\varphi \in \mathcal{L}_{PL}$, the propositional sublanguage of $\mathcal{L}$, such that:

(1) $M \models (B\varphi \equiv B\omega) \wedge (B\neg\varphi \equiv B\neg\omega)$,

(2) $M \models \varphi \Rightarrow \omega$, and

(3) $M \models \neg\omega \Rightarrow \neg\varphi$,

where $B\alpha \equiv_{def} true \Rightarrow \alpha$, i.e., α is believed.

(1) means that an observation ω is accepted exactly when an explanation φ is. (2) formulates the predictive condition that belief in φ induces belief in ω, and (3) states that φ is sufficient for ω, as suspending the belief in ω induces the belief in $\neg\varphi$. This condition is needed in the case where both φ and ω are already believed and conditions (1) and (2) are thus trivially fulfilled.

According to the epistemic status of ω, an explanation can be further classified and the above requirements can be simplified accordingly. For a *factual explanation*, i.e., ω is believed, and $\omega, \varphi \in K$, it suffices to show $\neg\omega \Rightarrow \neg\varphi$ as stated above. *Hypothetical explanations* cover the cases were the observation ω is *rejected* or *indeterminate*. If we have $\neg\omega, \neg\varphi \in K$, then φ is an explanation for ω if it can be shown that $\varphi \Rightarrow \omega$. If ω is indeterminate, i.e., $\omega, \neg\omega \notin K$, it suffices to show for $\varphi, \neg\varphi \notin K$ that $\varphi \Rightarrow \omega$ iff $\neg\omega \Rightarrow \neg\varphi$, which means that belief in φ induces belief in ω if and only if the absence of ω would induce the absence of φ.

Remarks:

- Preference between explanations is defined with respect to the plausibility of the explanations. Most-plausible explanations are preferred.
- The semantics even allows for trivial or irrelevant explanations. To discard these the authors suggest application-dependant filters for the selection of explanations.

- Non-predictive explanations which don't induce belief in the observation but just "make it reasonable" can also be captured by employing a weaker notion of revision.

This weaker notion is also used to show how the abductive approach of THEORIST (cf. [Poole, 1988], [Poole, 1989a] and section 5.1) and the preferred subtheories of Brewka (see [Brewka, 1991]) can be embedded in the presented logical approach. Apart from showing how to augment THEORIST with *preferred explanations*, the authors also show how consistency-based diagnosis (cf. [Reiter, 1987], [de Kleer *et al.*, 1990] and [Console and Torasso, 1991]) is captured by THEORIST and thus also by the presented conditional framework.

By viewing abduction in the context of belief revision as presented above, a very general framework relying on the epistemic state of an agent has been defined. It allows for different notions of explanation and preference between explanations is determined by the application. Further investigation is needed (see [Boutilier and Becher, 1995]) to discriminate between evidential and causal explanations, or to allow for explanations which are themselves causal laws. Another point are explanations which simply make an observation more plausible as it was before.

A Modal Logic for Abduction

Mayer and Pirri (see [Mayer and Pirri, 1996]) use the semantics of multimodal structures to formalize the abductive relation in a modal system.

Assuming an ordering relation $<$ on propositional formulae which is compatible with logical equivalence, i.e., if $\models A \equiv B$ and $\models C \equiv D$ then $A \leq C$ iff $B \leq D$, a *preferred explanation* φ of Γ, denoted φ pref-expl Γ, is defined as follows:

$$\varphi \text{ pref-expl } \Gamma \text{ iff } \models_{PC} \varphi \supset \Gamma \text{ and } \nexists\, \varphi' < \varphi \text{ with } \varphi' \supset \Gamma,$$

with the propositional logical entailment $\models_{PC}$. By choosing Γ to be $\mathcal{T} \supset \omega$ for a finite propositional theory $\mathcal{T}$, we can reformulate the entailment condition $\mathcal{T} \cup \{\varphi\} \models_{PC} \omega$ of abduction as $\models_{PC} \varphi \supset \Gamma$.

The idea is now to define a formula $\varphi \rightsquigarrow \Gamma$ over a modal language $\mathcal{L}$ which captures preference

$$\varphi \text{ pref-expl } \Gamma \text{ iff for every } P\text{-model } M \text{ of } <,\ M \models \varphi \rightsquigarrow \Gamma,$$

where a P-model of $<$ is a P-structure which captures $<$ in terms of modal operators defined over a transitive relation $\succ$.

In more detail, a P-structure is a multimodal structure $M = (k_0, K, \succ, t)$ with K being a set, $k_0 \in K$, $\succ$ being a transitive relation, and t a mapping from K into a set of so-called k-structures, i.e., each k_i is associated with a set of k-structures. k-structures are tuples (w_0, W, R, V) with W being a set, $w_0 \in W$ (the "actual world"), $R \subseteq W^2$, and $V : W \mapsto S$ maps the worlds into propositional interpretations of the finite language $\mathcal{L}$ respecting totality and canonicity (cf. [Mayer and Pirri, 1996]).

$\mathcal{L}$ is defined as usual with the modal operators $\Box$ and $\overleftarrow{\Box}$ which should roughly be interpreted as truth in accessible worlds and truth in all non-accessible worlds, respectively. With the aid of the derived modality $\bigcirc$ for "only knowing" ($\bigcirc F$ means F holds in all accessible and $\neg F$ in all non-accessible worlds), it can be shown that every k-structure can uniquely up to equivalence be associated with a formula $\bigcirc A$ for a propositional A, and furthermore that φ is an explanation for Γ in the propositional language, i.e., it holds $\models_{PC} \varphi \supset \Gamma$, if and only if we have in $\mathcal{L}$ that $\models \bigcirc\varphi \supset \Box\Gamma$, with $\models$ being entailment in $\mathcal{L}$.

In order to catch preference the language is augmented with the modality $[\star]$, which is not allowed to occur inside other modalities and is interpreted over P-structures M. The truth of $[\star]$ is defined with respect to the relation $\succ$ between k-structures, and by the above thus between uniquely identifiable formulae of the propositional language:

$$M \models_k [\star]\alpha \ \text{ iff } \ \forall k', \text{if } k \succ k', M \models_{k'} \alpha.$$

With the aid of this it becomes possible to express preference of explanations in the modal language as follows

$$\varphi \leadsto \Gamma \equiv_{def} [\star](\bigcirc\varphi \supset (\Box\Gamma \wedge [\star]\neg\Box\omega)).$$

This implies $[\star](\bigcirc\varphi \supset \Box\Gamma) \wedge [\star](\bigcirc\varphi \supset [\star]\neg\Box\Gamma)$. By the first conjunct $\models_{PC} \varphi \supset \Gamma$. The second conjunct captures the fact that for a formula $\varphi' < \varphi$ with associated k-structure k, i.e., $k \models \bigcirc\varphi'$, we have $M \models_k \bigcirc\varphi' \wedge \Box\neg\Gamma$ which is equivalent to $\not\models_{PC} \varphi' \supset \Gamma$. This means that φ' doesn't explain Γ.

With respect to P-structures the sound and complete system $AK4$ is defined by appropriate axioms and inference rules. Apart from showing some general structural properties of the abductive relation $\leadsto$, the authors show how the axiomatization of special preference relations in $AK4$ can be used to derive abductive relations of the form $\varphi \leadsto \Gamma$ in $AK4$.

As shown above, this approach captures the classical definition of abduction in terms of monotonic derivability in a logic whose semantics is defined in terms of a preference relation. As the authors also state, among others, further research is needed to extend the approach to an infinite language and theory, or to first-order logic, and to investigate automated deduction methods.

3 HYPOTHESES GENERATION

The most important part of an abductive procedure is the formation of a new theory that explains the observed data. In section 2.1 one such method was described in connection with the set-cover-based model. This, however, was not hypotheses generation but hypotheses assembly. Out of a set of predefined hypotheses, the feasible ones are selected with the aid of a computable, predefined mapping e. The mapping e goes from hypotheses to data sets and determines accountability.

In this section we will investigate algorithms that do not rely on such a mapping, i.e., all relevant information is encoded in the underlying theory. We distinguish between the following methods for hypotheses generation that strongly depend on the respective formal model:

- generation by a form of linear resolution and
- generation with an Assumption-based Truth Maintenance System.

The generation by methods of probability theory is discussed in section 4 in the context of the selection of the "best" hypotheses.

3.1 Generation by Resolution

Many authors have shown that forms of linear resolution can be used for the generation of hypotheses in logic-based models, e.g. [Pople, 1973; Eshghi and Kowalski, 1989; Kakas *et al.*, 1998; Console *et al.*, 1991a; Cox and Pietrzykowski, 1986]. In this context, we will outline the basic algorithm of Pople (cf. [Pople, 1973]), the algorithm of Cox and Pietrzykowski (cf. [Cox and Pietrzykowski, 1986]), the IFF proof procedure of Fung and Kowalski (cf. [Fung and Kowalski, 1997]), and the abductive reasoning method of Shanahan (cf. [Shanahan, 1989]) that incorporates a generalization of SLD-resolution.

Linear Resolution and Skolemization

The procedure of Cox and Pietrzykowski is based on linear resolution and reverse skolemization. This allows the explanation of facts that are described by arbitrary formulae, not just variable-free clauses (cf. [Eshghi, 1991]).

Let $\mathcal{T}$ be a logical theory and let ω be a formula (the fact to be explained). $\mathcal{T}$ also denotes the conjunction of all formulae in $\mathcal{T}$. Furthermore, we assume $\mathcal{T} \nvdash \omega$, as otherwise ω is already explained by $\mathcal{T}$. A cause φ of ω in $\mathcal{T}$ is a formula determined by the following two conditions

- $\mathcal{T} \wedge \varphi \vdash \omega$ and
- φ is consistent with $\mathcal{T}$.

To compute the cause φ, do the following:

(1) Convert $\mathcal{T}$ in clausal form.

(2) Negate ω and perform linear resolution with input clauses from $\mathcal{T}$ and some element of $\neg\omega$ as top clause.

(3) We have assumed $\mathcal{T} \nvdash \omega$. Thus, we distinguish two cases:

(4) the deduction does not terminate,

(5) the deduction terminates with some dead-end, the clause d.

(6) Let D be the set of all dead-ends. Do the following steps for each dead-end d in D.

Negate d, thus obtaining a conjunction $p_1 \wedge \ldots \wedge p_n$ of literals.

Apply the reverse skolemization algorithm, i.e., replace all skolem terms in p_i by universally quantified variables. This yields the formula

$$\varphi :\equiv Q_1 x_1 \ldots Q_k x_k (q_1 \wedge \ldots q_n),$$

where the Q_i stand for the universal quantifiers. Then, φ is a cause for ω in $\mathcal{T}$.

Remarks:

- The algorithm incorporates a selection criterion. All computed causes φ are *basic*, which means that every cause φ' of φ itself is trivial, i.e., φ' implies φ directly. In most cases a minimality criterion is also satisfied. If φ is a computed cause then there is also no more specific cause x in the sense $x \supset \varphi$. This is a consequence of the resolution algorithm that works until no further resolution step can be made. An exception occurs only if a dead-end is subsumed by another.
- The algorithm is not complete, i.e., there exist cases where not all basic and minimal causes are found. For an example consult [Cox and Pietrzykowski, 1986].

The criteria introduced for good explanations, in particular *basicness* and *minimality* will be presented in more detail in section 4.

Example:

(1) Take as given the example theory of section 2. Transforming the theory in clausal form yields:

$$\{\neg bird(x), ab(x), flies(x)\} \tag{9}$$

$$\{\neg ufo(x), flies(x)\} \tag{10}$$

$$\{\neg penguin(x), ab(x)\}\{\neg ostrich(x), ab(x)\} \tag{11}$$

$$\{\neg songbird(x), bird(x)\} \tag{12}$$

$$\{\neg songbird(x), eats\_insects(x)\} \tag{13}$$

$$\{\neg frog(x), eats\_insects(x)\} \tag{14}$$

$$\{\neg frog(x), green(x)\}\{\neg frog(x), croaks(x)\} \tag{15}$$

$$\{\neg frog(x), ab(x)\} \tag{16}$$

In addition, we assume as in section 2 that the different animal types are disjoint corresponding to our intuition, e.g., $\{\neg frog(x), \neg ostrich(x)\}$.

(2) Let the observation ω be given by

$$\omega \equiv \exists x(flies(x) \wedge eats\_insects(x))$$

We start resolution with some element of

$$\{\neg flies(F), \neg eats\_insects(F)\}$$

as top clause.

(3)-(5) Resolution with top clause $\neg flies(F)$ yields

$$\{\neg ufo(F), \neg eats\_insects(F)\} \qquad (*)$$
$$\{\neg bird(F), ab(F), \neg eats\_insects(F)\} \qquad (**)$$

Further resolving (*) does not lead to a dead-end, as the resolvents become inconsistent. A dead-end is reached only if (**) is resolved with (12) and (13). We get

$$\{\neg songbird(F), ab(F)\}$$

(6)

$$D = \{\{\neg songbird(F), ab(F)\}\}$$

Negating the elements of D and applying reverse skolemization gives the solution

$$\forall x(songbird(x) \wedge \neg ab(x)).$$

Problem Reduction by Linear Resolution

Instead of adding the negation of the observed data ω to the logical theory, Pople (cf. [Pople, 1973]) tries to directly show $\mathcal{T} \supset \omega$. The observation is converted into disjunctive form. By regarding each disjunct separately the problem of proving ω is split into an equivalent set of subproblems. A clause in the logical theory is interpreted as a kind of rewrite rule and by backward chaining one arrives at nodes that will not create successor nodes, i.e., that cannot be proved from the axiom set. These are candidates for an explanatory hypothesis. If all different subproblems are considered, a set of possibly competing hypotheses is generated.

In detail Pople proceeds as follows:

(1) Convert the theory $\mathcal{T}$ to quantifier-free conjunctive normal form.

(2) Use skolemization of universally quantified variables to convert the observation ω in disjunctive normal form:

- Eliminate implication.
- Reduce the scope of negation.
- Replace each *universally* quantified variable by a skolem function that has as arguments the variables of any existential quantifiers occuring before the universal quantifier.
- Drop existential quantifiers.
- Transform the resulting expression into disjunctive normal form.

(3) By considering all possible combinations for satisfying the disjunction ω, the problem can be split into n different but equivalent formulations, where n is the number of disjuncts. To see this, assume that

$$\begin{aligned} \omega &\equiv d_1 \vee \ldots \vee d_{n-1} \vee d_n \\ &\equiv (\neg d_1 \wedge \ldots \wedge \neg d_{n-1}) \supset d_n \end{aligned}$$

If the conjunction $\neg d_1 \wedge \ldots \wedge \neg d_{n-1}$ is added to $\mathcal{T}$ the problem is reduced to showing d_n. This process can be repeated with each d_i in turn as the right-hand side of the implication so that n different problem formulations are generated.

(4) Clauses in $\mathcal{T}$ are interpreted as productions with a single literal on the left-hand side. Every clause with l disjuncts allows the formation of l different rules in which each disjunct constitutes the left-hand side in turn.

(5) Try to show each subproblem by backward chaining. This can yield up to n different sets of dead-ends.

(6) Note that for each subproblem the conjecture d_i consists of a possibly unary conjunction. Now, those dead-ends are preferred that give the most coherent explanation of the observations, i.e., that account for the most conjuncts of those to be explained. This heuristics is incorporated in the reasoning mechanism by a selection process called *synthesis* or *factoring across partial trees*. It states that, whenever possible, literals occuring in the different proof trees built for each conjunct of the observation should be unified. The resulting unifier accounts for both top literals of the corresponding proof trees, i.e., both observations are explained. Among several feasible hypotheses those are always preferred that account for the most observations.

Example:

(1) Let $\mathcal{T}$ be given by

$$\{\neg bird(x), ab(x), flies(x)\} \tag{17}$$
$$\{\neg airplane(x), ab(x), flies(x)\} \tag{18}$$
$$\{\neg ufo(x), flies(x)\} \tag{19}$$
$$\{\neg penguin(x), ab(x)\}\{\neg ostrich(x), ab(x)\} \tag{20}$$
$$\{\neg songbird(x), bird(x)\} \tag{21}$$
$$\{\neg songbird(x), eats\_insects(x)\} \tag{22}$$
$$\{\neg frog(x), eats\_insects(x)\} \tag{23}$$
$$\{\neg frog(x), croaks(x)\} \tag{24}$$
$$\{\neg frog(x), ab(x)\} \tag{25}$$

(2) Let the observation ω be described by

$$\exists x((flies(x) \wedge eats\_insects(x)) \vee (eats\_insects(x) \wedge croaks(x)))$$

By skolemization we get

$$(flies(F) \wedge eats\_insects(F)) \vee (eats\_insects(F) \wedge croaks(F))$$

(3) The reformulation yields two subproblems, namely

(a) Add $\neg flies(F) \vee \neg eats\_insects(F)$ to the theory and show

$$eats\_insects(F) \wedge croaks(F).$$

(b) Add $\neg eats\_insects(F) \vee \neg croaks(F)$ to the theory and show

$$flies(F) \wedge eats\_insects(F).$$

(4) Assume $\mathcal{T}$ is reformulated as demanded, e.g., the first clause

$$\{\neg bird(x), ab(x), flies(x)\}$$

allows for three "production rules":

$$\begin{aligned} bird(x) &\supset ab(x) \vee flies(x) \\ \neg ab(x) &\supset \neg bird(x) \vee flies(x) \\ \neg flies(x) &\supset \neg bird(x) \vee ab(x) \end{aligned}$$

(5) Thus, the dead-ends for (a) are $\{frog(F)\}$ and for (b) $\{songbird(F) \wedge \neg ab(F)\}$

Remarks:

- With the synthesis process, Pople implements a selection procedure that obeys Ockham's Razor in the sense that only those hypotheses are regarded that are strongly confirmed by several observable data.
- For diagnostic tasks, Pople suggests the choice of one abducible predicate, e.g., "$presence(X, Y)$," that explains the observed malfunction. In this case, synthesis is only regarded with respect to the occurrence of this predicate.
- Other heuristics for the controlled search for abductive explanations are estimated as useful, but are not yet implemented in the medical diagnosis system described by Pople.

A Proof Procedure for Abductive Logic Programming

In [Fung and Kowalski, 1997], the IFF proof procedure for ALP (see section 2.2) is presented which is sound and complete with respect to the three-valued semantics of the completion of an abductive logic program. It generalizes the work of Console et al. (see [Console *et al.*, 1991a]) and is closely related to the SLDNFA-procedure in [Denecker and de Schreye, 1992]. SLDNFA gives the more general answers, but as the authors argue IFF seems to be the simpler procedure and yields a stronger completeness.

Given an abductive logic framework $(\mathcal{T}, I, A)$, $\mathcal{T}$ consists of the completion of a logical program in if-form: $P(x_1, \ldots x_n) \leftrightarrow D_1 \vee \ldots \vee D_m$, where P is neither $=$, $true$, $false$ or an abducible in A, each D_i is a conjunction of literals and the variables x are implicitly universally quantified. The integrity constraints I are given as a set of implications

$$a_1 \vee \ldots \vee a_n \longleftarrow b_1 \wedge \ldots \wedge b_m, \quad m, n \geq 0.$$

A contains abducible predicate symbols. An *abductive solution* to a query q is a pair (A', σ) with ground abducible atoms in A' and a substitution of ground terms for the variables of the query such that

- $\mathcal{T} \cup Comp(A') \cup CET \models q\sigma$, and
- $\mathcal{T} \cup Comp(A') \cup CET \models I$

with $Comp(A')$ being the completion of A' in terms of iff-definitions as above and CET being Clark equality theory. All other abducibles in A are equivalent to false.

Note: In section 2.2 the consistency-view of integrity constraints, i.e., $\mathcal{T} \cup Comp(A') \cup CET \cup I$ is satisfiable, has been adopted. By requiring here semi-decidable logical entailment a complete proof procedure with respect to the three-valued semantics can be defined. The answers are also answers with respect to the consistency-view (cf. [Fung and Kowalski, 1997]).

IFF Proof Procedure

Starting with a query q

- construct an initial node by conjoining q with the set of integrity constraints;
- apply the following inference rules as long as possible:

unfolding: replace an atom which occurs as a conjunct in a node or in the condition of an implication by the body of its definition;

propagation: resolve an atom with an implication in the same node;

splitting: use distributivity laws to create two successor nodes;

case analysis: if $x = t$ occurs in the condition of an implication, create one successor node for $x = t$ and one for $x \neq t$;

factoring: if two abducibles $P(x)$ and $P(y)$ occur in a node, create two successor nodes, one in which the arguments are identical and one where they differ;

rewrite rules for equality: apply to simulate unification;

logical simplifications: apply in order to reformulate implications and to simplify expressions containing $true$ and $false$.

Remarks:

- The IFF top-down derivation consists of two phases: Unfolding, splitting, and simplification with rewrite rules and case analysis wherever possible and then propagation and factoring again using simplifications and case analysis.
- The authors give no strategy for the application of the inference rules, but they expect completeness for every "fair" strategy.

Example: We reformulate the example of the preceding sections as an abductive framework $(\mathcal{T}, I, A)$ with definitions in iff-form.

$$\begin{aligned} \mathcal{T} = \{ & flies(x) \leftrightarrow (bird(x) \wedge \neg ab(x)) \vee (airplane(x) \wedge \neg ab(x)), \\ & eats\_insects(x) \leftrightarrow frog(x) \vee songbird(x)\} \end{aligned}$$

$$\begin{aligned} I = \{ & bird(x) \leftarrow songbird(x), \\ & ab(x) \leftarrow frog(x), \\ & ab(x) \leftarrow penguin(x), \\ & ab(x) \leftarrow airplane(x) \wedge defective(x)\} \end{aligned}$$

together with implications stating the disjointness of animals corresponding to our intuition, e.g., $\{false \leftarrow frog(x) \wedge bird(x), \ldots\}$.

$$A = \{bird(x), airplane(x), ab(x)\}$$

For the query $q = eats\_insect(Sam) \wedge flies(Sam)$ the initial node is obtained by conjoining the integrity constraints:

$$\begin{array}{c} eats\_insect(Sam) \wedge flies(Sam) \wedge \ (bird(x) \leftarrow songbird(x)) \wedge \\ (ab(x) \leftarrow penguin(x)) \wedge (ab(x) \leftarrow frog(x)) \wedge \\ (false \leftarrow frog(x) \wedge bird(x)) \wedge \ldots \end{array}$$

By unfolding and splitting $eats\_insects$ we get

$$\begin{array}{rcl} (frog(Sam) \wedge flies(Sam & \wedge & IC) \vee \\ (songird(Sam) \wedge flies(Sam) & \wedge & IC) \end{array}$$

where we use IC as an abbreviation for the conjunction of integrity constraints as stated above.

Unfolding and splitting $flies$ yields

$$\begin{array}{rcl} (frog(Sam) \wedge bird(Sam) \wedge (false \leftarrow ab(Sam)) & \wedge & IC) \vee \\ (frog(Sam) \wedge airplane(Sam) \wedge (false \leftarrow ab(Sam)) & \wedge & IC) \vee \\ (songbird(Sam) \wedge bird(Sam) \wedge (false \leftarrow ab(Sam)) & \wedge & IC) \vee \\ (songbird(Sam) \wedge airplane(Sam) \wedge (false \leftarrow ab(Sam)) & \wedge & IC) \end{array}$$

Propagation with the implications stating disjointness of different animals and airplanes results in one non-failure node, which gives the answer

$$songbird(Sam) \wedge bird(Sam) \wedge \neg ab(Sam)$$

This entails the definition of the abducibles *songbird*, *bird*, and *ab* corresponding to the semantics above

$$\begin{array}{rcl} sonbird(x) & \leftrightarrow & x = Sam \\ bird(x) & \leftrightarrow & x = Sam \\ ab(x) & \leftrightarrow & x \neq Sam \end{array}$$

Remarks:

- As the authors state, the constructed answer N is minimal, i.e., there is no answer N' with $N' \subset N$. Minimality here doesn't mean that as few abducibles as possible are used in the answer, e.g., by preferring $songbird$ $(Sam) \wedge \neg ab(Sam)$ which entails $bird(Sam)$, but corresponds to the number of definitions for abducibles in $Comp(A')$ as given in the semantics. If Sam is the only songbird, i.e., $songbird(x) \leftrightarrow x = Sam$, then by the implication in the integrity constraints the extension of $bird$ cannot be empty. So every minimal answer defining $songbird$ also has to define $bird$.

- In [Console *et al.*, 1991a] as well, predicate completion of the domain theory is used to find abductive explanations. The authors emphasize the importance of an explicit representation of the incompleteness of a theory in order to avoid undesired results. Furthermore, they remark that the completion has to be done only for the general knowledge, not the observations (and integrity constraints in the above case). The minimization of predicate completion alone does not allow for the derivation of new knowledge.

Abduction and Default Persistence

The approach of Shanahan (cf. [Shanahan, 1989]) will be briefly reviewed, as it incorporates a resolution-based abductive mechanism in a temporal reasoning system, in which prediction and explanation are realized by deduction respectively abduction within the same logical theory $\mathcal{T}$. This shows possible applications of abductive algorithms working with resolution.

The approach is based on Kowalski's and Sergot's Event Calculus (see [Kowalski and Sergot, 1986]). The prediction problem consists of the search for a set of causal consequences ω of a set of events φ such that $\mathcal{T} \cup \varphi \models \omega$. If in turn ω describes the observed events, then abduction is used to find a feasible set φ of explanations satisfying $\mathcal{T} \cup \varphi \models \omega$.

As usual, the abductive algorithm requires that all causal relationships are expressed by formulae of the form "effect if cause." Furthermore, it is assumed that all properties known to be true have an explanation in terms of events. This embodies a kind of *default persistence*. To see this, assume that in order to explain $prop_1$ at time t_1 an event e_1 is introduced that must occur before t_1, i.e., at a time $t < t_1$. Properties can only be changed by another event, so $prop_1$ holds also at time $t_2 > t_1$, if no such event has been observed. As Shanahan points out, this combination of abduction and persistence allows for the correct handling of some hard problems not solvable with other approaches to default persistence.

The abductive algorithm works with a kind of linear resolution. If ω_0 are the facts to be explained, then the problem can be formulated as follows: Find a set of unit clauses φ_n, called the *residue*, such that $\mathcal{T} \cup \varphi_n \models \omega_0$ and φ_n contains only abducible predicates. As usual, the set of abducible predicate symbols is predefined. As the procedure corresponds in general to the algorithm of Cox and Pietrzykowski described in section 3.1, it will not be outlined further.

In order to handle default persistence correctly, the mechanism must be extended. All negated assumptions derived by negation as failure must be recorded and checked in every later proof step, as abduction allows the addition of facts to the theory that may invalidate them. Shanahan also describes a method to cope with nested negation as failure. Essentially, it consists of a combination of normal SLD-Resolution that does not allow addition to the theory, and abduction.

In [Boutilier, 1996], Boutilier also uses an abductive model for the problem of updating belief due to a changing world. In his approach no particular action model is adopted and explaining events are judged corresponding to the plausibility of the event. In addition, conditional explanations are allowed. Plausibility is determined by an event ordering with respect to a world w.

The use of abduction in a temporal context is also described in [Brusoni *et al.*, 1997] and [Ribeiro and Porto, 1994]. Brusoni et al. develop an efficient nondeterministic algorithm for computing temporally consistent explanations for observations with associated temporal constraints using a temporal constraint network (cf. [Dechter *et al.*, 1991]). Temporally inconsistent hypotheses are thereby pruned as early as possible by using local propagation of constraints to the atoms occuring in a formula F which is used in an explanation step.

A query-oriented approach is presented in [Ribeiro and Porto, 1994]. It relies on a language of maximal time intervals. Instead of adding the constraints and the axiomatization of the constraint relations to the theory, a constraint machine and a separate reasoning machine, incorporating abduction to cope with incomplete knowledge, are used.

3.2 An Assumption-based Truth Maintenance System as Abductive Procedure

In section 2.3 the basic ideas of the knowledge-level account for abduction were described. We have

$$EXPLAIN_{\lambda}[\![e, \omega]\!] = \|\{\ \varphi \mid \varphi\ min\_expl_{\lambda}\ \omega \text{ with respect to } e\ \}\|.$$

$EXPLAIN$ gives a semantic characterization of explanation. So in order to develop an abductive algorithm, we need a syntactic counterpart that allows the actual computation of the simplest explanations.

In the following, we will give such a syntactic definition of explanation, and show how an abductive procedure can be developed, depending on a belief type λ. It will turn out that in the case of implicit belief, abductive reasoning can be modelled with an Assumption-based Truth Maintenance System (ATMS).

Levesque (cf. [Levesque, 1989]) defines the function ∇ for two sets of clauses Σ and Γ. If Γ is the set of clauses to be explained and Σ contains the believed clauses with respect to a belief of type λ, then $\nabla(\Sigma, \Gamma)$ determines the corresponding minimal explanations.

DEFINITION 15. ([Levesque, 1989]) $\nabla(\Sigma, \Gamma) = \mu(\Phi)$, where Φ is a set defined by

$$\Phi = \left\{ \neg z \;\middle|\; \begin{array}{l} \forall y \in \Sigma, y \not\subseteq z \\ \forall x \in \Gamma, \exists y \in \Sigma, x \cap y \neq \emptyset \text{ and } (y - x) \subseteq z \end{array} \right\}$$

The function μ determines the smallest set with respect to subsets.

The function ∇ can be interpreted as follows: First, the clauses z searched for are not allowed to be expansions of some $y \in \Sigma$, the set of believed clauses. This is required in order to avoid inconsistencies, if $\neg z$ is later added to the belief set as an explanation. Additionally, z must satisfy another condition. If it is actually possible to explain all observations in Γ, then there must exist for every observation $x \in \Gamma$ a belief $y \in \Sigma$ that, at least partially, accounts for it, i.e., x and y have common literals, $x \cap y \neq \emptyset$. The clauses z are chosen in such a way that they contain every literal of such a y that is not a part of the corresponding observation x. If, in the following, z is negated and added to the belief set Σ, this has the consequence that the observations of Γ are forced to be true. The formula $\neg z$ denies exactly those disjuncts of the clauses y that do not account for the observations. Thus, the remaining disjuncts of each y that are part of the observations have to be true. If different clauses z can be found, they constitute alternative explanations. Finally, the function μ that is applied to the set of clauses $\neg z$ determines the minimal set with respect to set inclusion. This obeys the principle that an explanation should not contain more literals than necessary.

Example: Let Σ be given by

$$\begin{array}{l}\{\neg bird(x) \vee ab(x) \vee flies(x)\}\\ \{\neg songbird(x) \vee bird(x)\}\\ \{\neg songbird(x) \vee eats\_insects(x)\}\\ \{\neg frog(x) \vee eats\_insects(x)\}\\ \{\neg frog(x) \vee green(x)\}\\ \{\neg frog(x) \vee ab(x)\}\end{array}$$

As usual, we assume that corresponding to our intuition the extension of predicates that model animals are disjoint, i.e., we also have axioms of the form $\{\neg frog(x), \neg songbird(x)\}$.

The proposition ω to be explained is given by

$$\omega \equiv \{eats\_insects(F), flies(F) \vee green(F)\}$$

The set $y - x$ of definition 11 is determined by

$$\begin{array}{rll} y - x = & \{ & \{\neg songbird(F) \vee \neg bird(F) \vee ab(F)\},\\ & & \{\neg frog(F) \vee \neg bird(F) \vee ab(F)\},\\ & & \{\neg frog(F)\},\\ & & \{\neg songbird(F) \vee \neg frog(F)\} \;\; \} \end{array}$$

The second and fourth clause are already elements of Σ, but this contradicts $\forall y \in \Sigma, y \not\subseteq z$ as $(y - x) \subseteq z$. Thus, Φ is given by

$$\begin{array}{rll} \Phi = & \{ & \{songbird(F) \wedge bird(F) \wedge \neg ab(F)\},\\ & & \{frog(F)\} \;\; \} \end{array}$$

This can be simplified to

$$\Phi = \{ \ \{songbird(F) \wedge \neg ab(F)\}, \\ \{frog(F)\} \ \}$$

As Φ is already minimal with respect to subsets, we get $\nabla(\Sigma, \Gamma) = \Phi$.

∇ is only defined for clauses, so it can not be used in developing a general abductive procedure. But Levesque has shown that for a certain class of beliefs the simplest explanations are in fact the result of an application of ∇. We will first characterize the beliefs that can be handled.

DEFINITION 16. ([Levesque, 1989]) (regular beliefs) A type of belief λ is *regular*, iff for every epistemic state the following sentences of $\mathcal{L}^*$ are true:

- $B_\lambda \neg \Box$, where $\Box$ denotes the empty clause;
- $(B_\lambda \alpha \vee B_\lambda \beta) \supset B_\lambda(\alpha \vee \beta)$;
- $(B_\lambda \alpha \wedge B_\lambda \beta) \supset B_\lambda(\alpha \wedge \beta)$;
- $B_\lambda(\alpha \wedge \beta) \supset (B_\lambda \alpha \wedge B_\lambda \beta)$;
- $B_\lambda \alpha \equiv B_\lambda \alpha^*$, if α^* is α in conjunctive or disjunctive normal form, or if α is the result of replacing any subformula β in α by β^*, where (recursively) $B_\lambda \beta \equiv B_\lambda \beta^*$ is always true.

For regular beliefs, the simplest explanations can be computed by applying ∇ to the belief set and the transformation of the observations in conjunctive normal form (CNF).

THEOREM 17 ([Levesque, 1989]). *For regular belief*

$$EXPLAIN_\lambda[\![e, \omega]\!] = \|\nabla(\{y \mid e \models B_\lambda y\}, CNF(\omega))\|.$$

We will now separately consider implicit and explicit belief and the resulting connection between the operation ∇ for the respective belief type and an Assumption-based Truth Maintenance System (ATMS).

Implicit Belief

Implicit belief will be denoted by the belief operator B_I. A corresponding epistemic state e is determined by a set of assignments r. Implicitly believed formulae are characterized by

$$e \models B_I \alpha \text{ iff for every assignment } r \in e,\ r \models \alpha,$$

i.e., if α is believed in an epistemic state e, then it must be true in all assignments r that characterize e. This implies that exactly those formulae are believed that are logical consequences of the underlying theory:

$$e \models B_I\alpha \text{ iff } \mathcal{T} \models \alpha,$$

where $e = \mathcal{R}_I(\mathcal{T})$ (see section 2.3). In section 2.3 we defined explanation with respect to an arbitrary belief type as follows:

$$\varphi \text{ explains } \omega \text{ in } e \text{ iff } e \models B_\lambda(\varphi \supset \omega) \wedge \neg B_\lambda \neg\varphi.$$

If we apply this to the above defined implicit belief, we get

$$\varphi\ expl_I\ \omega \text{ with respect to } e \text{ iff } \mathcal{T} \models (\varphi \supset \omega) \text{ and } \mathcal{T} \not\models \neg\varphi.$$

This is equivalent to the definition of abduction in logic-based approaches:

$$\mathcal{T} \cup \{\varphi\} \models \omega \text{ and } \mathcal{T} \cup \{\varphi\} \text{ is consistent.}$$

With the function ∇ we already indicated a computational procedure for the generation of abductive explanations. Thus, the above shows that the further development that is based on the function ∇ will also be applicable to logic-based approaches.

The result of ∇ is reminiscent of the support sets used in an ATMS, and Levesque shows that in fact "an ATMS can be understood as computing all simplest explanations with respect to implicit belief" (cf. [Levesque, 1989]). To see this, we first show that an ATMS procedure with respect to a set of clauses Σ and a symbol p can be defined in terms of ∇.

THEOREM 18 ([Levesque, 1989]).

$$atms[\Sigma, p] = \nabla(Th(\Sigma), \{\{p\}\}).$$

$Th(\Sigma)$ is the deductive closure of theory Σ.

This result can be generalized in order to treat clauses as second argument:

DEFINITION 19. ([Levesque, 1989]) (generalized ATMS) A generalized ATMS procedure for a set of clauses Σ and a clause β is defined by

$$gatms[\Sigma, \beta] = \nabla(Th(\Sigma), CNF(\beta)).$$

Implicit belief is regular. Thus, if Σ is taken as the set of beliefs in an epistemic state e, theorem 17 yields the following for a correct abductive procedure:

LEMMA 20 ([Levesque, 1989]).

$$\|EXPLAIN_I[\![\mathcal{R}_I(\Sigma), \omega]\!]\| = \|gatms(\Sigma, \omega)\|.$$

Proof. For a correct abductive procedure we have

$$EXPLAIN_I[\![\mathcal{R}_I(\Sigma), \omega]\!] = EXPLAIN_I[\![e, \omega]\!].$$

According to theorem 17, this is equivalent to

$$\|\nabla(\{y \mid e \models B_I y\}, CNF(\omega))\|.$$

Implicit belief is defined in such a way that $e \models B_I\alpha$ iff $\Sigma \models \alpha$, when $\mathcal{R}_I(\Sigma)$ determines the epistemic state e. So, the above is equivalent to

$$\|\nabla(\{Th(\Sigma)\}, CNF(\omega))\|.$$

The definition of a generalized ATMS completes the proof. ∎

The lemma proves that an ATMS can be used to generate explanations in a model for implicit belief and, as the equivalence was shown, an ATMS can also be used in logic-based models. But in section 5.2 the exponential time of the ATMS procedure will be discussed. This suggests the consideration of another kind of belief that might be computationally more tractable.

For a general discussion of the relationship between abductive logic programming and Truth Maintenance Systems and in particular the abductive procedures presented in [Satoh and Iwayama, 1991] and [Inoue, 1990] see [Kakas *et al.*, 1998].

Explicit Belief

If α is implicitly believed, then all logical consequences of α are believed as well. In contrast, explicit belief in α only sanctions the belief in all tautological consequences in the sense of *relevance logic*, which allows contradictory information in the knowledge base.

An epistemic state for explicit belief is defined with the aid of *situations*. Situations are total functions that assign a truth value to literals p, where at least p or $\neg p$ is assigned the value 1, but possibly both. Thus, a situation is in general not a valid assignment. A literal follows from a situation s, iff it is assigned the value 1: $s \models p$ iff $s(p) = 1$ and $s \models \neg p$ iff $s(\neg p) = 1$. Connectors and negation are defined as follows:

$$\begin{aligned} s \models (\alpha \wedge \beta) \quad &\text{iff} \quad s \models \alpha \text{ and } s \models \beta; \\ s \models \neg(\alpha \wedge \beta) \quad &\text{iff} \quad s \models \neg\alpha \text{ or } s \models \neg\beta; \\ s \models \neg\neg\alpha \quad &\text{iff} \quad s \models \alpha. \end{aligned}$$

An epistemic state is determined by a set of situations. This yields the following characterization for explicit belief:

$$e \models B_E\alpha \text{ iff for every } s \in e,\ s \models \alpha.$$

The function $\mathcal{R}_E$ is defined similar to $\mathcal{R}_I$ (see section 16). For a given knowledge base $\mathcal{T}$, it determines the set of all situations that satisfy $\mathcal{T}$, i.e., the corresponding epistemic state. For $e = \mathcal{R}_E(\mathcal{T})$ we get

$$e \models B_E\alpha \text{ iff } \mathcal{T} \cup \mathcal{S} \text{ tautologically entails } \alpha \text{ in the sense of relevance logic,}$$

where $\mathcal{S}$ is the set of all clauses of the form $\{p, \neg p\}$. This set is added to ensure that at least one of p or $\neg p$ is assigned the value 1.

The abductive explanations for explicit belief can also be described with the function ∇. But instead of using the logical closure of Σ as belief set, we now need the set $EXPS(\Sigma) = \{y \mid y \text{ is tautologous or } \exists y^* \in \Sigma, y^* \subseteq y\}$.

DEFINITION 21. ([Levesque, 1989]) (explanation with respect to explicit belief) Let Σ be a theory and β a clause.

$$abd[\Sigma, \beta] = \nabla(EXPS(\Sigma), CNF(\beta)).$$

The next theorem states that *abd* in fact yields the simplest explanations with respect to explicit belief.

THEOREM 22 ([Levesque, 1989]).

$$EXPLAIN_E[\![\mathcal{R}_E(\Sigma), \beta]\!] = \|abd[\Sigma, \beta]\|.$$

Levesque shows that generating the explanation of a single clause with respect to explicit belief is computationally easier than the corresponding ATMS procedure for implicit belief. A theorem that relates explicit to implicit beliefs allows the use of *abd* in those cases to compute the explanations for implicit beliefs by recursively computing explicit beliefs.

It has been shown that an ATMS is one possible tool to generate abductive explanations. If we take the results of Selman and Levesque into account, presented in section 5.2, then we will see that an ATMS is computationally also the best procedure that can be expected. This suggests lowering the demands in favour of higher efficiency, e.g., by finding *some* explanation instead of all possible ones. The knowledge-level account suggests further ideas that point in this direction. Considering explicit belief, a computationally simpler procedure can be found that also allows one to recursively compute implicit belief. In addition, this is a very general approach that permits the investigation of further belief types.

4 SELECTION OF HYPOTHESES

The result of an abductive procedure is a *set* of possible explanations. As this set can be quite extensive, it seems reasonable to discard the "less interesting" hypotheses in order to gain efficiency in the following computational process. Most abductive procedures described in the previous sections already incorporate such a selection criterion (e.g., [Cox and Pietrzykowski, 1986], [Allemang *et al.*, 1987]), to prevent, for instance, the generation of the observations themselves as trivial explanations.

Selection principles are based on heuristics that try to determine good explanations. What seems promising depends partly on the application domain, e.g., fault diagnosis systems should yield the most specific explanation for a malfunction, whereas other tasks may require more abstract hypotheses. But syntactic criteria also exist that can be applied to both domains, e.g., in order to exclude trivial explanations.

The philosopher Peirce (see [Goudge, 1950]) claims that those hypotheses should be selected that correspond to Ockham's Razor in the sense that they are the psychologically simplest, i.e., the most intuitive explanations. This seems hard to realize, for which reason Ockham's Razor is normally interpreted as meaning logical and syntactical simplicity. Hypotheses containing superfluous literals or hypotheses that are subsumed by others are unwanted. The negative results of Levesque (cf. [Levesque, 1989]) confirm that no more can be expected, because he shows that it is impossible to formulate a selection criterion on purely semantic grounds.

In the following, we will investigate different approaches for finding the most promising explanations. The greatest part is formed by methods that realize in some way Ockham's Razor. An alternative has been proposed that is based on a metric of coherence. This approach will also yield a method for determining an appropriate level of specificity of an explanation.

4.1 Simplicity Criteria

Independent of the respective criteria that determine a level of specificity, i.e., the level of abstraction of an explanation, in general, all accepted hypotheses have to satisfy some basic conditions that formalize syntactical simplicity. Representative are the ones that Cox (see also section 3.1) demands for the hypotheses generated by his algorithm.

Let $\mathcal{T}$ be a first-order knowledge base and Φ a set of hypotheses for the observation ω. If $\varphi \in \Phi$ is interesting, it should satisfy the following conditions:

Consistency: $\mathcal{T} \wedge \varphi$ is satisfiable. This is required in all logic-based models.

Non-Triviality: $\neg\varphi \supset \omega$, i.e., the observation is not a direct consequence of the hypotheses. In particular, this excludes that ω itself is synthesized as a feasible explanation.

Basicness: Every consistent explanation of φ itself is trivial. This favours the most specific explanation, in the sense that there is no non-trivial explanation for the explanation itself.

Minimality: For all hypotheses φ' of ω: $\varphi \supset \varphi'$ implies $\varphi' \equiv \varphi$, i.e., there exists no more general hypothesis for ω than φ, in the sense that all superfluous literals or unnecessary universal quantifications are omitted from explanations.

Given, for example, $croaks(x) \supset frog(x)$ as explanation, $croaks(x) \wedge green(x) \supset frog(x)$ is not minimal.

Note:

- If $\varphi \supset \psi$ is a part of the theory, then ψ will not be among the generated hypotheses. The reason for this lies in the resolution algorithm that chains backward as far as possible, thus reaching only φ as possible dead-end. This means that such a resolution-based method always guarantees basicness, and non-minimality can only occur in the cases mentioned.
- Basicness already realizes a strategy for determining an appropriate level of specificity of the explanation set that is called *most specific abduction* (see also below).

Explanations satisfying consistency, minimality and non-triviality realize Ockham's Razor, if it is interpreted syntactically. But the set of interesting hypotheses selected this way can still be quite extensive and usually an additional selection is made to prefer hypotheses of a certain level of specificity. Appelt and Pollack consider in [Appelt and Pollack, 1990] two different sorts of criteria:

- global criteria, and
- local criteria.

Global Criteria

Selection methods based on global criteria consider the assumption set as a whole. We distinguish (cf. [Appelt and Pollack, 1990]):

- cardinality comparisons,
- least presumptive or least specific abduction,
- most specific abduction, and
- minimal abnormality.

Cardinality comparisons: These are used only in diagnostic systems and ensure that those hypotheses are assumed that imply the failure of the smallest number of components. Thus, the application system has to be specified in terms of distinguishable components whose input and output behaviour is completely determined (see [Appelt and Pollack, 1990]). As Appelt and Pollack state, this is not the case for natural language understanding or planning resp. plan recognition.

Least and most presumptive explanations: Less presumptive or less specific is defined as follows: Let h_1 and h_2 be hypotheses and $\mathcal{T}$ a logical theory, then h_1 is less presumptive or less specific than h_2 iff $\mathcal{T} \cup h_2 \models h_1$. Thus, the least presumptive or specific explanations are those that provide the most general explanation. An abductive process that realizes this form of selection was proposed by Stickel (cf. [Stickel, 1990]).

Alternatively, the *most specific explanations* can also be chosen, i.e., those that themselves have no more non-trivial explanations (see also the definition of basicness above). Above all, this strategy seems to be useful for fault diagnosis, where very detailed knowledge about the origin of the failure is required.

Appelt suggests that for plan recognition and mental state ascription, a combination of most and least specific abduction seems adequate. Nevertheless, there is still the problem of deciding which strategy should be applied in which particular case. Furthermore, it remains possible that there exists no single least or most specific explanation which forces a further selection.

In [Dupré and Rossotto, 1995], it is shown how abstraction axioms together with an abductive theory can be used to compute the most abstract, i.e., the least presumptive explanations. The use of an abstraction hierarchy increases the efficiency of the computation of explanations by avoiding the construction of unpreferred hypotheses instead of adopting a selection criterion afterwards (see also section 5). Abstractions can also be used to focus the search for more detailed explanations.

Minimal abnormality: In [Poole, 1989b], Poole defines an abductive framework that relies on the specification of abnormality and normality assumptions. As Appelt claims, an inherent problem of this approach is the fact that minimally abnormal assumptions may be inconsistent with the "best" explanation.

Local Criteria

To avoid the shortcomings of those approaches local criteria have been introduced that allow one to single out the "best" hypotheses, prior to computing the space of all possible explanations. We have

- Bayesian statistical methods,
- weighted abduction, and
- cost-based abduction.

Bayesian statistical methods: Standard probability theory can be used to single out the "most probable" hypothesis. In [Poole, 1993], Poole describes the pragmatically motivated framework of *probabilistic Horn abduction*. It is defined over theories consisting of a set of definite clauses, representing the general knowledge, and a number of "disjoint-declarations" which determine the set of disjoint and covering hypotheses with associated probabilities. Each disjoint-set stands for a set of instances of independent and mutually exclusive hypotheses whose probabilities sum up to 1. Poole shows that any probabilistic knowledge representable in a discrete Bayesian belief network can be represented in this framework. The approach thus yields a simple logical formulation which is semantically transparent and in addition allows for simple implementations. But therefore it requires, among other assumptions, a partitioning of the space of hypotheses into independent alternatives. As Appelt and Pollack state (see [Appelt and Pollack, 1990]), e.g., for the problem of mental state ascription it is hard to give alternatives and establish their independence.

In [O'Rorke, 1994], a generalization is described, which allows not only for the generation of the most probable explanations but also the most important ones. Importance is measured using decision-theory.

Weighted abduction: Appelt and Pollack adapt in [Appelt and Pollack, 1990] a method developed by Hobbs, Stickel et al. (see [Hobbs *et al.*, 1989]). To guide the application of rules, they assign weighting factors to all literals in the premise. Thus, rules have the form

$$\varphi_1^{\mu_1} \wedge \ldots \wedge \varphi_n^{\mu_n} \supset \psi.$$

The weights μ_i are used to compute the assumption cost of literals. This is done by multiplying the assumption cost of the consequent by the weighting factor of the considered literal in the premise, e.g., if ψ has cost c then φ_1 can be assumed at cost $c * \mu_1$. In the abductive procedure the assumption set with the lowest cost is preferred.

A model-theoretic semantics of the approach is based on model preference (cf. also [Appelt, 1990]). By weighted abduction the models of a theory $\mathcal{T}$ are restricted in such a way that those models are filtered out that are inferior according to the model preference constraints. Thereby, an underlying partial preference order on the models of $\mathcal{T}$ is assumed. The weights in the rules are interpreted as additional constraints on this order. If the rule $\varphi^{\alpha} \supset \psi$ is given with weight $\alpha < 1$, this means that every model satisfying $\varphi \wedge \psi$ is preferred to some model satisfying $\neg\varphi \wedge \psi$.

Charniak and Shimony (cf. [Charniak and Shimony, 1990]) note that there is no semantics given for the case $\alpha > 1$, so only a kind of most-specific abduction can be modelled. Furthermore, the rule weightings must result in a consistent ordering without cycles, which makes it even more difficult to determine the weights of newly added rules. Weighted abduction has been used, e.g., in the approach of Wærn ([Wærn, 1992]) to reactive diagnosis.

Cost-based abduction: Charniak and Shimony (cf. [Charniak and Shimony, 1994], [Charniak and Shimony, 1990]) present an alternative approach based on a probabilistic semantics for cost-based abduction. They try to find the best explanation of an observation by finding a minimal cost proof of it as is done in [Hobbs *et al.*, 1989], but with the advantage that a suitable semantics is given in terms of a Boolean belief network. In more detail, they represent the weighted abduction rules as a directed acyclic graph (DAG) and show that finding a model of the DAG with minimal costs corresponds to finding the least-cost set of abductive assumptions. The proposed algorithm relies on a search heuristics on the DAG for finding partial explanations which are further expanded.

In order to define a semantics a Boolean belief network is constructed based on the weighted AND/OR DAG (WAODAG). It is shown that finding the minimal cost model, i.e., the least-cost assumptions, corresponds to finding the maximum a-posteriori probability (MAP) assignment for given facts to the belief network.

In [Santos, 1994], Santos presents an alternative approach for the computation of minimal-cost explanations by using linear constraints between the nodes of the WAODAG which represent causal relationships. Minimizing a cost function associated with the WOADAG with linear programming techniques gives the least-cost explanations. In some cases an additional search is needed in order to find the minimal solutions. With an empirical evaluation and by a comparison with the MAP-approach of Charniak and Shimony the efficiency of the presented algorithm is demonstrated.

The above described ideas show that if broader applications are intended, the assignment of cost or weighting factors seems a promising strategy. But, as also becomes obvious, the respective application domain must be formalized very carefully in order to avoid unintuitive results, i.e., the probabilistic knowledge must be available. In addition, the domain knowledge and in particular the relationship between explanations and observations must be representable in terms of rules which can be transformed into a directed acyclic graph.

In the following, we will investigate another approach that judges the quality of explanations with a *coherence metric*. This method also yields an alternative to determine an appropriate level of specificity.

4.2 Explanatory Coherence as Selection Criterion

Instead of implementing Ockham's Razor as syntactical simplicity, Ng and Mooney (see [Ng and Mooney, 1990]) have determined a metric of coherence that helps to find the most intuitive explanations. The search for explanations is guided by heuristics (beam search), in order to increase efficiency. The metric also determines the level of specificity.

As the authors claim, Ockham's Razor is insufficient for text understanding and plan recognition. They suggest selecting those explanations that best "tie together" all observations. This means that they prefer "those with more connections between any pair of observations," where connections are interpreted as directed paths in the proof graph. More precisely, the metric is defined as follows:

DEFINITION 23. ([Ng and Mooney, 1990]) (coherence metric) The *coherence metric* C is defined by

$$C = \frac{\sum_{1 \leq i < j \leq l} N_{i,j}}{N\binom{1}{2}},$$

where $\mathbf{l}$ is the number of observations, $\mathbf{N}$ the number of nodes in the proof graph, and $\mathbf{N_{i,j}}$ the number of distinct nodes in the proof graph such that there exists a sequence of directed edges from n_k to n_i and from n_i to n_j, where n_i and n_j are observations. The sequences may be empty.

Remarks:

- $N\binom{1}{2}$ is a scaling factor ensuring that the final value lies between 0 and 1.
- The numerator is the sum of the number of nodes in a graph that simultaneously support the interpretation represented by this graph. To determine a "good" explanation all possible connections of nodes, i.e., several different proof graphs, are regarded. The interpretation represented by the graph with the highest coherence is chosen.
- The metric can be computed in time $O(lN + e)$ by using depth-first search.

As the authors point out, this approach has several advantages. On the one hand, coherent explanations are usually also syntactically simpler explanations. This is the case because unification is favoured by preferring tight connections. Furthermore, this method does not allow "too many degrees of freedom," as is the case, e.g., with weighted abduction where the rule weights can be chosen arbitrarily. In addition, the metric helps to determine a level of specificity, since only the proof of a subgoal is attempted by backward chaining if this will increase the overall coherence.

An algorithm using this metric is implemented in the system ACCEL (Abductive Calculation of Causal Explanations for Language) (see [Ng and Mooney, 1992]). To gain efficiency, the authors use ATMS-like caching of partial explanations and a fixed-size set of current partial hypotheses for computation.

Explanatory Coherence appears to be very promising for systems that incorporate in some way human reasoning capabilities, i.e., systems whose behaviour cannot be determined for all cases in advance. The authors show that it can be applied successfully in the field of natural language understanding, plan recognition and diagnosis. But, as they also suggest, the most promising approach is one that uses both coherence and likelihood information in order to be able to cope with incomplete or vague information.

5 THE COMPLEXITY OF ABDUCTION

Abduction is the process of finding plausible explanations for some observed events. If the reasoner becomes more experienced, i.e., more facts become known, it is possible that previously assumed explanations have to be rejected. Thus, a non-monotonic mechanism is needed to keep track of the "plausible" hypotheses.

Default logic seems to be a non-monotonic formalism that can be quite easily adapted to the problem of explanation finding. It is possible to interpret defaults as predefined hypotheses and reasoning with them as a simple kind of theory formation. This view of default logic corresponds to the abductive approach of Poole in THEORIST (see [Poole, 1988]) that is further investigated in the next section. In the following section, the comparison of abductive and default reasoning yields complexity results for both reasoning mechanisms.

5.1 The Relationship between Abduction and Default Reasoning

DEFINITION 24. [Poole, 1988] (scenario) Let $\mathcal{T}$ be a set of closed first-order formulae and Δ an arbitrary set of first-order formulae. A *scenario* of $\mathcal{T}$ and Δ is a set $D \cup \mathcal{T}$ where D is a set of ground instances of elements of Δ such that $D \cup \mathcal{T}$ is consistent.

DEFINITION 25. [Poole, 1988] (explanation) A closed formula ω is *explainable* from a scenario of $\mathcal{T}$ and Δ if there is a set of ground instances D of elements from Δ such that

- $\mathcal{T} \cup D \models \omega$, and
- $\mathcal{T} \cup D$ is consistent.

The set $\mathcal{T} \cup D$ is called an *explanation* of ω.

The correspondence to the logic-based model for abductive reasoning is obvious at first sight. A set of hypotheses out of Δ is accepted as an explanation if it can be consistently added to the theory $\mathcal{T}$ and the observation can then be proved. The sentences in $\mathcal{T}$ are used as a kind of constraint for the hypotheses in Δ. They determine which elements of Δ are feasible hypotheses.

DEFINITION 26. [Poole, 1988] (extension) An *extension* of $\mathcal{T}$ and Δ is the set of logical consequences of a maximal scenario of $\mathcal{T}$ and Δ with respect to set inclusion.

This definition can be related to the notion of extension in default logic. Poole shows that if for the default

$$\frac{\varphi : \psi}{\psi}$$

the formula $\varphi \supset \psi$ is added to the set Δ, his notion of extension corresponds to the one defined by Reiter (cf. [Reiter, 1980]). For general defaults there is no exact translation, but Poole argues that in those cases where the approaches give different results, Reiter's method yields nonintuitive conclusions. This pathological case occurs if the underlying theory contains only disjunctive information about the prerequisites of defaults. Consider, e.g., the theory $\{p(a) \vee q(a)\}$ and the defaults

$$\frac{p(x) : r(x)}{r(x)}, \frac{q(x) : r(x)}{r(x)}.$$

Then, no default is applicable as no single disjunct can be proved. However, Poole claims that the knowledge that the disjunction of the prerequisites is true is enough to sanction the belief that the common conclusion $r(a)$ of the defaults is also true. Thus, in his system the corresponding inference is allowed.

But as lined out in [Kakas *et al.*, 1998], the system doesn't give the right answer if it is confronted with the Yale Shooting Problem, whereas interpreting negation as failure provides the correct solution. This is also the case for the abductive framework as defined in section 2.2.

If abductive approaches are compared with default reasoning systems in more detail some differences are striking. In general, defaults yield several extensions and in order to be able to use them for further derivations, an attempt is made to rule some of them out by assigning priorities. Abductive systems use heuristics to constrain the set of feasible hypotheses but nevertheless multiple hypotheses are not always undesirable. The origin of this discrepancy lies in the different tasks of the two approaches. Hypotheses generated by abduction are used for explanation whereas defaults are used in connection with incompletely specified knowledge bases, in order to allow at least non-monotonic conclusions. Thus, each alternative extension of a default theory determines a maximal extension of the knowledge base. In contrast, abductive hypotheses are in general not mutually exclusive and only partially determine the underlying theory (cf. [Eshghi, 1991]). The fact that the approaches of Poole and Reiter coincide for normal defaults has its origin in Poole's special definition of an extension. An extension is a *maximally consistent* scenario and hence different sets of hypotheses are in fact orthogonal, i.e., inconsistent.

As a conclusion, one could point out that default reasoning can be classified within the framework of theory formation if defaults are viewed as hypotheses. But abduction constitutes the more general approach in the sense that it allows more flexibility by admitting partial extensions.

5.2 *The Complexity of the Abductive Task*

Some results concerning the complexity of abduction can be received by filtering out a common subtask of abductive procedures and methods for computing default extensions.

In section 3.2 an Assumption Based Truth Maintenance System (ATMS) was shown to constitute one possibility for generating abductive hypotheses in a logic-based model. Selman and Levesque prove in [Selman and Levesque, 1996] (see also [Selman and Levesque, 1989]) that the exponential time needed by the ATMS-procedure is not a consequence of a possibly exponential number of hypotheses. Even when the explanations are restricted to those of a predefined set A of sentences, the task is NP-complete. The crucial point is the so-called *support selection task*, where support of the observations has to be found. This task is shown to also be a subtask of every algorithm determining default extensions that contain a given propositional letter P, i.e., for so-called *goal-directed reasoning*.

Assume that a propositional Horn clause theory $\mathcal{T}$, a set of predicate symbols A, and a symbol ω are given. An ATMS computes all explanations φ for ω with respect to the theory $\mathcal{T}$ in such a way that the symbols of φ are in A and φ is a minimal set of literals satisfying

$$\mathcal{T} \cup \varphi \models \omega \text{ and } \mathcal{T} \cup \varphi \text{ is consistent.}$$

It is shown (cf. [Selman and Levesque, 1996]) that finding such an assumption-based explanation is NP-complete, as well as generating an explanation containing a given propositional letter P, i.e., goal-directed reasoning, whereas *some* non-trivial explanation can be computed in polynomial time. This means that even for Horn-clause theories there is probably no improvement in efficiency for abduction. The responsible subproblem can be singled out:

Support Selection Task:

Let $\mathcal{T}$ be a set of Horn clauses, A a predefined set of predicates and ω a predicate symbol. Find a set of literals φ, the so-called *support set* such that

- $\mathcal{T} \cup \varphi \models \omega$;
- $\mathcal{T} \cup \varphi$ is consistent;
- φ contains only symbols from A.

An assumption-based explanation for an observation ω is a minimal support set of ω. But the support selection task can also be used to describe default extensions. Let a default theory Δ be defined by $(\mathcal{T}, D)$, where all defaults in D have the form

$$\frac{:p}{p}$$

with $p \in A$. Then $Th(\mathcal{T} \cup \varphi)$, the deductive closure of $\mathcal{T}$ augmented with φ, is an extension of Δ if and only if φ is a maximal support set of ω (see [Selman and Levesque, 1989]).

Both minimality and maximality do not add to the computational difficulty. In both cases the support selection task is responsible for the intractability. Concerning the computation of solutions, an improved algorithm can be found if not all explanations are computed, but only an arbitrary one. As stated above some non-trivial explanation can be found in polynomial time. Restricting the theory to definite Horn clauses, i.e., clauses with one positive literal, gives another tractable subclass: an assumption-based explanation can be computed in polynomial time. Further results for definite Horn theories can be found in [Bylander *et al.*, 1991].

Thus, to retain efficiency for non-tractable classes of theories *credulous reasoning*, i.e., generating one extension and one set of compatible hypotheses instead of all possible ones, should be favoured over goal-directed reasoning. This result applies both to abduction and default reasoning, since they have the support selection task in common.

In [Eiter and Gottlob, 1995], Eiter and Gottlob generalize and complete the above results. They treat the decision problems of

- determining the existence of an explanation;
- determining the relevance of a hypothesis with respect to an explanation, i.e., whether it contributes to the explanation;
- determining the necessity of a hypothesis for all explanations, and the computational cost of computing explanations with respect to different preference criteria.

Apart from subset-minimality, as outlined above for propositional Horn clause theories, minimum cardinality, priorities and weights for hypotheses are considered.

As the authors state, "the different variations of abduction (...) populate all mayor complexity classes between P and Σ_3^P, Π_3^P", and thus abduction is harder than deduction. For the general case, to decide the existence of an explanation as well as the relevance is Σ_2^P-complete, while determining the necessity is Π_3^P-complete. Subset-minimality in contrast to minimal cardinality doesn't add to the complexity. Using weights for minimal-cost proofs makes the decision problem Δ_3^P-complete (for weights whose value is exponential in the input size n). If priorities are employed with respect to subset-minimal explanations in a priority level, the relevance and necessity checking get Σ_3^P and Π_3^P-complete, respectively. For clausal theories there is no efficiency gain, but for Horn-theories. For details on the complexity classes and an extensive overview see [Eiter and Gottlob, 1995].

These results are rather discouraging for concrete abductive algorithms. To gain efficiency in the application of abductive reasoning it is suggested in [Eiter and Gottlob, 1995] and [Bylander *et al.*, 1991] to use tractable subclasses whenever possible for an application (see also [Eshghi, 1993]). Apart from that, lowering the quality of explanations in terms of allowing explanations which don't explain all the observations might increase efficiency. This implies that a suitable notion of approximation is defined. If possible, a concrete algorithm could also use heuristic knowledge for guiding the search process.

For the case of abductive diagnosis, Console et al. (cf. [Console *et al.*, 1991b]) present an approach which uses compiled knowledge to prune the search process but retains completeness of the process, i.e., only useless search is avoided. This is done by associating the entities of a fault model with necessary conditions which are "easy to compute". The unsatisfiability of such conditions, which can be automatically synthesized from the fault model, allows to focus on the model without the associated entities thus avoiding extensive traversal.

Another point for efficiency gain (cf. [Eiter and Gottlob, 1995]) is the use of polynomial algorithms, if there exist any for abductive tasks, which are neither complete nor correct, but yield the correct result with a sufficiently high probability.

6 APPLICATIONS TO PLANNING AND PLAN RECOGNITION

Planning or plan recognition is one possible application of abductive reasoning that requires a quite general and sophisticated abductive method because of the wide range of possible application domains. In this section, we will investigate two logic-based approaches to planning and plan recognition that use abduction as problem solving strategy. The model presented by Esghi (cf. [Eshghi, 1991]) uses abduction with metalevel integrity constraints instead of negation as failure (see also section 2). The second approach to be described presents a theory for plan recognition as abduction and relevance (cf. [Helft and Konolige, 1990]).

6.1 Abduction and the Event Calculus

Instead of following the usual approach and employing situation calculus and deduction, Eshghi (see [Eshghi, 1991]) develops a planning algorithm with Event Calculus and abduction. His approach solves the frame problem, allows the generation of non-linear plans, and is able to incorporate an ATMS and a least-commitment strategy for efficient search.

The context of abduction is the same as described in section 2. An abductive framework $(\mathcal{T}, I, A)$ consists of a Horn clause theory $\mathcal{T}$ without denials, a set I of integrity constraints, and a set A of abducible predicate symbols. A set of sentences Φ is an abductive solution for the framework $(\mathcal{T}, I, A)$ and the query ω, iff

- Φ consists of a set of variable free abducible atoms,
- $\mathcal{T} \cup \Phi \vdash \omega$,
- $\mathcal{T} \cup \Phi \cup I$ is satisfiable.

Backward chaining with resolution is used for the generation of assumptions. A dead-end R consisting exclusively of abducible predicates is called a *residue*. $\mathcal{T} \cup \neg R$ entails the formula to be explained.

In general, $\neg R$ is existentially quantified and thus needs skolemization. In order to be able to substitute for the skolem constants in the further reasoning process an equality theory is required. Eshghi argues that if all clauses in $\mathcal{T}$ are *homogenised* (see below), then a restricted equality theory that contains only equality axioms and no schemata for inequality is sufficient. The schemata should be avoided, because they lead to an explosion of the search space and render the procedure inefficient.

DEFINITION 27. ([Eshghi, 1991]) (homogeneous clauses) A clause is homogeneous iff

- there are no constant symbols in its head atom, and
- no variable symbols occur more than once in the head.

For example, the homogenised form of the clause $P(a)$ is $\forall x((x = a) \supset P(x))$.

Remark: Eshghi gives an algorithm for how to transform arbitrary clauses into homogenised form.

The final algorithm for finding the set of explanations Δ for a goal G with integrity constraints I of the form $\mathcal{T} \cup \Delta \vdash R_1(x) \leftarrow \mathcal{T} \cup \Delta \vdash R_2(x)$ consists of five phases. We will first outline the complete procedure and in the following give some explanations and an example.

(1) Abductive phase:

$GS \longleftarrow \{G\}; \Delta \longleftarrow \{\};$

repeat until $\Delta' = \{\}$

find Δ' such that $\mathcal{T} \cup \Delta \cup \Delta' \vdash GS;$

By this algorithm a residue is found.

(2) $\Delta \longleftarrow \Delta \cup \Delta'$

(3) Consistency checking phase:
Test with an arbitrary checking algorithm whether $\mathcal{T} \cup \Delta \vdash R_1(x) \leftarrow R_2(x)$ is consistent. If the answer is no, remove clauses from Δ until the clause is consistent.

(4) Precondition determination phase:
if no inconsistency was found in (3)

then determine the instantiated preconditions (bindings $B_1, B_2, \ldots$) for $R_2(x)$ in $\mathcal{T} \cup \Delta'$, i.e., find the bindings for x such that $\mathcal{T} \cup \Delta' \vdash R_2(x)$

else determine these bindings $B_1, B_2, \ldots$ in Δ. (Note that clauses from Δ have been removed in order to guarantee consistency.)

(5) **if** no inconsistency was found in (3).

then $GS \leftarrow \{R_1(B_1), R_1(B_2), \dots\}$

else $GS \leftarrow \{G, R_1(B_1), \dots\}$

This algorithm follows from a theorem about the unsatisfiability of the integrity constraints:

THEOREM 28 ([Eshghi, 1991]). *An integrity constraint* $\mathcal{T} \cup \Delta \vdash R_1(x) \leftarrow \mathcal{T} \cup \Delta \vdash R_2(x)$ *is unsatisfiable iff*

a) $\mathcal{T} \cup \Delta \cup \{R_1(x) \leftarrow R_2(x)\}$ *is inconsistent, or*

b) *for some* β*:* $\mathcal{T} \cup \Delta \vdash R_2(\beta) \wedge \mathcal{T} \cup \Delta \not\vdash R_1(\beta)$.

Remarks:

- As a consequence, case a) is first tested in (3). If no inconsistency is found, the preconditions can be determined in $\mathcal{T}$, plus the assumptions generated by abduction. We proceed then to ensure case b) and try to find a hypothesis for $R_1(B_1), R_1(B_2), \dots$ in the next iteration.
- If inconsistencies occur in step (3), then clauses have to be removed from Δ. (4) is executed with respect to this new theory that guarantees consistency. In the next iteration besides $R_1(B_1), R_1(B_2), \dots$ the goal G still has to be explained.
- As the author points out, an ATMS will be of great use for recording the relationship between assumptions and goals and for determining the assumptions to be rejected in step (3).

For the purpose of planning, Eshghi combines this algorithm with a variant of Kowalski's and Sergot's Event Calculus (cf. [Kowalski and Sergot, 1986]). Events are the basis of the ontology and represent points in time. Actions are associated with events with the predicate "$action$." The predicate "$holds$" is defined as usual, and "$postcon$" describes the postconditions of actions. The calculus will not be introduced in more detail; it is assumed that the meaning of the predicates is intuitively clear from their names. The relationship between facts in the domain are described with the predicates $holds$, $=$ and $\neq$. The theory has Horn clause format with denials. We have the following domain-independent axioms:

$$initiates(t, prop) \leftarrow action(com, t), postcon(com, effect) \quad (26)$$

$$holds(prop, c(t)) \leftarrow initiates(t, prop) \quad (27)$$

$$t \leq c(t) \quad (28)$$

$$c(t) \leq t_1 \leftarrow t < t_1 \quad (29)$$

where $c(t)$ denotes the earliest time point after the execution of an action at which the postcondition is true.

Preconditions are expressed as integrity constraints stressing the fact that they must be believed to be true by the planning agent. This is formalized by the axiom

$$\begin{aligned} \mathcal{T} \cup \Delta \vdash holds(prop, t) \quad \leftarrow \quad & \mathcal{T} \cup \Delta \vdash action(com, t) \wedge \\ & precon(com, prop) \end{aligned} \tag{30}$$

Furthermore, a partial order on events is defined where the initial state is characterized by a special time point $INITIAL$. The predicate $persists(prop, t_1, t_2)$ is introduced to formalize persistence. An axiom states that a proposition $prop$ is true at time t_2 if $prop$ is initiated at time t_1 lying before t_2 and if $prop$ persists through time until time t_3 lying after t_2 or being equal to t_2.

$$\begin{aligned} holds(prop, t_2) \quad \leftarrow \quad & initiates(t_1, prop) \\ & t_1 < t_2 \\ & persists(prop, t_1, t_3) \\ & t_2 \leq t_3 \end{aligned} \tag{31}$$

In this framework the abducible predicates are $action$, $<$, $=$ and $persists$. A least commitment strategy requests that objects are not determined until necessary. To avoid unconstrained search through looping during the consistency checking phase, the predicates $=$ and $<$ are treated specially (as in Constraint Logic Programming (cf. [Eshghi, 1991])).

We will comment on the planning algorithm with a small example coming from the UNIX Mail System. A plan will be provided that one allows to read and then delete the mail from sender "kurt." As domain specific axioms we assume

$$precond(delete(x), \neg first(flag(x)) = 1) \tag{32}$$
$$precond(read(x), \neg first(flag(x)) = 1) \tag{33}$$

Delete and read flag are represented as pair (x, y), where x denotes the delete flag and y the read flag. (32) and (33) state that a message can only be read or deleted if the delete flag is not set.

$$postcon(read(x), second(flag(x) = 1)) \tag{34}$$
$$postcon(delete(x), first(flag(x) = 1)) \tag{35}$$

The initial state is described by

$$\begin{aligned}
initiates(t, member(mbox(x))) &\leftarrow t = INITIAL, x = mail_1\\
initiates(t, member(mbox(x))) &\leftarrow t = INITIAL, x = mail_2\\
&\vdots\\
initiates(t, member(mbox(x))) &\leftarrow t = INITIAL, x = mail_6\\
initiates(t, sender(x) = \text{“}kurt\text{”}) &\leftarrow t = INITIAL, x = mail_4\\
initiates(t, sender(x) = \text{“}kurt\text{”}) &\leftarrow t = INITIAL, x = mail_5\\
initiates(t, flag(x) = (0,0)) &\leftarrow t = INITIAL, x = mail_5\\
initiates(t, flag(x) = (1,1)) &\leftarrow t = INITIAL, x = mail_4
\end{aligned}$$

The goal of the plan is specified by

$$\begin{aligned}
\leftarrow \quad & holds(flag(mail_4) = (1,1), FINAL),\\
& holds(flag(mail_5) = (1,1), FINAL) \qquad (36)
\end{aligned}$$

We will now briefly sketch how the algorithm works with this example.

(1) In the abductive phase, a residue is searched for. (36) is resolved with (31). By the ordering of the events that states $\forall t\ (t < FINAL)$ we get

$$\begin{aligned}
\leftarrow \quad & initiates(t_3, flag(mail_4) = (1,1)) \wedge\\
& persists(flag(mail_4) = (1,1), FINAL) \wedge\\
& initiates(t_3, flag(mail_5) = (1,1)) \wedge\\
& persists(flag(mail_5) = (1,1), FINAL)
\end{aligned}$$

We apply our knowledge about the actions *read* and *delete*

$$\begin{aligned}
\leftarrow \quad & action(read(mail_4), t_3) \wedge persists(second(flag(mail_4) = 1), FINAL)\\
\wedge \quad & action(delete(mail_4), t_4) \wedge persists(first(flag(mail_4) = 1), FINAL)\\
\wedge \quad & t_3 < t_4\\
\wedge \quad & action(read(mail_5), t_3) \wedge persists(second(flag(mail_5) = 1), FINAL)\\
\wedge \quad & action(delete(mail_5), t_4) \wedge persists(first(flag(mail_4) = 1), FINAL)
\end{aligned}$$

We omit the residue where $t_3 > t_4$ as this will lead to inconsistencies.

(2) This residue is added to the theory in homogenised form.

(3) Our integrity constraints, i.e., the preconditions of our actions, were formulated in (30). We have to test whether

$$\mathcal{T} \cup \Delta \vdash holds(prop, t) \leftarrow action(com, t) \wedge precon(com, prop)$$

holds. This is inconsistent, and we have to reject some assumptions. We remove

$$\begin{aligned} & action(read(mail_4), t_3) \quad \text{and} \\ & action(delete(mail_4, t_4) \end{aligned}$$

from Δ.

(4) We have to find the bindings satisfying

$$\mathcal{T} \cup \Delta \vdash action(com, t) \wedge precon(com, prop)$$

We get the binding $< read(mail_5), t_3, \neg first(flag(mail_5) = 1) >$.

(5) The goal can then be reduced to

$$\begin{aligned} \leftarrow \quad & precon(read(mail_5), \neg first(flag(mail_5) = 1)) \wedge \\ & precon(delete(mail_5), \neg first(flag(mail_5) = 1)) \end{aligned}$$

This is already satisfied by $\mathcal{T}$. Thus, the result of the algorithm is

$$\begin{aligned} \Delta \; = \quad & action(read(mail_5), t_3), persists(second(flag(mail_5) = 1), FINAL) \\ \wedge \quad & action(delete(mail_5), t_4), persists(first(flag(mail_5) = 1), FINAL) \end{aligned}$$

The result Δ of the abductive algorithm can be quite intuitively interpreted as a plan: the predicate "*action*" specifies the actions to be executed and "*persists*" in connection with $<$ the temporal ordering and persistence of the facts. The algorithm enforces an order only when necessary and as a consequence actions may also be carried out in parallel. As the author claims, this is a great advantage over "classical" situation calculus. Nevertheless, one should not forget to take into account orderings that follow logically from given ones. One further advantage of the approach follows from the persistence assumption. This assumption makes it possible to handle certain kinds of unforeseen events, namely those that do not violate persistence.

This method also allows the circumvention of the frame problem. Eshghi distinguishes two aspects: the *epistemological frame problem* that consists in writing down all frame axioms, and the *computational frame problem* that concerns the repeated application of the frame axioms. Since there are no frame axioms, the first problem is irrelevant. Persistence of propositions through time is modelled by the abducible predicate "*persists*." Thus, the computational frame problem only consists of checking the consistency of persistence assumptions. Eshghi points out that this can be reduced to checking the consistency of clauses of the form $\leftarrow holds(prop, t_2), t_1 < t_2, t_2 \leq t_3$ with time points t_1, t_2, t_3. Then, consistency is determined by testing the satisfiability of the negative conjunction and for this problem special purpose algorithms exist.

The above shows that abduction through deduction is a very promising problem solving technique for logic-based planning. In particular, it provides an elegant and intuitive solution to the frame problem, in contrast to most deductive approaches. By handling persistence with the aid of an abducible predicate that is itself part of the plan, the approach can even cope with unforeseen events, as long as they do not violate those assumptions. By the use of a variant of the event calculus in connection with abduction it furthermore becomes possible to specify the ordering of actions only partially.

6.2 *Abduction and Relevance*

Helft and Konolige present in [Helft and Konolige, 1990] a theory for plan recognition that accounts for both acting and planning. Plan recognition is interpreted as "the recovery of the hidden parts of the plan/act process."

This approach distinguishes between the world knowledge $\mathcal{T}_a$ of the agent and the theory $\mathcal{T}_\omega$ that describes the consequences of actions. To achieve a goal, the agent constructs a plan using solely the knowledge in $\mathcal{T}_a$. The plan P in its simplest form consists of a sequence of basic actions that may be observed as well as their effects. Observed events are denoted by Ω. For a goal g, this can be formally described by:

(1) $\mathcal{T}_a \cup \{P\} \vdash g$ and

(2) $\mathcal{T}_\omega \cup \{P_0\} \vdash \Omega$ with $P_0 \subseteq P$.

The combined planning and plan recognition process gets as input the observations Ω and the theories $\mathcal{T}_a$ and $\mathcal{T}_\omega$. It produces a plan P and a goal g such that (1) and (2) are satisfied. For finding P and g, as well as for explaining the observations, abduction is used. Relevance serves as a constraint mechanism in the sense that it is assumed that an agent only executes actions that help to achieve his goal.

This model has several advantages over deductive approaches. The "dual nature of causality" is recognized: the agent's intentions cause the basic actions of plan P in the sense that he wants to achieve his goals with the plan. The basic actions in turn cause the observations that can be made. Furthermore, the approach incorporates a "rich model of intention." It is possible to distinguish between intended consequences (goals) and side-effects by considering two processes, namely planning and acting.

Helft and Konolige define a formal model that incorporates the ideas presented above.

DEFINITION 29. ([Helft and Konolige, 1990]) (planning theory) A *planning theory* is a tuple $(A, G, \mathcal{T}_a, \mathcal{T}_\omega)$, where A is a set of basic actions, G is a set of predefined goals and $\mathcal{T}_a$ and $\mathcal{T}_\omega$ are first-order theories.

DEFINITION 30. ([Helft and Konolige, 1990]) (plan recognition problem) A *solution to the plan recognition problem* is a tuple (A_0, P, g) such that

(1) A_0 is a minimal subset of A, consistent with $\mathcal{T}_\omega$ with

$$\mathcal{T}_\omega \cup A_0 \vdash \Omega$$

(2) P is a subset of A consistent with $\mathcal{T}_a$, $A_0 \subseteq P$, $g \subseteq G$ with

$$\begin{array}{ll} \text{(a)} & \mathcal{T}_a \cup \{P\} \vdash g \\ \text{(b)} & \mathcal{T}_a \cup \{P - e\} \not\vdash g, \quad \forall e \in A_0 \end{array}$$

Remarks:

- (1) explains the observations with respect to $\mathcal{T}_\omega$, the "physical" world description.
- In (2) the intentions and beliefs of the agent are considered. With (a) a plan P containing the actions in A_0 is found that achieves the goal g, if the agent's theory $\mathcal{T}_a$ is taken into account. (b) ensures that every action out of A_0 is in fact relevant, i.e., essential for the plan in order to accomplish g.

DEFINITION 31. ([Helft and Konolige, 1990]) (side-effects) Propositions that are entailed by $\mathcal{T}_\omega \cup P$ but not by $\mathcal{T}_\omega$ alone are called side-effects.

The abductive algorithm that solves the plan recognition problem relies on the function NEW defined by

DEFINITION 32. ([Helft and Konolige, 1990]) (function NEW)

$$NEW_P(\Sigma, \varphi) = (Th(\Sigma \cup \{\varphi\}) - Th(\Sigma)) \cap P,$$

i.e., NEW specifies what a formula φ adds to a theory Σ with respect to a set P, where the intersection with P means that only formulae containing symbols of P should be regarded.

The function is computed by a linear resolution algorithm. A solution (A_0, P, g) is found by several applications of NEW:

Suppose that the knowledge is in conjunctive normal form.

(1) Compute $\Sigma := NEW_B(\mathcal{T}_\omega, \neg\Omega)$, where B contains the symbols denoting the basic actions.

NEW_B computes the relevant consequent actions of $\mathcal{T}_\omega \cup \{\neg\Omega\}$. Thus, for every $\sigma \in \Sigma$, $\neg\sigma := \alpha_1 \wedge \ldots \wedge \alpha_n$ is a candidate for A_0. To see this, consider the definition of NEW. We have

$$\mathcal{T}_\omega \cup \{\neg\Omega\} \vdash \sigma \qquad (37)$$

$$\mathcal{T}_\omega \not\vdash \sigma \qquad (38)$$

By (38), $\neg\sigma$ is consistent with $\mathcal{T}_\omega$. With the aid of the deduction theorem, (37) can be transformed into

$$\begin{aligned} \mathcal{T}_\omega &\vdash \neg\Omega \supset \sigma \\ \mathcal{T}_\omega &\vdash \neg\sigma \supset \Omega \\ \mathcal{T}_\omega \cup \{\neg\sigma\} &\vdash \Omega \end{aligned}$$

Thus, $\neg\sigma$ explains Ω in $\mathcal{T}_\omega$.

(2) Compute $\Gamma := NEW_{B\cup G}(\mathcal{T}_a, \alpha_1 \wedge \ldots \wedge \alpha_n)$. G is the set of instantiated goals. The elements of Γ have the form $\beta_1 \wedge \ldots \wedge \beta_m \supset \gamma$.

Verify as follows that the β_i are possible elements of $P\backslash A_0$ and that γ is a goal candidate:
The conjunctions $\alpha_1 \wedge \ldots \wedge \alpha_n$ are elements of A_0, therefore we abbreviate an arbitrary conjunction by a_0. The definition of NEW yields

$$\mathcal{T}_a \cup \{a_0\} \vdash \beta_1 \wedge \ldots \wedge \beta_m \supset \gamma$$

If the agent considers a_o as action, his theory tells him that a_0 together with the actions $\beta_1 \wedge \ldots \wedge \beta_m$ implies the goal γ. So a plan P to achieve the goal γ consists of the actions a_0 and $\beta_1 \wedge \ldots \wedge \beta_m$:

$$\mathcal{T}_a \cup \{a_0\} \cup \{\beta_1 \wedge \ldots \wedge \beta_m\} \vdash \gamma.$$

As a_0 is also considered as a part of the plan, the requirement $A_0 \subseteq P$ is fulfilled.

(3) Check if $\mathcal{T}_a \wedge \beta_1 \wedge \ldots \wedge \beta_m$ is consistent. If the answer is yes, a solution is found.

(4) Compute the side-effects in G by $NEW_G(\mathcal{T}_\omega, \alpha_1 \wedge \ldots \alpha_n \wedge \beta_1 \wedge \ldots \wedge \beta_m)$ for all elements $\alpha_1 \wedge \ldots \wedge \alpha_n \wedge \beta_1 \wedge \ldots \wedge \beta_m$ in P.

This function call computes exactly the side-effects, as NEW determines those elements of G for which holds

$$\mathcal{T}_\omega \cup p \vdash s \text{ but } \mathcal{T}_\omega \not\vdash s,$$

for an arbitrary element p of P. But this is exactly the definition of a side-effect.

Thus, this framework can also account for observations that are merely side-effects of actions. This overcomes a drawback of many current approaches, that is, that they demand a direct causal link between the actions and all observable effects.

This feature also comments on the different demands on diagnosis and plan recognition systems. In diagnosis, all observed symptoms are caused by faults that have to be determined. In plan recognition this need not be the case, as the occurring side-effects prove. An effect need not be caused by the agent's intention.

The separation of the world knowledge and the agent's knowledge is the reason for a further nice property. Plans are constructed with respect to the knowledge $\mathcal{T}_a$ of the agent, thus it is also possible to determine plans that are correct with respect to the user knowledge but ill-formed with respect to the world knowledge.

The two described approaches to abductive planning and plan recognition have shown how the previously presented general methods can be successfully adapted to special applications. Hard problems for deductive approaches, e.g., the frame problem, can be solved in an intuitive and elegant manner if abduction is augmented with additional principles, for instance with default persistence, that are especially suited for problems in which commonsense reasoning is used.

7 CONCLUSION

In the previous sections an overview of existing abductive reasoning methods has been given, focusing in the last part on applications in plan recognition and planning systems.

Repair or diagnostic problems have shown to be one obvious possibility for applying abductive reasoning. They possess the great advantage that the underlying theory usually has a very simple structure concerning causality. Cause and effect relationships can be represented in a straightforward way, as implications. Furthermore, the search space for hypotheses can often be limited or even completely determined. This is used by the set-cover-based model described in section 2.1 and results in more efficient concrete algorithms than the more general case can provide.

It was shown that such strong assumptions are not suited for more general tasks. The logic-based model (see section 2.2) is the most widely spread approach as it allows more flexibility and thus seems to be adequate for a greater range of applications. But, if the generation and selection methods in sections 3 and 4 are regarded in more detail, it becomes obvious that those approaches as well implicitly rely on a special knowledge representation. Causal relationships must be determinable, since they are responsible for choosing the right explanations.

Levesque (cf. section 2.3) goes one step further and defines a model for abduction in dependence of a belief type. He shows that his approach implies the logic-based model for implicit belief. Hence, further generality has been gained. The explanation operator is defined semantically and thus ensures the independence of the respective knowledge representation. A semantic account for explanations is also given with the modal logic for abduction of Mayer and Pirri (see section 14).

The defeasability of abduction is taken into account in the approaches of Inoue and Sakama, and Boutilier and Becher. Abduction is embedded in the broader context of theory revision or update and the retraction of hypotheses which turn out to be no longer valid is formalized. This leads one step further into the direction of Peirce's theory of abduction where it is claimed that the verification of hypotheses is an integral part of explaining (cp. section 1.1). While Inoue and Sakama follow the classical logic-based paradigm with respect to entailment and preference, Boutilier and Becher define explanations in terms of revision-models of an epistemic state of an agent. Thereby preference depends on the application and the framework allows for a detailed distinction of different types of explanations.

Nevertheless, it remains necessary to represent causality on the computational level. So, in order to gain further insight into the foundations of abductive reasoning, the role of causality should be incorporated into the theoretical investigations. In addition, it seems profitable to consider abduction in the context of the other reasoning mechanisms induction and deduction, as has already been stressed by Peirce.

In section 3, different methods for hypotheses generation were described, in particular resolution-based algorithms and an ATMS procedure.

In the actual realizations of abductive procedures, one suffers from the problem that in most cases one abductive explanation has to be singled out. As a consequence, selection criteria have to be defined that allow one to find, in some sense, the "best" hypothesis. Apart from syntactic criteria and different strategies to determine an appropriate level of specificity, a method based on a coherence metric was described (see section 4). The metric was shown to be a first step in the direction of preferring the psychologically simplest explanation, not the logically simplest. Nevertheless, in most cases a combination of some of the described approaches seems to be adequate. A unifying theory is a further research topic.

A comparison of default and abductive reasoning in section 5.1 has on the one hand clarified the close connection of both approaches by showing that defaults can be interpreted as a kind of predefined hypotheses. Furthermore, it became possible to derive some results concerning the complexity of abductive and default reasoning. Both approaches were shown to be NP-complete. It was further shown that in the general case abduction is harder than deduction. Tractable subclasses were discussed, and a consequence for concrete applications should be the choice of a sufficient algorithm, and the corresponding knowledge representation, not the most general one in order to gain efficiency.

The first steps for wider applications of abduction have already been made. If the field of planning is considered, it becomes obvious that the classical models for abduction are not sufficient. They have to be augmented with features that allow the handling of intentions, persistence, time, etc. The applications presented in section 6 show how the formal models have to be changed for special applications. A survey of the main discussed approaches is given in the following table.

A survey of main presented approaches

Formal Model	Paper	Context of Abduction	Generation Method described or used	Selection Method described or used	Comment
set-cover-based	[Allemang *et al.*, 1987]	best explanation for observations	predefined function	plausibility, parsimony	restricted applicability: observation - explanation relationship is predefined
logic-based	[Konolige, 1990]	diagnosis	—	subset-minimality	transformation to the consistency-based method of Reiter (s. [Reiter, 1987])
	[Eshghi and Kowalski, 1989] [Kakas *et al.*, 1998]	generation of answers to queries	resolution-based; generalizes [Console *et al.*, 1991a], [Denecker and de Schreye, 1992]	integrity constraints	abduction as generalization of logic programming
	[Cox and Pietrzykowski, 1984]	find explanations	linear resolution and reverse skolemization	basicness; minimality	algorithm is incomplete w.r.t. basicness and minimality
	[Pople, 1973]	find explanations	linear resolution	factoring across partial proof trees	suggestion: combination of abduction, induction, and deduction
	[Shanahan, 1989]	temporal reasoning	linear resolution	explanation postulating fewest events	based on Event Calculus (s. [Kowalski and Sergot, 1986])
	[Appelt and Pollack, 1990]	recognition of plans and goals	theorem prover + control	weighting factors; lowest cost explanation	how are weighting factors determined? models only most-specific abduction
	[Poole, 1988]	defaults as pre-defined hypotheses	PROLOG	allows different methods, e.g., definition of priorities	general tool for reasoning with defaults
	[Helft and Konolige, 1990]	determination of plans and corresponding goals	linear resolution	—	distinguishes between side-effects and intended consequences
	[Console *et al.*, 1991b]	abductive diagnosis	resolution-based	—	use of compiled knowledge for efficient explanation generation
	[Ng and Mooney, 1990]	deep causal explanations for natural language text	linear resolution	coherence	heuristic search (beam search) to improve efficiency

knowledge-level	[Levesque, 1989]	find explanations	function yielding disjunction of feasible explanations; ATMS for implicit belief	explanations with fewest propositional symbols	more general definition; allows for exchange of underlying belief model
	[Inoue and Sakama, 1995]	nonmonotonic update of autoepistemic theories	—	subset-minimality	captures defeasibility by formalizing retraction of hypotheses
logics for abduction	[Boutilier and Becher, 1995]	abduction as belief revision	—	can be determined by the application	general definition, explanations depend on epistemic state
	[Mayer and Pirri, 1996]	bi-modal logic for abduction	deduction in the defined logic	ordering relation on formulae	defined for propositional case, finite language and theory
probabilistic methods	[Charniak and Shimony, 1994]	cost-based abduction	computation of MAP assignment for belief network	least-cost assumptions	probabilistic knowledge must be available
	[Santos, 1994]	cost-based abduction	linear programming techniques	least-cost assumptions	empirically more efficient than [Charniak and Shimony, 1994]
	[Poole, 1993]	probabilistic Horn abduction	—	probabilities over hypotheses	PROLOG-like rules form Bayesian network

Thus, the presented ideas are valuable as a first sketch of the possibilities of abductive reasoning. What seems profitable is a theory unifying abduction with other logical operations or decision procedures such as revision, retraction or consistency, which is general enough to allow for different, maybe nonmonotonic, underlying logics in order to cover also applications that involve commonsense reasoning. By this means the definition of generally applicable systems becomes possible in which abduction is one reasoning possibility among others. This is presented in the next chapter by D. M. Gabbay [1999].

ACKNOWLEDGEMENTS

This paper is an updated version of [Paul, 1993].

German Research Center for AI (DFKI GmbH), Saarbrücken, Germany.

REFERENCES

[Adé and Denecker, 1994] H. Adé and M. Denecker. RUTH: An ILP theory revision system. In *Proceedings of the 8th International Symposium on Methodologies for Intelligent Systems*, 1994.

[Allemang *et al.*, 1987] D. Allemang, M. Tanner, T. Bylander, and J. Josephson. Computational complexity of hypothesis assembly. In *Proceedings of the 10th International Joint Conference on Artificial Intelligence*, pages 1112–1117, 1987.

[Appelt and Pollack, 1990] D.E. Appelt and M. Pollack. Weighted abduction for plan ascription. Technical report, Artificial Intelligence Center and Center for the Study of Language and Information, SRI International, Menlo Park, California, 1990.

[Appelt, 1990] D.E. Appelt. A theory of abduction based on model preference. In *Proceedings of the AAAI90 Spring Symposium on Abduction*, 1990.

[Bauer and Paul, 1993] M. Bauer and G. Paul. PHI—a logic-based tool for intelligent help systems. In *Proceedings of the 13th International Joint Conference on Artificial Intelligence*, pages 460–466, 1993.

[Bauer *et al.*, 1993] M. Bauer, S. Biundo, D. Dengler, J. Koehler, and G. Paul. Logic-based plan recognition for intelligent help systems. In *Current Trends in AI Planning, Proceedings of the 2nd European Workshop on Planning*, pages 60–73, 1993.

[Boutilier and Becher, 1995] C. Boutilier and V. Becher. Abduction as belief revision. *Artificial Intelligence*, 77:43–94, 1995.

[Boutilier, 1996] C. Boutilier. Abduction to plausible causes: an event-based model of belief update. *Artificial Intelligence*, 83:143–166, 1996.

[Brewka and Konolige, 1993] G. Brewka and K. Konolige. An abductive framework for general logic programs and other non-monotonic systems. In *Proceedings of the 13th International Joint Conference on Artificial Intelligence*, pages 9–15, 1993.

[Brewka, 1991] G. Brewka. Preferred subtheories: an extended logical framework for default reasoning. In *Proceedings of the 12th International Joint Conference on Artificial Intelligence*, pages 1043–1048, 1991.

[Brusoni *et al.*, 1997] V. Brusoni, L. Console, P. Terenziani, and D.T. Dupré. An efficient algorithm for temporal abduction. In *Proceedings of the 5th Congress of the Italian Association for Artificial Intelligence*, pages 195–206, 1997.

[Bürckert and Nutt, 1992] H.J. Bürckert and W. Nutt. On abduction and answer generation through constrained resolution. Technical report, German Reseach Center for AI (DFKI GmbH), 1992.

[Bylander *et al.*, 1991] T. Bylander, D. Allemang, M. Tanner, and J. Josephson. The computational complexity of abduction. *Artificial Intelligence*, 49:25–60, 1991.

[Charniak and McDermott, 1985] E. Charniak and P. McDermott. *Introduction to Artificial Intelligence*. Addison Wesley, Menlo Park, California, 1985.

[Charniak and Shimony, 1990] E. Charniak and S.E. Shimony. Probabilistic semantics for cost based abduction. In *Proceedings of the 8th National Conference on Artificial Intelligence*, pages 106–111, 1990.

[Charniak and Shimony, 1994] E. Charniak and S.E. Shimony. Cost-based abduction and MAP explanation. *Artificial Intelligence*, 66(2):345–374, 1994.

[Console and Torasso, 1991] L. Console and P. Torasso. A spectrum of logical definitions of model-based diagnosis. *Computational Intelligence*, 7:133–141, 1991.

[Console *et al.*, 1991a] L. Console, D.T. Dupré, and P. Torasso. On the relationship between abduction and deduction. *Journal of Logic and Computation*, 1(5):661–690, 1991.

[Console *et al.*, 1991b] L. Console, L. Portinale, and D.T. Dupré. Focussing abductive diagnosis. *AI Communications*, 4(2/3):88–97, 1991.

[Cox and Pietrzykowski, 1984] P. Cox and T. Pietrzykowski. A complete nonredundant algorithm for reversed skolemization. *Theoretical Computer Science*, 28:239–261, 1984.

[Cox and Pietrzykowski, 1986] P. Cox and T. Pietrzykowski. Causes for events: Their computation and application. In *Proceedings CADE 86*, pages 608–621, 1986.

[de Kleer *et al.*, 1990] J. de Kleer, A.K. Mackworth, and R. Reiter. Characterizing diagnoses. In *Proceedings of the 8th National Conference on Artificial Intelligence*, pages 324–330, 1990.

[Dechter *et al.*, 1991] R. Dechter, I. Meiri, and J. Pearl. Temporal constraint networks. *Artificial Intelligence*, 49:61–95, 1991.

[Denecker and de Schreye, 1992] M. Denecker and D. de Schreye. SLDNFA: An abductive procedure for normal logic programs. In *ICSLP*, pages 868–700, 1992.

[Dragoni and Puliti, 1994] A.F. Dragoni and P. Puliti. Mental states recognition from speech acts through abduction. In *Proceedings of the 11th European Conference on Artificial Intelligence*, pages 183–187, 1994.

[Dung, 1991] P.M. Dung. Negation as hypothesis: an abductive foundation for logic programming. In *Proceedings of the 8th International Conference on Logic Programming*, pages 3–17, 1991.

[Dung, 1993] P.M. Dung. On the acceptability of arguments and its fundamental role in nonmonotonic reasoning and logic programming. In *Proceedings of the 13th International Joint Conference on Artificial Intelligence*, pages 852–859, 1993.

[Dupré and Rossotto, 1995] D.T. Dupré and M. Rossotto. The different roles of abstraction in abductive reasoning. In *Proceedings of the Italian Association for Artificial Intelligence on Topics in Artificial Intelligence*, volume 992 of *LNAI*, pages 211–216, 1995.

[Eiter and Gottlob, 1995] T. Eiter and G. Gottlob. The complexity of logic-based abduction. *Journal of the Association for Computing Machinery*, 42(1):3–42, 1995.

[Eshghi and Kowalski, 1988] K. Eshghi and R. Kowalski. Abduction through deduction. Technical report, Imperial College of Science and Technology, Department of Computing, 1988.

[Eshghi and Kowalski, 1989] K. Eshghi and R. Kowalski. Abduction compared with negation by failure. In *Proceedings of the 6th International Conference on Logic Programming*, 1989.

[Eshghi, 1991] K. Eshghi. Abductive planning with Event Calculus. In *Logic Programming: Proc. of the Fifth International Conference and Symposium (Volume 1)*, pages 562–579, 1991.

[Eshghi, 1993] K. Eshghi. A tractable class of abduction problems. In *Proceedings of the 13th International Joint Conference on Artificial Intelligence*, pages 3–8, 1993.

[Fung and Kowalski, 1997] T.H. Fung and R.A. Kowalski. The IFF proof procedure for abductive logic programming. *The Journal of Logic Programming*, 1997.

[Gabbay, 1999] D. M. Gabbay. Abduction in labelled deductive systems. This volume, pp. 101–156, 2000.

[Goudge, 1950] Th.A. Goudge. *The Thought of C.S. Peirce*. Dover Publications Inc., New York, 1950.

[Helft and Konolige, 1990] N. Helft and K. Konolige. Plan recognition as abduction and relevance. Draft version, Artificial Intelligence Center, SRI International, Menlo Park, California, 1990.

[Hobbs *et al.*, 1989] J. R. Hobbs, M. Stickel, P. Martin, and D. Edwards. Interpretation as abduction. Draft version, SRI International, Artificial Intelligence Center, Menlo Park, CA, 1989.

[Inoue and Sakama, 1995] K. Inoue and C. Sakama. Abductive framework for nonmonotonic theory change. In *Proceedings of the 14th International Joint Conference on Artificial Intelligence*, pages 204–210, 1995.

[Inoue, 1990] K. Inoue. An abductive procedure for the CMS/ATMS. In *Proceedings of the Workshop on Truth Maintenance Systems (ECAI-90)*, volume 515 of *LNAI*, pages 34–53, 1990.

[Jung, 1998] C.G. Jung. Situated abstraction planning by abductive temporal reasoning. In *Proceedings of the 13th European Conference on Artificial Intelligence*, pages 383–387, 1998.

[Kakas and Mancarella, 1990a] A.C. Kakas and P. Mancarella. Database updates through abduction. In *Proceedings of the 16th Conference on Very Large Databases*, 1990.

[Kakas and Mancarella, 1990b] A.C. Kakas and P. Mancarella. Generalized stable models: a semantics for abduction. In *Proceedings of the 9th European Conference on Artificial Intelligence*, pages 385–391, 1990.

[Kakas and Riguzzi, 1997] A.C. Kakas and F. Riguzzi. Learning with abduction. In *Proceedings of the 7th International Workshop on Inductive Logic Programming*, volume 1297 of *LNAI*, pages 181–188, 1997.

[Kakas *et al.*, 1998] A.C. Kakas, R.A. Kowalski, and F. Toni. The role of abduction in logic programming. In D.M. Gabbay, C.J. Hogger, and J.A. Robinson, editors, *Handbook of Logic in Artificial Intelligence and Logic Programming*, pages 235–324. Oxford University Press, 1998.

[Konolige, 1990] K. Konolige. Closure + minimization implies abduction. In *PRICAI-90, Nagoya, Japan*, 1990.

[Kowalski and Sergot, 1986] R. Kowalski and M. Sergot. A logic based calculus of events. *New Generation Computing*, 4:67–95, 1986.

[Lamma *et al.*, 1997] E. Lamma, P. Mello, M. Milano, and F. Riguzzi. Introducing abduction into (extensional) inductive logic programming systems. In *Proceedings of the 5th Congress of the Italian Association for Artificial Intelligence : Advances in Artificial Intelligence (AI*IA-97)*, volume 1321 of *LNAI*, pages 183–194, 1997.

[Leake, 1993] D.B. Leake. Focusing construction and selection of abductive hypotheses. In *Proceedings of the 13th International Joint Conference on Artificial Intelligence*, pages 24–29, 1993.

[Levesque, 1989] H. Levesque. A knowledge-level account of abduction. In *Proceedings of the 11th International Joint Conference on Artificial Intelligence*, pages 1061–1067, 1989.

[Li and Pereira, 1996] R. Li and L.M. Pereira. Temporal reasoning with abductive logic programming. In *Proceedings of the 12th European Conference on Artificial Intelligence*, pages 13–17, 1996.

[Mayer and Pirri, 1996] M. Cialdea Mayer and F. Pirri. A study on the logic of abduction. In *Proceedings of the 12th European Conference on Artificial Intelligence*, pages 18–22, 1996.

[McIlraith, 1994] S. McIlraith. Generating tests using abduction. In *Proceedings of the 4th International Conference on Principles of Knowledge Representation and Reasoning*, pages 449–460, 1994.

[Morgan, 1971] Ch.G. Morgan. Hypothesis generation by machine. *Artificial Intelligence*, 2:179–187, 1971.

[Ng and Mooney, 1990] H.T. Ng and R.J. Mooney. On the role of coherence in abductive explanation. In *Proceedings of the 8th National Conference on Artificial Intelligence*, pages 337–342, 1990.

[Ng and Mooney, 1992] H.T. Ng and R.J. Mooney. Abductive plan recognition and diagnosis: a comprehensive empirical evaluation. In *Proceedings of the 3rd International Conference on Principles of Knowledge Representation and Reasoning*, pages 499–508, 1992.

[O'Rorke, 1994] P. O'Rorke. Focusing on most important explanations: decision-theoretic Horn abduction. In *Proceedings of the 12th National Conference on Artificial Intelligence*, pages 275–280, 1994.

[Paul, 1993] G. Paul. Approaches to abductive reasoning—an overview. *Artificial Intelligence Review*, 7:109–152, 1993.

[Pearl, 1988] J. Pearl. *Probabilistic Reasoning in Intelligent Systems: Networks of Plausible Inference*. San Mateo, 1988.

[Peirce, 1931-1958] C.S. Peirce. *Collected Papers of Charles Sanders Peirce (eds. C. Hartshorne et al.)*. Harvard University Press, 1931-1958.

[Poole, 1988] D. Poole. A logical framework for default reasoning. *Artificial Intelligence*, 36:27–47, 1988.

[Poole, 1989a] D. Poole. Explanation and prediction: an architecture for default and abductive reasoning. *Computational Intelligence*, 5:97–110, 1989.

[Poole, 1989b] D. Poole. Normality and faults in logic-based diagnosis. In *Proceedings of the 11th International Joint Conference on Artificial Intelligence*, pages 1304–1310, 1989.

[Poole, 1991] D. Poole. Representing diagnostic knowledge for probabilistic Horn abduction. In *Proceedings of the 12th International Joint Conference on Artificial Intelligence*, pages 1129–1135, 1991.

[Poole, 1993] D. Poole. Probabilistic Horn abduction and Bayesian networks. *Artificial Intelligence*, 64:81–129, 1993.

[Pople, 1973] H.E. Pople, Jr. On the mechanization of abductive logic. In *Proceedings of the 3rd International Joint Conference on Artificial Intelligence*, pages 147–151, 1973.

[Reggia, 1988] J. Reggia. Diagnostic expert systems based on a set covering model. *International Journal of Man-Machine Studies*, November 83:437–460, 1988.

[Reiter, 1980] R. Reiter. A logic for default reasoning. *Artificial Intelligence*, 13(2):81–132, 1980.

[Reiter, 1987] R. Reiter. A theory of diagnosis from first principles. *Artificial Intelligence*, 32:57–95, 1987.

[Ribeiro and Porto, 1994] C. Ribeiro and A. Porto. Abduction in temporal reasoning. In *Proceedings of the 1st International Conference on Temporal Logic*, pages 349–364, 1994.

[Santos, 1994] E. Santos, Jr. A linear constraint satisfaction approach to cost-based abduction. *Artificial Intelligence*, 65(1):1–28, 1994.

[Satoh and Iwayama, 1991] K. Satoh and N. Iwayama. Computing abduction by using the TMS. In *Proceedings of the 8th International Conference on Logic Programming*, pages 505–520, 1991.

[Satoh and Iwayama, 1992] K. Satoh and N. Iwayama. A correct top-down proof procedure for general logic programs with integrity constraints. In *Proceedings of the 3rd International Workshop on Extensions of Logic Programming*, pages 19–34, 1992.

[Satoh, 1998] K. Satoh. Using two level abduction to decide similarity of cases. In *Proceedings of the 13th European Conference on Artificial Intelligence*, pages 398–402, 1998.

[Selman and Levesque, 1989] B. Selman and H.L. Levesque. Abductive and default reasoning: A computational core. In *Proceedings of the 8th National Conference on Artificial Intelligence*, pages 343–348, 1989.

[Selman and Levesque, 1996] B. Selman and H.J. Levesque. Support set selection for abductive and default reasoning. *Artificial Intelligence*, 82:259–272, 1996.

[Shanahan, 1989] M. Shanahan. Prediction is deduction but explanation is abduction. In *Proceedings of the 11th International Joint Conference on Artificial Intelligence*, pages 1055–1060, 1989.

[Smith, 1985] J.W. Smith. *RED: A Classificatory and Abductive Expert System*. PhD thesis, Ohio State University, Laboratory for Artificial Intelligence Research, Department for Computer Science, 1985.

[Stickel, 1990] M. Stickel. Rationale and methods for abductive reasoning in natural-language interpretation. In R. Studer, editor, *Natural Language and Logic*, pages 331–352. Springer-Verlag, Berlin, Heidelberg, New York, 1990.

[Stone, 1998] M. Stone. Abductive planning with sensing. In *Proceedings of the 15th National Conference on Artificial Intelligence*, pages 631–636, 1998.

[Thompson and Mooney, 1994] C.A. Thompson and R.J. Mooney. Inductive learning for abductive diagnosis. In *Proceedings of the 12th National Conference on Artificial Intelligence*, pages 664–669, 1994.

[Toni and Kowalski, 1995] F. Toni and R.A. Kowalski. Reduction of abductive logic programs to normal logic programs. In *Proceedings of the 12th International Conference on Logic Programming*, pages 367–382, 1995.

[Wærn, 1992] A. Wærn. Reactive abduction. In *Proceedings of the 10th European Conference on Artificial Intelligence*, pages 159–163, 1992.

[Wetzel and Toni, 1998] G. Wetzel and F. Toni. Semantic query optimization through abduction and constraint handling. In *Proceedings of the 3rd International Conference on Flexible Query Answering Systems (FQAS-98)*, volume 1495 of *LNAI*, pages 366–381, 1998.

D. M. GABBAY

ABDUCTION IN LABELLED DEDUCTIVE SYSTEMS

1 ABDUCTION PRINCIPLES: AN OVERVIEW

The chapter by G. Paul in this Volume gives an overview of major existing abductive reasoning methods in Artificial Intelligence. The purpose of this chapter is to introduce a major case study: abduction in implicational labelled deductive systems, and demonstrate a general methodology of abduction in logical systems.

Given a logical system $\mathbf{L}$ ($\mathbf{L}$ may be a monotonic or non-monotonic logic [1]) a theory Δ in $\mathbf{L}$ and a wff Q there are several possible interactions to consider between $\mathbf{L}, \Delta$ and Q.

1. *Consequence*
   We can ask whether $\Delta \vdash_{\mathbf{L}} Q$

2. *Retraction*
   If $\Delta \vdash_{\mathbf{L}} Q$ holds, we can ask how do we revise Δ to a theory Δ' (if $\mathbf{L}$ is a monotonic logic we require that $\Delta' \subseteq \Delta$) such that $\Delta' \nvdash_{\mathbf{L}} Q$.

3. *Consistency*
   If $\Delta \nvdash_{\mathbf{L}} Q$, we can ask whether $\Delta \cup \{Q\}$ is consistent in $\mathbf{L}$.

4. *Revision*
   If $\Delta \cup \{Q\}$ is not consistent, we can ask how do we revise $\Delta \cup \{Q\}$ to a Δ' preferably satisfying $\Delta' \subseteq \Delta \cup \{Q\}$, such that $\Delta' \vdash_{\mathbf{L}} Q$ and Δ' is consistent.

5. *Abduction*
   If $\Delta \nvdash_{\mathbf{L}} Q$, we can ask what X can we add to Δ to get $\Delta' = \Delta \cup \{X\}$ such that $\Delta \cup \{X\} \vdash_{\mathbf{L}} Q$. (If $\mathbf{L}$ is non-monotonic we may delete X from Δ to get a Δ' such that $\Delta' \vdash Q$).

   We would, of course, like X to be as logically weak as possible.

[1] The consequence relation $\vdash_{\mathbf{L}}$ of a monotonic logic satisfies

1. *Reflexivity*: $\Delta \vdash_{\mathbf{L}} A$ if $A \in \Delta$
2. *Monotonicity*: $\Delta \vdash_{\mathbf{L}} A$ implies $\Delta, \Delta' \vdash_{\mathbf{L}} A$
3. *Cut*: $\Delta, A \vdash_{\mathbf{L}} B$ and $\Delta' \vdash_{\mathbf{L}} A$ imply $\Delta, \Delta' \vdash_{\mathbf{L}} B$.

For a non-monotonic logic (2) is replaced by the weaker

2′. *Restricted monotonicity*: $\Delta \vdash_{\mathbf{L}} A$ and $\Delta \vdash_{\mathbf{L}} B$ imply $\Delta, A \vdash_{\mathbf{L}} B$.

D. M. Gabbay and R. Kruse (eds.), Abductive Reasoning and Learning, 99–154.

A possible answer to (1) is a proof theory system for $\mathbf{L}$. Such a system would be an algorithm $\Pi_{\mathbf{L}}$ which for a given Δ and Q can be activated and which, if terminating (it might loop), may give an answer yes ($\Pi_{\mathbf{L}}(\Delta, Q) = 1$) or no ($\Pi_{\mathbf{L}}(\Delta, Q) = 0$).

The answer to (2) may be through a contraction mechanism which tells us what to take out to make Q unprovable.

The answer to (3) can be a semantical interpretation or a model building process.

The answer to (4) may be a revision mechanism which tells us how to revise a theory to accept an input Q.

Note that the cases (2) and (4) can be quite complex in the case of non-monotonic theories.

The answer to (5) is an abduction mechanism, which is the concern of this chapter.

The above considerations used the notions 'logic', (the notion of 'logic' includes the notions of 'consistency' and 'consequence') 'theory', and 'formula'. To answer any of the above questions in a general setting, we need to have general settings for these notions.

Our view of the notion of logic is that in order to present a logic $\mathbf{L}$ it is not enough to specify its consequence relation (what proves what in $\mathbf{L}$), but a specific proof procedure $\Pi_{\mathbf{L}}$ must also be given. Thus, according to this view, classical logic with the tableaux proof theory is not the same 'logic' as classical logic with resolution proof theory. This view was motivated abundantly by the author on many occasions. See Chapter 1 of [Gabbay, 1996].

Our view of the notion of a theory (or database) is that of a structured family of wffs, (data items). The structure of the data can be sets, multisets, sequences, or even a general algebraic structure. In such structured data both a formula and its negation may reside in different locations. The appropriate notion of 'consistency' for structured databases is that of *acceptability*. Some databases are acceptable to us and some are not.[2] Furthermore, a database Δ is not just a family of structured data items. Part of the notion of a database is a recipe of how the database receives input. Since databases are structured, there may be several options concerning where the input can be placed.[3] Our logic $\mathbf{L}$ must have a clear cut notion for each database and each input formula where the formula is to be placed. Similarly we need a clear concept of how to take a formula out of the database and what the new database is to be. Other required notions are how to join together two databases and how to decompose a database into several databases. The proper set-up for a general database and logic is that the framework of ındexlabelled deductive system*Labelled Deductive System*. In *LDS* the notion of a formula becomes the notion of *Declarative unit* of the form $t : A$, t is a label from an algebra and A a wff from a logic.

[2]In traditional logics, a theory Δ is acceptable iff it is consistent, i.e. $\Delta \nvdash \perp$. However, the more general notion of acceptability may be employed for traditional logics as well. We may find some theories unacceptable even though they are consistent.

[3]In traditional logics, where the theories are sets of formulas, input is done by set union.

Thus, in our view, the proper general set-up for answering questions (1)–(5) is the *LDS* methodology.

Let us now describe how we view the nature of the abductive mechanism in general.

The new ideas we shall put forward are:

- *Metalevel properties*:
  abduction is a metalevel notion and should be understood in the context of object level/metalevel interaction.

- *Dependence on proof theory*:
  abduction for a logic $\mathbf{L}_1$ depends on proof procedures $\Pi_{\mathbf{L}_1}$ for $\mathbf{L}_1$. Different proof procedures may give different abductive mechanisms.

- *Abduction mechanisms are data*:
  abductive principles can be part of the data. In other words, a declarative item of data can be either a (labelled) formula or a principle of abduction. [4]

- *Abduction can change the logic*:
  abductive principles can be a new principle of proof. In other words new rules can be abduced which can be used to prove more, i.e. the abductive mechanism for $\mathbf{L}_1$ can modify $\Pi_{\mathbf{L}_1}$. The effect of such modification can enrich the logic $\mathbf{L}$.

- *Abduction requries a second background logic*:
  Assume we have a logic $\mathbf{L}_1$ together with a proof theory $\Pi_{\mathbf{L}_1}$ for it. In order to give an abductive mechansim for $\mathbf{L}_1$ we need to provide two independent components: a procedure for the abduction (which will make use of the proof theory of $\mathbf{L}_1$, and can be thought of as the *explanation* component) together with an additional, possibly completely different, logic $\mathbf{L}_2$, (which can be thougth of as the *plausibility* component). The abductive procedure determines, using the logic $\mathbf{L}_1$ and its proof theory, possible candidates formulas to be abduced. The logic $\mathbf{L}_2$ is used to decide which candidates it is plausible to choose for abduction. The logic $\mathbf{L}_2$ itself may involve its own abduction process, and so on ....

- *Abduction may not succeed:*
  There may not be anything acceptable to abduce, (adding the goal as abducible may not be an option).

The structure of the chapter is as follows. Section 2 will give an *LDS* formulation for logics with $\Rightarrow$ only. This kind of *LDS* is general enough to cover in a uniform way a variety of implications: classical, intuitionistic, strict (modal), resource/substructural, many-valued and conditional. This system can therefore

[4] We have argued elsewhere [Gabbay, to appear] that databases should contain the other mechanisms such as actions as part of the data. See also [Gabbay and Woods, forthcoming].

serve as a general case study for abduction. The later part of the section will give many examples of proofs from resource logics.

Section 3 will give a goal-directed proof theory for the $\Rightarrow$ of Section 2. Although we are presenting a specific *LDS* and its proof theory, our example is general enough to illustrate how abduction can be done in general, for any logic.

Section 4 discusses intuitively various options of how to do abduction.

Section 5 gives the general abstract theory of abduction for logical systems.

Section 6 presents further interesting case studies.

Section 7 concludes with a final discussion.

2 INTRODUCING LDS

We have stated in Section 1 that we believe the best way of describing the abduction mechanism in its general form is to present it within the framework of *Labelled Deductive Systems* (*LDS* [Gabbay, 1996]). This framework is general enough to contain as special cases most, if not all, well known logics, whether classical, non-classical, monotonic or non-monotonic logics.

We begin by introducing a typical *LDS* formulation for implication $\Rightarrow$ and within such a system we discuss the principles of abduction.

2.1 LDS for $\Rightarrow$

To present an *LDS* for implication $\Rightarrow$ we need an algebra $\mathcal{A}$ of labels of the form $(\mathbb{A}, f, \varphi)$, where $\mathbb{A}$ is a set of labels, f is a binary operation on $\mathbb{A}$ and φ is a binary compatibility relation on $\mathbb{A}$.[5]

Our notion of a well-formed declarative unit is defined as a pair $\alpha : B$, where α is a label i.e. $\alpha \in \mathbb{A}$ and B is a traditional wff in the propositional language with $\Rightarrow$ and with atomic propositional variables $\{q_1, q_2, q_3, \ldots\}$.

Using the above algebra we can present the (f, φ) $\Rightarrow$-elimination rule as follows:

- $(f, \varphi) \Rightarrow E$ Rule:

$$\frac{\alpha : A; \beta : A \Rightarrow B; \varphi(\beta, \alpha)}{f(\beta, \alpha) : B}$$

We need to assume that $\varphi(\beta, \alpha)$ is a decidable predicate over $\mathbb{A}$ and that f is a total recursive function on A. This makes the rule effective. This rule is a general labelled *Modus Ponens* rule. We shall see later in Definition 1 how to use it.

[5]This is not the most general definition, but it is sufficient for having a general system in which abduction principles are explained. f, φ are necessary predicates in any *LDS*. Different logics may have additional relations and functions on $\mathbb{A}$, besides the compulsory f and φ. The most common additions are an ordering $<$ and the constant d. φ is in the language with $f, <$ and d.

We need one more notion. Let $t(x)$ be a term of the algebra built up using the function symbol f and the variable x ranging over $\mathbb{A}$. The function $\lambda x t(x)$ is a unary function over $\mathbb{A}$ which may or may not be definable in $\mathcal{A}$.

We say that $\lambda x t(x)$ is *φ-realisable* by an element $\alpha \in \mathbb{A}$ iff the following holds

$$\forall x[\varphi(\alpha, x) \rightarrow f(\alpha, x) = t(x)].$$

We denote this α, if it exists, by $(\eta x)t(x)$.

We must assume that our algebra is such that it is decidable to check whether $(\eta x)t(x)$ exists and if it does exist we have algorithms available to effectively produce it.

Given an (f, φ) $\Rightarrow$-elimination rule, we can define the (f, φ) $\Rightarrow$-introduction rule as follows:

- To show $\gamma : A \Rightarrow B$, assume $x : A$ and $\varphi(\gamma, x)$ and prove $f(\gamma, x) : B$,[6] where x is a new atomic variable.

The traditional way of writing this rule is as follows:

Show $\gamma : A \Rightarrow B$ from subcomputation:

	show $f(\gamma, x) : B$
Assume	$x : A; \varphi(\gamma, x)$
⋮	
$f(\gamma, x) : B$	

Exit with $\gamma : A \Rightarrow B$

DEFINITION 1.

1. A database Δ is a set of declarative sentences (labelled formulas) together with a set of compatibility relations of the form $\varphi(\alpha_i, \beta_i)$ on the labels of Δ. Part of the notion of a database is the notion of where to place additional inputs. Thus, if we want to add the declarative unit $x : B$ to the database Δ to get a new database Δ', we need an insertion operation. We can assume that with the database there is a metapredicate $\Psi_2(\Delta, \Delta', x : B)$ saying Δ' is the result of inserting $x : B$ into Δ. We also need a notion of deleting $x : B$ from Δ. We can use the metapredicate $\Psi_3(\Delta, \Delta', x : B)$. Of course Ψ_3 and Ψ_2 satisfy some obvious properties. These predicates need not be definable in the algebra $\mathcal{A}$, but could be higher order. We can write $\Delta' = \Delta + (x : B)$ for insertion and $\Delta' = \Delta - (x : B)$ for deletion, provided we understand

[6] In practice one proves $t(x) : B$ for some term $t(x)$ and then shows that $\gamma = (\eta x)t(x)$.

$+$ and $-$ as referring to Ψ_2 and Ψ_3. Ψ_2 and Ψ_3 must be computable. In practice we can express Ψ_3 and Ψ_2 in the algebra.[7]

2. We define the notion $\Delta \vdash \alpha : A$ by induction.

   (a) $\Delta \vdash_0 \alpha : A$ iff there exists a sequence of labelled wffs $\alpha_1 : A_1, \ldots, \alpha_k : A_k$ such that each $\alpha_i : A_i$ is either an item of data in Δ or is obtained from previous two elements

   $$\alpha_m : A_m, \alpha_n : A_n, m, n < i$$

   via the $(f, \varphi) \Rightarrow$-elimination rule. This means that $\varphi(\alpha_n, \alpha_m)$ can be proved from Δ, $A_n = A_m \Rightarrow A_i$ and $\alpha_i = f(\alpha_n, \alpha_m)$. We also have $\alpha_k : A_k$ is $\alpha : A$.

   (b) We say $\Delta \vdash_{n+1} \alpha : B \Rightarrow C$ iff $\Delta + \{x : B; \varphi(\alpha, x)\} \vdash_m f(\alpha, x) : C$, where x is a new variable and $m \leq n$.

3. We write $\Delta \vdash_\infty \alpha : A$ if for some n, $\Delta \vdash_n \alpha : A$.

4. Let Ψ be a decidable meta-predicate on pairs of the form (S, α) where S is a set of labels and α is a label.[8] Given a database Δ let $S_\Delta = \{\alpha \in \mathbb{A}$ s.t. for some formula $B \mid \alpha : B \in \Delta\}$.

   We write $\Delta \vdash_\Psi B$ if for some α, $\Delta \vdash_\infty \alpha : B$ and $\Psi(S_\Delta, \alpha)$ holds.[9]

5. We say B is a theorem of the logic if $\varnothing \vdash_\Psi B$.

EXAMPLE 2. We give a simple example. Let $\mathcal{A}$ be a free non-commutative semigroup with operation $*$, i.e. it has the form $\mathcal{A} = (\mathbb{A}, *)$. $*$ will be our f. Consider the database Δ with

1. $a_1 : A \Rightarrow (A \Rightarrow B)$
2. $a_2 : A$

We can prove B as follows

3. $(a_1 * a_2) : A \Rightarrow B$, from (1) and (2)
4. $(a_1 * a_2) * a_2 : B$, from (3) and (2)

[7] We let $\Delta' = \Delta \cup \{x : B\}$ but by definition we also add some formulas about x, relating x to the labels y of Δ. For example, the labels in Δ may be linearly ordered by $<$. We may agree that all inputs $x : B$ receive the highest priority in the ordering, in which case the formula to add is $\forall y(y \leq x)$. We write $\Psi_2(\Delta, \Delta', x : B)$ to say Δ' is the result of proper input of $x : B$ into Δ. See Definition 11 below.

[8] Although Ψ is a predicate on the algebra it need not be first order. It need only be recursive.

[9] Note that the logic $\vdash_\Psi$ can turn out to be either monotonic or non-monotonic. It all depends on Ψ and the properties of the input function.

5. Thus we have $\Delta \vdash (a_1 * a_2) * a_2 : B$.

   We have several options for Ψ.

   (a) $\Psi(S_\Delta, \alpha)$ holds if all atoms in Δ appear in α. (We say that all assumptions are used.)
   (b) $\Psi(S_\Delta, \alpha)$ holds if all atoms in Δ appear exactly once, (we say that all assumptions in Δ are used exactly once).
   (c) $\Psi(S_\Delta, \alpha) = \top$ (we say we ignore the labels).
   (d) Assuming $*$ is associative, we can have $\Psi(S_\Delta, \alpha)$ holds iff $x_1 * x_2 * \ldots * x_n = \alpha$, where $S_\Delta = \{x_1, \ldots, x_n\}$.

2.2 *Examples of Resource LDS*

We shall illustrate the use of the labels to perform computations in substructural logics. This will show how different labelling algebras can give different implicational logics and justify our choice of the *LDS* forum to discuss general abduction principles. Our labels are multisets as defined in Definition 3 below. Multisets are like sets except that elements may appear in them more than once. The use of multisets allows us to keep track of how many times an assumption is used in *modus ponens*.

DEFINITION 3 (Multisets). Let $\mathbb{A}$ be a set of atoms. A multiset based on $\mathbb{A}$ is a function α on $\mathbb{A}$ giving for each element $a \in \mathbb{A}$ a natural number $\alpha(a) \geq 0$. $\alpha(a)$ tells us how many copies of a we have in the multiset α. Let $\gamma = \alpha \cup \beta$ be defined as the function $\gamma = \alpha + \beta$ (i.e. for each $a, \gamma(a) = \alpha(a) + \beta(a)$). Let $\gamma = \alpha \dot{-} \beta$ be the function defined for each a by $\gamma(a) = 0$, if $\alpha(a) \leq \beta(a)$, and $\gamma(a) = \beta(a) - \alpha(a)$, otherwise.

EXAMPLE 4.

1. The simplest example is the set resource labelling. The labels are sets $\mathbb{A}$ of atomic labels and $f(\beta, \alpha) = \beta \cup \alpha$. φ and Ψ depend on the logic under consideration. In many logics, however, we have that $\varphi(\beta, \alpha) = \top$, i.e. φ always holds.

2. Consider the following database Δ_1:

$$\begin{array}{ll} \{a_1\}: & A \\ \{a_2\}: & A \\ \{a_3\}: & A \Rightarrow B \end{array}$$

   This database can prove

$$\begin{array}{ll} \Delta_1 \vdash_1 \{a_3, a_2\}: & B \\ \Delta_1 \vdash_1 \{a_3, a_1\}: & B \end{array}$$

Let $\Psi(S, \alpha)$ be $\alpha = \bigcup_{\beta} \beta \in S$. Then we do *not* have that $\Delta_1 \vdash_\Psi B$.

3. Consider the database Δ_2

$$\begin{array}{ll} \{a_1\} & (A \Rightarrow B) \Rightarrow (A \Rightarrow B) \\ \{a_2\} & A \Rightarrow B \\ \{a_3\} & A. \end{array}$$

We have (using the last two items)

$$\Delta_2 \vdash \{a_2, a_3\} : B$$

we can also first use $\{a_1\}$ and $\{a_2\}$ and get

$$\Delta_2 \vdash \{a_1, a_2, a_3\} : B.$$

Thus we do have in this case

$$\Delta_2 \vdash_\Psi B.$$

Note the following three conventions:

1. Each new assumption is labelled by a new atomic label. An ordering on the labels can be imposed, namely $a_1 < a_2 < a_3$. This is to reflect the fact that the assumptions arose from our attempt to prove $A \Rightarrow (A \Rightarrow ((A \Rightarrow B) \Rightarrow B))$ and not for example from $(A \Rightarrow B) \Rightarrow (A \Rightarrow (A \Rightarrow B))$ in which case the ordering would be $a_3 < a_1 < a_2$. The ordering can affect the proofs in certain logics. Some logics allow us to bring in, anywhere in the proof, theorems of the logic with the empty label and allow their use in the proof. Other logics do not allow this.

2. If in the proof A is labelled by the multiset α and $A \Rightarrow B$ is labelled by β then B can be derived with a label $\alpha \cup \beta$ where $\cup$ denotes multiset union.

3. If B was derived using A as evidenced by the fact that the label α of A is a singleton $\{a\}$, a atomic and a is in the label β of B ($\alpha \subseteq \beta$) then we can derive $A \Rightarrow B$ with the label $\beta \dot{-} \alpha$ ('$\dot{-}$' is multiset subtraction).

In case our labels are sets, we use $\beta - \alpha$, where '$-$' is set subtraction.

EXAMPLE 5. Show $(B \Rightarrow A) \Rightarrow ((A \Rightarrow B) \Rightarrow (A \Rightarrow B))$

$$\underline{\{a_2, a_1\} : (A \Rightarrow B) \Rightarrow (A \Rightarrow B)}$$

(1) $\{a_1\} : B \Rightarrow A$
(2) $\{a_2, a_1\} : (A \Rightarrow B) \Rightarrow (A \Rightarrow B)$

$$\underline{\{a_2, a_1, a_2\} : A \Rightarrow B}$$

(2.1) $\{a_2\} : A \Rightarrow B$
(2.2) $\{a_2, a_1, a_2\} : A \Rightarrow B$

$$\underline{\{a_2, a_1, a_2, a_3\} : B}$$

(2.2.1) $\{a_3\} : A$
(2.2.2) $\{a_2, a_3\} : B$
(2.2.3) $\{a_1, a_2, a_3\} : A$
(2.2.4) $\{a_2, a_1, a_2, a_3\} : B$

The above is the box method of representing the deduction. Note that in leaving the inner box for $\{a_2, a_1, a_2\} : A \Rightarrow B$, multiset subtraction was used and only one copy of the label a_2 was taken out. The other copy of a_2 remains and cannot be cancelled, so that the entire computation finishes with the label of $\{a_2\}$. We thus have scope here to define different logics by saying when a labelled proof is acceptable. For linear logic, the final label at the end of the computation must be empty, signifying that formulae have only been used once. Hence this formula is not a theorem of linear logic because the outer box does not exit with label $\varnothing$. In relevance logic, the discipline uses sets and not multisets. Thus the label upon leaving the inner box in this case would be $\{a_1\}$ and that upon leaving the outermost box would be $\varnothing$.

Note that different conditions on labels correspond to different logics, given informally in the following table:

condition Ψ:	**logic:**
ignore the labels	intuitionistic logic
accept only the derivations which use all the assumptions	relevance logic
accept derivations which use all assumptions exactly once	linear logic

The conditions on the labels can be translated into reasoning rules.

The following are examples of further proofs. The examples speak for themselves.

EXAMPLE 6.

(1)	$\varnothing : (A \Rightarrow (A \Rightarrow B)) \Rightarrow (A \Rightarrow B)$	subcomputation

$\underline{\{a_1\} : (A \Rightarrow B)}$

(1.1)	$\{a_1\} : A \Rightarrow (A \Rightarrow B)$	assumption
(1.2)	$\{a_1\} : A \Rightarrow B$	subcomputation

$\underline{\{a_1, a_2\} : B}$

(1.2.1)	$\{a_2\} : A$	assumption
(1.2.2)	$\{a_1, a_2\} : A \Rightarrow B$	by (1.1), (1.2.1)
(1.2.3)	$\{a_1, a_2\} : B$	by (1.2.1), (1.2.2)

The proof is acceptable in relevance logic, but not in linear logic, because if labels are multisets we obtain (1.2.3) $\{a_1, a_2, a_2\} : B$ in the innermost box, which provides only (1.2) $\{a_1, a_2\} : A \Rightarrow B$ in the next box, and then (1) $\{a_2\} : (A \Rightarrow (A \Rightarrow B)) \Rightarrow (A \Rightarrow B)$ in the outermost box. This happens because the assumption (1.2.1) is used twice in this proof.

It is clear that for linear logic the labelling becomes more efficient if we introduce the condition $\varphi(\beta, \alpha) = \beta \cap \alpha = \varnothing$ (instead of $\varphi(\beta, \alpha) = \top$). This new condition prevents assumptions from being used more than once. Ψ then makes sure that all assumptions are used.

EXAMPLE 7.

(1) $\varnothing : (C \Rightarrow A) \Rightarrow ((B \Rightarrow C) \Rightarrow (B \Rightarrow A))$ subcomputation

> $\underline{\{a_1\} : (B \Rightarrow C) \Rightarrow (B \Rightarrow A)}$
>
> (1.1) $\{a_1\} : C \Rightarrow A$ assumption
> (1.2) $\{a_1\} : (B \Rightarrow C) \Rightarrow (B \Rightarrow A)$ subcomputation
>
> > $\underline{\{a_1, a_2\} : B \Rightarrow A}$
> >
> > (1.2.1) $\{a_2\} : B \Rightarrow C$ assumption
> > (1.2.2) $\{a_1, a_2\} : B \Rightarrow A$ subcomputation
> >
> > > $\underline{\{a_1, a_2, a_3\} : A}$
> > >
> > > (1.2.2.1) $\{a_3\} : B$ assumption
> > > (1.2.2.2) $\{a_2, a_3\} : C$ by (1.2.2.1), (1.2.1)
> > > (1.2.2.3) $\{a_1, a_2, a_3\} : A$ by (1.2.2.2), (1.1)

The proof is acceptable in linear logic. In fact, the proof is acceptable in Lambek logic, where a database is a sequence of formulas, input into the database is done by adding the input to the end of the sequence and modus ponens is done by order adjacent formulas, where the implication comes first, i.e. '$X \Rightarrow Y, X$' yields 'Y' and the result of the modus ponens replaces the participants in the position of the ordering.

EXAMPLE 8.

> (1) $\varnothing : (A \Rightarrow (B \Rightarrow C)) \Rightarrow ((A \Rightarrow B) \Rightarrow (A \Rightarrow C))$ subcomputation
>
> > $\underline{\{a_1\} : (A \Rightarrow B) \Rightarrow (A \Rightarrow C)}$
> >
> > (1.1) $\{a_1\} : A \Rightarrow (B \Rightarrow C)$ assumption
> > (1.2) $\{a_1\} : (A \Rightarrow B) \Rightarrow (A \Rightarrow C)$ subcomputation
> >
> > > $\underline{\{a_1, a_2\} : A \Rightarrow C}$
> > >
> > > (1.2.1) $\{a_2\} : A \Rightarrow B$ assumption
> > > (1.2.2) $\{a_1, a_2\} : A \Rightarrow C$ subcomputation
> > >
> > > > $\underline{\{a_1, a_2, a_3\} : C}$
> > > >
> > > > (1.2.2.1) $\{a_3\} : A$ assumption
> > > > (1.2.2.2) $\{a_1, a_3\} : B \Rightarrow C$ by (1.1), (1.2.2.1)
> > > > (1.2.2.3) $\{a_2, a_3\} : B$ by (1.2.1), (1.2.2.1)
> > > > (1.2.2.4) $\{a_1, a_2, a_3\} : C$ by (1.2.2.2), (1.2.2.3)

The proof is acceptable in relevance logic. It is not acceptable in linear logic, because the assumption (1.2.2.1) is used twice (cf. the solution to Example 6).

EXAMPLE 9.

> (1) $A \Rightarrow (B \Rightarrow A)$ subcomputation
>
> > $\underline{B \Rightarrow A}$
> >
> > (1.1) A assumption
> > (1.2) $B \Rightarrow A$ subcomputation
> >
> > > $\underline{A}$
> > >
> > > (1.2.1) B assumption
> > > (1.2.2) A by (1.1)

The proof is acceptable in intuitionistic logic. It cannot be transferred to relevance logic by introducing labels because the label for B in (1.2.1) does not disappear in the outermost box.

EXAMPLE 10.

(1) $\varnothing : ((A \Rightarrow A) \Rightarrow A) \Rightarrow A$ subcomputation

$\{a_1\} : A$

(1.1) $\{a_1\} : (A \Rightarrow A) \Rightarrow A$ assumption
(1.2) $\varnothing : A \Rightarrow A$ subcomputation

$\{a_2\} : A$

(1.2.1) $\{a_2\} : A$ assumption
(1.2.2) $\{a_2\} : A$ by (1.2.1)

(1.3) $\{a_1\} : A$ by (1.1), (1.2)

This proof is *not* in relevance logic, because (1.2) is proved as a 'lemma' with label $\varnothing$. Relevance logic does not allow for such labelling: you should not forget assumptions while making a proof, so $A \Rightarrow A$ should be proved with a label $\{a_1\}$, which is impossible.

Of course, by deleting all labels we get an intuitionistic proof.

3 GOAL DIRECTED ALGORITHM FOR $\Rightarrow$

One of our general abduction principles for general logics is that abductive procedures do not depend only on the consequence relation (i.e. on $\Delta \vdash A$ alone) but also on the proof procedure given for finding whether $\Delta \vdash A$ or not. In other words we abduce something in the context of our algorithmic search. To make this concrete, we offer, in this section a general goal directed proof procedure for our $\Rightarrow$ *LDS*. We will then show how abduction can be dependent on it.

In fact, we will show how abduction can be dependent on any general type proof theoretic procedure since our goal directed algorithm is a typical reduction algorithm.

3.1 The Algorithm

We now give a goal directed computation which seeks to find for a given Δ and B whether $\Delta \vdash_\Psi B$, namely, whether there exist any labels β such that $\Delta \vdash_\infty \beta : B$ and $\Psi(S_\Delta, \beta)$ hold.

We begin by observing that each declarative formula in the database has the form $\alpha : A_1 \Rightarrow (A_2(\Rightarrow \cdots \Rightarrow (A_n \Rightarrow q)\ldots))$ where q is atomic. q is called the *head* and $(A_1, \ldots, A_n)$ is called the *body* (sequence). Consider the question of whether $\Delta \vdash_k \delta : A$. A can have one of two forms:

1. A can be an atom q

2. A can have the form $A = (B \Rightarrow C)$.

Also recall that

3. $\Delta \vdash_\Psi A$ if for some $\delta, k, \Delta \vdash_k \delta : A$ and $\Psi(S_\Delta, \delta)$ holds.

In the first case, if $\Delta \vdash_k \delta : q$, then there must exist an $\alpha : A_1 \Rightarrow (A_2 \Rightarrow \cdots (A_n \Rightarrow q) \ldots)$ in Δ and labels $\alpha_1 \ldots, \alpha_n$ such that $\Delta \vdash_k \alpha_i : A_i$ and labels $\delta_1, \ldots, \delta_n = \delta$ such that the following all hold:

$$\begin{aligned} &\delta_1 = f(\alpha, \alpha_1); \varphi(\alpha, \alpha_1) \\ &\delta_2 = f(\delta_1, \alpha_2); \varphi(\delta_1, \alpha_2) \\ &\vdots \\ &\delta = \delta_n = f(\delta_{n-1}, \alpha_n); \varphi(\delta_{n-1}, \alpha_n) \end{aligned}$$

Let us abbreviate the above as

$$\delta = f[\alpha, (\delta_1, \ldots, \delta_n)]$$

and

$$\varphi[\alpha, (\delta_1, \ldots, \delta_n)].$$

Of course it may be the case that there is no item of data in the database with head q. In this case we are stuck and the computation clearly cannot continue. We say that we have finite failure at this point.

Another possibility of finite failure is that although a clause does exist, φ cannot be satisfied. Again we cannot go on.

In the second case, if $\Delta \vdash_k \delta : B \Rightarrow C$ holds, then if we add $x : B$ to Δ for a new variable x, we must have that

$$\Delta + \{x : B\} \vdash_m f(\delta, x) : C$$

for some $m \leq k - 1$.

Given the above, we can write a constraints logic programming program (the algebra of constraints is our algebra $\mathcal{A} = (\mathbb{A}, f, \varphi))$ for the metapredicate,

$$\textit{Success}\ (\Delta, \delta : A, \textit{constraints}, \theta) = 0 \text{ or } 1.$$

where Δ is a labelled database, (with labels containing variables), $\delta : A$ is the current goal, *constraints* is a set of constraints on the labels and θ is a possibly partial substitution for the label variables.

Success = 1 means the computation succeeds, and
Success = 0 means the computation finitely fails.

Of course the computation can loop or just continue forever for whatever reasons. So the two options are not exhaustive. The consequence relation meaning of *Success* $= 1$ means that $\Delta\theta \vdash \delta\theta : A$ and $\mathcal{A} \vdash$ *constraints* θ.

We now give a recursive definition. The definition is formulated using meta-predicates $\Psi_1, \Psi_2, \Psi_3, \Psi_4$ and Ψ. This way we can change the computation for different logics by varying the Ψs. (See [Gabbay and Olivetti, 2000].) In general the Ψs need to satisfy some conditions discussed in Section 5. The meaning of the Ψs is as follows. Ψ_1 tells us when an atomic query can succeed or fail from a database in *one step* (immediate success or failure). Ψ_2 is an insertion predicate. Ψ_3 is a deletion predicate and Ψ_4 is a decomposition predicate; given a database Δ, and databases $\Delta'_1, \ldots, \Delta'_n$, it may be that Δ'_i are a decomposition of Δ in some way. We write $\Psi_4(\Delta, \Delta'_1, \ldots, \Delta'_n)$ to express this relation. Note that n may vary and that the decomposition may not be disjoint. Ψ tells us when a proof is acceptable.

The choice of Ψs proposed below is for the simple resource logics case.

DEFINITION 11. We define the metapredicate *Success* $(\Delta, \delta : A$, *constraint*, $\theta)$ $= 1$ or 0, where θ is a substitution to label variables, as follows:

1. *Immediate Success Case: Success*$(\Delta, \delta : q$, *constraints*, $\theta) = 1$ if q is atomic and $\delta\theta = \alpha\theta$, for some label α such that $\Psi_1(\Delta\theta, \delta\theta : q)$ holds and the constraints are provable in the algebra $\mathcal{A}$. In the case of resource logics, $\Psi_1(\Delta, \delta : q)$ can be, for example, $\alpha : q \in \Delta$.

2. *Implication case: Success*$(\Delta, \delta : B \Rightarrow C$, *constraints*, $\theta) = x$ if for $\Delta' = \Delta{+}(a : B)$ (i.e. Δ' such that $\Psi_2(\Delta, \Delta', a : B)$) we have *Success*$(\Delta', f(\delta, a) : C$, *constraints*, $\theta) = x$, for a new atomic constant a. Recall that Ψ_2 says that Δ' is the (usually unique) result of inserting $a : B$ into Δ. In case of resource logics, $\Psi_2(\Delta, \Delta', a : B)$ is $\Delta' = \Delta \cup \{a : B\}$.

3. *Immediate Failure case: Success*$(\Delta, \delta : q$, *constraints*, $\theta) = 0$ for q atomic if either the constraints are not provable under the substitution θ or there is no clause in the database Δ with head q, such that $\Psi_1(\Delta\theta, \delta\theta : q)$ holds.

4. *Cut Reduction case: Success*$(\Delta, \delta : q$, *constraints*, $\theta) = 1$ (resp. $= 0$) if for some (resp. any) $E = (\alpha : A_1 \Rightarrow (A_2 \Rightarrow \ldots (A_n \Rightarrow q) \ldots))$ in Δ and for some (resp. any) new variables $\delta_1, \ldots, \delta_n$ and some (resp. all) choices of $\Delta'_1, \ldots, \Delta'_n$ and $\Delta'_0 = \{\alpha : A_1 \Rightarrow (A_2 \Rightarrow \ldots (A_n \Rightarrow q) \ldots)\}$ such that $\Psi_4(\Delta, \Delta'_0, \ldots, \Delta'_n)$ holds and substitution θ' extending θ to $\delta_1, \ldots, \delta_n$ we have for each $1 \leq i \leq n$ (resp. for some i) *Success*$(\Delta'_i, \delta_i : A_i$, *constraints*$', \theta') = 1$ (resp. =0) where *constraints*$'$ = *constraints*$\cup\{\delta\theta' = f[\alpha\theta', (\delta_1\theta', \ldots, \delta_n\theta'] \} \cup \{\varphi[\alpha\theta', (\delta_1\theta', \ldots, \delta_n\theta')]\}$.[10]

5. *Consequence:* We have $\Delta \vdash A$ if for some variable δ and some substitution θ for δ we have *Success*$(\Delta, \delta : A, \{\Psi(S_\Delta, \delta)\}, \theta) = 1$.

[10] Δ_0 appears in Ψ_4 as a parameter containing the formula we are using in this rule.

Note that the computation may not be decidable unless we assume we have recursive algorithms for all the conditions and metapredicates mentioned in it and all ranges of choice are computably finite.

The computation may loop, so a historical loop checker may be needed.

The above algorithm is a typical reduction algorithm. We start with an initial state involving $(\Delta, A,$ *parameters*) and rewrite or reduce it to other states. Some states (Γ, B, P) are reduced to *success* and some states may be such that no reduction rules apply, in which case we can say they are reduced to *(finite) failure.*

The typical abduction principle will consider the history of the reduction up to a point of failure and tell us what to do at that stage.

Of course the process may loop but in the propositional case a historical loop checker can take care of that and so we may assume that the parameters include a device that eliminates looping.

3.2 Examples

EXAMPLE 12. In any logic, consider $q \Rightarrow q \vdash ?q$.

Success$(q \Rightarrow q, q,$ *no constraints*$) = x$, simply loops.

EXAMPLE 13. Let us do Example 6 in a goal-directed way, for linear logic. The labels are multisets. Note the way the computation is displayed, as
Data $\vdash ?$ **Goal; constraints.**

1. $\varnothing \vdash ?\delta : (A \Rightarrow (A \Rightarrow B)) \Rightarrow (A \Rightarrow B); \delta = \varnothing$
2. $\{a_1\} : A \Rightarrow (A \Rightarrow B) \vdash ?\delta \cup \{a_1\} : A \Rightarrow B; \delta = \varnothing$
3. $\{a_1\} : A \Rightarrow (A \Rightarrow B), \{a_2\} : A \vdash ?\delta \cup \{a_1, a_2\} : B; \delta = \varnothing$
4. We split here into two parallel computations, listed below as (5a) and (5b). Using the clause $\{a_1\} : A \Rightarrow (A \Rightarrow B)$.
5. (a) $\{a_1\} : A \Rightarrow (A \Rightarrow B), \{a_2\} : A \vdash ?x_1 : A$
   (b) $\{a_1\} : A \Rightarrow (A \Rightarrow B), \{a_2\} : A \vdash ?x_2 : A$

The constraints for both 5a and 5b (done in parallel) are

$$\{a_1\} \cup \{x_1\} \cup \{x_2\} = \delta \cup \{a_1, a_2\} \text{ and } \delta = \varnothing$$

5a can succeed by substituting $\theta(x_1) = a_2$ and similarly 5b can succeed by substituting $\theta(x_2) = a_2$. Thus we get the constraints to be $\{a_1, a_2, a_2\} = \{a_1, a_2\}$.

Since for linear logic we are dealing with multisets, this computation fails, because the constraints cannot be satisfied and there are no other options for the computation. In the case of relevance logic, the labels are sets and the constraints are satisfied and the computation succeeds.

The above computation can be made more efficient. Our first optimisation move is to throw out of the database any clause 'used' by rule 3. This saves us time because we are allowed to use each clause at most once.

So condition 4 of the definition of the *Success* predicate can be modified by changing Ψ_4 and requiring it to say that $\Delta'_i \subseteq \Delta \dot{-} \{\alpha : (A_1 \Rightarrow \ldots \Rightarrow (A_n = q)\ldots)$ for all i.

Since also all the clauses in the database must be used, we can further modify Ψ_4 to be the conjunction of the two following conditions:

- $\bigcup_{i=1}^{n} \Delta'_i = \Delta \dot{-} \{\alpha : (A_1 \Rightarrow \ldots \Rightarrow (A_n \rightarrow q)\ldots)\}$
- $\Delta'_i \cap \Delta'_j = \varnothing$ for $i \neq j$.

We will need to change clause 1 of the algorithm to be

- $\mathit{Success}(\Delta, \delta : q, \mathit{constraints}, \theta) = 1$ if $\{\alpha : q\} = \Delta$ and $\alpha\theta = \delta\theta$, i.e. change Ψ_1 to the condition $\{\alpha : q\} = \Delta$.

The computation of our example will now be short and quick as follows (we don't even need to use labels):

$$\varnothing \vdash ?((A \Rightarrow (A \Rightarrow B)) \Rightarrow (A \Rightarrow B)$$

if

$$A \Rightarrow (A \Rightarrow B) \vdash ?A \Rightarrow B$$

if

$$A; A \Rightarrow (A \Rightarrow B) \vdash ?B$$

if

$$A \vdash ?A \text{ and } \varnothing ?A$$

if

success and *failure*.

We take this opportunity to use this example and indicate options for abduction. Our purpose is to succeed. The simplest option in the case of linear logic is to duplicate A (as we are missing one copy of A).

This would normally be done at the stage where $\{a_2\} : A$ was put into the database. At that stage $A \Rightarrow (A \Rightarrow B)$ was already in the database, so we need to add to the database the wff $(A \Rightarrow (A \Rightarrow B)) \Rightarrow A$, (rather than adding just A).

So in order for $(A \Rightarrow (A \Rightarrow B)) \Rightarrow (A \Rightarrow B)$ to succeed from the empty database, we need to abduce $(A \Rightarrow (A \Rightarrow B)) \Rightarrow A$ into the database. This will not help in our case, where the logic is linear logic, because to get A we need another copy of $A \Rightarrow (A \Rightarrow B)$. So we are better off putting A in the database. This can be more readily seen in the case of the more optimised computation where failure results from $\varnothing \vdash ?A$.

Another option for abduction is to regard the constraints as satisfied if they hold for the labels when regarded as sets. This means changing the proof procedures to those of relevance logic.

4 INTUITIVE THEORY OF LABELLED ABDUCTION

This section will introduce our intuitive theory of abduction within the framework of *Labelled Deductive Systems*, and give some simple examples.

The basic situation we are dealing with can be presented as

Δ	$\vdash ?!Q$
data	?query or ! input

It is a relationship between a database and a (labelled) formula. The relationship is either declarative (i.e. $?Q$, Q a query) or imperative ($!Q$, Q is an input or a demand to perform abduction or a demand for explanation etc). In the imperative case there is an interaction between Δ and Q and a new database Δ' emerges.

We have put forward in previous sections that the most general and useful database is the one where the data is labelled and the proof procedures use the labelling.

In this set up, the abduction rules are extra moves that help answer the query or help change the database as a result of the query or input.

Thus to do abduction we need more precise proof procedures or update procedures for labelled databases and then on top of that we can define the extra abductive rules.

The exact proof procedures can be conveniently formalised in the framework of *LDS* and the previous section gave one example procedure, namely the goal directed proof procedure. We shall give a sample abduction algorithm for the goal directed procedure in a later section. The reason why abduction should depend also on the proof procedures can be intuitively motivated by an example. Suppose I am looking for an object in my home which I need for some purpose. I search the house and keep track of where I looked of what I found along the way. If I fail to find the object, I look for a substitute. This is abduction. I may use something else I found during the search or I may add a new principle of supplementary searching in the hope of finding either the object or a new possible substitute. If I find several possible candidates for a substitute I may use other unrelated reasoning to decide which one to take.

Let us illustrate our ideas through a series of examples.

4.1 Abduction in Knowledge Bases

EXAMPLE 14. The database below is a Horn clause database. It is labelled in the sense that each clause is named. The query is D. The query does not follow from the database as it is. We are going to use it to illustrate principles of abduction.

	Data	**Query**
a_1	$I \Rightarrow (T \Rightarrow D)$	$?\,D$
a_2	$L \Rightarrow I$	
a_3	$L \Rightarrow (S \Rightarrow T)$	
a_4	$O \Rightarrow (P \Rightarrow T)$	
a_5	L	

The database literals have no meaning. Let us give them a meaning. In the Stanford University English Department, there are two main ways of getting a PhD title. One can either put forward a thesis, stay in the department for 4-5 years acquiring and displaying an immense breadth of knowledge and pass an interview, or one can write a very good publication and get a job offer from another university in the top ten in the country. The database then becomes:

	Data
a_1	Interview $\Rightarrow$ (Thesis $\Rightarrow$ Degree)
a_2	Lecture $\Rightarrow$ Interview
a_3	Lecture $\Rightarrow$ (Scholarly Survey $\Rightarrow$ Thesis)
a_4	(Job) Offer $\Rightarrow$ (Publications $\Rightarrow$ Thesis)
a_5	Lecture

Query

? Degree

Another interpretation for the same database is a component interpretation. To do the laundry (D) one needs a washing machine (T) and washing powder (I). For washing powder one can use dishwashing soap (L). For a washing machine one may use a dishwasher (S) and dishwashing soap or one may handwash (P) but then one at least needs a spinner (O).

We thus get in this case

Data

a_1 Washing Powder $\Rightarrow$ (Washing Machine $\Rightarrow$ Laundry)

a_2 Dishwashing Soap $\Rightarrow$ Washing Powder

a_3 Dishwashing Soap $\Rightarrow$ (Dishwasher $\Rightarrow$ Washing Machine)

a_4 Spinner $\Rightarrow$ (Handwash $\Rightarrow$ Washing Machine)

a_5 Dishwashing Soap.

Query

? Laundry

We now list several possible abductive principles for the query $?D$. The principles depend on the computation, so let us suppose that we compute the query prolog like, where the pointer always starts at the top clause (assume $a_1 > a_2 > a_3 > a_4 > a_5$.)

We note that in logic programming [Eshghi and Kowalski, 1989] abduction for Horn clause programs is done via a system of the form (Δ, I, A), where Δ

is the program, I is a set of integrity constraints and A is a set of literals which are *abducible*. Whenever an abducible literal is encountered in the computation (e.g. $?D$) it is immediately added to the database provided it does not violate the integrity constraints.

Let us now examine our options:

Possible principles of Abduction

1. The first option is to abduce on anything as soon as needed. This corresponds in our case, to no integrity constraints and every literal is abducible. In this case we add D to the database, i.e. the Abduction principle yields D. In the component example such abduction makes no sense. I want to know which parts are missing so that we can get them and wash our clothes.

2. The second option is to abduce on literals which are not heads of clauses. In this case, we add S. This is because S is the first literal encountered in the top down order of execution. Note that we do not use here a set of abducibles. The structure of the database determines what we add.

3. If our underlying logic is not classical logic but some other resource logic, we will not succeed by adding S to the database because that would require the 'use' of L twice: once to make I succeed in clause a_2 and once to make T succeed in clause a_3. In the component example we need more dishwashing soap if we use a dishwasher, and we have only one lot of it (i.e. a_5).

   Note that the database is structured and thus we can add
   $a_6 \quad L$
   and $\{a_1, \ldots, a_5\}$ is *not* the same database as $\{a_1, \ldots, a_6\}$.

   Anyway, if the underlying logic is a resource logic, the result of our abduction will be O and P, unless we are prepared to add another copy of L.

4. If we require the weakest logical assumption (in classical logic) which makes the goal succeed then we must add $S \vee (O \wedge P)$. This abduction principle is independent of the computation.

5. In co-operative answering, the abduction principle takes the top level clause. In this case the answer is T. To the query '$?D$' we answer 'yes if T'. Think of the thesis example. If an ordinary student wants to know what is missing to get a PhD, the obvious answer is 'a thesis' and not 'a paper and a job offer from Harvard'.

6. The power of our labelling mechanism can be easily illustrated by a more refined use of the labels. If atoms are labelled, for example, by cost (laundry example) the abduction principle can aim for minimal cost. One can also

'cost' the computation itself and aim to abduce on formulas giving maximal provability with a least number of modus ponens instances.

EXAMPLE 15. To show that the abduction depends on the computation let us change the computation to forward chaining or Gentzen like rules. From

$$\text{Data} \vdash ?D$$

we get

$$I \Rightarrow (T \Rightarrow D), I, S \Rightarrow T, O \Rightarrow (P \Rightarrow T) \qquad \vdash ?D$$

which reduces to

$$T \Rightarrow D, S \Rightarrow T, O \Rightarrow (P \Rightarrow T), \qquad \vdash ?D$$

which reduces to the following by chaining:

$$S \Rightarrow D, O \Rightarrow (P \Rightarrow D), \qquad \vdash ?D$$

As we see, not many abduction possibilities are left!

So far we have discussed the possibilities of abduction principles being added to proof rules. We now come to our second new idea, namely:

- Abduction principles are data!

EXAMPLE 16 (Abduction as explanation). Let Δ_1 be the following database:[11]

$$\Delta_1 =: \boxed{\begin{array}{l} \mathit{Holiday}(x,y) \text{ if } \mathit{Available}(y) \wedge \mathit{Afford}(x,y) \\ \mathit{Afford}(x,y) \text{ if } \mathit{Cost}(y,z) \wedge \mathit{Credit}(x,z) \\ \mathit{Credit}(x,z) \text{ if } \mathit{In\_limit}(x,r) \\ \mathit{Sunnyplace}(a_i), i = 1 \ldots 3 \\ \mathit{Cost}(a_1, 200) \\ \mathit{Cost}(a_2, 250) \\ \mathit{Cost}(a_3, 700) \\ \mathit{Sunnyplace}(b) \\ \mathit{Cost}(b, 500) \end{array}}$$

A customer accesses the database by inserting his visa card into the slot and typing the query (*holiday*, a_1, a_2, a_3). The database can automatically get from the visa card his name x_0, and other credit details, and asks itself the queries:

$$?\mathit{Holiday}(x_0, a_i), i = 1, 2, 3,$$

An ordinary non-intelligent database will simply check if each query succeeds and output the details if it does and output fail if not. A more intelligent database might output an explanation in case of failure, such as:

[11] Note that our language does not contain '$\wedge$'. However, X if $Y \wedge Z$ is logically equivalent, in all our examples, to $Z \Rightarrow (Y \Rightarrow X)$. The use of '$\wedge$' makes our examples more readable.

You can't have a_1 because it is not available

or:

You can't have a_3 because you cannot get credit

An even more intelligent database may recognise, for example, that none of the a_i are available but they are all sunny places and b is a sunny place within the nearest price range. So using a metaprogram of 'interest', 'view' etc, might ask the user (in whatever way it 'communicates'):

Would you like (holiday,b)

and output the details. We are sure many papers in this volume address many of the issues related to the above capabilities.

Ultimately, though, we would like the human to ask in natural language rather than type the 'agreed' form of input (e.g. (holiday, a)) and furthermore reply naturally to a computer query. This might transcend simple 'yes' and 'no' answers: for example, a 'natural' form of response to the computer-generated query above might be:

'b is expensive'.

The computer should understand the English and deduce (as a human would) that the answer to the query is 'no', *and* that any holiday costing more than $500 is not a candidate. Ideally, we would like to imbue the intelligent database with the same faculty, so that it could now reply:

There are no cheaper holidays

to which a possible form of response might be:

OK, I will take it.

which includes anaphora, *it* referring to b, and withdrawal of the previous denial of b as a suitable holiday.

There are a number of hidden problems here, but essentially we need to spell out two processes: one the reasoning underlying the parsing process, including the interpretation of anaphoric expressions, the access of related information, etc.; and the other the reasoning underlying question-answer dialogues, which may involve indirect answers, negotiation requiring changes in beliefs, plans and goals, and so on.

EXAMPLE 17. Abduction can be data: Consider the following database and query:

a_1: A

a_2: $A \Rightarrow (B \Rightarrow S)$

a_3: B

....

a_4: X, abduce on structure to find X.

a_5: $B \Rightarrow D$
The goal is $?S \otimes D$.

By writing $S \otimes D$ for the goal we are saying we want to partition the database, which is a list of assumptions, into two parts, the first part must prove S and the second part must prove D. This is done in resource logics, where one pays attention to the question of which part of the database proves what.

Such considerations arise in many areas for example in what is known as *parsing as logic*.

Consider the text:
Mary hit John. He cried.

The way this can be analysed is by assuming that each word is assigned a wff of some resource logic (actually concatenation logic, see [Gabbay and Kempson, 1991]) with a label. This assignment is done at the lexical level. Thus a noun n is assigned $n' : NP$. An intransitive verb v_1 is assigned $v_1' : NP \Rightarrow S$. A transitive verb v_2 is assigned $v_2' : NP \Rightarrow (NP \Rightarrow S)$. The pronoun 'he' is assigned an abduction principle. Our problem becomes:

Data

1. Mary$'$: NP
2. hit$'$: $NP \Rightarrow (NP \Rightarrow S)$
3. John$'$: NP
4. he: Abduce on structure. Take the first literal up the list.
5. cried$'$: $NP \Rightarrow S$.

Query

Prove $?S$ or $S \otimes S$ or $?S \otimes S \otimes S \ldots$ etc, in order to show we have a text of sentences.

We are thus saying that Anaphora resolution makes use of structural abduction.

The reader should note that anaphora resolution is a complex area and we are not making any linguistic claims here beyond the intuitive example that abduction principles can be treated as data. We do admit however that logical principles underlying database management do seem to be operative in natural language understanding and we are working full steam ahead in making our case.

Coming back to our view of abduction as data, we are really saying:

- A database can either display data items or give us pointers to where to get them.

Thus a labelled database can appear as below:

n_1: data m
$\vdots \quad \vdots$
n_k: get datum from ...
$\vdots \quad \vdots$

Abductive Labelled Database

4.2 Abduction in Planning and Natural Language

I would like to give next a combined example of planning and parsing, based on ideas of [Gabbay and Kempson, 1991].

EXAMPLE 18 (Planning). Consider the situation described by the diagram below. There are three languages involved

1. The database language containing the predicates $\text{On}(x, y)$ and $\text{Free}(x)$
2. The imperative (Input) command language with the predicates $\text{Move}(x, y)$.
3. The mixed metalanguage with the connectives '$\wedge$' for 'and' and '$\Rightarrow$' for 'precondition and action imply postcondition'.

a			t_1: $\text{On}(a, b)$	
b	c		t_1: $\text{On}(b, tab)$	$\leftarrow \text{Move}(a, c)$
			t_1: $\text{On}(c, tab)$	
	a		t_2: $\text{On}(a, c)$	$\leftarrow \text{Move}(a, tab)$
b	c			
b	c	a	t_3: $\text{On}\ (a, tab)$.	

$$\text{On}(x, y) \wedge \text{Free}(x) \wedge \text{Free}(z) \wedge \text{Move}(x, z) \Rightarrow \text{On}(x, z) \wedge \text{Free}(y) \wedge \text{Free}(x)$$

The diagram describes the initial layout of the blocks ready to respond to command. t_1 labels all data true at the initial situation and t_2 and t_3 the additional data after each of the actions. We have

$$t_1 < t_2 < t_3.$$

If we query the system with

$$?\ \text{On}(a, x)$$

we get three answers, with different labels, indicating where the answer was obtained in the database, namely:

$$\vdash t_1 : \text{On}(a, b)$$
$$\vdash t_2 : \text{On}(a, c)$$
$$\vdash t_3 : \text{On}(a, tab)$$

The reply to the user is determined by the system as the answer proved with the stronger label, namely:

$$\text{On}(a, tab)$$

Call the deductive system governing the planning consideration LDS_1.[12] This system involves proving where the blocks are after which action. This system accepts commands in logical form Move(x, y). It does not accept commands in English. If the command comes in English, which we can represent as *move* x *onto* y, it needs to be parsed into the LDS_1 language. This is done in a parsing logic LDS_0. The following diagram explains the scheme, see Figure 1:

English Input:

move a onto c. move it onto table

$\downarrow$

Parsing Logic LDS_0

$\downarrow$

move (a, c); move (a, tab)

Planning Logic LDS_1

Figure 1.

The following diagram describes the database-query problem of LDS_0:

[12]We remark in passing that this approach offers a *conceptual* (not computational) solution to the frame problem. Conceptually, given an initial labelled database and a sequence of actions to be performed, we model the sequence by another labelled database; the database obtained by adding the results of the actions to the initial database. We label the additions appropriately. This idea will be pursued elsewhere. There are several such 'non-monotonic' solutions in the literature. This is probably the most general

move′:	$NP \Rightarrow (NP \Rightarrow S)$	$?S \otimes S$
a':	NP	
c':	NP	
move′:	$NP \Rightarrow (NP \Rightarrow S)$	
it:	use abduction. First use structural abduction to get the first NP higher in the list, then use inferential abduction to try and get maximal inferential effects in LDS_1.	
tab′:	NP	

Notice that the abduction principle in LDS_0 also uses inferential effect in LDS_1. Intuitively we are trying to abduce on who 'it' refers to. If we choose 'it' to be a block which is already on the table, it makes no sense to move it onto the table. Thus the command when applied to the database will produce no change. The abduction principle gives preference to abduced formulas which give some effect.

From the logical point of view we are using the following principle, (see also Example 17):

- Abduction principles can serve as items of data in one database Δ_0, being a pointer to another database Δ_1 where some computation is carried out jointly involving both databases and the result is the abduced data item in Δ_0.

4.3 Abduction in Logic Programming

EXAMPLE 19 (Logic programming abduction). The abductive system in logic programming can be schematically put into our form by making use of the way the Prolog pointer scans the database. An abductive system has the form (Δ, I, A), where $\Delta = (C_1, \ldots, C_n)$ is a sequence of clauses and literals, where I are the integrity constraints and where A is the set of abducible atoms. This system can be translated into the following Horn clause database:

(0) C_0 = Abduce on the goal by checking whether the goal is in A and whether when added it satisfies the integrity constraints.

(1) C_1

$\vdots \quad \vdots$

(n) C_n

We are now ready for our next conceptual step. If an abductive principle is to be considered as a declarative item of data, say $\mathcal{Q}_{Abduce}$, then what would be the meaning of

$$\Delta \; ?! \; \mathcal{Q}_{Abduce}$$

For the imperative interaction $!\mathcal{Q}_{Abduce}$, the meaning is obvious, we simply apply the abductive principle to the database and get a new database. However the query meaning of the abductive principle is a bit more difficult to explain in general, and we may need to provide an explanation for each specific case. For example, if the abduction principle abduces a formula B, then $\Delta?\mathcal{Q}_{Abduce}$ would mean $\Delta?B$. This seems all right at first sight, however, the problem here can be that the abduction process by its nature tries to find B's which are not available in the database and so the answer to the query $\Delta?\mathcal{Q}_{Abduce}$ will always be no. This is clearly unacceptable. We must seek another meaning for the query.

Let us for the time being, accept only the imperative reading of $\Delta!\mathcal{Q}_{Abduce}$. We can immediately allow ourselves to write databases with clauses containing $\mathcal{Q}_{Abduce}$ in them. Let us see through a few examples what this means.

EXAMPLE 20. Let $\Delta = \{\mathcal{Q}_{Abduce} \wedge B \Rightarrow D\}$. Think of the above as a database, and assume the computation procedure to be Prolog-like. We consider the following query

$$\Delta?D$$

which reduces to

$$\Delta?(\mathcal{Q}_{Abduce}, B)$$

which reduces to

$$\Delta, B?B$$

which succeeds.

Here we assumed that $\mathcal{Q}_{Abduce}$ yields B. Our database is similar in this case to the Prolog database:

$$\mathit{Assert}\,(B) \wedge B \Rightarrow D$$

Indeed, asserting is a form of unconditional abduction.

EXAMPLE 21 (Abduction and negation by failure). From our point of view, negation by failure is abduction. This point has also been made in [Eshghi and Kowalski, 1989]. However, we want to make our position crystal clear to avoid confusion. We believe that abduction is a principle of reasoning of equal standing to deduction and that every logical system is comprised of both proof rules and various mechanisms including abductive rules. This view has developed through our interaction with the logics of common sense reasoning and our work in natural language understanding [Gabbay and Kempson, 1991]. Negation by failure is not central to the general abduction scheme, though it is an interesting example from our point of view.

We begin with a precisely specified proof system. The query $\Delta?Q$ can be algorithmically checked. If the algorithm succeeds, the answer is yes. The algorithm may loop or it may fail. We may be able to prove that for the particular $\Delta?Q$, the algorithm must fail (e.g. in a case where none of the rules can even be applied). In this case we can say $\Delta?Q$ *finitely fails* (relative to the algorithm). Thus the notion of *finite failure* can be defined for any proof theoretic system.

Given a system, we can consider the following abduction principles which we call $Fail(Q, B)$:

If $\Delta?Q$ *finitely fails then abduce (or assert)* B.

To make our example specific, let us choose a language and computation procedures. By an atomic literal let us understand either an atom $\{p, q, r, \ldots\}$ or an abduction principle $Fail(a, b)$, where a, b are atoms. By clauses let us understand Horn clauses of literals. Goals are conjunctions of literals. Thus we can write the following clauses:

1. $q \wedge Fail(a, b) \wedge c \Rightarrow p$
2. $Fail(q, r)$
3. $a \Rightarrow Fail(b, b)$.

To explain the computational meaning, we will translate into Prolog. Ordinary Prolog is not expressive enough for our purpose, so we use N-Prolog [Olivetti and Terracini, 1991; Gabbay and Reyle, 1984; Gabbay, 1985] with negation by failure, mainly because it allows hypothetical reasoning, i.e. embedded implications.

We translate:

$$Fail(a, b) \mapsto (\neg a \Rightarrow b).$$

After translation, the database becomes:

1. $q \wedge (\neg a \Rightarrow b) \wedge c \Rightarrow p$
2. $\neg q \Rightarrow r$
3. $a \wedge \neg b \Rightarrow b$

which is meaningful computationally in goal directed N-Prolog.

A Horn clause with negation by failure of the form:

$$a \wedge \neg b \Rightarrow c$$

can be translated back into our abductive language as

$$a \Rightarrow Fail(b, c)$$

A Prolog goal of the form $\neg a$ an be translated as $Fail(a, \varnothing)$, $\varnothing$ is truth.

N-Prolog is not as expressive as our abductive language. In our abductive language we also have the imperative meaning of

$$\Delta?\ Fail(a, b)$$

which means *apply* the abduction to Δ.

This would correspond to

$$Assert\ (\neg a \Rightarrow b)$$

in N-Prolog. N-Prolog does not allow for that. The syntax is defined in such a way that we do not get goals of the form $\Delta?(\neg a \Rightarrow b)$. The N-Prolog computation rule would require in this case to add $\neg a$ to Δ, which is not meaningful.

The connection between abduction and negation by failure was observed in [Eshghi and Kowalski, 1989]. Since their abductive systems have the restricted form (Δ, I, A) as described in Example 19, they need to rewrite the Horn clause program into a more convenient form, translating the Prolog $\neg a$ as a^* and adding the integrity constraint:

$$a \wedge a^* \Rightarrow .$$

Let us now do in detail the abduction for logic programming.

DEFINITION 22 (Logic programs).

1. Our language contains the connectives $\wedge, \neg, \Rightarrow$ the constant $\perp$ and atomic letters $p, q, \ldots$. A clause has the form

$$\bigwedge_i a_i \wedge \bigwedge_j \neg b_j \Rightarrow q$$

   where a_i, b_j and q are atoms. q is the Head of the clause and $\bigwedge_i a_i \wedge \bigwedge_j \neg b_j$ is the Body of the clause. The body may not appear in a clause. An integrity constraint has the same form except that $q = \perp$. An integrity constraint must have a body. A program is a set of clauses and integrity constraints.

2. Success or failure of an atom q from a program Δ is defined via the metapredicate $S(\Delta, q, x)$, for $x = 1$ (success) or $x = 0$ (failure) as follows:

   - $S(\Delta, q, 1)$ if $q \in \Delta$
   - $S(\Delta, q, 0)$ if q is not the head of any clause in Δ.
   - $S(\Delta, q, 1)$ if for some clause $\bigwedge_i a_i \wedge \bigwedge_j \neg b_j \Rightarrow q$ (or integrity constraint if $q = \perp$) we have $\bigwedge_i S(\Delta, a_i, 1) \wedge \bigwedge_j S(\Delta, b_j, 0)$ holds.
   - $S(\Delta, q, 0)$ if for every clause (every integrity constraint if $q = \perp$) $\bigwedge_i a_i \wedge \bigwedge_j \neg b_j \Rightarrow q$ there exists an i such that $S(\Delta, a_i, 0)$ or a j such that $S(\Delta, b_j, 1)$ holds.
   - We write $\Delta \vdash a$ if $S(\Delta, a, 1)$ holds and $\Delta \vdash \neg a$ if $S(\Delta, a, 0)$ holds.
   - We also write $\Delta?a = x$ for $S(\Delta, a, x)$.

3. A program is consistent if $\perp$ is not successful.

Let us consider an example. Consider the propositional logic program Δ below

$\neg a \Rightarrow b$
$\neg b \Rightarrow d$
$c \Rightarrow b$
c

clearly $\Delta \vdash c \wedge b$ and $\Delta \not\vdash a$.

If we add a to the database and delete c we get $\Delta + a - c \not\vdash b$, and therefore $\Delta \vdash d$. Δ is clearly non-monotonic. By adding a and deleting c we took out b and further added d.

Suppose we want to update Δ to Δ' by ensuring that the query $a \wedge \neg b$ succeeds from Δ'. Thus $\Delta' = \Delta + (a \wedge \neg b)$, where $+$ symbolises our revision process. For this we need an abduction algorithm that will follow the computation of the goal $\Delta?a \wedge \neg b$ and suggest what changes to make to Δ to ensure the success of this query. Δ' is the result of making these changes to Δ. There may be more than on way of doing the change so the suggestions will be disjunctive.

Let us see how it works for our particular example.

Try $\Delta?a = 1$. This fails, since a is not the head of any clause in Δ. So the suggestion here is to add a to Δ.

We write this as

$$Ab(\Delta, a, 1) = \{\{+a\}\}$$

$+a$ means add a to Δ.

We also want b to fail from Δ, so let us ask $\neg b$:
$\Delta?\neg b = 1$ iff $\Delta?b = 0$ iff $\Delta?\neg a = 0$ and $\Delta?c = 0$ iff both $\Delta?a = 1$ and $\Delta?c = 0$.

We know what to suggest here, namely, add a and take out c. Thus $Ab(\Delta, b, 0) = \{\{+a, -c\}\}$. Thus $Ab(\Delta, a \wedge \neg b, 1) = Ab(\Delta, a, 1) \curlywedge A(\Delta, b, 0) = \{\{+a, -c\}\}$ where $\curlywedge$ means that we choose the cases that are compatible. We need a formal definition.

DEFINITION 23 (of $\curlywedge$).

1. Let $S_1, \ldots, S_k$ be sets of signed literals of the form $\pm q_i$. We say that they are compatible if for no atom q do we have $+q$ in one of the S_i and $-q$ in another.

2. Let $\mathsf{A}^i = \{\mathsf{S}^i_1, \ldots, \mathsf{S}^i_{m_i}\}$. Define $\bigcurlywedge \mathsf{A}_i$ as the set containing all elements of the form $\mathsf{S} = \bigcup_i \mathsf{S}^i_{k_i}$, where $\mathsf{S}^i_{k_i} \in A_i$ are all compatible. If none exists then $\bigcurlywedge A_i = \varnothing$.

DEFINITION 24 (Abduction algorithm).

1. A propositional logic program Δ is a list, $C_1, \ldots, C_k$ of clauses of the form:

$$C_i = \bigwedge_{j=1}^{m_i} a^i_j \wedge \bigwedge_{j=1}^{n_i} \neg b^i_j \Rightarrow q_i \qquad i = 1, \ldots, k,$$

where a^i_j, b^i_j, q_i are atoms, with q_i possibly $\bot$.

2. Let q be a literal. Define $Ab(\Delta, q, 1)$ and $Ab(\Delta, q, 0)$ by simultaneous induction as follows

   - $Ab(\Delta, q, 1) = \{\{q\}\}$ if q is not the head of any clause.
   - $Ab(\Delta, q, 0) = \varnothing$ if q is not the head of any clause.

   Suppose q is the head of the first r clauses

   - $Ab(\Delta, q, 1) = \bigcup_{i=1}^{r}(\bigwedge\!\!\!\bigwedge_{j=1}^{m_i} Ab(\Delta, a^i_j, 1) \wedge\!\!\!\wedge \bigwedge\!\!\!\bigwedge_{j=1}^{n_i} Ab(\Delta, b^i_j, 0))$
   - $Ab(\Delta, q, 0) = \bigcup_\alpha A_\alpha$, where α ranges over all choice vectors such that for each i, $\alpha(i)$ is either some $a^i_{j(i)}, j \leq m_i$ or some $b^i_{j(i)}, j \leq n_i\}$ and
   $$\begin{array}{rcl} \mathsf{A}_\alpha & = & \bigwedge\!\!\!\bigwedge_{\{i|\alpha(i)=a^i_{j(i)}\}} Ab(\Delta, \alpha(i), 0) \wedge\!\!\!\wedge \\ & & \bigwedge\!\!\!\bigwedge_{\{i|\alpha(i)=b^i_{j(i)}\}} Ab(\Delta, \alpha(i), 1) \end{array}$$
   - $Ab(\Delta, \neg q, x) = Ab(\Delta, q, 1 - x)$, for $0 \leq x \leq 1$.
   - $Ab(\Delta, \bigwedge_i A_i, 1) = \bigwedge\!\!\!\bigwedge_i Ab(\Delta, A_i, 1)$.
   - $Ab(\Delta, \bigwedge_i A_i, 0) = \bigcup_i Ab(\Delta, A_i, 0)$.

Given a Δ and a goal G, $Ab(\Delta, G, 1)$ (resp. $Ab(\Delta, G, 0)$) gives all the possible options for adding/deleting literals that would make the goal succeed resp. fail). It is up to us to make a choice of what to add/delete. This may involve other considerations, such as ranking or whatever measure we choose to put on the options. Given an inconsistent theory Δ, (i.e. such that $\Delta \vdash \bot$) we can activate $Ab(\Delta, \bot, 0)$ and find out what to add/delete to restore consistency. $Ab(\Delta, A, 1)$ may be empty.

For example, let $A = a \wedge \neg b$ and $\Delta = a \Rightarrow b$. There is no way that $a \wedge \neg b$ can succeed. This happens because we are trying to leave the clauses of Δ untouched and revise Δ just by adding/subtracting literals. If $Ab(\Delta, A, 1)$ is empty we must be prepared to delete clauses from Δ. How can we do that?

Let us consider the example again

$$\Delta = a \Rightarrow b$$

move to Δ_{name}

$$\Delta_{name} = \{\neg name \wedge a \Rightarrow b\}$$

The clause $a \Rightarrow b$ is active as long as $name$ is not in the database. The minute $name$ is added to the data, the clause is practically out.

To allow deleting clauses from Δ we move to Δ_{name} and apply Ab.

Δ_{name} is obtained from Δ by replacing each clause i, of the form $B_i \Rightarrow q_i$ by the clause $name_i \wedge B_i \Rightarrow q_i$ for $i = 1, \ldots, r$.

Let us see what happens with our program.

Δ_n is : $\{\neg n \wedge a \Rightarrow b\}$.

The update is: $a \wedge \neg b$.

$$Ab(\Delta_n, a, 1) = \{\{+a\}\}$$
$$Ab(\Delta_n, \neg b, 1) = \{\{-a\}, \{+n\}\}$$
$$Ab(\Delta_n, a \wedge \neg b, 1) = Ab(\Delta_n, a, 1) \bigwedge\!\!\!\bigwedge Ab(\Delta_n, \neg b, 1) = \{\{+a, +n\}\}.$$

The new program $\Delta_n + (a \wedge \neg b)$ is $\{\neg n \wedge a \Rightarrow b, a, n\}$.

4.4 The plausibility logic component of abduction

This subsection will give an example to show how we choose to abduce depends on a second plausibility logic.

This subsection gives examples illustrating the need for the second logic for abduction, the plausibility logic which helps us decide which abduced option to actually choose.

We quote this example from a paper by S. Marc Cohen and David Keyt [1992]

The example is from Euclid Elements. In Book I, Euclid gives five postulates for geometry and his goal is to prove from these postulates how to construct an equilateral triangle on a given finite straight line. The proof has a gap in it; we need to abduce extra assumptions. The point we make is that the considerations and reasoning involved in what assumptions to add (abduce) to make the proof correct, have to do with the general body of knowledge of ancient Greek mathematics and culture rather than geometric proof theory.

Here is what happens. We are quoting directly from the paper [Cohen and Keyt, 1992, pp. 190–192], including the footnotes:

The proof, divided into discrete steps, runs as follows:[13]

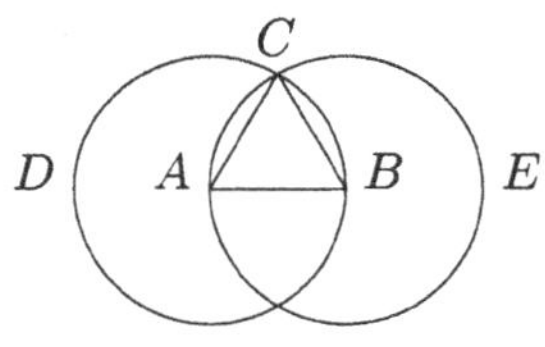

1. Let AB be the given finite straight line.
2. With centre A and distance AB, let the circle BCD be described. (Postulate 3)
3. With centre B and distance BA, let the circle ACE be described. (Postulate 3)
4. The circles cut one another at point C.
5. From the point C to the points A and B let the straight lines CA and CB be joined. (Postulate 1)

[13] Thomas L. Heath, *The Thirteen Books of Euclid's* Elements, 2nd edn, Cambridge 1926. i. 241–2.

6. Since the point A is the centre of the circle CDB, AC is equal to AB, (Definition 15[14])
7. Since the point B is the centre of the circle CAE, BC is equal to BA, (Definition 15)
8. Therefore, AC is equal to BC. (From 6 and 7 by axiom 1[15])
9. Hence the three straight lines AC, AB, and BC are equal to each other. (From 6, 7 and 8)
10. Therefore, the triangle ABC is equilateral. (Definition 20).[16]

An obvious flaw in this proof and one that has often been pointed out[17] is that no justification is given for step (4). What guarantees that the two circles intersect at a point and do not slip through each other without touching? Step (4) tacitly assumes that Euclidean circles are continuous. Can this be proved? Is there anything in Euclid's postulates that ensures that Euclidean lines have no gaps?

Here are Euclid's five postulates:

1. Let it be postulated to draw a straight line from any point to any point,
2. and to produce a limited straight line in a straight line,
3. and to describe a circle with any center and distance,
4. and that all right angles are equal to one another,
5. and that, if one straight line falling on two straight lines makes the interior angle s in the same direction less than two right angles, the two straight lines, if produced *ad infinitum*, meet one another in that direction in which the angles less than two right angles are.[18]

One way to settle the question of continuity is to see if there is an interpretation of the concepts that enter into these five postulates and into the above proof under which the postulates are true but under which step (4) is false. Such an interpretation can be found by exploiting the techniques of analytic geometry and set theory. The basic idea is to give the relevant terms an arithmetic interpretation in the domain of rational numbers. The domain of rational numbers is chosen since a line whose points correspond to rational numbers, though everywhere dense (between any two points there is a third), is not continuous. (There is a gap, for example, between all the points of such a line greater than the square root of 2 and all the points less than the square root of 2.) Following this strategy each of the relevant terms is assigned an arithmetic meaning that corresponds by way of a Cartesian (or rectangular) co-ordinate system to the intended geometric meaning of the term.

Under this arithmetic interpretation the word 'point' means 'ordered pair of rational numbers'; 'straight line' means 'set of points that satisfy an equation of the form $ax + by + c = o$', and 'circle' means 'set of points that satisfy an equation of the form $x^2 + y^2 + ax + by + c = o$'. (In these equations and in those that follow 'x' and 'y' are variables and all other letters are constants.) These two equations are chosen since the graph of the first, using a Cartesian coordinate system, is a geometric straight line and that of the second is a geometric circle. Finally, 'line AB intersects line CD' means 'the intersection of the set of points identified with line AB and the set of points identified with line CD is not the null set'.

Consider now Euclid's five postulates. Postulate Four is provable,[19] so it can be set aside. The other four are all true under the foregoing arithmetic interpretation when it is elaborated in an obvious way.

---

[14] A circle is a plane figure contained by one line such that all straight lines falling upon it from one particular point among those lying within the figure are equal.

[15] Things which are equal to the same thing are also equal to one another.

[16] Of the three trilateral figures, an *equilateral triangle* is the one which has its three sides equal ....

[17] Heath, *Euclid's* Elements, i. 25, 242; Felix Klein, *Elementary Mathematics from an Advanced Standpoint: Geometry*, trans E. R. Hedrick and C. A. Noble (New York, 1939), 197; Howard Eves, *A Survey of Geometry*, revised edn. (Boston, 1972), 321; Ian Mueller, *Philosophy of Mathematics and Deductive Structure in Euclid's* Elements (Cambridge, Mass., 1981), 28.

[18] Mueller's translation, *Philosophy of Mathematics*, 318–19.

[19] The proof, which goes back at least to Proclus, is given in Mueller, ibid. 22.

If the points mentioned in Postulate One are the two ordered pairs $\langle h_1, k_1\rangle$ and $\langle h_2, k_2\rangle$, the following set is the straight line through these points (where Ra is the set of rational numbers):

$$\{\langle x,y\rangle \mid x,y \in \mathrm{Ra}\&(k_1-k_2)x+(h_2-h_1)y+h_1k_2-h_2k_1=\mathrm{o}\}.$$

Under the arithmetic interpretation Postulate One makes the true assertion that this set has members. Postulate Two makes the true assertion that if $h_1 \neq h_2$, then for any given rational number, n, the set contains an ordered pair whose first element is larger than n and a second ordered pair whose first element is smaller than n, and it makes a similar true assertion for the second element if $k_1 \neq k_2$. (If $h_1 = h_2$, the line is parallel to the y-axis; and if $k_1 = k_2$, it is parallel to the x-axis.)

In order to interpret Postulate Three it is necessary to specify that by 'any centre' is meant 'any point as centre' and by 'distance' 'distance between two points'. If this is laid down, then under the arithmetic interpretation Postulate three makes the true assertion that the following set, which is determined by the equation for a circle with the centre $\langle h,k\rangle$ and the radius r is not null:

$$\{\langle x,y\rangle \mid x,y \in \mathrm{Ra}\&(x-h)^2+(y-k)^2=r^2\}.$$

(In the equation determining this set r^2 must always be a rational number though r can be irrational.)

The fifth, or parallel, postulate is the most complex. Consider the two following sets:

$$\begin{aligned}\mathrm{A} &= \{\langle x,y\rangle \mid x,y \in \mathrm{Ra}\&a_1x+b_1y+c_1=\mathrm{o}\}\\ \mathrm{B} &= \{\langle x,y\rangle \mid x,y \in \mathrm{Ra}\&a_2x+b_2y+c_2=\mathrm{o}\}\end{aligned}$$

Postulate Five asserts that if A and B each have at least two members and if $a_1/a_2 \neq b_1/b_2$, then $A \cap B \neq \mathrm{o}$. This amounts to the claim that the equations that determine A and B have a simultaneous solution in the domain of rational numbers. It is established by showing that a linear equation that is satisfied by two distinct ordered pairs of rational numbers has rational coefficients[20] and that the simultaneous solution of two linear equations with rational coefficients must be a pair of rational numbers.[21]

All of Euclid's postulates are true under the proposed arithmetic interpretation, but under this interpretation the key step in Euclid's proof of his first theorem and the theorem itself are false. For suppose that $\langle -1,\mathrm{o}\rangle$ and $\langle 1,\mathrm{o}\rangle$ are the end-points of the finite straight line upon which the equilateral triangle is to be constructed. Then under the arithmetic interpretation the two circles are:

$$\begin{aligned}&\{\langle x,y\rangle \mid x,y \in \mathrm{Ra}\&(x-1)^2+y^2=2^2\}\\ &\{\langle x,y\rangle \mid x,y \in \mathrm{Ra}\&(x+1)^2+y^2=2^2\}\end{aligned}$$

But these sets do not intersect. (In the domain of real numbers the points of intersection are $\langle \mathrm{o}, \sqrt{3}\rangle$ and $\mathrm{o}, -\sqrt{3}\rangle$.) Thus Euclid's first proof cannot be repaired, and the reason is that his postulates do not guarantee that the lines of his geometry are continuous.

Obviously we have several options for abduction. The first one is to assume that all lines are continuous. This means we are working in a model of the plane of all points (x, y) where x, y are real numbers. It is plausible for us to assume that Euclid has these intuitions at the time. The second option is to assume, based

[20]The equation for a straight line through the two points $\langle h_1, k_1\rangle$ and $\langle h_2, k_2\rangle$ is:

$$(k_1-k_2)x+(h_2-h_2)y+h_1k_2-h_2k_1=\mathrm{o}.$$

Since h_1, h_2, k_1 and k_2 are all rational, the coefficients in the equation are rational.

[21]A system of two linear equations can be solved by successive uses of addition and multiplication. Hence, if all of the coefficients of each equation are rational, the solution must also be rational.

on our knowledge of the extensive use and attention the Greeks gave to ruler and compass constructions, that Euclid had in mind that all ruler and compass construction points are available. In modern terms this means that we are looking at all points (x, y) such that $x, y \in \mathbb{E}$, where $\mathbb{E}$ is the extension field of the rationals closed under the additional operation of getting square roots, i.e. if $x \in \mathbb{E}$ and $x > 0 \rightarrow \sqrt{x} \in \mathbb{E}$ (i.e. the set of all numbers we can get from rationals by the operations of addition $+$, multiplication $\times$, subtraction $-$, division $/$ and squareroot $\sqrt{\cdot}$.)

Which abduction hypothesis do we adopt? Do we add the axiom that the plane is continuously full of points or do we add the axiom that the plane includes all rational as well as rule and compass points?

No geometrical abductive process can decide for us. Only our second logic, our reasoning based on ancient Greek culture can decide.

This example shows our second logic at play!

By the way, Plato's work is full of arguments in need of abductive 'help' and the Cohen and Keyt paper discusses the methodology of 'helping' Plato. In our terminology, the paper discusses the options for a second logic for supporting abduction for Enthymemes[22] in Plato.

5 GENERAL PRINCIPLES OF ABDUCTION

The abduction process is a recursive metafunction *Abduce* which follows the computation steps of the metapredicate *Success* and yields modification to the database at every step where the *Success* metapredicate seems to fail at its task. The suggested modifications are designed to help *Success* achieve its aim.

We begin our study with a discussion of the ideas involved in the definition of *Abduce*, for the *LDS* goal directed system for $\Rightarrow$. We then define the abduction algorithm for this system, and proceed to give a simplified version of it for the case of intuitionistic $\Rightarrow$ (where no labels are used).

5.1 Discussion

Let our starting point be Definition 11. This definition gives the recursive rules for the metapredicate

$$\textit{Success}\,(\Delta, \delta : A, \textit{constraints}, \theta) = x.$$

for $x = 1$ we want the computation to succeed and for $x = 0$ we want it to fail. We now examine what kind of difficulties we encounter in defining the abduction algorithm using Definition 11 as a basis:

[22]Enthymemes are arguments that need additional premisses, (which cannot be deduced from the immediate context) to make them work.

Case of immediate success or failure
We need to examine what can happen in the case of clauses 1 and 3 of Definition 11.

Case $x = 1$
Clause 1 says that *Success* $(\Delta, \delta : q, \mathit{constraints}, \theta) = 1$ (i.e. $\Delta\theta \vdash_{constraints} \delta\theta : q$), if two conditions hold:

(a) The constraints mentioned in *constraints* are all provable in $\mathcal{A}$ for the substitution θ.

(b) $\Psi_1(\Delta\theta, \delta\theta : q)$ holds.

If either (a) or (b) cannot be shown to hold then *success* cannot do its job and our abduction process can be activated to help.

If (a) cannot be shown to hold, we may choose to add axioms to our algebra $\mathcal{A}$ so that the new algebra $\mathcal{A}'$ can prove the constraints. This amounts to a change of the logic (compare with the last paragraph of Section 3). Our abduction policy is not to change the logic (at least not in this manner). Thus, in this case our abduction will not help. So let us consider the case where the constraints can be proved but where (b) fails, that is $\Psi_1(\Delta\theta, \delta\theta : q)$ does not hold. We can look for a Δ' such that $\Psi_1(\Delta'\theta, \delta\theta : q)$ does hold.

Whether we can find such a reasonable Δ' depends on Ψ_1. For example, for the case of resource logics and the Ψ_1 suggested in Definition 11, namely the Ψ_1 saying that $\delta\theta : q \in \Delta\theta$, we can always find a Δ'; we simply add (input) $\delta : q$ into Δ, i.e. let $\Delta' = \Delta + (\delta : q)$. In other words, using the predicate Ψ_2, we find the Δ' such that $\Psi_2(\Delta, \Delta', \delta : q)$ holds.

For a general Ψ_1, we do not know whether a Δ' can be found. We are going to have to stipulate some propertes of Ψ_1:

- Ψ_1 *Abduction axiom*:
  For every finite Δ and any $\delta : q$ and any $x \in \{0, 1\}$ such that $\Psi_1(\Delta, \delta : q) = x$ ($\Psi_1 = 1$ means Ψ_1 holds, $\Psi_1 = 0$ means Ψ_1 does not hold), there is an effective procedure to produce a finite set (which may be empty) of databases $\{\Gamma_1, \Gamma_2, \ldots, \Gamma_k\}$ such that $\Psi_1(\Gamma_i, \delta : q) = 1 - x$. We denote this set by $\mathbf{Ab}_1(\Delta, \delta : q, 1 - x)$.

Let us see what this algorithm can be for some of the examples we know. Note that the nature of the abduction depends on the application area it is being used. In the application area all of these algorithms have a meaning, and the choice of algorithm for Ψ_1 and indeed for any other Ψ will be dictated by the needs of the application.

- For $\Psi_1(\Delta, \delta : q) = (\delta : q \in \Delta)$ the algorithm is to let Γ be $\Delta + (\delta : q)$, as already mentioned.

- For the condition $\Psi_1(\Delta, \delta : q) = (\{\delta : q\} = \Delta)$, let the algorithm be to delete all other data from Δ except $\delta : q$ if $\delta : q \in \Delta$ and if $\delta : q \not\in \Delta$ let the algorithm give us nothing. (We do not want to add $\delta : q$ to Δ for reasons having to do with the application area.)

Case $x = 0$
This case is the opposite of the case $x = 1$. We want the computation to fail. So if the computation succeeds, this means that the *constraints* can be proved as well as that Ψ_1 holds. We will not fiddle with the logic and make the constraints unprovable. We will concentrate on Ψ_1. We have $\Psi_1(\Delta\theta, \delta\theta : q)$ holds and we want to make it not hold. Again we need to stipulate an algorithm that yields for each Δ and $\delta : q$ such that $\Psi_1(\Delta, \delta : q)$ holds a set (possibly empty) of $\{\Gamma_1, \ldots, \Gamma_k\}$ options such that $\Psi_1(\Gamma_i, \delta : q)$ fails to hold. We use $\mathbf{Ab}_1(\Delta, \delta : q, 0)$ to produce the set.

For example, we can adopt the algorithm that takes $\delta : q$ out of Δ (i.e. let $\Gamma = \Delta - \{\delta : q\}$), i.e. we solve $\Psi_3(\Delta, \Gamma, \delta : q)$.

See Definition 24 for an example.

Case of $\Rightarrow$:
Let us now consider the case of $\Rightarrow$.

In order to deal with abduction for a goal of the form $\delta : B \Rightarrow C$ we are going to need some additional machinery. We know that *Success*$(\Delta, \delta : B \Rightarrow C,$ *constraints*$,\theta) = x$ iff *success*$(\Delta + (a : B), f(\delta, a) : C,$*constraints*$,\theta) = x$, where a is a new atomic label and $+$ is the insertion operation. The abduction process should replace Δ by a new Γ which will facilitate the success/failure of the computation. Suppose now that using abduction Γ_a replaces $\Delta + (a : B)$ in giving the desired success/failure value x to the goal $?f(\delta, a) : C$. We now ask ourselves, what is the appropriate replacement Γ of Δ? We want a theory Γ such that upon adding $a : B$ to it (i.e. forming $\Gamma + (a : B)$) we get Γ_a. So we want (using the metaoperator *The* x *such that* $F(x)$ *holds*):

The Γ *such that* $(\Gamma + (a : b) = \Gamma_a)$

let $\mathbf{Ab}_{\Rightarrow}(\Gamma_1, a : B)$ be a metafunction giving *the* Γ *such that* $(\Gamma + (a : B) = \Gamma_1)$, when it exists and $\varnothing$ otherwise.

Let us get a rough idea what this Γ can be.

If we ignore the labels for a moment, we really want the database $\Gamma = (B \Rightarrow \Gamma_1)$, because when we add B to Γ we can get Γ_1.

Thus (ignoring labels) we get

$$\mathbf{Ab}_{\Rightarrow}(\Gamma_1, B) = \{B \Rightarrow X \mid X \in \Gamma_1\}.$$

Since Γ_1 is supposed to be *Abduce*$(\Delta + B, C)$ we get the 'rough' equation:

- *Abduce*$(\Delta, \delta : B \Rightarrow C) = (a : B) \Rightarrow$ *Abduce* $(\Delta + (a, B), f(\delta, a) : C)$

where a is a new constant and the operation $(a : B) \Rightarrow \Gamma_a$ has some meaning for theories Γ_a and units $a : B$. We are not defining this operation but are using it intuitively to explain the kind of operation we need to do abduction in the case of $\Rightarrow$.

If we don't have labels, say in intuitionistic logic, let us see what we get.

Assume the abduction problem is $\mathit{Abduce}(\{A \Rightarrow (B \Rightarrow q)\}, B \Rightarrow q)$.

This reduces to

$$B \Rightarrow \mathit{Abduce}(\{B, A \Rightarrow (B \Rightarrow q)\}, q)$$

$\mathit{Abduce}(\{B, A \Rightarrow (B \Rightarrow q\}, q)$ reduces to $\mathit{Abduce}(\{B, A \Rightarrow (B \Rightarrow q)\}, A)$ and $\mathit{Abduce}\{B, A \Rightarrow (B \Rightarrow q)\}, B\}$.

The second computation, $\mathit{Success}(\{B, A \Rightarrow (B \Rightarrow q)\}, B) = 1$ is successful and so *Abduce* does not change the database but the first one recommends that we add A to the database. So it is intuitively clear that

$$\mathit{Abduce}(\{B, A \Rightarrow (B \Rightarrow q)\}, q) = \{\{B, A \Rightarrow (B \Rightarrow q), A\}\}$$

Therefore our new theory is:

$$\begin{aligned} &\mathit{Abduce}(\{A \Rightarrow (B \Rightarrow q)\}, B \Rightarrow q) = \\ &\quad \{\{B \Rightarrow B, B \Rightarrow (A \Rightarrow (B \Rightarrow q)), B \Rightarrow A\} = \\ &\quad \{\{B \Rightarrow A, B \Rightarrow q\}\} = \{\Delta'\}. \end{aligned}$$

Let us take another point of view of abduction. Since intuitionistic logic is monotonic, let us look at $\mathit{Abduce}(\Delta, A)$ as telling us *what to add* to Δ to obtain a Δ' which makes A succeed. In this case $\mathit{Abduce}(\{A \Rightarrow (B \Rightarrow q\}, B \Rightarrow q) = B \Rightarrow \mathit{Abduce}(\{A \Rightarrow B \Rightarrow q), B\}, q) = B \Rightarrow A$.

So $B \Rightarrow A$ needs to be added to Δ to get Δ'. Thus

$$\begin{aligned} \Delta' &= \{A \Rightarrow (B \Rightarrow q), B \Rightarrow A\} \\ &= \{B = q, B \Rightarrow A\}. \end{aligned}$$

We are getting the same result!

Let us see if we can be more specific about the operation $(a : B) \Rightarrow \Gamma$, of the labels.

Consider the database $\Gamma_1 = \{\alpha : A\}$. We want to turn it into $\Gamma = \{\beta : B \Rightarrow A\}$ such that $\Gamma + (a : B)$ is 'equivalent' to $\alpha : A$.

Let us try the modus ponens:

$$\frac{\beta : B \Rightarrow A; a : B}{f(\beta, a) : A}.$$

thus we have to solve *the* β *such that* $(f(\beta, a) = \alpha)$, and let

$$(a : B) \Rightarrow \{\alpha : A\} = \{\beta : B \Rightarrow A\}.$$

Thus $\mathbf{Ab}_{\Rightarrow}(\{\alpha : A\}, a : B) = \{\{\beta : B \Rightarrow A\}\}$.

Of course in a general *LDS* database the $\mathbf{Ab}_{\Rightarrow}$ function can be more complex to define.

Case of database decomposition:
We now consider the case where an atomic query $\delta : q$ unifies with a head of a clause in the database. This is rule 4 of Definition 11.

This case decomposes the database into several databases using the Ψ_4 predicate. This decomposition forces us to consider the question of what to do with the replacement databases proposed by *Abduce* after the decomposition. We will get output for the abduction metafunction at the points of difficulty of the *Success* predicate but these replace databases deep into some iterated decomposition. How do we propagate the replacements upwards?

Let us illustrate the problem by being more specific. The abductive algorithm for the Ψ_1 case, both for $x = 0$ and $x = 1$, for example, replaces Δ with a new Γ. This Γ is proposed in the middle of an overall computation. We need to examine the 'replacement' operation and how it interacts with decomposition. A simple example will illustrate our problem (we will ignore the constraints). Suppose we have a clause $E = (q_1 \Rightarrow (q_2 \Rightarrow a)) \in \Delta$. We ask $Success(\Delta, \delta : a, constraints, \theta) = 1$. Rule 4 of Definition 11 tells us to look at partitions databases Δ'_1 and Δ'_2 such that $\Psi_4(\Delta, \{E\}, \Delta'_1, \Delta'_2)$ holds and labels δ_1, δ_2 such that $success(\Delta'_i, \delta_i : q_i, constraints', \theta_i) = 1$ for $i = 1, 2$ hold. E appears in Ψ_4 as a parameter. It may be that to make these succeed we abduce Γ_1, Γ_2 for the two subcomputations.

Our question is how does the replacement of Δ'_i by Γ_i yield a replacement of Δ by Γ? What is Γ? How do we construct it?

Recall that the idea of abduction is to ask for the initial database Δ and a goal $\delta : A$ the predicate $Success(\Delta, \delta : A, constraints, \theta) = x$ and we expect our abduction metafunction to answer: $Abduce(\Delta, \delta : A, constraints, \theta, x) = \{\Gamma\}$. Thus Γ is the result of abduction and then we are supposed to be assured that $Success(\Gamma, \delta : A, constraints, \theta) = x$ succeeds!

Let us present our problem clearly. We have $\Delta'_1, \Delta'_2, \Gamma_1, \Gamma_2$ and Δ and we are looking for a Γ. What we know is the following:

- $\Psi_4(\Delta, \{E\}, \Delta'_1, \Delta'_2)$
- Γ_1 is abduced to replace Δ'_1
- Γ_2 is abduced to replace Δ'_2
- Ψ_2 holds for Γ_i and $\delta_i : q_i$.

We need to know a corresponding Γ to replace Δ such that $\Psi_4(\Gamma, \{E\}, \Gamma_1, \Gamma_2)$ holds.

To find the Γ we need to know more about the Ψs.

We need some basic axioms relating the properties of the Ψs. We perhaps need to stipulate inverse functions that allow one to retrieve Δ from any $\Delta_1, \ldots, \Delta_n$, i.e. we need

- $\forall n \forall \Delta_1, \ldots, \Delta_n \exists ! \Delta \Psi_4(\Delta, \Delta_1, \ldots, \Delta_n)$.
  Let us introduce a new function $\mathbf{Ab}_+$ with $\mathbf{Ab}_+(\Delta_1, \ldots, \Delta_n) = \Delta$.

- $\bigwedge_{j=1}^{n} \Psi_4(\Delta_n, \Delta_n^1, \ldots, \Delta_n^{k(n)}) \wedge \Psi_4(\Delta, \Delta_1, \ldots, \Delta_n) \rightarrow$
  $\Psi_4(\Delta, \Delta_1^1, \ldots, \Delta_1^{k(1)}, \ldots, \Delta_n^1, \ldots, \Delta_n^{k(n)})$.

- $\forall \Delta \forall n \exists \Delta_1, \ldots, \Delta_n \Psi_4(\Delta, \Delta_1, \ldots, \Delta_n)$. (Note that the decomposition need not be disjoint and Δ_i can be empty.)

- $\Psi_2(\Delta, \Delta', a : B) \rightarrow \Psi_3(\Delta', \Delta, a : B)$

We also want some compatibility of Ψ_4 with Ψ_2 and Ψ_3. If we regard Ψ_4 as a decomposition of Δ into $\Delta_1, \ldots, \Delta_n$ and regard $\Psi_2(\Delta, \Delta', a : B)$ as insertion of $a : B$ into Δ to form Δ' then do we expect $\Psi_4(\Delta', \Delta, \{a : B\})$ to hold?

We can have an even stronger condition, namely that Δ can be generated by successive inputs of its own elements, namely:

- $\forall \Delta \exists n \exists (\alpha_1 : A_1, \ldots, \alpha_n : A_n) \exists \Delta_1, \ldots, \Delta_{n-1} \bigwedge_{i=1}^{n-1} \Psi_2(\Delta_{i+1}, \Delta_i, \alpha_{i+1} : A_{i+1}) \wedge \Psi_2(\varnothing, \Delta_1, \alpha_1 : A_1)$.

The problem of soundness

This is not all that is needed. We started from a decomposition Δ_1', Δ_2' of Δ. We abduced and got Γ_1, Γ_2 and formed Γ. Now we must assume the original clause $E = (q_1 \Rightarrow (q_2 \Rightarrow a))$ is in Γ. In such a case, it is obvious that we can use the clause to decompose Γ back into Γ_1, Γ_2, and ensure that the computation from Γ succeeds.[23]

What happens if we want to fail? In this case for every clause, say $q_1 \Rightarrow (q_2 \Rightarrow a)$ and every decomposition, say Δ_1', Δ_2' the *abduce* predicate will choose some replacement, say Γ_1 for Δ_1' for which the computation $\Gamma_1 ? q_1$ fails. Thus when we form Γ to replace Δ, we do not care if $q_1 \Rightarrow (q_2 \Rightarrow a) \in \Gamma$. If it is in Γ, however, we must worry whether it gives rise to *new* decompositions Γ_1', Γ_2' such that $\Gamma_i' ? q_i$ succeed. In fact we must worry about new clauses in Γ with heads q for which there are successful decompositions. This is again the problem of ensuring soundness

There are two ways for going around this difficulty. The first is to make some strong assumptions on the decomposition and $\mathbf{Ab}_+$ predicates. One such possible assumption is that Γ has the same clauses with head q as Δ and the changes are only in addition or deletion of atoms. Such an assumption may not be enough, however, without knowing the specific properties of Ψ_4 we may not be able to formulate additional assumptions.

The second method may be in principle more general. We compose $\Gamma = \mathbf{Ab}_+(\Gamma_1, \Gamma_2)$ to replace Δ but we are not assured (by any strong assumptions) that $\Gamma ? q$ fails. We compute *Success*(Γ, q). If it fails, then fine, we take this Γ as

[23]This means that the function $\mathbf{Ab}_+(\Delta_0, \Delta_1, \ldots, \Delta_n) = \Delta$ must satsify $\Delta_0 \subseteq \Delta$. Δ_0 is the parameter $\{E\}$ in our case.

the result of the abduction. If it succeeds, then we either say that the abduction process produces nothing or we apply abduction again to $\Gamma?q$. This will produce a Γ_1, and now we iterate the process. To ensure that this process does not loop forever, we could introduce some complexity assumptions either on the number of iterations or on Ψ_4 and $\mathbf{Ab}_+$ which will tell us that abduction involves adding or deleting wffs from a finite set of wffs constructed from the initial database and query. This assumption is enough to allow for a historical loop checker to work.

5.2 *Abduction algorithm for* $\Rightarrow$

Following our discussion, we are now ready to write the abduction algorithm

DEFINITION 25 (Abduce metafuction).

1. $\mathit{Abduce}(\Delta, \delta : q, \mathit{constraints}, \theta, 1) = \{\Delta\}$ if $\mathit{Success}(\Delta, \delta : q, \mathit{constraints}, \theta) = 1$.

2. $\mathit{Abduce}(\Delta, \delta : q, \mathit{constraints}, \theta, 1) = \varnothing$ if $\mathcal{A} \not\vdash \mathit{constraints}\theta$

3. $\mathit{Abduce}(\Delta, \delta : q, \mathit{constraints}, \theta, 1) = \{\Gamma_1, \ldots, \Gamma_k\}$ if $\mathcal{A} \vdash \mathit{constraints}\theta$ and $\Psi_1(\Delta\theta, \delta\theta : q)$ does not hold and $\mathbf{Ab}_1(\Delta\theta, \delta\theta : q, 1) = \{\Gamma_1, \ldots, \Gamma_k\}$. Note that each $\Gamma \in \mathbf{Ab}_1$ satisfies $\Psi_1(\Gamma\theta, \delta\theta : q)$. $\mathbf{Ab}_1$ may yield $\varnothing$ if the abduction is not possible.

4. $\mathit{Abduce}(\Delta, \delta : q, \mathit{constraints}, \theta, 0) = \{\Delta\}$ if $\mathit{Success}(\Delta, \delta : q, \mathit{constraints}, \theta) = 0$.

5. $\mathit{Abduce}(\Delta, \delta : q, \mathit{constraints}, \theta, 0) = \{\Gamma_1, \ldots, \Gamma_k\}$ if $\mathcal{A} \vdash \mathit{constraints}\theta$ and $\Psi_1(\Delta\theta, \delta\theta : q)$ does hold and $\mathbf{Ab}_1(\Delta\theta, \delta\theta : q, 0) = \{\Gamma_1, \ldots, \Gamma_k\}$.

   Note that for each $\Gamma \in \mathbf{Ab}_1$ we have that $\psi_1(\Gamma\theta, \delta\theta : q)$ does not hold. $\mathbf{Ab}_1$ may give $\varnothing$ if the abduction is not possible.

6. $\mathit{Abduce}(\Delta, \delta : B \Rightarrow C, \mathit{constraints}, \theta, x) = \{\Gamma_1, \ldots, \Gamma_k\}$, where for each Γ_i there exists a Γ'_i such that $\Gamma'_i \in \mathit{Abduce}(\Delta + (a : B), f(\delta, a) : C, \theta, x)$ and $\Gamma_i = \mathbf{Ab}_{\Rightarrow}(\Gamma'_i, (a : B))$.

We now consider the case of decomposition. We need some notation first. Let Δ be a database and $\delta : q$ a goal. By a decomposition $\mathbb{D}(\Delta, \delta : q)$ we mean a choice of a clause E of the form $\alpha : A_1 \Rightarrow \ldots \Rightarrow (A_n \Rightarrow q)\ldots)$ in Δ called the clause of $\mathbb{D}$ and of databases $\Delta'_1, \ldots, \Delta'_n$ such that $\Psi_4(\Delta, \{E\}, \Delta'_1, \ldots, \Delta'_n)$ holds. n is referred to as the length of the decomposition and depends on E. To introduce the rules for *Abduce* recall that by rule 4 of Definition 11, $\mathit{Success}(\Delta, \delta : q, \mathit{constraints}, \theta) = 1$ (resp. $= 0$) if for some $\mathbb{D}$ (resp. all $\mathbb{D}$) and some $\delta_1, \ldots, \delta_n$ we have for all i (resp. some i) $\mathit{Success}(\Delta'_i, \delta_i : A_i, \mathit{constraints}', \theta') = 1$ (resp. $= 0$).

7. *Abduce*$(\Delta, \delta : q, constraints, \theta, 1)$ wants to make success out of failure so it will give all options for success. So for each $\mathbb{D}$, it will give a set of $\Gamma^{\mathbb{D}}_{\mathbf{e}_1}, \Gamma^{\mathbb{D}}_{\mathbf{e}_2}, \ldots$ that will make this choice of $\mathbb{D}$ succeed. For the choice of $\mathbb{D}$ to succeed we need all *Success*$(\Delta'_i, \delta_i : A_i, constraints', \theta') = 1$.

   Let *Abduce*$(\Delta'_i, \delta_i : A_i, constraints', \theta', 1) = \{\Gamma^{\mathbb{D}}_{i,1}, \Gamma^{\mathbb{D}}_{i,2} \ldots \Gamma^{\mathbb{D}}_{i,k(i)}\}$, i.e. the set of options for replacing Δ'_i and succeeding.

   The replacement for Δ will be all possible choices $\Gamma^{\mathbb{D}}_{\mathbf{e}} = \{\mathbf{Ab}_+(\{E\}, \Gamma^{\mathbb{D}}_{1,\mathbf{e}(1)}, \ldots, \Gamma^{\mathbb{D}}_{n,\mathbf{e}(n)}) \mid \mathbf{e}$ a function giving each $1 \leq i \leq n$ a value $1 \leq \mathbf{e}(i) \leq k(i)\}$. We therefore define: *Abduce*$(\Delta, \delta : q, constraints, \theta, 1) = \{\Gamma^{\mathbb{D}}_{\mathbf{e}} \mid \mathbb{D}, \mathbf{e}$ as above and such that the clause of $\mathbb{D}$ is in $\Gamma_{\mathbf{e}}\mathbb{D}\}$.

8. Consider now *Success*$(\Delta, \delta : q, constraints, \theta) = 0$. This fails if for some $\mathbb{D}$ we have for all $i = 1, \ldots, n$, *Success*$(\Delta'_i, \delta_i : A_i, constraints'\theta') = 1$.

   So for every $\mathbb{D}$ we choose an index $1 \leq \mathbf{c}(\mathbb{D}) \leq n$ and replace $\Delta'_{\mathbf{c}(\mathbb{D})}$ by $\Gamma'_{\mathbf{c}(\mathbb{D})}$ from the set *Abduce*$(\Delta'_{\mathbf{c}(\mathbb{D})}, \delta_{\mathbf{c}(\mathbb{D})} : A_{\mathbf{c}(\mathbb{D})}, constraints', \theta', 0)$.

   Let $\Gamma^{\mathbb{D}}_{\mathbf{c}} = \mathbf{Ab}_+(\{E\}, \Delta'_1, \ldots, \Delta'_{\mathbf{c}(\mathbb{D})-1}, \Gamma'_{\mathbf{c}(\mathbb{D})}, \Delta'_{\mathbf{c}(\mathbb{D})+1}, \ldots, \Delta'_n)$.

   Let *Abduce*$(\Delta, \delta : q, constraints, \theta, 0) = \{\Gamma^{\mathbb{D}}_{\mathbf{c}} \mid \mathbb{D}$ and $\mathbf{c}$ as above and *Success*$(\Gamma^{\mathbb{D}}_{\mathbf{c}}, \delta : q, constraints, \theta) = 1\}$.

9. In case there are no strong assumptions on $\mathbf{Ab}_+$ and Ψ_4 ensuring that the $\Gamma^{\mathbb{D}}_{\mathbf{c}}$ produced in the preceding item 8 above are sound, then we iterate the *Abduce* computation for each candidate $\Gamma^{\mathbb{D}}_{\mathbf{c}}$ until we either find such candidates or give up and produce $\varnothing$.

THEOREM 26 (Soundness of abduction). *If* $\Gamma \in \text{Abduce}(\Delta, \delta : A, \text{constraints}, \theta, x)$, *then* $\text{Success}(\Gamma, \delta : A, \text{constraints}, \theta) = x$. *Of course,* $\text{Abduce}(\Delta, \delta : A, \text{constraints}, \theta, x)$ *may be empty.*

Proof. By induction on the definition of *Abduce*.

We follow the clauses of Definition 25 and itme 3 of Definition 28

1. clear

2. nothing to prove

3. By definition $\mathbf{Ab}_1(\Delta\theta, \delta\theta : q, 1)$ produces Γ_i which satisfy Ψ_2 and so *Success* $= 1$ holds.

4. Clear

5. By definition $\mathbf{Ab}_1(\Delta\theta, \delta\theta : q, 0)$ produces Γ_i for which there is failure of Ψ_1 and so *Success* $= 0$ holds.

6. Clear from the definitions.

7. The abduction replaces Δ by Γ. To check the success predicate for Γ we had to assume that $\mathit{Success}(\Gamma, \delta : q, \mathit{constraints}, \theta) = 1$.

8–9. Again measures were introduced in this clause to ensure that the choices of Γ produced by the *Abduce* function do the job. ∎

5.3 Example: Abduction for intuitionistic logic

Intuitionistic logic is monotonic and requires no labels and therefore no constraints. Doing abduction for it becomes simple. In Definition 11, $\Psi_1(\Delta, q)$ becomes $q \in \Delta$. The decomposition $\Psi_4(\Delta, \{E\}, \Delta_1, \ldots, \Delta_n)$ does not need the parameter Δ_0 and can be taken as $\bigwedge_{i=1}^{n} \Delta_i = \Delta$. $\Psi_2(\Delta, \Delta', A$ is $\Delta' = \Delta \cup \{A\}$ and $\Psi_3(\Delta, \Delta', A)$ is $\Delta' = \Delta - \{A\}$.

The computation rules for $\mathit{Success}(\Delta, A) = x$, $x \in \{0, 1\}$ become the following:

DEFINITION 27 (Success for intuitionistic logic).

1. *Immediate success case*
   $\mathit{Success}(\Delta, q) = 1$, for q atomic if $q \in \Delta$.

2. *Implication case*
   $\mathit{Success}(\Delta, B \Rightarrow C) = x$ if $\mathit{Success}(\Delta \cup \{B\}, C) = x$.

3. *Immediate failure case*
   $\mathit{Success}(\Delta, q) = 0$, q atomic, if q is not the head of any clause in Δ.

4. *Cut reduction case*
   $\mathit{Success}(\Delta, q) = 1$ (resp. $\mathit{Success}(\Delta, q) = 0$) for q atomic, if for some (resp. all) clauses of the form $A_1 \Rightarrow (A_2 \Rightarrow \ldots \Rightarrow (A_n \Rightarrow q) \ldots) \in \Delta$ we have that for all i (resp. for some i) we have $\mathit{Success}(\Delta, A_i) = 1$ (resp. $\mathit{Success}(\Delta, A_i) = 0$).

5. *Consequence*
   We have $\Delta \vdash A$ iff $\mathit{Success}(\Delta, A) = 1$.

We can therefore see $\mathit{Abduce}(\Delta, A, x)$ as yielding a set of wffs to add (for $x = 1$) or delete (for $x = 0$) from Δ to make A succeed or respectively fail.

We can therefore expect the following:

- $\Delta + \mathit{Abduce}(\Delta, A, 1) \vdash A$
- $\Delta - \mathit{Abduce}(\Delta, A, 0) \nvdash A$

We can simplify our notation and write $\Delta?Q = x$ for $\mathit{Success}(\Delta, Q, \mathit{constraints}, \theta) = x$ and $\mathit{Abduce}^{\pm}(\Delta, Q) = \{\Gamma_1, \ldots, \Gamma_k\}$, where Γ_i are all alternative sets of wffs to be added or taken out of Δ to yield the desired results.

We therefore have the following connection between the old and new notations:

$$\mathit{Abduce}(\Delta, Q, \mathit{constraints}, \theta, 1/0) = \{\Delta \pm \Gamma \mid \Gamma \in \mathit{Abduce}^{\pm}(\Delta, Q)\}$$

DEFINITION 28 (Abduce for intuitionistic logic).

1. $\mathit{Abduce}^{+}(\Delta, Q) = \{\varnothing\}$ if $\Delta?Q = 1$
2. $\mathit{Abduce}^{-}(\Delta, Q) = \{\varnothing\}$ if $\Delta?Q = 0$
3. $\mathit{Abduce}^{+}(\Delta, q) = \{q\}$ if q is atomic and is not the head of any clause in Δ
4. $\mathit{Abduce}^{-}(\Delta, q) = \{q\}$ if $q \in \Delta$, q atomic[24]
5. $\mathit{Abduce}^{\pm}(\Delta, A_1 \Rightarrow (A_2 \Rightarrow \ldots (A_n \Rightarrow q)\ldots)\{A_1 \Rightarrow (A_2 \Rightarrow \ldots (A_n \Rightarrow X)\ldots) \mid X \in \mathit{Abduce}^{\pm}(\Delta \cup \{A_1, \ldots, A_n\}, q)\}$

For clauses (6), (7) below, we need to assume that

$$B^j = (B_1^1 \Rightarrow \ldots \Rightarrow (B^j_{n(j)} \Rightarrow q)\ldots), j = 1, \ldots, m,$$

lists all clauses of heads q in Δ.

We also need the notion of a choice function $\mathbf{c}$. For each $1 \leq j \leq m$, $\mathbf{c}(j)$ is a theory $\Gamma = \mathbf{c}(j)$ such that for some $1 \leq i \leq n(j), \Gamma \in \mathit{Abduce}^{-}(\Delta, B^j)$.

6. $\mathit{Abduce}^{+}(\Delta, q) = \{\bigcup\{X_i \in \mathit{Abduce}^{+}(\Delta, B_i^j) \mid i = 1, \ldots, n(j)\} \mid j = 1, \ldots, m\}$.
7. $\mathit{Abduce}^{-}(\Delta, q) = \{\Gamma_{\mathbf{c}} \mid \mathbf{c}$ is a choice function as explained above and $\Gamma_{\mathbf{c}} = \bigcup_{j=1}^{m} \Gamma_{\mathbf{c}(j)}\}$.

The explanation for clause (7) is the following: For each clause B^j as above, we want to choose an $1 \leq i \leq n(j)$ such that $\Delta?B_i^j = 0$. To ensure that we look to $\mathit{Abduce}^{-}(\Delta, B_i^j)$.

Our choice functions are functions $\mathbf{c}$ choosing for each $j = 1, \ldots, m$ an $1 \leq i \leq j$ and a theory $\mathbf{c}(j) \in \mathit{Abduce}^{-}(\Delta, B_i^j)$.

Let $\Gamma_{\mathbf{c}} = \bigcup_{j=1}^{m} \Gamma_{\mathbf{c}(j)}$.

REMARK 29 (Warning on *Abduce*$^{-}$).

1. Note that *Abduce*$^{+}$ and *Abduce*$^{-}$ are defined independently of each other. The definition of *Abduce*$^{-}$ above is formally incorrect. To explain what is needed consider $\Delta, X \vdash q$ and assume that the abudced^{-} formula to delete is X, i.e. we delete X and $\Delta \nvdash q$. The situation is equivalent to $\Delta \vdash X \Rightarrow q$. So we want to abduce some Y so that $\Delta - \{Y\} \nvdash X \Rightarrow q$.

[24]*Abduce*$^{-}$ is not properly defined. See Remark 29 below.

This may not be possible (e.g. $\Delta = \varnothing$). The only way is to add a commitment to delete X, the instant it is added to Δ. Let us denote this commitment by $\textcircled{X}$ and think of it as a *blackhole* consuming X. Thus we have

$$\Delta, X, \textcircled{X} = \Delta.$$
$$\textcircled{\textcircled{X}} = X$$

and for every A, Δ, X we have $\Delta \vdash A$ iff $\Delta, \textcircled{X} \vdash X \Rightarrow A$.

We therefore have for our example above
$Abduce^{-}(\Delta, X \Rightarrow q) = X \Rightarrow Abduce^{-}(\Delta \cup \{X\}, q) = X \Rightarrow \textcircled{X}$.

Item (4) in Definition 28 should be:

4′. $Abduce^{-}(\Delta, q) = \textcircled{q}$, if $q \in \Delta$, and indeed $\textcircled{q}$, when added to Δ will delete q.

If all *Abduce*$^{-}$ sets are blackholes then the definition works.

This will be studied in our new book [Gabbay and Woods, forthcoming].

2. Note that in the previous definition if $B^{j} = q$ then clause 4 of the definition of *Abduce*$^{-}$ gets activated.

EXAMPLES 30.

1. Consider $\Delta = \{a\}?q$.
   The *Abduce*$^{+}$ wff is q.
   The new abduced theory for success is $\{a, q\}$.

2. Consider $\Delta = \{a \Rightarrow (b \Rightarrow q), a, c \Rightarrow (d \Rightarrow q)\}?q$.
   The *Abduce*$^{+}$ candidates are $\{\{b\}, \{c \Rightarrow d\}\}$. There are two new abduced theories for success $\Delta \cup \{b\}$ and $\Delta \cup \{c, d\}$.

3. Consider $\Delta = \{q, a \Rightarrow q\}?q$. The *Abduce*$^{-}$ candidate is $\{q\}$. The new abduced theory for failure is $\Delta - \{q\} = \{a \Rightarrow q\}$.

THEOREM 31. $\Delta \cup \Gamma \vdash Q$ *for any* $\Gamma \in$ Abduce$^{+}(\Delta, Q)$.

Proof. The proof is by induction on the definition of the *Abduce* predicate.

1. If the set *Abduce*(Δ, Q) contains $\varnothing$ then this means $\Delta \vdash Q$.

2. Assume that $Q = q$ is atomic and that it is not the head of any clause. Then the abduced formula is q and clearly $\Delta, q \vdash q$.

3. Assume Q has the form

$$Q = (A_1 \Rightarrow \ldots \Rightarrow (A_n \Rightarrow q) \ldots).$$

Then the abduced set is

$$A_1 \Rightarrow \ldots \Rightarrow (A_n \Rightarrow \mathit{Abduce}^+(\Delta \cup (A_1, \ldots, A_n), q), \ldots).$$

where $A_1 \Rightarrow \ldots (A_n \Rightarrow \{\Gamma_1, \ldots, \Gamma_k\}) \ldots)$ denotes the set of all theories of the form $\Gamma_i^{\Rightarrow}$

$$\Gamma_i^{\Rightarrow} = \{A_1 \Rightarrow (A_2 \ldots \Rightarrow (A_n \Rightarrow X) \ldots) \mid X \in \Gamma_i\}.$$

We need to show $\Delta \cup \Gamma_i^{\Rightarrow} \vdash Q$ for any $\Gamma_i^{\Rightarrow} \in A_1 \Rightarrow \ldots \Rightarrow (A_n \Rightarrow \mathit{Abduce}^+(\Delta \cup \{A_1 \ldots, A_n\}, q) \ldots)$.

By the induction hypothesis

$$\Delta \cup \{A_1, \ldots, A_n\} \cup \Gamma_i \vdash q \text{ for any } \Gamma_i \in \mathit{Abduce}^+(\Delta \cup (A_1, \ldots, A_n), q)$$

hence

$$\Delta \cup \{A_1, \ldots, A_n\} \cup \Gamma_i^{\Rightarrow} \vdash q$$

hence

$$\Delta \cup \Gamma_i \vdash Q$$

4. Assume $Q = q$ is atomic and let $B^j = (B_1^j \Rightarrow \ldots \Rightarrow (B_{n(j)}^j \Rightarrow q) \ldots), j = 1, \ldots, m$, be all the clauses in Δ with head q.

Then we need to show for

$$\Delta \cup \Gamma \vdash q \text{ for } \Gamma \in \mathit{Abduce}^+(\Delta, q)$$

where $\mathit{Abduce}^+(\Delta, q) = \{\Gamma_1^j \cup \ldots \cup \Gamma_{n(j)}^j \mid \Gamma_i^j \in \mathit{Abduce}^+(\Delta, B_i^j), j = 1, \ldots, m\}$.

By the induction hypothesis for j fixed, we have

$$\Delta \cup \Gamma_i^j \vdash B_i^j$$

for $i = 1, \ldots, n(j)$.

Hence for each j, since B^j is in Δ, we have:

$$\Delta \cup \bigcup_{i=1}^{n(j)} \Gamma_i^j \vdash q. \qquad \blacksquare$$

6 CASE STUDY: A CONVERSATION BETWEEN TWO INTELLIGENT DATABASES

We have the logical means to allow for two *LDS* databases to negotiate and reach an understanding. Imagine two databases S and H exchanging formulas continuously. At time n, the databases have evolved through the sequences

$$\Delta_1^S, \ldots, \Delta_n^S$$

and

$$\Delta_1^H, \ldots, \Delta_n^H.$$

At time n, database S sends a logical input I_n^S to database H and conversely, database H sends an input I_n^H to database S. The two databases use abduction to update themselves. Thus

$$\Delta_{n+1}^H = \mathit{Abduce}(\Delta_n^H, I_n^S)$$

and

$$\Delta_{n+1}^S = \mathit{Abduce}(\Delta_n^S, I_n^H).$$

To continue and communicate we need action principles for sending the next input. This is also part of our abduction scheme as hinted at in Example 18

Let us now consider an extended example:

Given a set of wffs Δ, taken as assumptions, and a formula or query Q, we can pose the logical question $\Delta?Q$. The meaning of this question is: Does Q follow from Δ in the logic we are dealing with, *i.e.* is $\Delta \vdash Q$ true? There are three possibilities:

1. $\Delta \vdash Q$ holds
2. $\Delta \vdash {\sim}Q$ holds
3. Neither $\Delta \vdash Q$ nor $\Delta \vdash {\sim}Q$ hold.

We do not adopt any additional conventions such as $\Delta \not\vdash Q$ implies $\Delta \vdash \sim Q$, where Q is atomic (i.e. the closed-world assumption). We thus want a definite proof of the negation $\sim Q$ in all cases. We denote by $\Delta?!Q = 1$ the situation where logically the database is consistent and is able to prove either Q or $\sim Q$. In other words $\Delta?!Q = 1$ means that the database Δ decides the truth of Q in a definite way.

The situation we are going to study arises when there is an exchange of information between an asker, A and a responder R. At the outset of the exchange, A has a set of assumptions Δ_A: so too does R, i.e. Δ_R. Note that Δ_A may include A's beliefs about the content of Δ_R, and vice versa.

Suppose now that A asks R the query Q. If Δ_A were sufficient to prove or disprove Q then we can assume that A could have figured out Q for himself (unless, of course, he wanted to find out if R knew that Q). However, we will begin by assuming that $\Delta_A?!Q \neq 1$.

R should now respond by supplying some new information B to be added to Δ_A to enable $\Delta_A \cup \{B\}?!Q = 1$. The answer could be:

- $B = Q$ (i.e. 'yes'), in which case $\Delta_A \cup \{Q\}?!Q = 1$;
- $B = {\sim}Q$, (i.e. 'no'), in which case $\Delta_A \cup \{{\sim}Q\}?!Q = 1$;

- some B such that $\Delta_A \cup \{B\}?!Q = 1$;
- some B such that $\Delta_A \cup \{B\}?!Q \neq 1$.

In the first three cases things are fine for A. The situation becomes more interesting when the new data B is still insufficient for A to (logically) determine an answer to Q.

R may have her own reasons for not answering directly and saying Q or $\sim Q$. Perhaps she wants to forestall any future questions. Perhaps she wants to give the reason for Q (or for $\sim Q$), namely B. For whatever reason, and assuming that R is trying to be co-operative rather than evasive, A expects the answer B to logically decide the query Q, i.e. for $\Delta_A \cup \{B\}?!Q = 1$ but nevertheless finds this not to be the case. At this point we invoke what we call *the Rule of Relevant Assumptions*; and require that further data B' should be abduced such that $\Delta_A \cup \{B, B'\}?!Q = 1$. We assume by relevance that R gives A the minimal input B which allows him to decide the query Q. To find B' we need to know and use the computation procedure we have for $\vdash$.

We call the relevance-theoretic principle which allows us to find the enrichment B' *The Rule of Relevant Assumptions (RRA): the principle of consistent least enrichment relative to a proof procedure for* $\vdash$, whereby:

1. Given a proof procedure for finding answers to queries of the form $\Delta?Q$, then if Δ is a partial database and we ask a query $\Delta?Q$ and we get the input B, then we try and prove Q from $\Delta \cup \{B\}$ and assume that any information $B_1, B_2, \ldots$ required for the success of the proof was intended to be added to Δ along with the input B. $\Delta \cup \{B, B_1, B_2, \ldots\}$ must emerge as consistent. When we are talking about proof or deduction, we do not necessarily restrict ourselves to monotonic deduction. The deduction rules used in trying to prove $\Delta \cup \{B\} \vdash Q$ may be defeasible or non-monotonic. These rules correspond better to common sense reasoning. The nature of the logic involved is as yet unspecified by us.

2. If there are several ways of adding $\{B_i\}$ leading to success, we choose the one which involves the least deductive effort for some inferential effect. (These notions will be made precise in later work: they require specification of links between databases so that we can describe for example some concept of closeness between a pair of databases).

We note that the process of finding the necessary B_i requires a given proof procedure for the logic licensing the addition of such assumptions to the database. Thus for different proof procedures we may get different B_i. We also note that the asker A may get his enriched database $\Delta_A \cup \{B, B_i\}$ wrong, in which case further interaction would be required to clarify the matter.

6.1 An Example

Suppose we have a database Δ for an asker A and his query Q to a responder R. Δ could contain the fact that A has apples, and Q could be 'would you like [to eat] an apple?' Here we assume success in the parsing process is achieved and we have some outcome to be evaluated against the database, i.e.:

$$\begin{aligned} \Delta &\ni \exists x[\text{Apple}(x)] \\ Q &= \exists x[\text{Apple}(x) \wedge \text{Eat}(x)] \end{aligned}$$

The answer A gets from the responder R may be 'yes' (Q) or 'no' ($\sim Q$), or A may get another answer B which allows A to conclude Q or $\sim Q$ (e.g. 'I'm not hungry'). Consider instead the answer: B = 'I don't eat South African apples'. Our proposed formal mechanism for parsing must bring the content of the sentence out, i.e.:

$$B = \forall x[\text{SA}(x) \wedge \text{Apple}(x) \Rightarrow \sim\text{Eat}(x)]$$

A has asked on the basis of Δ whether Q and got the answer B. We can therefore assume by the principle of relevance that $\Delta \cup \{B\}?!Q = 1$, i.e. the extra input information B is expected to (logically) decide the question Q. However, adding B to Δ does not decide Q in this case. Now A can invoke the principle of consistent least enrichment and try to abduce some further data B' such that $\Delta \cup \{B, B'\}?!Q = 1$.

We will now examine one way to compute the enrichment of Δ with some B' to give the desired effect. In this case, we will use as the RRA for this $\vdash$ the (propositional) abduction algorithm in Definition 28, which we have also implemented in Prolog, whereby:

1. $\mathit{Abduce}(\Delta, Q) = \top$ if $\Delta?Q = 1$;
2. $\mathit{Abduce}(\Delta, q) = q$, for q atomic such that q is not the head of any clause in Δ;
3. $\mathit{Abduce}(\Delta, Q_1 \wedge Q_2) = \mathit{Abduce}(\Delta, Q_1) \wedge \mathit{Abduce}(\Delta, Q_2)$;
4. $\mathit{Abduce}(\Delta, A_1 \Rightarrow (A_2 \Rightarrow \ldots (A_n \Rightarrow q) \ldots)) = A_1 \Rightarrow (A_2 \Rightarrow \ldots (A_n \Rightarrow \mathit{Abduce}(\Delta \cup \{A_1, A_2, \ldots A_n\}, q)) \ldots)$;
5. Let q be atomic and let $B^j = B^j_1 \Rightarrow (B^j_2 \Rightarrow \ldots (B^j_{n_j} \Rightarrow q) \ldots)$ for $j = 1, \ldots, m$ be all clauses in Δ with head q. Then $\mathit{Abduce}(\Delta, q) = \bigvee_{j=1}^{m} \bigwedge_{i=1}^{n_j} \mathit{Abduce}(\Delta, B^j_i)$.

If we introduce Skolem constants and Herbrand constants so that we are dealing only with propositional variables, put Δ in clausal form, and rewrite $\sim p$ as $p \Rightarrow \bot$, we can apply this abduction algorithm to the example. We also focus on the relevant parts of Δ, i.e. $\exists x[\text{Apple}(x)]$.

We want to know whether the answer to $\Delta \cup \{B\}?Q$ is yes or no. We therefore try and prove both Q and $\sim Q$, abducing any further assumptions along the way. We have:

1. a (i.e. Δ Skolemised)
2. $a \Rightarrow (sa \Rightarrow (e \Rightarrow \bot))$ (i.e. B Herbrandised in clausal form)

We want to prove $Q = a \wedge e$ and $\sim Q = (a \wedge e) \Rightarrow \bot$. We will examine each in turn.

For Q we have:

$$\begin{aligned} \mathit{Abduce}(\Delta, a \wedge e) &= \mathit{Abduce}(\Delta, a) \wedge \mathit{Abduce}(\Delta, e) \\ &= \top \wedge \mathit{Abduce}(\Delta, e) \\ &= \mathit{Abduce}(\Delta, e) \end{aligned}$$

To compute $\mathit{Abduce}(\Delta, e)$ we have two policies. Firstly, if we don't unify e with $\bot$, then the result of $\mathit{Abduce}(\Delta, e) = e$. Since we are assuming that the RRA yields some new information B' and $Q \vdash e$ anyway, we ignore this result.

Secondly, we do let e unify with $\bot$. Then the algorithm gives:

$$\begin{aligned} \mathit{Abduce}(\Delta, e) &= \mathit{Abduce}(\Delta, a) \wedge \mathit{Abduce}(\Delta, sa) \wedge \mathit{Abduce}(\Delta, e) \\ &= sa \wedge \mathit{Abduce}(\Delta, e) \end{aligned}$$

We therefore need to solve the following 'equations' for the unknown $x =$ Abduce (Δ, e):

$\Delta \cup \{x\} \vdash e$

$x = sa \wedge x$

x is minimal in satisfying the two previous properties

The only solution is $x = sa \wedge e$. This, however, will yield an inconsistent database when added to Δ. Therefore the RRA 'fails' for Q.

For $\sim Q$ we have:

$$\begin{aligned} \mathit{Abduce}(\Delta, (a \wedge e) \Rightarrow \bot) &= (a \wedge e) \Rightarrow \mathit{Abduce}(\Delta \cup \{a, e\}, \bot) \\ &= (a \wedge e) \Rightarrow sa \end{aligned}$$

Since, in effect, this came from the Herbrandised query $\text{Apple}(c) \wedge \text{Eat}(c) \Rightarrow \bot$ (from $\forall y[\text{Apple}(y) \wedge \text{Eat}(y) \Rightarrow \bot]$, itself rewritten from $\sim \exists y[\text{Apple}(y) \wedge \text{Eat}(y)]$, i.e. $\sim Q$), we unHerbrandise to give:

$$B' = \forall x[\text{Apple}(x) \wedge \text{Eat}(x) \Rightarrow \text{SA}(x)]$$

which is added to A's database Δ and so yields an answer to $\sim Q$. We now have $\Delta \cup \{B, B'\}?!Q = 1$, as required.

It should be noted that although adding B' is the minimal information enrichment for success (using an implicit ordering relation such that $p \leq (\rightarrow p) \Rightarrow q$), it does cost in deductive effort. This additional cost must be offset by some additional inferential effect for A, the details of which are not necessarily accessible to R. In this case, one natural such additional conclusion is that R is telling A that she believes that all the apples he has to eat are South African (and is inviting A to refute that belief before she will eat one of his apples). There may also be other conclusions in respect of other fruit R wouldn't eat for the same reason or, more vaguely, of R's political beliefs. A is encouraged to derive inferences in either of these directions given R's indirect mode of expression, but the only inference he is forced to make is that which induces an answer to his question in one step (this in itself is a reflection of the cognitive cost intrinsic to relevance).

Note that the need to add B' arose directly from the proof procedure involved. We 'hit' on the need to succeed with B' and we decided to add. Different proof procedures may give slightly different results especially when considerations like cost of deduction are involved. But given the question as input, no further enrichment (such as 'If the apples are Polish, R might not eat them either') is even attempted until it is established whether the indicated enrichment itself yields a definitive answer.

6.2 *An Extended Dialogue*

There are many issues to clarify here, but this provides an indication of how we think such abduction resolution should be addressed. We now turn to the analysis of an extended dialogue, in which indirect answers play a prominent part. The example shows a continuing temporal interaction which can lead to different enrichments as time progresses. Furthermore, the process of utterance interpretation in the dialogue will be embedded within an overall planning framework, in which the participants have certain desired outcomes in mind and plan what they say with these outcomes in mind, but during the course of the conversation have to renegotiate the source of conflicts between these outcomes as they become manifest in the discourse scenario.

Consider a situation in which a customer wants to hire a car 'today', and has a fair idea of the type of car she wants. She approaches a salesman who, understandably, wants to maximise his income by hiring and selling the biggest and most expensive cars. The following dialogue, let us suppose, ensues:

CUSTOMER:	*I want to hire a car.*	(1)
SALESMAN:	*When for?*	(2)
CUSTOMER:	*Today.*	(3)
SALESMAN:	*Would you like to hire this Sierra?*	(4)

CUSTOMER:	*I don't drive big cars.*	(5)
SALESMAN:	*The small cars are reserved for members only.*	(6)
CUSTOMER:	*Can I join?*	(7)
SALESMAN:	*It will take a week to process the papers.*	(8a)
	Would you like to reconsider?	(8b)
CUSTOMER:	*Big cars are not easy for me to drive. I can't control them.*	(9)
SALESMAN:	*But the Sierra has power-assisted steering!*	(10)
CUSTOMER:	*So?*	(11)
SALESMAN:	*Power-assisted steering makes a car easy to control.*	(12)

The dialogue is one which requires successive steps of abduction, both 'inferential' and 'structural'. We shall analyse this dialogue closely to try and bring this out, using formal notation quite loosely. We are assuming, *inter alia*, a logic with sorts, modal operators of wants and beliefs, agents who have goals and plans, temporal labels on formulas which we can reason with, etc. Because of the complexity of and interaction between these phenomena, we will gloss over many details, but aim for the exposition to establish the principles involved.

At step (1) the customer C sets out her goal, the hiring of a car. At step (2) the salesman, S, seeks to pinpoint this at some particular point in the temporal flow, and at step (3), C provides S with this information. From the salesman's point of view, we then have:

$$\begin{array}{lcl} G_S & = & W_S W_C \exists x_{car}[\text{hire}(C, x_{car}) \wedge \text{profitable}(x_{car})] \\ \Delta_S & \ni & t_{now} : W_C \exists x_{car}[\text{hire}(C, x_{car})] \end{array}$$

i.e. his goal G_S is to hire out a car 'profitably', and from (1) and (3) he has added to his database Δ_S the customer's goal.

From now on, the conversation is a successive attempt by each speaker to push the conversation towards satisfaction of their own ends - S to hire out an expensive car, C to hire a small car. We see immediately the point of indirect answers in dialogue. S offers (4). Rather than reply 'no', C provides an indirect answer (5) which is intended to be pre-emptive. She is maximising to S the relevance of her utterance, since the indirect answer, albeit *a priori* more costly (cognitively), obviates a whole series of similar questions.

This pattern of question and answer is similar to that discussed in Subsection 6.1, and it is instructive to observe the effect of applying the RRA in this case.

First we assume that in Δ_S there is the 'common knowledge' that if someone can't drive a car then they won't hire it, and that there is some correlation between the size of a car and its profitability. So 'formally' we now have:

$$\begin{array}{lcl} G_S & = & W_S W_C \exists x_{car}[\text{hire}(C, x_{car}) \wedge \text{profitable}(x_{car})] \\ \Delta_S & \ni & t_{now} : W_C \exists x_{car}[\text{hire}(C, x_{car})] \\ & & \forall x_{car}[\sim \text{drive}(C, x_{car}) \Rightarrow \sim \text{hire}(C, x_{car})] \\ & & \forall x_{car}[\text{big}(x_{car}) \Leftrightarrow \text{profitable}(x_{car})] \end{array}$$

$$B_C \forall x_{car}[\text{big}(x_{car}) \Rightarrow \sim \text{drive}(C, x_{car})]$$
$$Q \quad = \quad W_C \text{hire}(C, \text{Sierra}_{car})$$

Using (intuitively) the abduction algorithm of subsection 6.1, *Abduce*$(\Delta_S, Q) = Q$, which is ignored. On the other hand, *Abduce*$(\Delta_S, \sim Q) = B_C\text{big}(\text{Sierra}_{car})$, which is enough to prove $\sim Q$. This then is added to Δ_S.

At this point the salesman should be able to detect a refinement of the customer's goal, i.e.:

$$t_{now} : W_C \exists x_{car}[\text{hire}(C, x_{car}) \wedge \text{small}(x_{car})]$$

and that there is a conflict between the salesman's goal and the customer's goal, caused by the customer's beliefs. Rather than relaxing his own goal, (6) is an attempt by the salesman to revise the customer's belief so that she will revise her own goal and, in turn, communicate this to him.

So he tells her that the set of cars in which she is interested are for members only. In processing (6), C has to abduce *structurally* the existence of some association (formally A_{assoc}, say) to which these members belong. Making this step, she asks (7), requiring her hearer, S, in turn to abduce structurally the second argument of the relation 'join' as being this same association. In our model, we expect that S's 'parsing database' may look something like:

⋮

prove $\{\text{I}:\text{e}, \text{join}: e \Rightarrow e \Rightarrow t\} \vdash \gamma : \text{t}$

n:	I : e — abduce: set $I = C$
$n+1$:	join : $e \Rightarrow e \Rightarrow t$
$n+2$:	abduce: ? : e is required to exit
$n+3$:	abduce on structure: set ? $= A_{assoc}$ as right sort for second argument of join

exit $\gamma = \text{join}(C, A_{assoc})$

⋮

We would also expect, in context, that S can work out the implication of (7), namely that if C joins this association, then she will become a member (and so eligible to hire small cars).

S, presuming C able to reason that joining an association involves filling out, sending off, and processing application papers, replies *indirectly* with (8a). She is required to abduce, again structurally, that the papers so indicated are the papers relevant to her application of this presumed association, and that if she wants to join the papers need to be processed before she becomes a member. Furthermore, the indirect answer requires 'inferential' abductive reasoning on the customer's part. 'Formally', we have:

$$\Delta_C \ni \forall x_{car} \forall t[t : \text{member}(C, A_{assoc}) \Rightarrow t : (\text{hire}(C, x_{car}) \wedge \text{small}(x_{car}))]$$
$$Q = [t_1 \leq t_2] : \exists t_1, t_2[t_1 : \text{join}(C, A_{assoc}) \Rightarrow t_2 : \text{member}(C, A_{assoc})]$$
$$B = [t_3 \leq t_4, t_4 = t_5 - 1\text{week}] : \exists t_3, t_4, t_5[t_3 : \text{join}(C, A_{assoc}) \Rightarrow t_4 : \text{process}(\text{papers}) \Rightarrow t_5 : \text{member}(C, A_{assoc})]$$

Applying the abduction algorithm of Section 6.1 to an intuitively Skolemised/ Herbrandised version of the above (and assuming the ability to propagate temporal constraints (i.e. labels of the form $[t_1 \leq t_2]$)), we get:

$$\begin{aligned}
&Abduce(\Delta, [t_1 \leq t_2] : t_1 : join \Rightarrow t_2 : member) \\
&= [t_1 \leq t_2] : t_1 : join \Rightarrow Abduce(\Delta \cup t_1 : join, t_2 : member) \\
&= [t_1 \leq t_2, t_2 = t_5, t_3 \leq t_4, t_4 = t_5 - 1\text{week}] : t_1 : join \Rightarrow \\
&\qquad \bigwedge(Abduce(\Delta \cup t_1 : join, t_3 : join), \\
&\qquad\qquad Abduce(\Delta \cup t_1 : join, t_4 : process)) \\
&= [t_1 \leq t_2, t_2 = t_5, t_3 \leq t_4, t_4 = t_5 - 1\text{week}, t_1 = t_3] : t_1 : join \Rightarrow \\
&\qquad (\top \wedge t_4 : process) \\
&= [t_1 \leq t_4] : t_1 : join \Rightarrow t_4 : process
\end{aligned}$$

Although the RRA has succeeded, C is also able to reason that by adding this to the database and asking herself $\Delta_C ?t_{now} : \text{join}(C, A_{assoc})$ she can conclude:

$$\forall x_{car}[[t_{now} \leq t - 1\text{week}] : t : (\text{hire}(C, x_{car}) \wedge \text{small}(x_{car}))]$$

but the time constraint on when she can hire a small car, requiring this time t to be greater than t_{now} by at least one week, renders this conclusion inconsistent with her goal.

This, according to relevance, is what the salesman intends her to abduce, and since he knows that C cannot reconcile her goal with the given information he continues with (8b). With this question, he reminds her of the first steps in their negotiation (indeed we posit that the lexical content of reconsider forces her to set up as its second argument some step in the 'consideration' so far to hand). In (9) C now provides the reason for her earlier judgment (5) and in so doing gives S the information he needs to subvert her belief and to fulfill his own goal. He responds with (10). The keyword here is *but*. We intend the procedural specification of *but* to be formulated so that it instigates a search for a set of premises that derive inconsistency at some level of inference - the inference to be derived from the second conjunct to replace the inference to be derived from the first. Since in (9) it is the customer's assertion that big cars are not safe which is the reason for her belief of (5) it is this assertion that S will want her to retract. Accordingly he utters (10) requiring her to recover premises that combine with this to replace the inference from C's utterance (9).

Despite recognising this, the customer refuses to provide the requisite premise herself. However, the word *so* provides the salesman with an invitation to build just this premise for her. We can specify *so* to be an anaphoric expression precisely for this purpose. S duly provides C with the premise needed. The customer now has to reason (non-monotonically) from (12) to the point where she can relax her goal and inform the salesman that she is prepared to hire a big car, presuming it has power-assisted steering.

7 CONCLUDING REMARKS AND FURTHER IDEAS

We conclude this chapter by emphasising the metalevel nature of abduction and giving some pointers to a more general point of view for abduction.

Consider the database comprising of $\{a \Rightarrow q, b \Rightarrow q\}$ and suppose our goal is $?q$.

Let us examine how the properties of abduction mentioned in Section 1 (page 101) manifest themselves in this example.

1. *Abduction is metalevel*
   The very idea of what we want to do is metalevel. We want to add data to make q provable.

2. *Dependence on proof theory*
   We find what options we have to add to the database by following the proof theory and extracting the additional data at the point of failure. Possibilities for our case are $\{a\}, \{b\}, \{a \vee b\}, \{a \wedge b\}, \{q\}$.

   If we adopt a goal directed apporach our options are $\{a\}, \{b\}$.

3. *Abduction mechanisms are data*
   This is clear since they yield additional data.

4. *Abduction requires a background logic*
   Given the choice of whether to add $\{a\}$ or $\{b\}$, a background plausibility logic may help us choose.

5. *Abduction may not succeed*
   The plausibility logic may tell us not to add anything or the abduction algorithm may yield nothing.

We now address the metalevel aspect of abduction. From the metalevel point of view an abduction algorithm tells us what to add to the data at the point of failure (see for example item 3 of Definition 25 and item 3 of Definition 27).

In fact, we recognise a point of failure by the meta-predicate $\mathit{Success}(\Delta, A) = 0$. The abduction algorithm just tells us what to do in such a case. Let us write it explicitly in English (serving as a metalangauge).

- If $\mathit{Success}(\Delta, q) = 0$, then Do add $\mathit{Abduce}^{+}(\Delta, q)$ to Δ.

Viewed in this way, we can think of a general rule of the form

- If $\mathit{Success}(\Delta, q) = 0$ then take some new proof theoretic step.

The proof theoretic step can be any of the following:

- Replace the goal q by another goal; (for example the Restart Rule: See [Gabbay and Reyle, 1984; Gabbay and Olivetti, 2000].

- Add the data item p. (This is what *negation by failure* does. The rule can be written in the object language as, say, $fail(q) \to p$. See [Eshghi and Kowalski, 1989]).
- Take an action (see the Time-Action-Revision chapter in [Gabbay, to appear].

The above considerations are special cases of *fibring* the metalevel with the object level which in itself is a special case of fibring two systems. We can think of this kind of research as *the logic of getting unstuck*! Namely, what to do whenever the algorithm can no longer proceed under existing instructions.

ACKNOWLEDGEMENTS

This chapter is an expansion of Chapter 11 of [Gabbay, 1996], which is itself based on [Gabbay, Kempson and Pitt, 1994]. I would like to thank Oxford University Press for allowing me to use the material beginning on page 130. I am grateful to John Woods and Gaby Paul for carefully commenting on the text.

King's College, London.

REFERENCES

[Cohen and Keyt, 1992] S. M. Cohen and D. Keyt. Analysing Plato's argument: Plato and Platonism. In *Oxford Studies in Ancient Philosophy, Supplementary Volume 1992, Methods of Interpreting Plato and his Dialogues*, James C. Clagge and Nicholas D. Smith, eds. pp. 173–200. Oxford University Press, 1992.

[Eshghi and Kowalski, 1989] K. Eshghi and R. A. Kowalski. Abduction compared with negation by failure. In *Proc. 6th ICLP 89*, MIT Press, 1989.

[Gabbay, 1985] D. M. Gabbay. N-Prolog, Part II, *Journal of Logic Programming*, 5: 251–283, 1985.

[Gabbay, 1996] D. M. Gabbay. *Labelled Deductive Systems*, Oxford University Press, 510 p., 1996.

[Gabbay, to appear] D. M. Gabbay. What is a logical system, part 2. To appear.

[Gabbay and Kempson, 1991] D. M. Gabbay and R. Kempson. Natural language content: a proof theoretic perspective. In *Proceedings of the Eighth Amsterdam Colloquium*, pp. 173–196, 1991.

[Gabbay, Kempson and Pitt, 1994] D. M. Gabbay, R. Kempson and J. Pitt. Labelled abduction and relevance. In *Nonstandard Queries and Nonstandard Answers*, R. Demolombe, ed, pp. 155–186. Studies in Logic and Computation, Oxford University Press, 1994.

[Gabbay and Reyle, 1984] D. M. Gabbay and U. Reyle. N-Prolog, An extension of Prolog with hypothetical implication. *Journal of Logic Programming*, 4: 319–355, 1984.

[Gabbay and Olivetti, 2000] D. M. Gabbay and N. Olivetti. *Goal Directed Proof Theory*, Kluwer, 2000.

[Gabbay and Woods, forthcoming] D. M. Gabbay and J. Woods. *The Reach of Abduction*, forthcoming.

[Olivetti and Terracini, 1991] N. Olivetti and L. Terracini. N-Prolog and equivalence of logic Programs. *Journal of Logic, Language and Information*, 1: 253–340, 1992.

[Paul, 1999] G. Paul. Abductive logic — an overview. This Volume.

PETER A. FLACH

LOGICAL CHARACTERISATIONS OF INDUCTIVE LEARNING

1 INTRODUCTION

This chapter aims to develop a logical account of inductive reasoning, one of the most important ways to synthesise new knowledge. Induction provides an idealised model for empirical sciences, where one aims to develop general theories that account for phenomena observed in controlled experiments. It also provides an idealised model for cognitive processes such as learning concepts from instances. The advent of the computer has suggested new inductive tasks such as program synthesis from examples of input-output behaviour and knowledge discovery in databases, and the application of inductive methods to artificial intelligence problems is an active research area, which has displayed considerable progress over the last decades.

On the foundational side, however, our understanding of the essentials of inductive reasoning is fragmentary and confused. Induction is usually defined as inference of general rules from particular observations, but this slogan can hardly count as a definition. Clearly some rules are better than others for given observations, while yet other rules are totally unacceptable. A logical account of induction should shed more light on the relation between observations and hypotheses, much like deductive logic formalises the relation between theories and their deductive consequences.

This is by no means an easy task, and anyone claiming to provide a definitive solution should be approached sceptically. This chapter suggests a novel perspective by combining older work in philosophy of science with a methodology suggested by recent work in formalising nonmonotonic reasoning. This perspective provides us with a *descriptive* – rather than prescriptive – account of induction, which clearly indicates both the opportunities for and limitations of logical analysis when it comes to modelling induction.

We will take it for granted that there exists a distinct and useful form of reasoning called induction. As a model for this form of reasoning we may take the approaches to learning classification rules from examples that can be found in the machine learning literature, or the work on inducing Prolog programs and first-order logical theories from examples in the recently established discipline of inductive logic programming. By taking this position we will avoid the controversies abounding in philosophy of science as to whether or not science proceeds by inductive methods. This is not to say that philosophical considerations will be completely ignored – in fact, my approach has been partly motivated by works from the philosophers Charles Sanders Peirce and Carl G. Hempel, as I will explain shortly.

D. M. Gabbay and R. Kruse (eds.), Abductive Reasoning and Learning, 155–196.

The main question addressed in this chapter is the following: *Can we develop a logical account of induction that is sufficiently similar to the modern account of deduction?* By 'the modern account of deduction' I mean the approach, developed in the first half of this century and standard by now, of defining a logical language, a semantical notion of deductive consequence, and a proof system of axioms and inference rules operationalising the relation of deductive consequence. By the stipulation that the logical account of induction be 'sufficiently similar' to this modern account of deduction I mean that the former should likewise consist of a semantical notion of *inductive consequence*, and a corresponding proof system.

Those perceiving logic as the 'science of correct reasoning' will now object that what I am after is a deductive account of induction, and it is known already since Hume that inductive hypotheses are necessarily defeasible. My reply to this objection is that it derives from a too narrow conception of logic. In my view, *logic is the science of reasoning*, and it is the logician's task to develop formal models of every form of reasoning that can be meaningfully distinguished. In developing such formal models for nondeductive reasoning forms, we should keep in mind that deduction is a highly idealised and restricted reasoning form, and that we must be prepared to give up some of the features of deductive logic if we want to model reasoning forms that are less perfect, such as induction.

The fundamental question then is: which features are inherent to logic *per se*, and which are accidental to deductive logic? To illustrate this point, consider the notion of truth-preservation: whenever the premisses are true, the conclusion is true also. It is clear that truth-preservation must be given up as soon as we step out of the deductive realm. The question then arises whether a logical semantics is mainly a tool for assessing the truth of the conclusion given the truth of the premisses, or whether its main function is rather to define what property is preserved when passing from premisses to conclusion. We will address this and similar fundamental questions in this chapter.

Another objection against the approach I propose could be that deductive logic is inherently prescriptive: it clearly demarcates the logical consequences one *should* accept on the basis of given premisses, from the ones one *should not* accept. Clearly, our understanding of induction is much too limited to be able to give a prescriptive account of induction. My reply to this objection is that, while such a demarcation is inherent to logic, its interpretation can be either prescriptive or descriptive. The inductive logics I propose distinguish between hypotheses one *should not* accept on the basis of given evidence, relative to a certain goal one wants the hypothesis to fulfill, and hypothesis one *might* accept. Put differently, these inductive logics formalise the logic of inductive hypothesis generation rather than hypothesis selection, which I think is the best one can hope to achieve by purely logical means.

The objective pursued in this chapter, then, is to develop semantics and proof systems for inductive hypothesis generation. What is new here is not so much this objective, which has been pursued before (see e.g. [Delgrande, 1987]), but the meta-theoretical viewpoint taken in this chapter, which I think greatly benefits our understanding of the main issues. This meta-theoretical viewpoint has been inspired by the seminal paper of Kraus, Lehmann and Magidor [1990], where it is employed to unravel the fundamental properties of nonmonotonic reasoning. Readers familiar with the paper of Kraus *et al.* may alternatively view the present chapter as a constructive proof of the thesis that their techniques in fact establish a *methodology*, by demonstrating how these techniques can be successfully applied to analyse a rather different form of reasoning.

The chapter is structured as follows. In Section 2 the philosophical, logical, and machine learning backgrounds of this chapter are surveyed. Section 3 introduces the main logical tool employed in this chapter: the notion of a meta-level consequence relation. Sections 4 and 5 form the technical core of this chapter, stating representation theorems characterising sets of meta-level properties of explanatory induction and confirmatory induction, respectively. In Section 6 we discuss the implications of the approach taken and results obtained in this chapter. Section 7 reviews the main conclusions in a wider context.

2 BACKGROUNDS

This section reviews a number of related approaches from the philosophical, logical, and machine learning literature. With such a complex phenomenon as induction, one cannot hope to give an overview that can be called complete in any sense – I will restrict attention to approaches that are most closely related to the one I propose in this chapter.

In Section 2.1 I discuss the idea of defining inductive or subjective probability as the cornerstone of an inductive logic, as pioneered by Rudolf Carnap. I will however argue that this approach is unsuitable for the kind of logical analysis that I am pursuing. Section 2.2 discusses Hempel's qualitative notion of confirmation that can be seen as a precursor to my approach. In Section 2.3 I relate this notion of confirmation to explanation as introduced by Peirce. Finally, in Section 2.4 relevant approaches in inductive machine learning are reviewed.

2.1 Inductive probability

By now it is commonplace to draw a connection between inductive reasoning and probability calculus. Inductive or subjective probability assesses the degree to which an inductive agent is willing to accept a hypothesis on the basis of available evidence. A so-called posterior probability of the hypothesis after observing the evidence is obtained by applying Bayes' theorem to the probability of the hypothesis prior to observation. Rudolf Carnap has advocated the view that inductive probability gives rise to a system of inductive logic [Carnap, 1950]. Briefly, Carnap defines a function $c(H, E)$ assigning a *degree of confirmation* (a number between 0 and 1) to a hypothesis H on the basis of evidence E. This function generalises the classical notion of logical entailment – which can be seen as a 'confirmation function' from premisses and conclusion to $\{0, 1\}$ – to an inductive notion of 'partial entailment':

> 'What we call inductive logic is often called the theory of nondemonstrative or nondeductive inference. Since we use the term 'inductive' in the wide sense of 'nondeductive', we might call it the theory of inductive inference... However, it should be noticed that the term 'inference' must here, in inductive logic, not be understood in the same sense as in deductive logic. Deductive and inductive logic are analogous in one respect: both investigate logical relations between sentences; the first studies the relation of [entailment], the second that of degree of confirmation which may be regarded as a numerical measure for a partial [entailment]... The term 'inference' in its customary use implies a transition from given sentences to new sentences or an acquisition of a new sentence on the basis of sentences already possessed. However, only deductive inference is inference in this sense.' [Carnap, 1950, §44B, pp. 205–6]

This citation succinctly summarises why inductive probability is *not* suitable, in my view, as the cornerstone of a logic of induction. My two main objections are the following.

Inductive probability treats all nondeductive reasoning as inductive. This runs counter to one of the main assumptions of this chapter, namely that induction is a reasoning form in its own right, which we want to characterise in terms of properties it enjoys rather than properties it lacks. A more practical objection is that a single logical foundation for all possible forms of nondeductive reasoning is likely to be rather weak. Indeed, I would argue that in many forms of reasoning the goal that is to be fulfilled by the hypothesis, such as explaining the observations, is not reducible to a degree of confirmation.[1]

Inductive probability, taken as partial entailment, leads to a degenerated view of logic. This is essentially what Carnap notes when he states that his inductive logic does not establish inference in the same sense as deductive logic (although he would not call it a degeneration). This means that, for instance, the notion of a proof reduces to a calculation of the corresponding degree of confirmation. A possible remedy is to define and axiomatise a qualitative relation of confirmation, such as the relation defined by $qc(H, E) \Leftrightarrow c(H, E) > c(H, \mathbf{true})$. However, such a qualitative relation of confirmation can also be postulated without reference to numerical degrees of confirmation, which would give us much more freedom to investigate the relative merits of different axiom systems. In fact, this is the course of action taken by Hempel, as we will see in the next section.

I should like to stress that it is not inductive probability or Bayesian belief measures as such which are criticised here – on the contrary, I believe these to be

[1]Note that degree of confirmation, in the sense defined by Carnap, is not a quantity that is simply to be maximised, since this would lead us straight back into deductive logic.

significant approaches to the important problem of how to update an agent's beliefs in the light of new information. Since belief measures express the agent's subjective estimates of the truth of hypotheses, let us say that inductive probability and related approaches establish a *truth-estimating procedure*. My main point is that such truth-estimating procedures are, generally speaking, *complementary* to logical systems. Truth-estimating procedures answer a type of question which nondeductive logical systems, in general, cannot answer, namely: how plausible is this hypothesis given this evidence? The fact that deductive logic incorporates such a truth-estimating procedure is accidental to deductive reasoning; the farther one moves away from deduction, the less the logical system has to do with truth-estimation. For instance, the gap between logical systems for nonmonotonic reasoning and truth-estimating procedures is much smaller than the gap between the latter and logical systems for induction. Indeed, one may employ one and the same truth-estimating procedure for very different forms of reasoning.

2.2 *Confirmation as a qualitative relation*

Carl G. Hempel [1943; 1945] developed a qualitative account of induction that will form the basis of the logical system for what I call confirmatory induction (Section 5). Carnap rejected Hempel's approach, because he considered a quantitative account of confirmation as more fundamental than a qualitative account. However, as explained above I think that the two are conceived for different purposes: a function measuring degrees of confirmation can be used as a truth-estimating procedure, while a qualitative relation of confirmation can be used as the cornerstone for a logical system. I also consider the two as relatively independent: a qualitative confirmation relation that cannot be obtained from a numerical confirmation function is not necessarily ill-conceived, as long as the axioms defining the qualitative relation are intuitively meaningful.

Hempel's objective is to develop a material definition of confirmation. Before doing so he lists a number of adequacy conditions any such definition should satisfy. Such adequacy conditions can be seen as meta-level axioms, and we will discuss them at some length. The following conditions can be found in [Hempel, 1945, pp. 103–106, 110]; logical consequences of some of the conditions are also stated.

(H1) *Entailment condition*: any sentence which is entailed by an observation report is confirmed by it.

(H2) *Consequence condition*: if an observation report confirms every one of a class K of sentences, then it also confirms any sentence which is a logical consequence of K.

(H2.1) *Special consequence condition*: if an observation report confirms a hypothesis H, then it also confirms every consequence of H.

(H2.2) *Equivalence condition*: if an observation report confirms a hypothesis H, then it also confirms every hypothesis which is logically equivalent with H.

(H2.3) *Conjunction condition*: if an observation report confirms each of two hypotheses, then it also confirms their conjunction.

(H3) *Consistency condition*: every logically consistent observation report is logically compatible with the class of all the hypotheses which it confirms.

(H3.1) Unless an observation report is self-contradictory, it does not confirm any hypothesis with which it is not logically compatible.

(H3.2) Unless an observation report is self-contradictory, it does not confirm any hypotheses which contradict each other.

(H4) *Equivalence condition for observations*: if an observation report B confirms a hypothesis H, then any observation report logically equivalent with B also confirms H.

The entailment condition (H1) simply means that entailment 'might be referred to as the special case of *conclusive* confirmation' [Hempel, 1945, p. 107]. The consequence conditions (H2) and (H2.1) state that the relation of confirmation is closed under weakening of the hypothesis or set of hypotheses (H_1 is weaker than H_2 iff it is logically entailed by the latter). Hempel justifies this condition as follows [Hempel, 1945, p. 103]: 'an observation report which confirms certain hypotheses would invariably be qualified as confirming any consequence of those hypotheses. Indeed: any such consequence is but an assertion of all or part of the combined content of the original hypotheses and has therefore to be regarded as confirmed by any evidence which confirms the original hypotheses.' Now, this may be reasonable for single hypotheses (H2.1), but much less so for sets of hypotheses, each of which is confirmed separately. The culprit can be identified as (H2.3), which together with (H2.1) implies (H2). A similar point can be made as regards the consistency condition (H3), about which Hempel remarks that it 'will perhaps be felt to embody a too severe restriction'. (H3.1), on the other hand, seems to be reasonable enough; however, combined with the conjunction condition (H2.3) it implies (H3).

We thus see that Hempel's adequacy conditions are intuitively justifiable, except for the conjunction condition (H2.3) and, *a fortiori*, the general consequence condition (H2). On the other hand, the conjunction condition can be justified by a completeness assumption on the evidence, as will be further discussed in Section 5. We close this section by noting that Hempel's material definition of the relation of confirmation of a hypothesis by evidence roughly corresponds to what we would nowadays call 'truth of the hypothesis in the truth-minimal Herbrand model of the evidence' (see [Lachiche, 2000] for a discussion of closed-world assumptions in induction and abduction). We will return to material definitions of qualitative confirmation in Section 2.4.

2.3 *Confirmation and explanation*

Preceding Hempel's work on confirmation by almost half a century is the work of Charles Sanders Peirce on abduction: the process of forming explanatory hypotheses. For a full account of Peirce's theories of abduction and induction the reader is referred to the introductory chapter to this volume; here, a short summary suffices.[2]

In Peirce's view, inductive reasoning consists of two steps: (i) formulating a conjecture, and (ii) evaluating the conjecture. Both steps take the available evidence into account, but in quite different ways and with different goals. The first step requires that the conjectured hypothesis *explains* the observations; having a definite logical form, it represents a form of inference. The second step evaluates how well predictions offered by the hypothesis agree with reality; it is not inference, but assigns a numerical value to a hypothesis. The first step corresponds to what I call explanatory hypothesis generation, and the second to hypothesis evaluation.

Leaving a few details aside, Peirce's definition of explanatory hypothesis generation can be formalised as the inference rule

$$\frac{C \quad , \quad A \models C}{A}$$

I propose to generalise Peirce's definition by including the relation of 'is explained by' as a parameter. This is achieved by lifting the explanatory inference from C to A to the meta-level, as follows:

$$\frac{A \models C}{C \mathrel{|\!\!<} A}$$

The symbol $\mathrel{|\!\!<}$ stands for the explanatory consequence relation. Axiom systems for this relation will be considered in Section 4.

We have now encountered two fundamental notions that play a role in inductive hypothesis generation: one is that the hypothesis should be *confirmed* by the evidence (Hempel), the other that the hypothesis should *explain* the evidence (Peirce). Couldn't we try and build the requirement that the hypothesis be explanatory into our definition of confirmed hypothesis?

The problem is that an unthoughtful combination of explanation and confirmation can easily lead into paradox. Let H_1 and H_2 be two theories such that the latter includes the former, in the sense that everything entailed by H_1 is also entailed by H_2. Suppose E is confirming evidence for H_1; shouldn't we conclude that it confirms H_2 as well? To borrow an example of Hempel: 'Is it not true, for example, that those experimental findings which confirm Galileo's law, or Kepler's laws,

[2]In this chapter we are mainly interested in Peirce's later, inferential theory, which views abduction and induction in the context of scientific theory formation. See [Flach and Kakas, 2000] for a collection of papers on the relation between induction and abduction.

are considered also as confirming Newton's law of gravitation?' [Hempel, 1945, p. 103]. This intuition is formalised by the following condition:

(H5) *Converse consequence condition*: if an observation report confirms a hypothesis H, then it also confirms every formula logically entailing H.

The problem is, however, that this rule is incompatible with the special consequence condition (H2.1). This can be seen as follows: in order to demonstrate that E confirms H for arbitrary E and H, we note that E confirms E by (H1), so by the converse consequence condition E confirms $E \wedge H$; but then E confirms H by (H2.1). Thus, we see that the combination of two intuitively acceptable conditions leads to a collapse of the system into triviality, a clearly paradoxical situation.

Hempel concludes that one cannot have both (H2.1) and (H5), and drops the latter. His justification of this decision is however unconvincing, which is not surprising since neither is *a priori* better than the other: *they formalise different intuitions*. While (H2.1) formalises a reasonable intuition about confirmation, (H5) formalises an equally reasonable intuition about *explanation*:

(H5′) if an observation report is explained by a hypothesis H, then it is also explained by every formula logically entailing H.

In this chapter I defend the position that Hempel was reluctant to take, namely that with respect to inductive or scientific hypothesis generation there is more than one possible primitive notion: the relation 'confirms' between evidence and hypothesis, and the relation 'is explained by'. Each of these primitives gives rise to a specific form of induction. This position is backed up by recent work in machine learning, to which we will turn now.

2.4 *Inductive machine learning*

Without doubt, the most frequently studied induction problem in machine learning is concept learning from examples. Here, the observations take the form of descriptions of instances (positive examples) and non-instances (negative examples) of an unknown concept, and the goal is to find a definition of the concept that correctly discriminates between instances and non-instances. Notice that this problem statement is much more concrete than the general description of induction as inference from the particular to the universal: once the languages for describing instances and concepts are fixed, the desired relation between evidence and hypothesis is determined. A natural choice is to employ a predicate for the concept to be learned, and to use constants to refer to instances and non-instances. In this way, a classification of an instance can be represented by a truth value, which can be obtained by setting up a proof. We then obtain the following general problem definition:

Problem: Concept learning from examples in predicate logic.

Given:
(1) A predicate-logical language.
(2) A predicate representing the target concept.
(3) Two sets P and N of ground literals of this predicate, representing the positive and negative examples.
(4) A background theory T containing descriptions of instances.

Determine: A hypothesis H within the provided language such that

(i) for all $p \in P : T \cup H \models p$;
(ii) for all $n \in N : T \cup H \not\models n$.

Notice that condition (ii) is formulated in such a way that the hypothesis only needs to contain *sufficient conditions* for concept membership (since a negative classification is obtained by negation as failure). This suggests an analogy between concept definitions and Horn clauses, which can be articulated by allowing (possibly recursive) logic programs as hypotheses and background knowledge, leading us into the field of inductive logic programming (ILP) [Muggleton and De Raedt, 1994]. Furthermore, P and N may contain more complicated formulae than ground facts. The general problem statement then becomes: given a partial logic program T, extend it with clauses H such that every formula in P is entailed and none of the formulae in N.

The potential for inductive methods in artificial intelligence is however not exhausted by classification-oriented approaches. Indeed, it seems fair to say that most knowledge implicitly represented by extensional databases is non-classificatory. Several researchers have begun to investigate non-classificatory approaches to knowledge discovery in databases. For instance, in previous work I have demonstrated that the problem of inferring the set of functional and multivalued attribute dependencies satisfied by a database relation can be formulated as an induction problem [Flach, 1990; Flach, 1993; Flach, 1995; Flach and Savnik, 1999]. Furthermore, De Raedt and Bruynooghe have generalised the classificatory ILP-setting in order to induce non-Horn clauses from ground facts [De Raedt and Bruynooghe, 1993]. Both approaches essentially employ the following problem statement.

Problem: Knowledge discovery

Given:
(1) A predicate-logical language.
(2) Evidence E.

Determine: A hypothesis H within the provided language such that

(i) H is true in a model m_0 constructed from E;
(ii) for all g within the language, if g is true in m_0 then $H \models g$.

Essentially, the model m_0 employed in the approaches of De Raedt and Bruynooghe and myself is the truth-minimal Herbrand model of the evidence.[3] The hypothesis

[3]An alternative approach is to consider the information-minimal partial model of the evidence [Flach, 1995].

is then an axiomatisation of all the statements true in this model, including non-classificatory statements like 'everybody is male or female' and 'nobody is both a father and a mother'.

The relation between the classificatory and non-classificatory approaches to induction is that they both aim at extracting similarities from examples. The classificatory approach to induction achieves this by constructing a single theory that entails all the examples. In contrast, the non-classificatory approach achieves this by treating the examples as a model – a description of the world that may be considered *complete*, at least for the purposes of constructing inductive hypotheses. The approach is justified by the assumption that *the evidence expresses all there is to know about the individuals in the domain.* Such a completeness assumption is reminiscent of the Closed World Assumption familiar from deductive databases, logic programming, and default reasoning – however, in the case of induction its underlying intuition is quite different. As Nicolas Helft, one of the pioneers of the non-classificatory approach, puts it:

> 'induction assumes that the similarities between the observed data are representative of the rules governing them (...). This assumption is like the one underlying default reasoning in that a priority is given to the information present in the database. In both cases, some form of "closing-off" the world is needed. However, there is a difference between these: loosely speaking, while in default reasoning the assumption is "what you are not told is false", in similarity-based induction it is "what you are not told looks like what you are told".' [Helft, 1989, p. 149]

There is a direct connection between Peirce's conception of abduction as generation of explanatory hypotheses and the classificatory induction setting, if one is willing to view a theory that correctly classifies the examples as an *explanation* of those examples. I suggest to draw a similar connection between Hempel's conception of confirmation as a relation between evidence and potential hypotheses and the non-classificatory induction setting outlined above. Non-classificatory induction aims at constructing hypotheses that are *confirmed* by the evidence, without necessarily explaining them. Rather than studying material definitions of what it means to explain or be confirmed by evidence, as is done in the works referred to above, in the following sections I will be concerned with the logical analysis of the abstract notions of explanation and confirmation.

3 INDUCTIVE CONSEQUENCE RELATIONS

In the sections to follow I employ the notion of a consequence relation, originating from Tarski [1994] and further elaborated by Gabbay [1985], Makinson [1989], and Kraus, Lehmann and Magidor [1990; 1992]. In this section I give an introduction to this important meta-logical tool that is largely self-contained. The basic definitions are given in Section 3.1. In Section 3.2 I consider some general properties of consequence relations that arise when modelling the reasoning behaviour of inductive agents. Section 3.3 is devoted to a number of considerations regarding the pragmatics of consequence relations in general, and inductive consequence relations as used in this chapter in particular.

3.1 Consequence relations

We distinguish between the language L in which an inductive agent formulates premisses and conclusions of inductive arguments, and the meta-language in which statements about the reasoning behaviour of the inductive agent are expressed. Throughout the chapter L is a propositional language[4] over a fixed countable set of proposition symbols, closed under the usual logical connectives. We assume a set of propositional models U, and a satisfaction relation $\models \subseteq U \times L$ that is well-behaved with respect to the logical connectives and compact. As usual, we write $\models \alpha$ for $\forall m \in U : m \models \alpha$, for arbitrary $\alpha \in L$. Note that U may be a proper subset of the set of all truth-assignments to proposition symbols in L, which would reflect prior knowledge or *background knowledge* of the inductive agent. Equivalently, we may think of U as the set of models of an implicit background theory T, and let $\models \alpha$ stand for 'α is a logical consequence of T'.

The meta-language is a restricted predicate language built up from a unary meta-predicate $\models$ in prefix notation (standing for validity with respect to U in L) and a binary meta-predicate $\mid\!<$ in infix notation (standing for inductive consequence). In referring to object-level formulae from L we employ a countable set of meta-variables $\alpha, \beta, \gamma, \delta, \ldots$, the logical connectives from L (acting like function symbols on the meta-level), and the meta-constants **true** and **false**. Meta-level literals are atomic formulae or their negation; instead of $\neg(\models \alpha)$ we write $\not\models \alpha$, and instead of $\neg(\alpha \mid\!< \beta)$ we write $\alpha \mid\!\not< \beta$. Formulae of the meta-language, usually referred to as *rules* or *properties*, are of the form $P_1, \ldots, P_n/Q$ for $n \geq 0$, where $P_1, \ldots, P_n$ and Q are literals (atomic formulae or their negation). Intuitively, such a rule should be interpreted as an implication with *antecedent* $P_1, \ldots, P_n$ (interpreted conjunctively) and *consequent* Q, in which all variables are implicitly universally quantified. An example of such a rule, written in an expanded Gentzen-style notation, is

$$\frac{\models \alpha \wedge \beta \rightarrow \gamma \quad , \quad \alpha \mid\!< \beta}{\alpha \wedge \neg\gamma \mid\!\not< \beta}$$

This is a rule with two positive literals in its antecedent, and a negative literal in its consequent. Intuitively, it expresses that an inductive hypothesis β, previously inferred from evidence α, should be withdrawn if the negation of a consequence of

[4]One may argue that in induction the distinction between statements about individuals (ground facts) and statements about sets of individuals (universal sentences) is crucial, which calls for a predicate language. However, since all predicates are unary this can be translated into a propositional language. See [Flach, 1999].

α and β together is added to the evidence. A *rule system* is a set of such meta-level rules, denoted by abbreviations in boldface capitals.

Consequence relations provide the semantics for this meta-language, by fixing the meaning of the meta-predicate $\mid\!\!<$. Formally, a consequence relation is a subset of $L \times L$. They will be used to model part or all of the reasoning behaviour of a particular reasoning agent, by listing a number of *arguments* (pairs of *premiss* and *conclusion*) the agent is prepared to accept.[5] A consequence relation satisfies a rule whenever it satisfies all instances of the rule, and violates it otherwise, where an instance of a rule is obtained by replacing the variables of the rule with formulae from L. A consequence relation satisfies an instance of a rule if, whenever it satisfies the literals in the antecedent of the rule, it also satisfies the consequent. Finally:

- a ground literal $\models \alpha$ is satisfied whenever the propositional formula from L denoted by α is true in every model in U;

- a ground literal $\alpha \mid\!\!< \beta$ is satisfied whenever the pair of propositional formulae from L denoted by α and β is an element of the consequence relation.

For instance, the consequence relation $\{\langle p, q\rangle, \langle p \wedge \neg p, q\rangle\}$ violates the rule above. It is customary to ignore the distinction between the meta-language and its semantics by referring to a particular consequence relation as $\mid\!\!<$ and write $p \mid\!\!< q$ instead of $\langle p, q\rangle \in \mid\!\!<$.

A useful analogy with clausal logic is revealed by noting that the object language L establishes a Herbrand universe built up from the proposition symbols in L as constants and the connectives from L as function symbols. Consequence relations then correspond to Herbrand interpretations[6] of the meta-language, whose rules can be easily transformed to clausal notation. The following terminology is borrowed from logic programming:

(i) if all of $P_1, \ldots, P_n$ and Q are positive literals the rule is *definite*;

(ii) if at least one of $P_1, \ldots, P_n$ is a negative literal and Q is a positive literal the rule is *indefinite*;

(iii) if all of $P_1, \ldots, P_n$ are positive literals and Q is a negative literal the rule is a *denial*.[7]

For instance, all the rule systems of [Kraus *et al.*, 1990] are made up of definite rules only. Logic programming theory teaches us that the set of consequence relations satisfying a set of definite rules is closed under intersection. In contrast,

[5] This is not to say that the agent actually draws the conclusion when observing the premisses, but merely that she considers it one of the possible conclusions.

[6] These Herbrand interpretations are restricted to the meta-predicate $\mid\!\!<$; $\models$ is treated as a built-in predicate.

[7] This exhausts all the possibilities: the case that at least one of $P_1, \ldots, P_n$ is a negative literal and Q is a negative literal can be rewritten to (i) or (ii).

the rule system **R** characterised in [Lehmann and Magidor, 1992] contains one indefinite rule, *viz.* Rational Monotonicity. Consequently, the set of rational consequence relations is not closed under intersection.

3.2 Some properties of inductive consequence relations

After having stated the main definitions concerning consequence relations I will now list some properties generally obeyed by inductive consequence relations. In this section we will not distinguish between explanatory or confirmatory induction, and simply interpret $\alpha \mathrel{|\!<} \beta$ as 'β is a possible hypothesis given evidence α'.

The first two rules state that the logical form of evidence and hypotheses is immaterial:

Left Logical Equivalence $$\frac{\models \alpha \leftrightarrow \beta \quad , \quad \alpha \mathrel{|\!<} \gamma}{\beta \mathrel{|\!<} \gamma}$$

Right Logical Equivalence $$\frac{\models \beta \leftrightarrow \gamma \quad , \quad \alpha \mathrel{|\!<} \beta}{\alpha \mathrel{|\!<} \gamma}$$

Left Logical Equivalence states, for instance, that if the evidence is expressed as a conjunction of ground facts, the order in which they occur is immaterial. From the viewpoint of practical induction algorithms this may embody a considerable simplification – however, the framework represented in this chapter is intended to provide a model for inductive reasoning in general rather than particular algorithms.[8]

The following two rules express principles well-known from philosophy of science:

Verification $$\frac{\models \alpha \wedge \beta \rightarrow \gamma \quad , \quad \alpha \mathrel{|\!<} \beta}{\alpha \wedge \gamma \mathrel{|\!<} \beta}$$

Falsification $$\frac{\models \alpha \wedge \beta \rightarrow \gamma \quad , \quad \alpha \mathrel{|\!<} \beta}{\alpha \wedge \neg\gamma \mathrel{|\!\not<} \beta}$$

In these two rules γ is a *prediction* made on the basis of hypothesis β and evidence α. Verification expresses that if such a prediction is indeed observed, hypothesis β remains a possible hypothesis, while if its negation is observed, β may be considered refuted according to Falsification.[9] One might remark that typically the

[8] Practical algorithms establish a *function* from evidence to hypothesis rather than a relation, i.e. also Right Logical Equivalence would be invalidated by an induction algorithm (of all the logically equivalent hypotheses only one would be output).

[9] Notice that, contrary to the previous two rules, Verification and Falsification happen to be meaningful also when modelling the behaviour of an induction algorithm: Verification expresses that the current hypothesis should not be abandoned when the next observation is a predicted one (in the terminology of [Angluin and Smith, 1983] the algorithm is conservative), while Falsification expresses that the current hypothesis must be abandoned when the next observations run counter to the predictions of the algorithm (called consistency by [Angluin and Smith, 1983]). However, in the context of the present chapter these are not the intended interpretations of the two rules.

inductive hypothesis will entail the evidence, so that the first condition in the antecedent of Verification and Falsification may be simplified to $\models \beta \rightarrow \gamma$. However, this is only the case for certain approaches to explanatory induction; generally speaking inductive hypotheses, in particular those that are confirmed without being an explanation, may not contain all the information conveyed by the evidence. The formulation above represents the general case.

Falsification can be simplified in another sense, as shown by the following lemma.

LEMMA 1. *In the presence of Left Logical Equivalence, Falsification is equivalent with the following rule:*

Consistency $$\frac{\alpha \mathrel{|\!\!<} \beta}{\not\models \beta \rightarrow \neg\alpha}$$

Proof. To derive Falsification, suppose $\models \alpha \wedge \beta \rightarrow \gamma$, i.e. $\models \beta \rightarrow \neg(\alpha \wedge \neg\gamma)$, then by Consistency $\alpha \wedge \neg\gamma \not\mathrel{|\!\!<} \beta$. To derive Consistency from Falsification, suppose $\alpha \mathrel{|\!\!<} \beta$ and $\models \beta \rightarrow \neg\alpha$, i.e. $\models \alpha \wedge \beta \rightarrow$ **false**, then by Falsification $\alpha \wedge \neg$**false** $\not\mathrel{|\!\!<} \beta$, and by Left Logical Equivalence $\alpha \not\mathrel{|\!\!<} \beta$, a contradiction. ■

Falsification and Consistency rule out inconsistent evidence and hypotheses. The way inconsistent evidence is handled is merely a technicality, and we might have decided to treat it differently – for instance, Hempel's entailment condition (H1) implies that in his framework inconsistent evidence confirms arbitrary hypotheses. The case of inconsistent hypotheses is different however: it is awkward to say, for instance, that arbitrary evidence induces an inconsistent hypothesis. Furthermore, in inductive concept learning often negative examples are included, that are not to be classified as belonging to the concept, which requires consistency of the induced rule. Also, the adoption of Consistency is the only way to treat explanatory and confirmatory induction in a unified way as regards the consistency of evidence and hypothesis.

In the presence of Consistency a number of other principles have to be formulated carefully. For instance, we have reflexivity only for consistent formulae. In the light of Consistency a formula is consistent if it occurs in an inductive argument, either as evidence or as hypothesis, so we have the following weaker versions of reflexivity:[10]

Left Reflexivity $$\frac{\alpha \mathrel{|\!\!<} \beta}{\alpha \mathrel{|\!\!<} \alpha}$$

Right Reflexivity $$\frac{\alpha \mathrel{|\!\!<} \beta}{\beta \mathrel{|\!\!<} \beta}$$

[10]One might argue that induction is inherently non-reflexive if the hypothesis is to generalise the evidence. This point will be taken up in Section 6.

If a consequence relation contains an argument $\alpha \mathrel{|\!\!<} \alpha$, this signals that α is consistent with the reasoner's background theory. We will call such an α *admissible* (with respect to the consequence relation), and use conditions of this form whenever we require consistency of evidence or hypothesis in a rule.[11]

The final rule mentioned in this section is a variant of Verification that allows to add any prediction to the hypothesis rather than the evidence:

Right Extension

$$\frac{\models \alpha \wedge \beta \rightarrow \gamma \quad , \quad \alpha \mathrel{|\!\!<} \beta}{\alpha \mathrel{|\!\!<} \beta \wedge \gamma}$$

Further rules considered in this chapter are specific to either explanatory or confirmatory induction, and are therefore to be discussed in later sections.

3.3 The pragmatics of consequence relations

Before proceeding with the technical results of the chapter, we may spend a few thoughts on the exact nature of consequence relations. As defined above, a consequence relation is an extensional specification of the behaviour of a reasoning agent. The symbol $\mathrel{|\!\!<}$ is introduced in order to reason about consequence relations and reasoning behaviour, and functions as a binary predicate in the metalanguage. Rules describing properties of $\mathrel{|\!\!<}$ express boundaries of rationality of inductive reasoning: a consequence relation violating such a rule would be considered irrational.

Now suppose $\mathbf{X}$ is a set of such rationality postulates, and let A be a set of inductive arguments. Clearly, by means of $\mathbf{X}$ we could derive additional inductive arguments A', the significance of which is: if an inductive agent accepts arguments A, and it behaves rationally according to $\mathbf{X}$, it should also accept arguments A'. For instance, if $\mathbf{X}$ includes the rule of Verification, A contains the argument

$$\texttt{chad\_is\_black} \mathrel{|\!\!<} \texttt{crows\_are\_black}$$

and the background knowledge includes

$$\models \texttt{crows\_are\_black} \rightarrow \texttt{chevy\_is\_black}$$

then A' contains the additional inductive argument

$$\texttt{chad\_is\_black} \wedge \texttt{chevy\_is\_black} \mathrel{|\!\!<} \texttt{crows\_are\_black}$$

implying that if the agent would not accept the latter, it would behave irrationally (w.r.t. $\mathbf{X}$).

One may now ask: what is the smallest set of arguments containing A and satisfying the rules of $\mathbf{X}$? Such a set, if it exists, would represent the closure of A under $\mathbf{X}$, denoted $A^{\mathbf{X}}$. The significance of such a closure is that if two agents

[11] Readers with a background in inductive learning may interpret $\alpha \mathrel{|\!\!<} \alpha$ as 'hypothesis α does not cover any negative example'.

start from the same set of arguments A, they cannot possibly disagree about any other argument if they both act rationally according to **X**. Clearly, the existence of such a closure operation depends on the rules in **X**. As has been remarked before, if all the rules in **X** are definite (having only positive literals in their antecedents and consequent) the closure of A under **X** is unique for arbitrary A. All the rule systems in [Kraus *et al.*, 1990] consist solely of definite rules.

However, the situation changes drastically if not all rules are definite. In particular, rule systems containing indefinite rules (having at least one negative literal in their antecedent and a positive literal as consequent) will not have an associated closure operation. An indefinite rule represents a rationality principle of the following kind: 'if you accept this argument, you should accept at least one of those'. One example is Rational Monotonicity as studied in [Lehmann and Magidor, 1992]; other examples will be found in this chapter. The upshot of such rules is that even if two agents agree on the initial set of arguments they accept, they can disagree about some other arguments without violating the rationality postulates. Put differently, if we have two consequence relations both satisfying the rules of **X**, and we take their intersection to find out on what arguments they agree, this intersection itself may not satisfy the rules of **X**.

Lehmann and Magidor are not content with this indefiniteness: they define an operation of *rational closure* which selects, among the many supersets of A that satisfy **X**, one that has certain desirable properties [Lehmann and Magidor, 1992, p. 33]. This seems to be motivated by the tacit assumption that *rationality postulates* **X** *should lead to a closure operation* $A \rightarrow A^{\mathbf{X}}$. Such a closure operation can be seen as a consequence operation in the sense of Tarski [1936], mapping any set of premisses to the set of its consequences under some inference system. However, the notion of inference represented by such closure operations should be clearly distinguished from the notion of inference represented by consequence relations – if the latter operate on a meta-level, closure operations establish a *meta-meta-level*.

In this chapter we will not be concerned with inference on this meta-meta-level. This choice is motivated on methodological rather than technical grounds: I don't believe that rationality postulates for induction necessarily establish an unequivocal closure operation. If we intersect two inductive consequence relations to see what two inductive agents agree upon, the resulting set of arguments may, as a consequence relation, not satisfy some rationality postulate, but this is, it seems to me, just to be expected.

A further criticism of the meta-meta-level view is that it obscures the status of the consequence relation symbol. The relevant question on the meta-meta-level is: what expressions of the form $\alpha{\mid\!\sim}\beta$ are entailed by a set of expressions of the same form? In other words: on the meta-meta-level ${\mid\!\sim}$ acts as a *connective*.[12] This is also apparent from Kraus, Lehmann and Magidor's terminology: they call $\alpha{\mid\!\sim}\beta$ a *conditional assertion*, a set of such conditional assertions is called a *conditional*

[12]This raises the question why conditional assertions cannot be nested, as in $(\alpha{\mid\!\sim}\beta){\mid\!\sim}\gamma$. Note that the answer to this question is perfectly clear on the meta-level, since this expression makes as little sense as, say, $(\alpha \models \beta) \models \gamma$.

knowledge base, and they ask themselves the question: *what does a conditional knowledge base entail*?, i.e. what other conditional assertions can be deduced from it? It seems to me that the meta-meta-level perspective is at odds with the meta-level perspective – indeed, it is my conjecture that the theory of entailment of conditional assertions can be developed without reference to an intermediate level of consequence relations.

I will now proceed with a technical analysis of the process of forming an explanatory hypothesis from evidence *à la* Peirce (Section 4) and the process of forming a confirmed hypothesis *à la* Hempel (Section 5).

4 EXPLANATORY INDUCTION

In this section we will study abstract properties and semantics for explanatory consequence relations. Throughout the section $\alpha \mathrel{|\!\!<} \beta$ is to be read as 'evidence α is explained by hypothesis β' or 'hypothesis β is a possible explanation of evidence α'. What counts as a possible explanation will initially be left unspecified – the framework of consequence relations allows us to formulate, in Section 4.1, abstract properties of hypothesis generation, without fixing a particular material definition. In Section 4.2 we will then single out a particular set of properties (the system **EM**) and characterise it semantically by means of strong explanatory structures.

4.1 Properties of explanatory induction

A natural requirement for explanatory induction is that every consistent hypothesis that entails the evidence counts as a possible explanation. As explained above the condition that hypothesis β be consistent is expressed by $\beta \mathrel{|\!\!<} \beta$, which gives us the following rule:

Admissible Converse Entailment $$\frac{\models \beta \rightarrow \alpha \quad , \quad \beta \mathrel{|\!\!<} \beta}{\alpha \mathrel{|\!\!<} \beta}$$

Another requirement for explanations has been discussed above as (H5′): possible explanations may be logically strengthened, as long as they remain consistent. This is expressed as follows:

Admissible Right Strengthening $$\frac{\alpha \mathrel{|\!\!<} \beta \quad , \quad \models \gamma \rightarrow \beta \quad , \quad \gamma \mathrel{|\!\!<} \gamma}{\alpha \mathrel{|\!\!<} \gamma}$$

We may note that Admissible Converse Entailment can be derived from Admissible Right Strengthening if we assume Consistency and the following rule:

Explanatory Reflexivity $$\frac{\alpha \mathrel{|\!\!<} \alpha \quad , \quad \neg\beta \not\mathrel{|\!\!<} \alpha}{\beta \mathrel{|\!\!<} \beta}$$

This rule represents a weakening of reflexivity especially tailored for explanatory induction. It is best understood by rewriting it into its contrapositive: from $\alpha \mid\!< \alpha$ and $\beta \not\mid\!< \beta$ infer $\neg\beta \mid\!< \alpha$, which states that if β is inadmissible, i.e. too strong a statement with regard to the background knowledge, its negation $\neg\beta$ is so weak that it is explained by arbitrary admissible hypotheses α.

LEMMA 2. *In the presence of Consistency and Explanatory Reflexivity, Admissible Right Strengthening implies Admissible Converse Entailment.*

Proof. Suppose $\models \beta \rightarrow \alpha$, then by Consistency $\neg\alpha \not\mid\!< \beta$. Suppose furthermore $\beta \mid\!< \beta$, then by Explanatory Reflexivity $\alpha \mid\!< \alpha$. The desired result follows by Admissible Right Strengthening. ■

While the rules above express properties of possible explanations, the following two rules concentrate on the evidence. The underlying idea is a basic principle in inductive learning: if the evidence is a set of instances of the concept, we can partition the evidence arbitrarily and find a single hypothesis that is an explanation of each subset of instances. This principle is established by the following two rules:[13]

Incrementality
$$\frac{\alpha \mid\!< \gamma \quad , \quad \beta \mid\!< \gamma}{\alpha \wedge \beta \mid\!< \gamma}$$

Convergence
$$\frac{\models \alpha \rightarrow \beta \quad , \quad \alpha \mid\!< \gamma}{\beta \mid\!< \gamma}$$

LEMMA 3. *If $\mid\!<$ is a consequence relation satisfying Incrementality and Convergence, then $\alpha \wedge \beta \mid\!< \gamma$ iff $\alpha \mid\!< \gamma$ and $\beta \mid\!< \gamma$.*

Proof. The *if* part is Incrementality, and the *only-if* part follows from Convergence. ■

Incrementality and Convergence are of considerable importance for computational induction, since they allow for an incremental approach. Incrementality states that pieces of evidence can be dealt with in isolation. Another way to say the same thing is that the set of evidence explained by a given hypothesis is conjunctively closed. By the rule of Consistency this set is consistent, which yields the following principle:

Left Consistency
$$\frac{\alpha \mid\!< \beta}{\neg\alpha \not\mid\!< \beta}$$

LEMMA 4. *In the presence of Right Reflexivity and Admissible Converse Entailment, Left Consistency implies Consistency.*

[13] In previous work [Flach, 1995] Incrementality was called Additivity, and Convergence was called Incrementality. The terminology employed here better reflects the meaning of the rules.

Proof. Suppose $\models \beta \rightarrow \neg\alpha$. Now, either $\beta \mathrel{|\!\!<} \beta$ or $\beta \mathrel{|\!\!\not<} \beta$; in the former case, $\neg\alpha \mathrel{|\!\!<} \beta$ by Admissible Converse Entailment, and we conclude by Left Consistency. In the latter case, we have $\delta \mathrel{|\!\!\not<} \beta$ for any δ by Right Reflexivity. ∎

It follows that Left Consistency and Consistency are equivalent in the presence of Right Reflexivity, Admissible Converse Entailment, and Incrementality.

Convergence expresses a *monotonicity* property of induction, which can again best be understood by considering its contrapositive: a hypothesis that is rejected on the basis of evidence β cannot become feasible again when stronger evidence α is available. In other words: the process of rejecting a hypothesis is not defeasible (i.e. based on assumptions), but based on the evidence only. This is the analogue of the monotonicity property of deduction, which states that the process of *accepting* a consequence is non-defeasible (note that this property of deductive monotonicity can be obtained by reversing the implication in the first condition of Convergence).

LEMMA 5. *The combination of Verification and Convergence is equivalent with the following rule:*

Predictive Convergence $$\frac{\models \alpha \wedge \gamma \rightarrow \beta \quad , \quad \alpha \mathrel{|\!\!<} \gamma}{\beta \mathrel{|\!\!<} \gamma}$$

Proof. To derive Predictive Convergence, suppose $\models \alpha \wedge \gamma \rightarrow \beta$ and $\alpha \mathrel{|\!\!<} \gamma$, then by Verification $\alpha \wedge \beta \mathrel{|\!\!<} \gamma$, and by Convergence $\beta \mathrel{|\!\!<} \gamma$.

Predictive Convergence implies Convergence, since $\models \alpha \rightarrow \beta$ implies $\models \alpha \wedge \gamma \rightarrow \beta$.

Predictive Convergence implies Verification, since $\models \alpha \wedge \beta \rightarrow \gamma$ implies $\models \alpha \wedge \beta \rightarrow \alpha \wedge \gamma$. ∎

Predictive Convergence can be seen as a strengthening of Convergence, in the sense that β is not merely a weakening of evidence α, but can be any set of predictions. Note that Right Reflexivity is an instance of Predictive Convergence (put $\gamma = \beta$).

The final postulate we consider in this section expresses a principle well-known from algorithmic concept learning: if α represents the classification of an instance and β its description, then we may either induce a concept definition from examples of the form $\beta \rightarrow \alpha$, or we may add β to the background theory and induce from α alone. Since in our framework background knowledge is included implicitly, β is added to the hypothesis instead.

Conditionalisation $$\frac{\alpha \mathrel{|\!\!<} \beta \wedge \gamma}{\beta \rightarrow \alpha \mathrel{|\!\!<} \gamma}$$

After having discussed various abstract properties of generation of explanatory hypotheses we now turn to the question of characterising explanatory induction semantically.

4.2 Strong explanatory consequence relations

As we have seen, Peirce's original idea was to define explanatory hypothesis generation as reversed deduction. I will amend Peirce's proposal in two ways. First, as explained above it is required that the hypothesis be consistent with respect to the background knowledge. Secondly, I reformulate reversed deduction as inclusion of deductive consequences. The main reason for the latter is that in this way the explanatory consequence relation is defined in terms of a property that is *preserved* by arguments (*viz.* explanatory power).

DEFINITION 6. An *explanation mechanism* is some consequence relation $\mid\!\sim$.[14] Let $C_{\sim}(\alpha) = \{\gamma \mid \alpha \mid\!\sim \gamma\}$, i.e. the set of formulae explained by γ under $\mid\!\sim$. The *explanatory consequence relation* $\mid\!<$ defined by $\mid\!\sim$ is defined as $\alpha \mid\!< \beta$ iff $C_{\sim}(\alpha) \subseteq C_{\sim}(\beta) \subset L$. A *strong* explanatory consequence relation is defined by a monotonic explanation mechanism.

Thus, an explanation is required to have at least the same consequences under the explanation mechanism as the premiss it is obtained from, without becoming inconsistent. It should be noted that, in the general case, the conditions $C_{\sim}(\alpha) \subseteq C_{\sim}(\beta)$ and $\beta \mid\!\sim \alpha$ are not equivalent.[15] However, for monotonic explanation mechanisms they are, which provides us with the following 'Peircean' definition of strong explanatory consequence relations.

DEFINITION 7. A *strong explanatory structure* is a set $W \subseteq M$. The consequence relation it defines is denoted by $\mid\!<_W$ and is defined by: $\alpha \mid\!<_W \beta$ iff (i) there is a $m_0 \in W$ such that $m_0 \models \beta$, and (ii) for every $m \in W, m \models \beta \rightarrow \alpha$.

The following system of rules will be proved to axiomatise strong explanatory structures.

DEFINITION 8. The system **EM** consists of the following rules: Admissible Right Strengthening, Explanatory Reflexivity, Incrementality, Predictive Convergence, Left Consistency, and Conditionalisation.

We note the following derived rules of **EM**: Convergence, Admissible Converse Entailment and Right Reflexivity (instances of Predictive Convergence) and Consistency (Lemma 4). The following derived rule will also prove useful.

LEMMA 9. *The following rule is a derived rule of* **EM**:

Consistent Right Strengthening $$\frac{\alpha \mid\!< \gamma \quad , \quad \neg\beta \not\mid\!< \gamma}{\alpha \mid\!< \beta \wedge \gamma}$$

Proof. Suppose $\neg\beta \not\mid\!< \gamma$; since $\models \neg(\beta \wedge \gamma) \wedge \gamma \rightarrow \neg\beta$, we have $\neg(\beta \wedge \gamma) \not\mid\!< \gamma$ by Predictive Convergence. Furthermore, suppose $\alpha \mid\!< \gamma$, then by Right Reflexivity

[14] An explanation mechanism reasons from an explanans to an explanandum, and thus should not be confused with an explanatory consequence relation which reasons from explanandum to a possible explanans.

[15] $C_{\sim}(\alpha) \subseteq C_{\sim}(\beta)$ implies $\beta \mid\!\sim \alpha$ if $\mid\!\sim$ is reflexive; $\beta \mid\!\sim \alpha$ implies $C_{\sim}(\alpha) \subseteq C_{\sim}(\beta)$ if $\mid\!\sim$ is transitive.

$\gamma \mathrel{|\!\!<} \gamma$, so by Explanatory Reflexivity we have $\beta \wedge \gamma \mathrel{|\!\!<} \beta \wedge \gamma$. We conclude by Admissible Right Strengthening. ■

Soundness of **EM** is easily checked.

LEMMA 10 (Soundness of **EM**). *Any strong explanatory consequence relation satisfies the rules of* **EM**.

Proof. Let $W \subseteq U$ be a strong explanatory structure; we need to demonstrate that $\mathrel{|\!\!<}_W$, as defined in Definition 7, satisfies the rules of **EM**.

Admissible Right Strengthening: if $\gamma \mathrel{|\!\!<}_W \gamma$, then some model in W satisfies γ. Furthermore, if $m \models \beta \rightarrow \alpha$ and $\models \gamma \rightarrow \beta$, then $m \models \gamma \rightarrow \alpha$.

Explanatory Reflexivity: we have that some model in W satisfies α, while not all models in W satisfy $\alpha \rightarrow \neg\beta$, i.e. there is a model in W satisfying $\alpha \wedge \beta$ and hence β.

Incrementality: if $m \models \gamma \rightarrow \alpha$ and $m \models \gamma \rightarrow \beta$ then $m \models \gamma \rightarrow (\alpha \wedge \beta)$.

Predictive Convergence: if $\models \alpha \wedge \gamma \rightarrow \beta$ and $m \models \gamma \rightarrow \alpha$ then $m \models \gamma \rightarrow \beta$.

Left Consistency: if some model in W satisfies β while all models in W satisfy $\beta \rightarrow \alpha$, then there is a model in W not satisfying $\beta \rightarrow \neg\alpha$.

Conditionalisation: trivial. ■

In order to prove completeness we build a strong explanatory structure W from a given consequence relation $\mathrel{|\!\!<}$ satisfying the rules of **EM**, such that $\alpha \mathrel{|\!\!<} \beta$ iff $\alpha \mathrel{|\!\!<}_W \beta$. For non-empty explanatory relations the following construction is used:

$$W = \{m \in U \mid \text{ for all } \alpha, \beta \text{ such that } \alpha \mathrel{|\!\!<} \beta : m \models \beta \rightarrow \alpha\}.$$

An empty explanatory relation signals inconsistent background knowledge, and is hence defined by the empty explanatory structure.

We need a few intermediate results. The following lemma states that every strong explanatory hypothesis is satisfiable in W.

LEMMA 11. *Let* $\mathrel{|\!\!<}$ *be a consequence relation satisfying the rules of* **EM**, *and let* W *be defined as above. If* $\alpha \mathrel{|\!\!<} \beta$ *then there is a model* $m \in W$ *such that* $m \models \beta$.

Proof. Let $\alpha \mathrel{|\!\!<} \beta$; we will prove that $\{\beta\} \cup \{\delta \rightarrow \gamma \mid \gamma \mathrel{|\!\!<} \delta\}$ is satisfiable. Suppose not, then by compactness there is a finite $\Delta \subseteq \{\delta \rightarrow \gamma \mid \gamma \mathrel{|\!\!<} \delta\}$ such that $\models \beta \rightarrow \neg\Delta$. Furthermore, since $\varphi \mathrel{|\!\!<} \psi$ for any $\psi \rightarrow \varphi \in \Delta$, we have $\psi \rightarrow \varphi \mathrel{|\!\!<} \mathbf{true}$ for any $\psi \rightarrow \varphi \in \Delta$ by Conditionalisation, $\Delta \mathrel{|\!\!<} \mathbf{true}$ by Incrementality, and $\Delta \mathrel{|\!\!<} \beta$ by Right Reflexivity and Admissible Right Strengthening. But then by Consistency $\not\models \beta \rightarrow \neg\Delta$, a contradiction. ■

Furthermore, we have that every inadmissible formula is unsatisfiable in W.

LEMMA 12. *Let* $\mathrel{|\!\!<}$ *be a non-empty consequence relation satisfying the rules of* **EM**, *and let* W *be defined as above. If* $\gamma \not\mathrel{|\!\!<} \gamma$ *then* γ *is unsatisfiable in* W.

Proof. Let $\alpha \mathrel{|\!<} \beta$, then **true** $\mathrel{|\!<}$ **true** by Convergence and Left Reflexivity. Furthermore, if $\gamma \mathrel{|\!\not<} \gamma$ then $\neg\gamma \mathrel{|\!<}$ **true** by Explanatory Reflexivity, hence $m \models$ **true** $\rightarrow \neg\gamma$ for every $m \in W$. ■

I will now show that W defines a consequence relation that is included in $\mathrel{|\!<}$.

LEMMA 13. *Let $\mathrel{|\!<}$ be a non-empty consequence relation satisfying the rules of* **EM**, *and let W be defined as above. If $\alpha \mathrel{|\!<}_W \beta$ then $\alpha \mathrel{|\!<} \beta$.*

Proof. Suppose that $\alpha \mathrel{|\!\not<} \beta$, we will show that either no model in W satisfies β, or there exists a model $m_0 \in W$ that does not satisfy $\beta \rightarrow \alpha$.

First of all, if $\beta \mathrel{|\!\not<} \beta$ then β is unsatisfiable in W according to Lemma 12. In the remainder of the proof we will assume that $\beta \mathrel{|\!<} \beta$. Define $\Gamma_0 = \{\neg\alpha\} \cup \{\delta \mid \delta \mathrel{|\!<} \beta\}$; we will first show that Γ_0 is satisfiable. Suppose not, then by compactness there is a finite $\Delta \subseteq \{\delta \mid \delta \mathrel{|\!<} \beta\}$ such that $\models \Delta \rightarrow \alpha$, i.e. $\models \beta \rightarrow (\Delta \rightarrow \alpha)$; by Admissible Converse Entailment $\Delta \rightarrow \alpha \mathrel{|\!<} \beta$ (recall that $\beta \mathrel{|\!<} \beta$). But by Incrementality $\Delta \mathrel{|\!<} \beta$; using Incrementality and Convergence, we obtain $\alpha \mathrel{|\!<} \beta$. Contradiction, so Γ_0 is satisfiable.

Let $m_0 \models \Gamma_0$; clearly $m_0 \not\models \alpha$ and, since $\beta \in \Gamma_0, m_0 \models \beta$. It remains to prove that m_0 is in W; i.e., that for all φ, ψ such that $\varphi \mathrel{|\!<} \psi$ we have $m_0 \models \psi \rightarrow \varphi$. Let $\varphi \mathrel{|\!<} \psi$; if $\neg\beta \mathrel{|\!\not<} \psi$, then by Consistent Right Strengthening $\varphi \mathrel{|\!<} \psi \wedge \beta$, and by Conditionalisation $\psi \rightarrow \varphi \mathrel{|\!<} \beta$; thus $\psi \rightarrow \varphi \in \Gamma_0$ and therefore $m_0 \models \psi \rightarrow \varphi$. On the other hand, if $\neg\beta \mathrel{|\!<} \psi$ then by Conditionalisation and Convergence $\beta \rightarrow \neg\psi \mathrel{|\!<}$ **true**, by Admissible Right Strengthening $\beta \rightarrow \neg\psi \mathrel{|\!<} \beta$, and by Incrementality and Convergence $\neg\psi \mathrel{|\!<} \beta$; thus $\neg\psi \in \Gamma_0$ and therefore $m_0 \not\models \psi$, hence $m_0 \models \psi \rightarrow \varphi$. ■

Armed with the previous three lemmas we can prove the completeness of **EM**.

THEOREM 14 (Representation theorem for strong explanatory consequence relations). *A consequence relation is strong explanatory iff it satisfies the rules of* **EM**.

Proof. The only-if part is Lemma 10. For the if part, let $\mathrel{|\!<}$ be an arbitrary non-empty consequence relation satisfying the rules of **EM**, and let

$$W = \{m \in U \mid \text{ for all } \alpha, \beta \text{ such that } \alpha \mathrel{|\!<} \beta : m \models \beta \rightarrow \alpha\}.$$

Suppose $\alpha \mathrel{|\!<} \beta$, then by the construction of $W, m \models \beta \rightarrow \alpha$ for all $m \in W$. Furthermore, by Lemma 11 there is a model in W satisfying β. We may conclude that $\alpha \mathrel{|\!<}_W \beta$. Conversely, if $\alpha \mathrel{|\!<}_W \beta$ then Lemma 13 proves that $\alpha \mathrel{|\!<} \beta$. We conclude that W defines a consequence relation that is exactly $\mathrel{|\!<}$.

For an empty consequence relation put $w = \varnothing$. ■

In this section we have studied axioms and semantic characterisations for explanatory induction. I have proposed a novel definition of explanatory hypothesis generation in terms of preservation of explanatory power with respect to an explanation mechanism. A representation theorem has been obtained for the special case of a monotonic explanation mechanism. Characterisation of explanatory induction with respect to other (e.g. preferential) explanation mechanisms is left as an open problem.

5 CONFIRMATORY INDUCTION

We will now switch from the explanatory (classification-oriented) viewpoint to the confirmatory (non-classificatory) perspective. Throughout this section $\alpha \mathrel{|\!<} \beta$ is to be read as 'evidence α confirms hypothesis β'. Our goals will be to find reasonable properties of $\mathrel{|\!<}$ under this interpretation, and to characterise particular sets of properties by a suitable semantics.

Our starting point will be Hempel's adequacy conditions, which we express in our framework in Section 5.1. In Section 5.2 we characterise a subset of these conditions by means of simple confirmatory structures. Section 5.3 extends these with a preference ordering in the spirit of Kraus *et al.*'s preferential models. Finally, in Section 5.4 we consider weak confirmatory consequence relations which model, in effect, the relation of logical compatibility between evidence and hypothesis.

5.1 Properties of confirmatory induction

In this section I will translate Hempel's set of adequacy conditions for confirmation (Section 2.2) into rules for confirmatory consequence relations. The conditions will be slightly modified, in order to keep the treatment of inconsistent evidence and hypothesis in line with the explanatory case: inconsistent evidence does not confirm any hypothesis, and inconsistent hypotheses are not confirmed by any evidence.

Entailment condition (H1) is translated into two rules:

Admissible Entailment $$\frac{\models \alpha \to \beta \quad , \quad \alpha \mathrel{|\!<} \alpha}{\alpha \mathrel{|\!<} \beta}$$

Confirmatory Reflexivity $$\frac{\alpha \mathrel{|\!<} \alpha \quad , \quad \alpha \not\mathrel{|\!<} \neg\beta}{\beta \mathrel{|\!<} \beta}$$

Admissible Entailment expresses that admissible evidence (i.e. evidence that is consistent with the background knowledge) confirms any of its consequences. In other words, consistent entailment is a special case of confirmation. Confirmatory Reflexivity is the confirmatory counterpart of Explanatory Reflexivity encountered in the previous section. It is added as a separate rule since, in its original formulation, (H1) includes reflexivity as a special case. As with its explanatory counterpart, Confirmatory Reflexivity is best understood when considering its contrapositive: if β is inadmissible, i.e. too strong a statement with regard to the background knowledge, its negation $\neg\beta$ is so weak that it is confirmed by arbitrary admissible formulae α.

Consequence condition (H2) cannot be translated directly, since in the language of consequence relations as defined here we have no means to refer to a set of confirmed sentences. However, a translation of the special consequence condition (H2.1) and the conjunction condition (H2.3) will suffice:

Right Weakening $$\frac{\models \beta \rightarrow \gamma \quad , \quad \alpha \mathrel{|\!<} \beta}{\alpha \mathrel{|\!<} \gamma}$$

Right And $$\frac{\alpha \mathrel{|\!<} \beta \quad , \quad \alpha \mathrel{|\!<} \gamma}{\alpha \mathrel{|\!<} \beta \wedge \gamma}$$

LEMMA 15. *If* $\mathrel{|\!<}$ *is a consequence relation satisfying Right And and Right Weakening, then* $\alpha \mathrel{|\!<} \beta \wedge \gamma$ *iff* $\alpha \mathrel{|\!<} \beta$ *and* $\alpha \mathrel{|\!<} \gamma$.

Proof. The if part is Right And, and the only-if part follows from Right Weakening. ■

Right Weakening expresses that any hypothesis entailed by a given hypothesis confirmed by α is also confirmed by α. Notice that Admissible Entailment is an instance of Right Weakening (put $\beta = \alpha$).

LEMMA 16. *The combination of Right Extension and Right Weakening is equivalent to the following rule:*

Predictive Right Weakening $$\frac{\models \alpha \wedge \beta \rightarrow \gamma \quad , \quad \alpha \mathrel{|\!<} \beta}{\alpha \mathrel{|\!<} \gamma}$$

Proof. In order to derive Predictive Right Weakening, suppose $\models \alpha \wedge \beta \rightarrow \gamma$ and $\alpha \mathrel{|\!<} \beta$, then by Right Extension $\alpha \mathrel{|\!<} \beta \wedge \gamma$, and the result follows by Right Weakening.

Predictive Right Weakening implies Right Weakening, since $\models \beta \rightarrow \gamma$ implies $\models \alpha \wedge \beta \rightarrow \gamma$.

Predictive Right Weakening implies Right Extension, since $\models \alpha \wedge \beta \rightarrow \gamma$ implies $\models \alpha \wedge \beta \rightarrow \beta \wedge \gamma$. ■

In words, Predictive Right Weakening expresses that given a confirmatory argument, any predicted formula is confirmed by the same evidence. Notice that by putting $\gamma = \alpha$ in Predictive Right Weakening we obtain Left Reflexivity.

Right And states that the set of all confirmed hypotheses (interpreted as a conjunction) is itself confirmed. The combination of Right And and Right Weakening implies Hempel's general consequence condition (H2): if E confirms every formula of a set K, then it also confirms the conjunction of the formulae in K (by Right And), and therefore also every consequence of this conjunction (by Right

Weakening)[16]. It has already been remarked that Right And is probably too strong in the general case, if we have inconclusive evidence that is unable to choose between incompatible hypotheses. In this respect it is perhaps appropriate to point at a certain similarity between Right And and Right Extension: the latter rule requires γ to be predicted rather than being confirmed by α.

Like the general consequence condition (H2), general consistency condition (H3) cannot be translated directly into a rule, since we have no means to refer to the set of confirmed formulae. However, in the light of Right And the conjunction of the formulae in this set is itself confirmed, and therefore it is sufficient to formulate a rule expressing the special consistency condition (H3.1), which is the rule of Consistency previously encountered:

Consistency $$\frac{\alpha \mathrel{|\!\!<} \beta}{\not\models \beta \rightarrow \neg\alpha}$$

Condition (H3.2) expresses that for any formula β, if β is in the set of confirmed hypotheses then $\neg\beta$ is not. This principle is expressed by the following rule:

Right Consistency $$\frac{\alpha \mathrel{|\!\!<} \beta}{\alpha \mathrel{|\!\!\not<} \neg\beta}$$

LEMMA 17. *In the presence of Admissible Entailment and Left Reflexivity, Right Consistency implies Consistency.*

Proof. Suppose $\models \beta \rightarrow \neg\alpha$, i.e. $\models \alpha \rightarrow \neg\beta$. Now, either we have $\alpha \mathrel{|\!\!<} \alpha$, or else $\alpha \mathrel{|\!\!\not<} \alpha$. In the former case, $\alpha \mathrel{|\!\!<} \neg\beta$ by Admissible Entailment, and we conclude by Right Consistency. In the latter case, we have $\alpha \mathrel{|\!\!\not<} \delta$ for any δ by Left Reflexivity. ■

Clearly, Consistency implies Right Consistency in the presence of Right And. As a corollary to Lemma 17, we have that Right Consistency and Consistency are equivalent in the presence of Left Reflexivity, Admissible Entailment, and Right And.

Finally, the equivalence condition for observations (H4) is translated into

Left Logical Equivalence $$\frac{\models \alpha \leftrightarrow \beta \quad , \quad \alpha \mathrel{|\!\!<} \gamma}{\beta \mathrel{|\!\!<} \gamma}$$

We now turn to the question of devising a meaningful semantics for Hempel's conditions as re-expressed in our framework of inductive consequence relations.

[16]This holds only for finite K, an assumption that I will make throughout.

5.2 *Simple confirmatory structures*

It has been suggested in Section 2.4 that a semantics for confirmatory reasoning be expressed in terms of satisfaction by an appropriately constructed model or set of models. More precisely, a confirmatory semantics is conceived as one in which certain *regular* models are constructed from the premisses, such that a hypothesis is confirmed if it is true in all those regular models. Since this requires some completeness assumptions regarding the evidence we may call this a *closed* confirmatory semantics. In Section 5.4 I will consider a variant which relaxes the assumptions regarding the evidence (*open* confirmatory semantics).

DEFINITION 18. A *confirmatory structure* is a triple $W = \langle S, [\cdot], [\![\cdot]\!]\rangle$, where S is a set of *semantic objects*, and $[\cdot] : L \rightarrow 2^S$ and $[\![\cdot]\!] : L \rightarrow 2^S$ are functions mapping formulae to sets of semantic objects. The *closed* confirmatory consequence relation defined by W is given by: $\alpha \mathrel{|\!\!<}_W \beta$ iff (i) $[\alpha] \neq \varnothing$, and (ii) $[\alpha] \subseteq [\![\beta]\!]$.

Intuitively, $[\alpha]$ denotes the set of regular models constructed from premisses α, each of which should satisfy hypothesis β. A similar semantics has been considered by Bell [1991], who calls it a pragmatic model. There are however two differences between Bell's approach and mine. First, in order to rule out inconsistent premisses I have added condition (i) in the definition of $\mathrel{|\!\!<}_W$. Furthermore, I allow the possibility that some of the regular models may not satisfy the premisses (i.e. $[\alpha] \not\subseteq [\![\alpha]\!]$). The rationale of this possibility is further discussed in Section 6. However, the characterisation of such anti-reflexive logics of confirmation is left as an open problem, and the results obtained below are based on the assumption that $[\alpha] \subseteq [\![\alpha]\!]$ for all $\alpha \in L$. These results differ from those reported in [Kraus *et al.*, 1990; Bell, 1991] because of the additional condition (i), as will be detailed in the present section and the next.

Bell proves [1991, Theorem 4.8] that his pragmatic models, including the condition $[\alpha] \subseteq [\![\alpha]\!]$, are axiomatised by the rule system **B** consisting of Reflexivity, Right Weakening, and Right And, if one additionally assumes that $[\![\cdot]\!]$ is well-behaved with respect to the logical connectives and logical entailment:

$$[\![\alpha \wedge \beta]\!] = [\![\alpha]\!] \cap [\![\beta]\!]$$
$$[\![\neg\alpha]\!] = S - [\![\alpha]\!]$$
$$[\![\alpha]\!] = S \text{ iff } \models \alpha$$

Let us call confirmatory structures which satisfy these conditions, as well as $[\alpha] \subseteq [\![\alpha]\!]$ for all $\alpha \in L$, *simple* confirmatory structures, defining simple (closed) confirmatory consequence relations. I will now demonstrate that simple confirmatory structures are axiomatised by the following rule system.

DEFINITION 19. The system **CS** consists of the following rules: Predictive Right Weakening, Right And, and Right Consistency.

Derived rules of **CS** include Right Weakening and Right Extension (Lemma 16), Left Reflexivity (an instance of Predictive Right Weakening), Admissible Entailment (an instance of Right Weakening), and Consistency (Lemma 17).

The soundness of the rules of **CS** is easily proved.

LEMMA 20 (Soundness of **CS**). *Any simple closed confirmatory consequence relation satisfies the rules of* **CS**.

Proof. For Predictive Right Weakening, first of all we have $[\alpha] \subseteq [\![\alpha]\!]$ for all $\alpha \in L$, hence $[\alpha] \cap [\![\beta]\!] \subseteq [\![\alpha]\!] \cap [\![\beta]\!]$ for all $\alpha, \beta \in L$. Now suppose $\models \alpha \land \beta \to \gamma$, then $[\![\alpha]\!] \cap [\![\beta]\!] \subseteq [\![\gamma]\!]$, hence $[\alpha] \cap [\![\beta]\!] \subseteq [\![\gamma]\!]$. Furthermore suppose $\alpha \mathrel{|\!\!<}_W \beta$, then $\varnothing \subset [\alpha] \subseteq [\![\beta]\!]$, hence $[\alpha] \cap [\![\beta]\!] = [\alpha] \subseteq [\![\gamma]\!]$, and we conclude $\alpha \mathrel{|\!\!<}_W \gamma$.

For Right And, if $[\alpha] \subseteq [\![\beta]\!]$ and $[\alpha] \subseteq [\![\gamma]\!]$ then $[\alpha] \subseteq [\![\beta]\!] \cap [\![\gamma]\!] = [\![\beta \land \gamma]\!]$.

For Right Consistency, if $\varnothing \subset [\alpha] \subseteq [\![\beta]\!]$ then $[\alpha] \cap [\![\beta]\!] \neq \varnothing$, hence $[\alpha] \not\subseteq [\![\neg\beta]\!]$. ∎

In order to prove completeness we need to build, for a given consequence relation, a simple confirmatory structure W that defines exactly that relation. The key concept is the following.

DEFINITION 21. Let $\mathrel{|\!\!<}$ be a confirmatory consequence relation. The model $m \in U$ is said to be *normal for* α iff for all β in L such that $\alpha \mathrel{|\!\!<} \beta, m \models \beta$.

For admissible formulae, normal models will play the role of the regular models in the simple confirmatory structure we are building (remember that, given a consequence relation $\mathrel{|\!\!<}$, a formula $\alpha \in L$ is admissible iff $\alpha \mathrel{|\!\!<} \alpha$).

LEMMA 22. *Suppose a consequence relation $\mathrel{|\!\!<}$ satisfies Right Weakening and Right And, and let α be an admissible formula. All normal models for a satisfy β iff $\alpha \mathrel{|\!\!<} \beta$.*

Proof. The if part follows from Definition 21.

For the only-if part, suppose $\alpha \mathrel{|\!\!<} \alpha$ and $\alpha \not\mathrel{|\!\!<} \beta$; I will show that there is a normal model for α that does not satisfy β. Let $\Gamma_0 = \{\neg\beta\} \cup \{\delta \mid \alpha \mathrel{|\!\!<} \delta\}$; it suffices to show that Γ_0 is satisfiable. Suppose not, then by compactness there is a finite $\Delta \subseteq \{\delta \mid \alpha \mathrel{|\!\!<} \delta\}$ such that $\models \Delta \to \beta$, i.e. $\models \alpha \to (\Delta \to \beta)$; by Right Weakening $\alpha \mathrel{|\!\!<} \Delta \to \beta$. But by Right And $\alpha \mathrel{|\!\!<} \Delta$; using Right And and Right Weakening we obtain $\alpha \mathrel{|\!\!<} \beta$, a contradiction. ∎

In the standard treatment of inconsistent premisses they have all formulae as consequences, hence no normal models – since in our treatment inconsistent premisses confirm no hypothesis and thus have all models in U as normal models, we have to treat them as a separate case. Given a consequence relation $\mathrel{|\!\!<}$, let $W = \langle U, [\cdot], [\![\cdot]\!] \rangle$ be defined as follows:

1. U is the set of models of L under consideration;

2. $[\alpha] = \{m \in U \mid m \text{ is a normal model for } \alpha\}$ if α is admissible, and $\varnothing$ otherwise;

3. $[\![\alpha]\!] = \{m \in U \mid m \models \alpha\}$.

The following completeness result demonstrates that the consequence relation defined by W coincides with the original one if the latter satisfies the rules of **CS**.

THEOREM 23 (Representation theorem for simple confirmatory consequence relations). *A consequence relation is simple confirmatory iff it satisfies the rules of* **CS**.

Proof. The only-if part is Lemma 20. For the if part, let $\mathrel{|\!\!<}$ be a consequence relation satisfying the rules of **CS** and let W be defined as above. We will prove that $\alpha \mathrel{|\!\!<} \beta$ iff $\alpha \mathrel{|\!\!<}_W \beta$, i.e. $\mathrel{|\!\!<}$ is simple confirmatory.

First suppose that $\alpha \mathrel{|\!\!<} \beta$, then by Left Reflexivity $\alpha \mathrel{|\!\!<} \alpha$ and by Right Consistency $\alpha \not\mathrel{|\!\!<} \neg\alpha$, so the proof of Lemma 22 constructs a model m normal for α, hence $[\alpha] \neq \varnothing$. Furthermore any normal model for α satisfies β by Definition 21, hence $[\alpha] \subseteq [\![\beta]\!]$. We conclude that $\alpha \mathrel{|\!\!<}_W \beta$. Now suppose that $\alpha \mathrel{|\!\!<}_W \beta$, then $[\alpha] \neq \varnothing$ hence α is admissible by the definition of W. Furthermore we have $[\alpha] \subseteq [\![\beta]\!]$, i.e. every model normal for α satisfies β, and the conclusion follows by Lemma 22. ■

One may note that two of the rules obtained in the previous section have not been mentioned above, *viz.* Confirmatory Reflexivity and Left Logical Equivalence. Each of these rules poses additional restrictions on simple confirmatory structures:

(Left Logical Equivalence)	If $\models \alpha \leftrightarrow \beta$, then $[\alpha] = [\beta]$.
(Confirmatory Reflexivity)	If $[\![\beta]\!] \neq \varnothing$, then $[\beta] \neq \varnothing$.

Additional properties may be obtained by being more explicit about the construction of regular models. The ILP methods referred to in Section 2.4 suggest to take the truth-minimal Herbrand model(s) of the evidence as the regular model(s). In the analysis of nonmonotonic reasoning it is customary to abstract this into a preference ordering on the set of models, such that the regular models are the minimal ones under this ordering. We will work this out in the next section.

5.3 *Preferential confirmatory consequence relations*

The main result of this section concerns an adaptation of Kraus *et al.*'s preferential semantics, recast as a confirmatory semantics. In other words, the regular semantic objects are those that are minimal with respect to a fixed preference ordering.

DEFINITION 24. A *preferential structure* is a triple $W = \langle S, l, < \rangle$ where S is a set of *states*, $l : S \rightarrow U$ is a function that labels every state with a model, and $<$ is a strict partial order[17] on S, called the *preference ordering*, that is smooth.[18]

[17] I.e. $<$ is irreflexive and transitive.

[18] I.e. for any $S' \subseteq S$ and for any $s \in S'$, or there is a $t \in S'$ such that $t < s$ and t is minimal in S'. This condition is satisfied if $<$ does not allow infinite descending chains.

W defines a *preferential confirmatory* structure $\langle S, [\cdot], [\![\cdot]\!]\rangle$ and a preferential confirmatory consequence relation as follows: $[\![\alpha]\!] = \{s \in S \mid l(s) \models \alpha\}$, and $[\alpha] = \{s \in [\![\alpha]\!] \mid \forall s' \in S : s' < s \rightarrow s' \notin [\![\alpha]\!]\}$.

Note that preferential confirmatory structures are simple confirmatory structures, and that they also satisfy the conditions associated above with Left Logical Equivalence and Confirmatory Reflexivity (by the smoothness condition, if $s \in [\![\alpha]\!]$ then either $s \in [\alpha]$ or there is a $t < s$ such that $t \in [\alpha]$ – hence $[\![\alpha]\!] \neq \varnothing$ implies $[\alpha] \neq \varnothing$).

In comparison with the preferential semantics of [Kraus *et al.*, 1990], the only difference is that in a preferential confirmatory argument the evidence is required to be satisfiable, in order to guarantee the validity of Consistency. The intermediate semantic level of *states* is mainly needed for technical reasons, and can be identified with the set of models the reasoning agent considers possible in that epistemic state.

The following set of rules will be proved to axiomatise preferential confirmatory consequence relations.

DEFINITION 25. The system **CP** consists of the rules of **CS**, Confirmatory Reflexivity, Left Logical Equivalence, plus the following rules:

Left Or
$$\frac{\alpha \mid\!< \gamma \quad , \quad \beta \mid\!< \gamma}{\alpha \vee \beta \mid\!< \gamma}$$

Strong Verification
$$\frac{\alpha \mid\!< \gamma \quad , \quad \alpha \mid\!< \beta}{\alpha \wedge \gamma \mid\!< \beta}$$

The first rule can be seen as a variant of Convergence, which is clearly invalid in the general case: if we weaken the evidence, there will presumably come a point where the evidence no longer confirms the hypothesis. However, Left Or states that pieces of confirming evidence for a hypothesis can be weakened by taking their disjunction. The second rule is a variant of Verification, which states that a predicted formula γ can be added to confirming evidence α for hypothesis β. Strong Verification states that this is also allowed when γ is confirmed by α. The way in which Strong Verification strengthens Verification is very similar to the way Right And strengthens Right Extension. The underlying intuition is that the evidence is strong enough to have all confirmations 'point in the same direction', as it were.[19]

LEMMA 26 (Soundness of **CP**). *Any preferential confirmatory consequence relation satisfies the rules of* **CP**.

Proof. The proof only needs to be carried out for the two new rules.

[19] In the context of nonmonotonic reasoning, Strong Verification is known as Cautious Monotonicity. For the purposes of this paper I prefer to use the first name, which expresses more clearly the underlying intuition in the present context.

For Left Or, note that $[\![\alpha \vee \beta]\!] = [\![\alpha]\!] \cup [\![\beta]\!]$; thus , if $[\![\alpha]\!]$ and $[\![\beta]\!]$ are non-empty then so is $[\![\alpha \vee \beta]\!]$, hence $[\alpha \vee \beta] \neq \varnothing$. Furthermore, $[\alpha \vee \beta] \subseteq [\alpha] \cup [\beta] \subseteq [\![\alpha \vee \beta]\!]$, since a state cannot be minimal in $[\![\alpha \vee \beta]\!]$ without being minimal in at least one of $[\![\alpha]\!]$ and $[\![\beta]\!]$.

For Strong Verification, suppose that $[\alpha] \neq \varnothing, [\alpha] \subseteq [\![\gamma]\!]$ and $[\alpha] \subseteq [\![\beta]\!]$ – clearly, $[\![\alpha \wedge \gamma]\!]$ is non-empty, hence $[\alpha \wedge \gamma] \neq \varnothing$. Now, let $s \in [\alpha \wedge \gamma]$, then $s \in [\![\alpha]\!]$; I will prove that $s \in [\alpha]$. Suppose not, then there is a $t \in [\![\alpha]\!]$ such that $t < s$ and $t \in [\alpha] \subseteq [\![\gamma]\!]$, hence $t \in [\![\alpha \wedge \gamma]\!]$. But this contradicts the minimality of s in $[\![\alpha \wedge \gamma]\!]$, hence $s \in [\alpha]$ and thus $s \in [\![\beta]\!]$. ■

In order to prove completeness, we need to build a preferential structure W from a given consequence relation $\mid\!<$ satisfying the rules of **CP**, such that $\alpha \mid\!< \beta$ iff $\mid\!<_W \beta$. This structure $W = \langle S, l, < \rangle$ is constructed as follows:

1. $S = \{\langle m, \alpha \rangle \mid \alpha$ is an admissible formula, and m is a normal model for $\alpha\}$;
2. $l(\langle m, \alpha \rangle) = m$;
3. $\langle m, \alpha \rangle < \langle n, \beta \rangle$ iff $\alpha \vee \beta \mid\!< \alpha$ and $m \not\models \beta$.

Thus, states are pairs of admissible formulae and normal models. The labelling function simply maps a state to the model it contains. Condition (3) defines the preference ordering between states: note that $\beta \mid\!< \alpha$ is a special case of $\alpha \vee \beta \mid\!< \alpha$ by means of Left Or, and the fact that α is admissible. The condition $m \not\models \beta$ is added to make the ordering irreflexive; note that as a consequence any $\langle m, \alpha \rangle \in S$ is minimal in $[\![\alpha]\!]$.

The main difference between the preferential consequence relations of Kraus *et al.* and my preferential confirmatory consequence relations is the way unsatisfiable formulae are treated. In Kraus *et al.*'s system **P** unsatisfiable formulae are characterised by the fact that they have every formula in L as a plausible consequence, which means that they don't have normal models. In my framework, unsatisfiable formulae confirm no hypotheses, and have all models in U as normal models. In both cases, the structure W that is used to prove completeness contains only satisfiable formulae in its states. This means that we can replicate most of Kraus *et al.*'s results about the structure W.

PROPOSITION 27.

1. *[Kraus et al., 1990, Lemma 5.13] The relation $<$ is a strict partial order.*
2. *[Kraus et al., 1990, Lemma 5.15] The relation $<$ is smooth: for any $s \in [\![\alpha]\!]$, either s is minimal in $[\![\alpha]\!]$ or there exists a state $t < s$ minimal in $[\![\alpha]\!]$.*
3. *[Kraus et al., 1990, Lemma 5.11] If $\alpha \vee \beta \mid\!< \alpha$ and m is a normal model for α that satisfies β, then m is a normal model for β.*
4. *[Kraus et al., 1990, Lemma 5.14] $\langle m, \alpha \rangle \in [\![\beta]\!]$ iff $m \models \beta$ and $\alpha \vee \beta \mid\!< \alpha$.*

The first two statements express that W is a preferential structure. The remaining two are used in the proof of the following lemma.

LEMMA 28. *Let* $\mathrel{|\!<}$ *be a consequence relation satisfying the rules of* **CP**, *and let* W *be defined as above. If* $\alpha \mathrel{|\!<} \beta$ *then* $\alpha \mathrel{|\!<}_W \beta$.

Proof. First suppose that $\alpha \mathrel{|\!<} \beta$, then by Left Reflexivity $\alpha \mathrel{|\!<} \alpha$ and by Right Consistency $\alpha \not\mathrel{|\!<} \neg\alpha$, so the proof of Lemma 22 constructs a model m normal for α, hence $[\alpha] \neq \varnothing$. Furthermore suppose $s = \langle n, \gamma \rangle \in [\alpha]$, then γ is an admissible formula, n is a normal model for γ that satisfies α, and $\gamma \vee \alpha \mathrel{|\!<} \gamma$ by Proposition 27 (4). By Proposition 27 (3) n is a normal model for α and thus satisfies β by Definition 21, hence $s \in [\![\beta]\!]$. We conclude that $\varnothing \subset [\alpha] \subseteq [\![\beta]\!]$, hence $\alpha \mathrel{|\!<}_W \beta$. ■

The following lemma proves the converse of Lemma 28, and completes the proof of the representation theorem.

LEMMA 29. *Let* $\mathrel{|\!<}$ *be a consequence relation satisfying the rules of* **CP**, *and let* W *be defined as above. If* $\alpha \mathrel{|\!<}_W \beta$ *then* $\alpha \mathrel{|\!<} \beta$.

Proof. Suppose $\alpha \mathrel{|\!<}_W \beta$, i.e. $\varnothing \subset [\alpha] \subseteq [\![\beta]\!]$; we will first prove that α is admissible. Let $\langle n, \gamma \rangle \in [\alpha]$, then $\gamma \mathrel{|\!<} \gamma$ and n is a normal model for γ. If $\alpha \not\mathrel{|\!<} \alpha$, then by Confirmatory Reflexivity $\gamma \mathrel{|\!<} \neg\alpha$ and thus $n \models \neg\alpha$, contradicting the assumption that $\langle n, \gamma \rangle \in [\alpha]$ – so α is admissible . Furthermore, given any model m normal for α, $\langle m, \alpha \rangle \in [\alpha] \subseteq [\![\beta]\!]$, hence m satisfies β, and the conclusion follows by Lemma 22. ■

We may now summarise.

THEOREM 30 (Representation theorem for preferential confirmatory consequence relations). *A consequence relation is preferential confirmatory iff it satisfies the rules of* **CP**.

Proof. The only-if part is Lemma 26. For the if part, let $\mathrel{|\!<}$ be a consequence relation satisfying the rules of **CP** and let W be defined as above. Lemma 28 and Lemma 29 prove that $\alpha \mathrel{|\!<} \beta$ iff $\alpha \mathrel{|\!<}_W \beta$, i.e. $\mathrel{|\!<}$ is preferential confirmatory. ■

We end this section by noting that **CP** represents a more restrictive form of reasoning than Kraus *et al.*'s system **P**, even though neither system entails the other (Reflexivity, a rule of **P**, is invalid in **CP**, while Consistency, a rule of **CP**, is invalid in **P**). This is so because from every preferential consequence relation $\mid\!\sim$ we can construct a preferential confirmatory consequence relation by removing all arguments $\{\alpha \mid\!\sim \beta \mid \alpha, \beta \in L \text{ and } \alpha \mid\!\sim \neg\alpha\}$, i.e. all arguments with a left-hand side that is inconsistent with respect to the background knowledge. For further details the reader is referred to [Flach, 1998].

5.4 *Weak confirmatory consequence relations*

Any semantics that is to obey rules like Right And and Strong Verification must be based on completeness assumptions with regard to the evidence. On the other hand, such strong assumptions cannot be made for all induction tasks. It seems reasonable, then, to investigate also an alternative approach, in which a confirmed hypothesis is required to be true in *some* of the regular models.[20] It is not difficult to see that such an alternative semantics, based on some notion of consistency, would invalidate both Right And and Strong Verification. On the other hand, by not making completeness assumptions one may again have the desirable property of Convergence.

DEFINITION 31. Let $W = \langle S, [\cdot], [\![\cdot]\!]\rangle$ be a confirmatory structure. The *open* confirmatory consequence relation defined by W is given by: $\alpha \mathrel{|\!\!<}_W \beta$ iff $[\alpha] \cap [\![\beta]\!] \neq \varnothing$.

We will characterise an extreme form of open confirmatory relations, which arises when $[\cdot]$ is identified with $[\![\cdot]\!]$, which in turn is well-behaved with respect to the connectives and entailment.

DEFINITION 32. A *classical confirmatory structure* is a simple confirmatory structure $\langle S, [\![\cdot]\!], [\![\cdot]\!]\rangle$. A consequence relation is called *weak confirmatory* iff it is the open consequence relation defined by a classical confirmatory structure.

From this definition it is clear that weak confirmatory consequence relations satisfy both Right Weakening and Left Weakening (i.e. Convergence), as well as Consistency. One additional rule is needed.

DEFINITION 33. The system **CW** consists of the following rules: Predictive Convergence, Predictive Right Weakening, Consistency, and

Disjunctive Rationality $$\frac{\alpha \vee \beta \mathrel{|\!\!<} \gamma \quad , \quad \beta \mathrel{|\!\!\not<} \gamma}{\alpha \mathrel{|\!\!<} \gamma}$$

Disjunctive Rationality has not been considered in this chapter before. The name has been borrowed from Kraus *et al.*, who identify it as a valid principle of plausible reasoning. In the context of confirmatory reasoning, Disjunctive Rationality is a rather strong rule, which states that if a hypothesis is confirmed by disjunctive observations it is confirmed by at least one of the disjuncts. The following theorem proves the equivalence of weak confirmatory structures and the system **CW**.

THEOREM 34 (Representation theorem for weak confirmatory consequence relations). *A consequence relation* $\mathrel{|\!\!<}$ *is weak confirmatory iff it satisfies the rules of* **CW**.

Proof. The only-if part involves demonstrating that $\mathrel{|\!\!<}_W$, as defined in Definition 32, satisfies the rules of **CW**, which is trivial.

[20]This is sometimes called *credulous* inference, in contrast with *skeptical* inference which requires truth in all regular models.

For the if part, let $\mathrel{|\!\!<}$ be an arbitrary consequence relation satisfying the rules of **CW**, and consider the weak confirmatory structure $W = \langle S, [\![\cdot]\!], [\![\cdot]\!]\rangle$ with:

$$S = \{m \in U \mid \text{ for all } \alpha, \beta \in L \text{ such that } m \models \alpha \wedge \beta : \alpha \mathrel{|\!\!<} \beta\}$$

$$[\![\alpha]\!] = \{m \in S \mid m \models \alpha\}$$

We will prove that $\alpha \mathrel{|\!\!<} \beta$ iff $\alpha \mathrel{|\!\!<}_W \beta$. The if part follows directly from the construction of W. Suppose that $\alpha \mathrel{|\!\!<} \beta$, we will show that there exists a model $m_0 \in S$ that satisfies $\alpha \wedge \beta$.

Define $\Gamma_0 = \{\alpha\} \cup \{\delta \mid \neg\delta \not\mathrel{|\!\!<} \beta\}$; we will first show that Γ_0 is satisfiable. Suppose not, then by compactness there is a finite $\Delta \subseteq \{\delta \mid \neg\delta \not\mathrel{|\!\!<} \beta\}$ such that $\models \neg(\Delta \wedge \alpha)$, i.e. $\models \beta \rightarrow \neg(\Delta \wedge \alpha)$; by Consistency $\Delta \wedge \alpha \not\mathrel{|\!\!<} \beta$. Furthermore, since $\neg\delta \not\mathrel{|\!\!<} \beta$ for $\delta \in \Delta$, we have $\neg\Delta \not\mathrel{|\!\!<} \beta$ by Disjunctive Rationality and $\neg\Delta \wedge \alpha \not\mathrel{|\!\!<} \beta$ by Convergence. Combining this with$\Delta \wedge \alpha \not\mathrel{|\!\!<} \beta$ we obtain $(\neg\Delta \wedge \alpha) \vee (\Delta \wedge \alpha) \not\mathrel{|\!\!<} \beta$ by Disjunctive Rationality and $\alpha \not\mathrel{|\!\!<} \beta$ by Incrementality. Contradiction, so Γ_0 is satisfiable.

Let $m_0 \models \Gamma_0$; clearly $m_0 \models \alpha$ and, since by Consistency $\beta \in \Gamma_0, m_0 \models \beta$. It remains to prove that $m_0 \in S$; i.e., that for all φ, ψ such that $m_0 \models \varphi \wedge \psi$ we have $\varphi \mathrel{|\!\!<} \psi$. Let $m_0 \models \varphi \wedge \psi$, then $\neg(\varphi \wedge \psi) \notin \Gamma_0$, hence $\varphi \wedge \psi \mathrel{|\!\!<} \beta$; by Predictive Right Weakening $\varphi \wedge \psi \mathrel{|\!\!<} \psi \wedge \beta$, by Convergence $\varphi \mathrel{|\!\!<} \psi \wedge \beta$, and by Right Weakening $\varphi \mathrel{|\!\!<} \psi$. ■

The system **CW** thus provides an axiomatisation of the relation of logical compatibility.

In this section we have studied abstract properties and semantics for confirmatory induction. In the first part of this analysis I have demonstrated that Hempel's original conditions axiomatise a rather general form of confirmatory inference, characterised by truth in the regular models of the evidence. A close link with nonmonotonic or plausible reasoning is obtained by identifying the regular models with the minimal models under a preference ordering. Finally, I have proposed a more liberal form of confirmatory reasoning based on some notion of consistency, which is more appropriate if the evidence cannot be considered complete. This more liberal form of confirmatory reasoning invalidates the strong rule of Right And. A new representation theorem has been obtained for the extreme form of logical compatibility. Open problems include dropping the condition that regular models be models of the premisses, and finding more meaningful forms of open confirmatory reasoning.

6 DISCUSSION

In this chapter I have combined and extended old and recent work in philosophy, logic, and machine learning, in an attempt to gain more insight in induction as a reasoning process. I believe that the approach followed has implications for each of these three disciplines, which I will address separately below.

6.1 Philosophy

Induction is one of the traditional problems of philosophy. In spite of this it is perhaps also one of the least understood, and certainly one that hasn't been satisfyingly solved. From the perspective from which this chapter is written there is not one, but two 'problems of induction': one concerned with the generation of possible hypotheses, and one concerned with the justification of accepted hypotheses. Traditionally, philosophers have been concerned with the justification problem:

> 'Why is a single instance, in some cases, sufficient for a complete induction, while in others, myriads of concurring instances, without a single exception known or presumed, go such a very little way towards establishing an universal proposition? Whoever can answer this question knows more of the philosophy of logic than the wisest of the ancients, and has solved the problem of induction.' [Mill, 1843, Book III, Chapter VIII, p. 314]

No attempt has been made, in the present chapter, to solve *this* problem of induction. As I have argued in Section 2.1, I don't think the justification problem manifests itself exclusively with inductive generalisations, but rather with all forms of nondeductive reasoning.

Furthermore, I don't think that the justification problem is a problem of logic. What I perceive as the *logical problem of induction*, and what has been the central problem of this chapter, is the problem of finding a sufficiently accurate description, in logical terms, of the process of inductive hypothesis generation. As Hanson observes, this logical problem of induction has been mostly ignored:

> 'Logicians of science have described how one might set out reasons in support of an hypothesis once it is proposed. They have said little about the conceptual considerations pertinent to the initial proposal of an hypothesis. There are two exceptions: Aristotle and Peirce. When they discussed what Peirce called "retroduction",[21] both recognized that the proposal of an hypothesis is often a reasonable affair. (...)
>
> Neither Aristotle nor Peirce imagined himself to be setting out a manual to help scientists make discoveries. There could be no such manual. Nor were they discussing the psychology of discoverers, or the sociology of discovery. There are such discussions, but they are not logical discussions. Aristotle and Peirce were doing logic. They examined characteristics of the reasoning behind the original suggestion of certain hypotheses.' [Hanson, 1958, pp. 1073–4]

The first philosophical contribution of this chapter, then, has been to reinforce the point that inductive hypothesis generation is a logical matter (and hypothesis

[21]Peirce's translation of Aristotle's term $\alpha\pi\alpha\gamma\omega\gamma\acute{\eta}$ – only later Peirce introduced the term 'abduction'.

selection is not). The second contribution lies in the distinction I have drawn between explanatory induction and confirmatory induction. More generally, I believe that a logical characterisation of inductive hypothesis generation is impossible unless one takes into account the *goal* which the hypothesis is intended to fulfil. Possible goals include providing classifications like in concept learning, or making implicit regularities explicit like in knowledge discovery in databases. Different goals lead to different forms of induction with different logical characteristics. For this reason I have taken a somewhat relativistic viewpoint, by setting up a general framework in which various logics of induction can be formulated, analysed, and compared, instead of fixing a particular logic of induction.

Two main families of logics of induction have been singled out, one portraying induction as explanation-preserving reasoning, the other as inference of confirmed hypotheses. As these families should be taken as starting points for further technical research, let us consider a number of possible improvements here. First of all, it has been remarked earlier that the restriction to a propositional object language L is counterintuitive in the context of induction, where the distinction between statements about individuals and statements about sets of individuals appears to be crucial. Upgrading the results of this chapter to predicate logic is certainly an important open problem. On the other hand, the propositional analysis of this chapter is not meaningless for the predicate logic case. First of all, in finite domains statements of predicate logic can be encoded in propositional logic.[22] Furthermore, I would expect each of the predicate logic rule systems to include the corresponding propositional rules, and the predicate logic semantics to be refinements of the propositional semantics.

A related point concerns the (restricted) reflexivity of the logics proposed here, which again runs counter to the intuition of induction as generalisation from instances to populations. We expect names of individuals to be present in the inductive premisses, and absent from the inductive hypothesis. As a consequence, changing the names in the premisses (in a way that does not distort the information they convey) would not invalidate the hypothesis. For instance, consider the observations

$$\text{Raven}(a)\wedge\text{Black}(a)\wedge\neg\text{Raven}(c)\wedge\text{Black}(c)\wedge\neg\text{Raven}(d)\wedge\neg\text{Black}(d)$$

Given these observations, $\forall x : \text{Raven}(x) \rightarrow \text{Black}(x)$ is a possible hypothesis. In the confirmatory setting this can be formalised by constructing a regular model from the evidence, and stipulating that any hypothesis be satisfied by that model. However, from that model we can construct another model by replacing a, c, d with (say) e, f, g. It makes sense to say that this constitutes another regular model that is to satisfy any hypothesis – this would rule out any hypothesis that talks about a, c or d, including the observations. In terms of the confirmatory structures of Section 5.2 this would mean to drop the condition that $[\alpha] \subseteq [\![\alpha]\!]$ for all $\alpha \in L$. In this way the perception of induction as reasoning from the particular to the

[22] This may require a substantial background theory, limiting the practical feasibility of this encoding.

general would be realised, not by posing syntactic restrictions (which is logically unattractive), but by wiring it into the semantics.

6.2 Logic

In my view, a major concern of logic is the systematic study of *reasoning forms* (deduction, induction, plausible reasoning, counterfactual reasoning, and so on). The emphasis on correct or mathematical reasoning is much more recent:

> The central process of reasoning studied by modern logicians is the accumulative *deduction*, usually explained semantically, as taking us from truths to further truths. But actually, this emphasis is the result of a historical contraction of the agenda for the field. Up to the 1930s, many logic textbooks still treated deduction, induction, confirmation, and various further forms of reasoning in a broader sense as part of the logical core curriculum. And moving back to the 19th century, authors like Mill or Peirce included various non-deductive modes of reasoning (induction, abduction) on a par with material that we would recognize at once as 'modern' concerns. Since these non-deductive styles of reasoning seemed irrelevant to foundational research in mathematics, they moved out quietly in the Golden Age of mathematical logic. But they do remain central to a logical understanding of ordinary human cognition. These days, this older broader agenda is coming back to life, mostly under the influence of Artificial Intelligence, but now pursued by more sophisticated techniques – made available, incidentally, by advances in mathematical logic... [van Benthem, 2000]

From a logical point of view this chapter can be seen as a contribution to the systematic study of reasoning forms, by trying to nail down the essence of induction in logical terms. By employing consequence relations as the central notion in this analysis, rather than, say, introducing non-truth-functional connectives or modalities, attention is focused on the underlying inference mechanism. Abstract properties of consequence relations can be used to classify reasoning forms: for instance, we could say that a reasoning form is deductive if it satisfies Monotonicity, quasi-deductive if it satisfies Cautious Monotonicity, explanatory if it satisfies (Admissible) Converse Entailment, and so on. These properties can be used to chart the whole map of reasoning forms, just like the properties of [Kraus *et al.*, 1990] chart the map of plausible reasoning. Thus, the present chapter can be seen as a constructive proof of the thesis that the method of analysis through consequence relations, pioneered by Gabbay [1985] Makinson [1989], and Kraus, Lehmann and Magidor [1990; 1992], constitutes in fact a *methodology*, that can be applied to analyse arbitrary forms of reasoning.

It is important to note that, by pursuing the analysis on the level of consequence relations, one is studying a class of logics, i.e. a reasoning form, rather than a particular logic. The formal analysis of a reasoning form is quite different from the material definition of a particular logic. The traditional picture of the latter process is like this. One starts with the semantics, which is designed to provide a precise meaning for the primitive symbols in the language, and formalises the relevant notion of consequence. Only then the proof-theoretic axiomatisation follows, accompanied by proofs of soundness and completeness. However, the abstract analysis of a class of logics may well proceed along different lines. Usually, a number of material definitions of specific logics are available (e.g. default logic, circumscription, negation as failure), and one tries to understand what these logics have in common. This extraction of commonalities may start on the semantic level (e.g. each of the logics selects among the models of the premisses) but also on the meta-theoretical level (e.g. each of the logics is nonmonotonic, and closed under conjunction on the right-hand side). Semantics does not necessarily come first anymore: the notion of Cautious Monotonicity may add as much to the understanding of plausible reasoning as the idea of a preference ordering on models.

Another tendency that can be observed when moving towards more abstract characterisations of reasoning forms is that semantics concentrates more on a characterisation of the notion of consequence, and less on the meaning of the primitive symbols in the language. This raises the question as to what constitutes a logical semantics. This is by no means a settled issue, but different authors have put forward the notion of *preservation* as playing a central role in semantics. For instance, Jennings, Chan and Dowad 'argue for a generalisation of inference from the standard account in terms of truth preservation to one which countenances preservation of other desirable metalinguistic properties':

> '(...) **truth is not the only inferentially preservable property**. A system of inference essentially provides procedures by which a set Σ of sentences (for example, the set of one's beliefs) having some complex of metalinguistic properties can be unfailingly extended to a larger set Σ' having the same complex of properties. By all means, we may regard *truth* as one of the properties to be preserved, but what other properties are to be preserved can depend upon our interests.' [Jennings *et al.*, 1991, p. 1047]

Furthermore, the view that abduction is the logic of preserving explanations has also been put forward by Zadrozny, who calls an inference rule abductive 'if it preserves sets of explanations' [Zadrozny, 1991, p. 1].

These points underline that many open problems of contemporary logic are conceptual in nature, rather than just technical. Motivated by problems from artificial intelligence, logic is widening its scope (again) to include forms of reasoning that are less and less similar to classical deduction. This development is far from nearing its completion. For instance, it seems commonplace among researchers studying consequence relations to assume that any consequence relation should minimally satisfy Reflexivity, Right Weakening, and Right And. Surely this makes sense if one limits attention to quasi-deductive reasoning, but the explanatory consequence relations studied in Section 4 of this chapter satisfy none of these properties. This chapter makes a case for a much more liberal perception of what is a consequence relation.

6.3 Machine learning

Whereas the main contributions of this chapter are logical and philosophical in nature, it also provides a novel perspective on inductive machine learning. First of all, the framework of inductive logic elaborated above provides a new logical foundation of inductive learning. As such, viewing inductive learning through logic is of course not new, as is witnessed by the subdiscipline of inductive logic programming (ILP) [Muggleton and De Raedt, 1994]. However, the usual logical perception of inductive learning differs from the one presented here. These two perceptions can be explained by considering the phrase 'inductive logic programming'. This phrase is ambiguous: it can be taken to mean – as it is usually done – 'doing logic programming inductively', but it can also be parsed as *programming in inductive logic* – which corresponds to the perception elaborated in this chapter. Let me explain why these interpretations are fundamentally different.

By identifying ILP with doing logic programming inductively, one effectively says that one's main goal is logic programming, i.e. answering queries by executing a declarative specification by means of its procedural semantics; however, since this declarative specification is only partly known through a number of examples, we should do some inductive patching before the real work can start. This results in a somewhat subsidiary view of induction as a subproblem that needs to be solved before we can do the main task. This can be seen from the problem specification (see Section 2.4), which defines induction as a sort of reversed deduction from positive examples p to hypothesis H. The slogan *ILP = inductive logic + programming* offers a different viewpoint, from which the inference from examples to a logic program is the main step. That is, the examples (and background theory) provide the declarative specification of the induction task, which is executed by applying the inference rules of an inductive logic (e.g. specialisation or generalisation operators). The hypothesis is an inductive consequence of the examples.

An immediate advantage of the latter viewpoint is that it provides an independent definition of induction, instead of defining it in terms of something else (e.g. reversed deduction). The following analogy may clarify this point. In mathematics, many concepts are introduced as inverses of other concepts: division as inverse of multiplication, roots as inverses of powers, integration as inverse of differentiation, and so on. However, once such a concept has been introduced in this way, it usually gets an independent treatment, providing further insight in and justification of the new concept. For instance, the definition of a definite integral as the limit of a Riemann sum formalises the idea that a definite integral calculates the area under a curve. The relationship between the new concept and previously defined concepts is then obtained as a theorem (cf. the fundamental theorem of calculus), rather than a seemingly arbitrary definition.

The framework of inductive consequence relations can be used to obtain soundness and completeness proofs of particular sets of operators employed by a particular induction algorithm. As an illustration, consider the inverse resolution operator absorption [Muggleton and Buntine, 1988]: from $p \leftarrow q, r$ and $s \leftarrow q, r, t$ infer $s \leftarrow p, t$. In our framework, a soundness proof of this induction step is established by the derivation, from an appropriate rule system such as **EM**, of the following statement:

$$\frac{\models p \leftarrow q, r}{s \leftarrow q, r, t \mid\!< s \leftarrow p, t}$$

which should be read as follows: if $p \leftarrow q, r$ is known from the background knowledge, then $s \leftarrow p, t$ is an inductive consequence of $s \leftarrow q, r, t$.

The framework also provides a well-defined *vocabulary* for reasoning about induction tasks and algorithms. Important notions like incrementality and convergence are linked to the underlying inductive consequence relation. This is important because, as we have seen, different induction tasks may have different characteristics – articulating these characteristics is a necessary and important first step in understanding the induction task at hand, and choosing or devising the right algorithm. Using such a vocabulary we can construct a taxonomy of induction tasks. Two families of induction tasks have been discerned in this chapter: explanatory induction, aimed at obtaining classification rules, and confirmatory induction, directed towards extracting structural regularities from the data. While explanatory induction corresponds to typical classification-oriented inductive machine learning tasks such as concept learning from examples, induction tasks belonging to the non-classificatory paradigm have only recently started to attract attention [Helft, 1989; Flach, 1990; De Raedt and Bruynooghe, 1993; Flach and Lachiche, forthcoming].indexBruynooghe, M. The contribution of the present chapter has been to propose Hempel's notion of qualitative confirmation as the unifying concept underlying these latter approaches. As Helft observed, inductive conclusions may be obtained by closed-world reasoning if the evidence may be considered complete; however, above I have proposed to distinguish an open variant of confirmatory reasoning, where the completeness assumptions on the evidence are considerably relaxed. While this may lead to an increase in the set of inductive hypotheses not refuted by given evidence, it has the distinct advantage of invalidating the rather strong property of Right And, and restoring the computationally attractive property of Convergence.

7 CONCLUSIONS

This chapter has been written in an attempt to increase our understanding of inductive reasoning through logical analysis. What logic can achieve for arbitrary forms of reasoning is no more and no less than a precise definition of what counts as a *possible* conclusion given certain premisses. Selecting the best (most useful, most plausible, etc.) hypothesis is an extra-logical matter. The logic of induction is the logic of inductive hypothesis generation.

There is not a single logic of induction. The logical relationship between evidence and possible inductive hypotheses depends on the task these hypotheses are intended to perform. Induction of classification rules such as concept definitions are based on a notion of explanation; an alternative logical account of induction starts from the qualitative relation of confirmation. Other forms of induction are conceivable. In this chapter I have proposed a meta-level framework for characterising and reasoning about different forms of induction. The framework does not fix a material definition of inductive hypothesis generation, but can be used to aggregate knowledge about classes of such material logics of induction.

A number of technical results have been obtained. The system **EM** axiomatises explanation-preserving reasoning with respect to a monotonic explanation mechanism. Characterisation of explanatory induction with respect to weaker (e.g. preferential) explanation mechanisms is left as an open problem. The systems **CS** and **CP** axiomatise the general and preferential forms of closed confirmatory reasoning, conceived as reasoning about selected regular models. They represent variations of earlier, differently motivated, characterisations by Bell [1991] and Kraus, Lehmann and Magidor [1990], with a different treatment of inconsistent premisses. An important open problem here is the axiomatisation of confirmatory structures where regular models may not be models of the premisses. Finally, the system **CW** represents an extreme form of open confirmatory reasoning (i.e. compatibility of premisses and hypothesis). Finding more realistic forms of open confirmatory reasoning remains an open problem.

ACKNOWLEDGEMENTS

I like to thank John-Jules Meyer and Daniel Lehmann for stimulating discussions. Part of this work was supported by Esprit IV Long Term Research Project 20237 (Inductive Logic Programming 2), and by the EU-funded Networks of Excellence ILPnet and ILPnet2.

University of Bristol, UK.

REFERENCES

[Angluin and Smith, 1983] D. Angluin and C.H. Smith. Inductive inference: theory and methods. *Computing Surveys*, **15**, 238–269, 1983.

[Bell, 1991] J. Bell. Pragmatic logics. In *Proceedings Second International Conference on Knowledge Representation and Reasoning KR'91*, pp. 50–60, Morgan Kaufmann, San Mateo, 1991.

[van Benthem, 2000] J. van Benthem. Reasoning in reverse. In [Flach and Kakas, 2000].

[Carnap, 1950] R. Carnap. *Logical Foundations of Probability*. Routledge & Kegan Paul, London, 1950.

[Delgrande, 1987] J. P. Delgrande. A formal approach to learning from examples. *Int. J. Man-Machine Studies* **26**:123–141, 1987.

[De Raedt and Bruynooghe, 1993] L. De Raedt and M. Bruynooghe. A theory of clausal discovery. In *Proceedings Thirteenth International Conference on Artificial Intelligence IJCAI'93*, pp. 1058–1063. Morgan Kaufmann, San Mateo, 1993.

[Flach, 1990] P.A. Flach. Inductive characterisation of database relations. In *Proceedings International Symposium on Methodologies for Intelligent Systems ISMIS'90*, Z.W. Ras, M. Zemankowa and M.L. Emrich (eds.), pp. 371-378. North-Holland, Amsterdam.

[Flach, 1993] P.A. Flach. Predicate invention in Inductive Data Engineering. In *Proceedings European Conference on Machine Learning ECML'93*, P.B. Brazdil (ed.), pp. 83–94. Lecture Notes in Artificial Intelligence 667, Springer-Verlag, Berlin, 1993.

[Flach, 1995] P.A. Flach. *Conjectures: an inquiry concerning the logic of induction.* PhD thesis, Tilburg University, 1995.

[Flach, 1996] P.A. Flach. Rationality postulates for induction. In *Proceedings Theoretical Aspects of Rationality and Knowledge TARK'96*, Yoav Shoham (ed.), pp. 267–281. Morgan Kaufmann, San Mateo, 1996.

[Flach, 1998] P.A. Flach. Comparing consequence relations. In *Proceedings Sixth International Conference on Knowledge Representation and Reasoning KR'98*, A.G. Cohn, L. Schubert and S.C. Shapiro (eds.), pp. 180–189. Morgan Kaufmann, San Mateo, 1998.

[Flach, 1999] P.A. Flach. Knowledge representation for inductive learning. In *Proceedings Symbolic and Quantitative Approaches to Reasoning and Uncertainty ECSQARU'99*, A. Hunter and S. Parsons (eds.), pp. 160–167. Lecture Notes in Artificial Intelligence 1638, Springer-Verlag, Berlin, 1999.

[Flach and Savnik, 1999] P.A. Flach and I. Savnik. Database dependency discovery: a Machine Learning approach. *AI Communications* **12**(3):139–160, 1999.

[Flach and Lachiche, forthcoming] P.A. Flach and N. Lachiche. Confirmation-guided discovery of first-order rules with Tertius. *Machine Learning*, accepted for publication.

[Flach and Kakas, 2000] P.A. Flach and A.C. Kakas, editors. *Abductive and inductive reasoning: essays on their relation and integration.* Kluwer Academic Publishers, 2000.

[Gabbay, 1985] D.M. Gabbay. Theoretical foundations for non-monotonic reasoning in expert systems. In *Logics and Models of Concurrent Systems*, K.R. Apt (ed.), pp. 439–457. Springer-Verlag, Berlin, 1985.

[Hanson, 1958] N.R. Hanson. The logic of discovery. *J. Philosophy* **55**:1073–1089, 1958.

[Helft, 1989] N. Helft. Induction as nonmonotonic inference. In *Proceedings First International Conference on Knowledge Representation and Reasoning KR'89*, pp. 149–156. Morgan Kaufmann, San Mateo, 1989.

[Hempel, 1943] C.G. Hempel. A purely syntactical definition of confirmation. *J. Symbolic Logic* **6**:122–143, 1943.

[Hempel, 1945] C.G. Hempel. Studies in the logic of confirmation. *Mind* **54**:1–26 (Part I); **54**:97–121 (Part II), 1945.

[Jennings *et al.*, 1991] R.E. Jennings, C.W. Chan and M.J. Dowad. Generalised inference and inferential modelling. In *Proceedings Twelfth International Joint Conference on Artificial Intelligence IJCAI'91*, pp. 1046–1051. Morgan Kaufmann, 1991.

[Kraus *et al.*, 1990] S. Kraus, D. Lehmann and M. Magidor. Nonmonotonic reasoning, preferential models and cumulative logics. *Artificial Intelligence* **44**:167–207, 1990.

[Lachiche, 2000] N. Lachiche. Abduction and induction from a non-monotonic reasoning perspective. In [Flach and Kakas, 2000].

[Lehmann and Magidor, 1992] D. Lehmann and M. Magidor. What does a conditional knowledge base entail? *Artificial Intelligence* **55**:1–60, 1992.

[Makinson, 1989] D. Makinson. General theory of cumulative inference. In *Proceedings Second International Workshop on Non-Monotonic Reasoning*, M. Reinfrank, J. de Kleer, M.L. Ginsberg and E. Sandewall (eds.), pp. 1–18. Lecture Notes in Artificial Intelligence 346, Springer-Verlag, Berlin, 1989.

[Mill, 1843] J. S. Mill. *A System of Logic*, Reprinted in *The Collected Works of John Stuart Mill*, J.M. Robson (ed.), Routledge and Kegan Paul, London, 1843.

[Muggleton and De Raedt, 1994] S. Muggleton and L. De Raedt. Inductive Logic Programming: theory and methods. *J. Logic Programming* **19-20**:629–679, 1994.

[Muggleton and Buntine, 1988] S. Muggleton and W. Buntine. Machine invention of first-order predicates by inverting resolution. In *Proceedings Fifth International Conference on Machine Learning*, J. Laird (ed.), pp. 339–352. Morgan Kaufmann, San Mateo, 1988.

[Tarski, 1936] A. Tarski. Über den Begriff der logischen Folgering, *Actes du Congrès Int. de Philosophie Scientifique* 7:1–11, 1936. Translated into English as On the concept of logical consequence. In *Logic, Semantics, Metamathematics*, A. Tarski, pp. 409–420, Clarendon Press, Oxford, 1956.

[Zadrozny, 1991] W. Zadrozny. On rules of abduction. IBM Research Report, August 1991.

F. BERGADANO, V. CUTELLO AND D. GUNETTI

ABDUCTION IN MACHINE LEARNING

1 INTRODUCTION

Both inductive learning and abductive reasoning start from specific facts or observations and produce some explanation of these facts. Both may be described as forms of defeasible reasoning from effects to causes. There are some differences, but they are minor and due to different understandings of the notions of observation and explanation (see for instance [Bergadano and Besnark, 1994]). We build on the general notions developed in the introductory Chapter, taking what was labeled there as the *syllogistic* view, in the sense that we isolate the differences between abduction and induction based on syntactic considerations.

Briefly, induction sees examples as *instances of a concept* and explanations as general *concept descriptions*, whereas abduction sees examples as specific observations and explanations as other specific facts that are true, and cause the observations to occur. As a typical case for induction, an example could be the description of a specific bird, and a concept description could be a rule such as

$$\text{bird}(X) \leftarrow \text{small}(X) \wedge \text{part\_of}(X, Y) \wedge \text{wing}(Y)$$

with the obvious intended meaning. For abduction, an observation could be the fact that some particular bird does not fly, and an explanation could be the fact that that particular bird has a broken wing.

In both cases, the relevant logical relation is that the explanation, together with the domain knowledge, implies the observation. This aspect is developed in depth in the introductory Chapter.

The Machine Learning literature has largely ignored such similarities, or has produced studies that emphasize the differences that are present [Console and Saitta, 1993; Flach, 1992]. On the other hand, abduction has been used as an effective technique within the underlying inductive framework of Machine Learning methods. Two different approaches, which we think are quite representative of the Machine Learning view of abductive reasoning, can be identified.

- First, abduction has been used to guide search in top-down relational learning. The idea has evolved from explanation-based learning methods, restricting the inductive hypotheses to be logical consequences of a given domain theory. This is then generalized by allowing inductive hypotheses to be obtained from the domain theory by either deductive or abductive reasoning.
- Second, abduction has been used to generate missing examples in relational learning. In fact, it is intrinsic to the nature of induction that the input data is incomplete: some examples are given, but not all. If some particular

D. M. Gabbay and R. Kruse (eds.), Abductive Reasoning and Learning, 197–229.

examples are missing, existing relational learning methods may however encounter serious problems, and abductive generation of these missing examples may be an effective solution.

Top-down relational learning algorithms suffer from the difficulty of searching a space of possible inductive hypotheses that is usually very large. In the first relational learning systems, and also in more recent approaches such as Foil [Quinlan, 1990], the problem was addressed with heuristics of a statistical nature. Top-down systems start from very general concept descriptions, and then try to obtain consistent rules with a number of specialization steps. Typically, heuristics would favor specializations that exclude negative examples while still covering a large number of positive examples. Such heuristics may be misleading, and may also be insufficient for an adequate pruning of the hypothesis space. The Explanation-Based Learning (EBL) paradigm restricts the concept descriptions that may be possibly produced to the logical consequences of a so-called *domain theory*, that is given as input to the learning system [Mitchell *et al.*, 1986; Bergadano and Giordana, 1988; Pazzani and Kibler, 1992]. Abduction has been used in this framework as follows: given the domain theory, the concept descriptions that may be obtained via abductive reasoning are considered as possibly true, and should be evaluated inductively on the basis of the available examples. Studies that follow this scheme may be found in [Bergadano *et al.*, 1989; Cohen, 1992; Cohen, 1994].

A well known problem in Machine Learning is to provide a learning method with an "adequate" set of examples of the target concept. Here "adequate" informally means that the training set should contain all those examples required to successfully complete the learning task, and no more. Obviously, this is a very hard condition to achieve, since usually it is not possible to know in advance exactly which examples are (and which are not) significant for learning a concept. As a consequence, the learning task may turn out to be too slow (if too many examples are given) and/or may fail (if the examples are not significant). This problem is particularly serious for an important class of learning methods: *Relational Learning Algorithms* based on an extensional interpretation of sub-predicates and recursion [Quinlan, 1990; Bergadano and Giordana, 1988; Pazzani and Kibler, 1992]. In these methods, the learning procedure can not only fail or be too slow, but also produce wrong results: the description of the target concept synthesized by the system may entail some of the negative examples given to the system.

It is our objective to show how abduction can be used to fix the above problem. An abductive procedure is used to query the user for any example that may be missing, depending on the hypothesis space that has been defined and the given examples. A similar technique has been used before, for example in [DeRaedt and Bruynooghe, 1991; Shapiro, 1983], to query the user for missing values allowing a single example to be covered. In our case, abduction will be systematically applied over the whole hypothesis space. As a result, the learning systems turns out to be correct and sufficient, in the sense that the learned description does not entail any of the negative examples and such a description can always be found if it exists. We also show how this technique can be adapted to the problem of program testing: a combination of induction and abduction can be used to generate an "adequate" test set for a program under testing.

2 ABDUCTION AND INDUCTION

Abductive and Inductive Reasoning in Artificial Intelligence are considered to be distinct and have generated separate fields of study. After a simple analysis, one finds in effect distinct inference schemes.

- For induction: $\dfrac{P(a)}{\forall x P(x)}$
- For abduction: $\dfrac{P(x) \rightarrow Q(x) \qquad Q(a)}{P(a)}$

As stated in the introductory Chapter, a deeper analysis, however, suggests that the difference between the two schemes is not always easy to state. For instance, using the tautology

$$\Phi \equiv (\forall x P(x)) \rightarrow P(a)$$

one gets:

$$\frac{\Phi \qquad P(a)}{\forall x P(x)}$$

as an abductive inference step, but it actually has the same premises and conclusions of the inductive inference scheme.

It would then seem that the abductive scheme includes simple forms of inductive reasoning. In [Bergadano and Besnark, 1994] the authors start from the above considerations and then isolate some minor differences of abductive and inductive reasoning within the same framework, that is determined by the above inference rules. In the same paper, a formalization based on non-monotonic logic is developed. Here, we simply note that induction and abduction are indeed very similar inference schemes. A deeper analysis is found in the introductory Chapter of the present Book, where general similarities are noted, and syntactic, inferential and semantic differences are considered.

However, the Machine Learning literature has not used abduction and induction as synonyms. The main keyword in learning is induction, and abductive reasoning is rather used as an additional technique for solving particular problems. We discuss in the following sections two of such uses of abduction in Machine Learning. In particular in the next section we survey its use as a form of bias for guiding the search in a top-down specialization. Then, we show how abductive reasoning and queries can solve some relevant problems in relational top-down learning, of both binary and fuzzy predicates.

3 GUIDING SEARCH WITH ABDUCTION

One use of abduction in Machine Learning is related to the problem of guiding search in a top-down specialization. As the number of concept descriptions that may possibly be generated is, in general, very large, and since even the number of descriptions that are consistent with the examples can be large, learning systems need extra-evidential criteria to prune the search space. Deductive inference with a domain theory has been used to this purpose in ML-SMART [Bergadano and Giordana, 1988] and FOCL [Pazzani and Kibler, 1992]. Similarly, a domain theory could be used abductively within the same framework as in [Bergadano *et al.*, 1989] and in [Cohen, 1992; Cohen, 1994].

Before we can suitably describe the details of abductive inference to prune the space of possible concept descriptions, we need to define a type of analytic learning called Explanation-based Learning (EBL), which has received much attention during the late eighties. A survey of EBL is found in [Ellman, 1989].

EBL needs as input one or more positive examples, as well as a so-called *domain theory*, which includes relevant prior knowledge. A resolution proof of the positive examples is produced, and the leaves of the proof tree are generalized and taken as the antecedent of a new rule for the target concept. Suppose, for instance, that the domain theory is the following

$$\begin{aligned} C(X) \quad &:- \quad B(X), A(X,Y). \\ B(X) \quad &:- \quad R(X,Y). \\ B(b). & \\ A(c,a). & \\ R(c,d). & \end{aligned}$$

and a positive example for C is c. It is then easy to see that $C(c)$ can be obtained deductively from the domain theory, where the leaves in the proof tree are $R(c,d)$ and $A(c,a)$. This produces the learned description for C:

$$C(X) \quad :- \quad R(X,Y), A(X,Z)$$

The learned description is actually a logical consequence of the given domain theory, and could be obtained from the domain theory even without looking at the examples, by resolving the first two clauses in the domain theory. The role of the example is that of suggesting some consequences of the domain theory rather than others: the ones that are useful to deduce the positive example are chosen.

One natural question arises: what is the purpose of this form of learning, if its output is just a logical consequence of something that was already known? The answer is that, although nothing really new is learned, the new form of knowledge may be more *operational*, that is to say either easier to use or leading to more efficient computations. In the above case, the positive example c, that is to say $C(c) =$ true may be deduced more efficiently from the learned clause than by the original domain theory, as one resolution step is saved. EBL may then be seen as a form of pre-compilation, or partial evaluation. The effectiveness of this form of learning speed-up depends on whether the given positive examples are representative of future cases: if this is not true, the learned clause may turn out to be useless, as the original domain theory would be necessary every time. Learned clauses that are useless in this sense will only take away some memory space without giving any computational advantage.

For this reason, systems soon started using EBL with many positive examples, so that only clauses that frequently proved to be useful would be kept. Other systems (e.g., [Bergadano and Giordana, 1988]) also introduced the possibility of using negative examples, in connection with the acceptance of a domain theory that might be partly incorrect. In this case, even clauses that follow deductively from the domain theory may be incorrect, and may cover negative examples. Clauses covering too many negative examples may then be discarded. When used in this way, the domain theory basically defines a hypothesis space. The concept descriptions that are logical consequences of the domain theory are descriptions that may potentially be learned. The description which is actually produced would also be required to perform well with respect to the data, i.e. cover many positive examples and few negative examples. EBL may then be considered as a way to guide search in top-down specialization: not all specializations are possible, but only the ones that produce deductive consequences of the domain theory. Legal specializations are obtained by resolving the clause to be specialized against some clause in the domain theory. Other specializations are not considered. In the previous example, the clause

$$C(X) \quad :- \quad R(X,Y), A(X,Z)$$

would be a legal specialization of the clause

$$C(X) \quad :- \quad B(X), A(X,Y).$$

while, e.g., the clause

$$C(X) \quad :- \quad B(X), A(X,Y), B(Y)$$

would not be considered. If the space of possible clauses is too large, a domain theory can be used effectively to prune such a space and suggest preferred specialization steps. ML-SMART [Bergadano and Giordana, 1988], and FOCL [Pazzani and Kibler, 1992] adopt similar goals and techniques.

Abduction comes into place if the consequences of the domain theory are not just taken to be deductive consequences, but are also obtained by means of an abductive theorem prover. For instance, if the above domain theory would contain the clause

$$D(X) \quad :- \quad A(X,Y)$$

then, an abductive/deductive consequence of the theory would be the clause

$$C(X) \quad :- \quad R(X,Y), D(X)$$

which would also be a legal specialization of the clause

$$C(X) \quad :- \quad B(X), A(X,Y).$$

Such an abductive use of a domain theory is found in [Bergadano *et al.*, 1989], with the purpose of restricting the hypothesis space without excluding concept descriptions that may be meaningful on the basis of abductive reasoning. In fact, such a use of abduction is just a means for specifying in a declarative way a form of *inductive bias*, a criterion for preferring an inductive hypothesis over another, or a rule for eliminating some description from the space of possible inductive generalizations. As such a criterion is heuristic in nature, one would not see why a deductive approach would be more justified than an abductive one. The deductive approach is the one found in ML-SMART and other approaches to the integration of inductive reasoning and EBL. The abductive approach follows exactly the same scheme, and is described, for example in [Bergadano *et al.*, 1989]. In this case the rule "accept generalizations that are a logical consequence of a given domain theory" is replaced by "accept generalizations that may be obtained from the domain theory via abductive reasoning". In both cases the obtained generalizations must *in primis* be consistent and complete with respect to the given examples, or satisfy such requirement *to a degree*. The notion of *abductive EBL* in a similar framework is introduced in [Cohen, 1992; Cohen, 1994].

4 RELATIONAL LEARNING ALGORITHMS BASED ON EXTENSIONALITY

Relational Learning Algorithms learn recursive relational descriptions from positive and negative examples, given as ground literals. Usually, a subset of the first-order predicate calculus, Horn clauses, is used for the concept description (see, e.g. [Birnbaum and Collins, 1991; Rouveirol, 1992; Cohen, 1992; Bergadano and Gunetti, 1993; Muggleton and Feng, 1990]). Learning definite clauses for multiple predicates is difficult, and systems tend to be slow. Many systems, such as Foil [Quinlan, 1990],

Focl [Pazzani and Kibler, 1992] and Golem [Muggleton and Feng, 1990] handle this problem by evaluating clauses extensionally. In this way candidate clauses can be generated directly from the examples one at a time and independently of one another.

Let us now introduce the concept of extensionally defined predicate, which, intuitively, will play the role of the basis case in the recursive definition of more complex predicates.

DEFINITION 1. A predicate $P(X)$ is said to be extensionally defined if it is defined over a given finite collection of elements $a_1, \ldots, a_n$, that is to say, if its characteristic set is given extensionally.

Examples of extensionally defined predicates are given by "mother" and "father" defined by the following table:

Parenthood Table

mother			father	
X	Y		X	Y
d	g		f	g
r	d		s	d
r	p		s	p
i	r		b	r
t	s		h	c
a	c			

Some other predicates can be intensionally defined over a collection of extensionally defined predicates by means of simple logical rules.

An example is given by the predicate *parent*:

$$\text{parent}(X,Y) \quad :- \quad \text{mother}(X,Y)$$
$$\text{parent}(X,Y) \quad :- \quad \text{father}(X,Y)$$

In this case, the extensional definition of the predicates *mother* and *father* provides as well an extensional definition of the predicate *parent*.

We formalize now the meaning of "extensional evaluation".

DEFINITION 2. Let $\alpha(X_1, X_2, \ldots, X_n)$ be a conjunction of predicates where there is at least one occurrence of each of the variables (not necessarily distinct) $X_1, \ldots, X_n$. We say that $\alpha(a_1, a_2, \ldots, a_n)$ is *extensionally true* if and only if

- in case $\alpha(X_1, \ldots, X_n) \equiv Q(X_1, \ldots, X_n)$, for some extensionally defined predicate Q, then $Q(a_1, \ldots, a_n)$ is a given positive example of Q;
- in case

  $$\alpha \equiv \gamma(X_{i_1}, \ldots, X_{i_h}, T_1, \ldots, T_l), Q(T_1, \ldots, T_l, X_{j_1}, \ldots, X_{j_k}),$$

  for some conjunction γ, predicate Q, $l \geq 0$ and

  $$\{X_{i_1}, \ldots, X_{i_h}, X_{j_1}, \ldots, X_{j_k}\} = \{X_1, \ldots, X_n\},$$

  then there must exist $e_1, \ldots, e_l$ such that both the conjunction $\gamma(a_{i_1}, \ldots, a_{i_h}, e_1, \ldots, e_l)$ and $Q(e_1, \ldots, e_l, a_{j_1}, \ldots, a_{j_k})$ must be extensionally true.

We say that the conjunct $\alpha(X_1, \ldots, X_n)$ *extensionally covers* a positive [resp. negative] example of P, $P(a_1, \ldots, a_n)$ [resp. $\neg P(a_1, \ldots, a_n)$] if and only if $\alpha(a_1, \ldots, a_n)$ is extensionally true.

The next subsection contains a simplified version of Foil, in order to illustrate the extensional approach and its drawbacks.

4.1 The Extensional Approach: Simplified Foil

Let P be the target concept and pos_examples(P) and neg_examples(P) the given positive and negative examples of P (in the following, α and γ represent generic conjunction of literals).

Algorithm 1 (Simplified Foil).

Let $E^+(P) =$ positive examples of P;

Let $E^-(P) =$ negative examples of P;

Let Π be the empty program;

while $E^+(P) \neq \emptyset$ **do**

Generate one clause $P\ :-\ \alpha$ and add it to Π;

Let $E^+(\alpha)$ be the positive examples of P covered by α;

$E^+(P) \leftarrow E^+(P) \setminus E^+(\alpha)$

end while

Output Π

Procedure **Generate one clause**

$\alpha \leftarrow$ true;

let $E^-(\alpha)$, $E^+(\alpha)$ be respectively the negative and positive examples of P covered by α;

while $E^+(\alpha) \neq \emptyset$ **do**

if $E^-(\alpha) = \emptyset$ then return($P\ :-\ \alpha$)

else choose a predicate Q and its arguments $\vec{X}$

$\alpha \leftarrow \alpha \wedge Q(\vec{X})$

end while

end Algorithm.

Every predicate Q can be defined by the user either intensionally by or extensionally. In particular, clauses can be recursive and, in this case, $Q = P$, and its truth value can only be determined by the available examples.

The choice of the literal $Q(\vec{X})$ to be added to the partial antecedent α of the clause being generated is guided by heuristic information. It might nevertheless be a wrong choice in some cases, in the sense that it may cause the procedure "Generate one clause" to fail by exiting the while loop without returning any clause. This problem can be fixed by making the choice of $Q(\vec{X})$ a backtracking point.

Suppose, for instance, that Foil is given the following positive and negative examples of the relation *ancestor*, which is the one to be learned:

$$+ < r, g >, < b, g >, < b, d >, < d, g >, < b, r >, < r, d >, < r, p >;$$

$$- < d, d >, < g, p >, < d, p >.$$

Let us suppose as well that we have the intensional definition for the relation *parent* given above along with the Parenthood Table which extensionally defines the predicates mother and father. Finally, we know that the logic program for *ancestor* depends on *parent* and on itself (i.e. it may be recursive). As there are at most 3 variables to be used, these are the possible literals:
parent(X, Y), parent(Y, X), parent(X, W),
parent(W, X), parent(Y, W), parent(W, Y)
ancestor(X, W), ancestor(W, X), ancestor(Y, W), ancestor(W, Y) [1].

The learning algorithm starts to generate the first clause - the antecedent α is initially equal to "true". We need to choose the first literal $Q(\vec{X})$ to be added to α. As we have left the heuristics unspecified, we will choose it so as to make the discussion short.

Let α = parent(W, Y); then all positive examples and the second and third negative examples are covered, so more literals need to be added.

Let α = parent$(W, Y) \wedge$ parent(W, X); in this case no positive examples are covered and the negative example $< d, p >$ is covered. Clause generation fails and we backtrack to the last literal choice.

Let α = parent$(W, Y) \wedge$ ancestor(X, W); no negative example is covered, and the first 3 positive examples are extensionally covered. A clause is generated and the covered positive examples are deleted.

We proceed to the generation of another clause; α is empty again. If we choose the first literal as parent(X, Y), the remaining positive examples are covered and the final solution is obtained:

$$\begin{aligned} \text{ancestor}(X,Y) &\ :-\ \text{parent}(W,Y), \text{ancestor}(X,W). \\ \text{ancestor}(X,Y) &\ :-\ \text{parent}(X,Y). \end{aligned}$$

Notice that the obtained definition of the relation "ancestor" is complete and consistent when completeness and consistency are characterized as follows.

DEFINITION 3. A definition Π of P is *complete* with respect to the positive examples E^+ of P if and only if

$$(\forall e \in E^+)\Pi \vdash P(e).$$

A definition Π of P is *consistent* with respect to the negative examples E^- if and only if

$$(\forall e \in E^-)\Pi \nvdash P(e).$$

[1] ancestor(X, Y) and ancestor(Y, X) are not listed because they may produce looping recursions

4.2 *Problems of extensionality*

The independence of the clauses in the generation of the definition of "ancestor" is given by the extensional interpretation of recursion and sub-predicates: when a predicate Q occurs in a clause antecedent α, it is evaluated as true when the arguments match one of the positive examples. For instance, the clause

$$\text{ancestor}(X,Y)\ :-\ \text{parent}(W,Y), \text{ancestor}(X,W).$$

was found to extensionally cover the positive example $< b, g >$ of ancestor because parent(d, g) is true (see Parenthood Table) and $< b, d >$ is also a positive example of ancestor. The previously generated logical definitions of Q are not used. The method is (partially) justified by the following result (see [Bergadano and Gunetti, 1993])

THEOREM 4. *Suppose Foil successfully exits its main loop and outputs a logic program* Π, *that always terminates for the given examples.*

Let $Q(X)\ :-\ \alpha$ *be a generated clause of* Π, *then, for any e positive example of* Q, *if* α *extensionally covers e then* $\Pi \vdash Q(e)$.

However, extensionality forces us to include many examples, which would otherwise be unnecessary. In fact other desirable properties, similar to the one given by Theorem 4, are not true, and two fundamental problems arise:

PROBLEM 5. For a logic program Π, it may happen that $\Pi \vdash Q(e)$, but none of its clauses extensionally cover e. As a consequence Foil would be unable to generate Π.

Consider for instance the program Π:

$$\begin{aligned}\text{ancestor}(X,Y)\ &:-\ \text{parent}(W,Y), \text{ancestor}(X,W).\\ \text{ancestor}(X,Y)\ &:-\ \text{parent}(X,Y).\end{aligned}$$

Let $< r, g >$ be the only positive example of ancestor. As it is easy to see, by using resolution on Π and the Parenthood Table this example logically follows from Π ($\Pi \vdash$ ancestor(r, g)). However, is not extensionally covered. Indeed, by inspecting the Parenthood Table, we can see that the second clause does not cover it because parent(r, g) is false and the first clause does not cover it extensionally because parent(d, g) is true, but $< r, d >$ is not given as a positive example of ancestor.

PROBLEM 6. Foil may generate definitions which are not consistent. In details, even if no clause of a definition Π extensionally covers a negative example e of Q, it may still happen that $\Pi \vdash Q(e)$.

Consider for instance the following definition Π for "ancestor":

$$\begin{aligned} \text{ancestor}(X, Y) &\ :- \ \text{parent}(W, X), \text{ancestor}(W, Y) \\ \text{ancestor}(X, Y) &\ :- \ \text{parent}(X, Y). \end{aligned}$$

It follows that $\Pi \vdash \text{ancestor}(g, p)$. Nevertheless, $< g, p >$ is not extensionally covered by the first clause: parent(d, g) is true, but $< d, p >$ is not a positive example of ancestor.

Since d is not an ancestor of p, it could not possibly be added as a positive example, and $< g, p >$ would not be extensionally covered even if all positive examples were given.

The solution for this problem differs from the one of problem 1: in this case Π will be ruled out by adding a negative example, namely $< d, p >$.

5 PROBLEMS OF INTENSIONAL METHODS

Giving up the extensional interpretation of predicates while keeping the basic computational structure of Foil produces unsolvable problems. In fact, the truth values of the missing examples could be obtained by means of the partial program generated at a given moment. But if the inductive predicates occurring in a clause being generated are evaluated by means of the clauses that were learned previously, then the order in which the clauses are learned becomes a major issue.

Suppose, for instance, that we are given as family tree the following subset of the Parenthood Table:

$$\text{parent}(i, r), \text{parent}(r, d), \text{parent}(d, g), \text{parent}(f, g)$$

as well as the following positive and negative examples of ancestor:

$$\begin{array}{ll} + & < i, d >, < i, r >, < i, g >; \\ - & < i, f > . \end{array}$$

Suppose also that the following clauses are generated, with the given order:

(1) ancestor(X, Y) :− parent(X, W), parent(W, Y).

(2) ancestor(X, Y) :− parent(Y, W), ancestor(X, W).

(3) ancestor(X, Y) :− parent(X, W), ancestor(W, Y).

Clause (1) does not use any inductive predicate and immediately covers the positive tuple $< i, d >$. Clause (2) contains ancestor(X, W), and this literal is evaluated with the clauses available at this stage, i.e. (1) and (2). With this kind of intensional interpretation, the only example covered is $< i, r >$, a positive example. When clause (3) is generated, it will cover the positive example $< i, g >$.

But clause (2) will now cover the negative example $< i, f >$, since parent(f, g) is true, and ancestor(i, g) may now be deduced by using the third clause.

Rules that seem consistent and useful at some stage may later be found to cover negative examples. We must then backtrack to the generation of the clause causing the inconsistency, e.g. clause (3), and to the generation of the clause that was later found to be inconsistent, e.g. clause (2). In general, we must give up our former assumption that clauses may be learned one at a time and independently. Alternatively, we may number the predicates occurring in the learned clauses every time an inconsistency is detected. For instance, the former program would be rewritten as

$$
\begin{array}{rcl}
\text{ancestor1}(X, Y) & :- & \text{parent}(X, W), \text{parent}(W, Y). \\
\text{ancestor2}(X, Y) & :- & \text{parent}(Y, W), \text{ancestor2}(X, W). \\
\text{ancestor3}(X, Y) & :- & \text{parent}(X, W), \text{ancestor3}(W, Y). \\
\text{ancestor}(X, Y) & :- & \text{ancestor2}(X, Y). \\
\text{ancestor2}(X, Y) & :- & \text{ancestor1}(X, Y). \\
\text{ancestor}(X, Y) & :- & \text{ancestor3}(X, Y). \\
\text{ancestor3}(X, Y) & :- & \text{ancestor1}(X, Y).
\end{array}
$$

Nevertheless, this technique does not totally avoid the need for backtracking, and explodes the number of possible clauses by multiplying the number of inductive predicates by the number of generated indexes. A solution based on abduction is presented in the next section, so that the advantages of extensionality are preserved, i.e. so that previously generated clauses do not need to be reconsidered.

6 COMPLETING EXAMPLES VIA ABDUCTION BEFORE LEARNING

There is no reason why particular positive (problem 1) or negative (problem 2) examples should have to be present. After all, the whole motivation of induction is that some information is missing. The important points are that

(1) if a definition Π consistent with the given examples exists, then it must be found; and

(2) the induced definition Π must not cover any negative examples.

The extensional approach guarantees neither, unless some specially determined positive and negative examples are given.

Here we show how abduction can be used to query the user for the missing examples, in order to preserve the computational advantages of extensional approaches, while guaranteeing that a correct solution be found.

Suppose that a learning system has to cover the positive example $P(a, b)$ and that the following candidate clause has been generated:

$$P(X, Y) \ :- \ \alpha(X, Y), Q(Y, Z).$$

Moreover, we know that $\alpha(a, b)$ is true (for example because every literal in α is extensionally evaluated to true). Then, that clause will extensionally cover $P(a, b)$ only if there exists some value c such that $Q(b, c)$ is a positive example (known to the system) of Q. Suppose that all such examples are missing. From the classical abductive inference rule:

$$\frac{\alpha \leftarrow \beta \qquad \alpha}{\beta}$$

we get:

$$\frac{P(a, b) \ :- \ \exists Z \alpha(a, b), Q(b, Z). \qquad P(a, b)}{\exists Z Q(b, Z)}$$

and then the user can be queried for a value of Z such that $Q(b, Z)$ is true. The new example of Q is added to the set of positive examples and $P(a, b)$ can now be covered. This is a *controlled* form of abduction, in the sense that the result of abductive inference is not asserted as true, but only proposed as a possible truth, which is then queried to the user. It is clear that if there is no value Z such that $Q(b, Z)$ is true then the clause must be discarded.

The above basic idea of generating abductive queries can be applied in a systematic way. That is, we may build chains of abductive queries, or chains of controlled abductive inference steps. A backward reasoning from an initial example that we know to be true (and that must be extensionally covered), up to what must be true in order for that initial example to be covered. The chain of abductive steps can be depicted as in Figure 1.

That is, a chain is produced because the queried examples are treated in the same way as the starting one. The added examples are used to generate other abductive queries again and again until no more examples can be added, and the backward reasoning can stop. We show this on a very simple example.

Let the following hypothesis space HS be given, containing only the correct clauses for the *append* concept (or program):

```
{ c1 = app(X,Y,Z) :- head(H,H),tail(X,T),app(T,Y,W),cons(H,W,Z).
  c2 = app(X,Y.Z) :- null(X),equal(Y,Z). }
```

Where all the predicates (except for *app*) are intensionally defined as usual. Let the only given positive example be: e_1 = "app([a,b],[c],[a,b,c])". A correct program for *append* cannot be extensionally learned from HS using only this example. However, let us apply the abductive queries as described above. Example e_1 is matched against the head of c1 producing the query "app([b],[c],W)?". That

$$\frac{\alpha \leftarrow \beta \qquad \alpha}{\beta}$$

$$\Downarrow$$

$$\frac{\beta \leftarrow \gamma \qquad \beta}{\gamma}$$

$$\Downarrow$$

$$\frac{\gamma \leftarrow \delta \qquad \gamma}{\delta}$$

$$\Downarrow$$

$$\vdots$$

Figure 1. A chain of abductive steps

is, the user is queried for a value of the variable W such that app([b],[c],W) is a positive example of *append*. Since the answer is "W = [b,c]", the new example e_2 = "app([b],[c],[b,c])" is added to the set of known examples of *append* and treated in the same way. It is matched against c1 producing the new example e_3 = "app([],[c],[c])". Now, e_3 is handled in the same way, but it cannot be extensionally covered by c1, as the evaluation of "head([],H)" fails. So, it is matched with the head of clause c2 that is found to cover the example. Since no more examples are added and all the examples are extensionally covered by some clause in HS, the abductive reasoning can stop and so also the learning task. Clauses c1 and c2 have been learned.

In fact, the effect of building a chain of controlled abductive steps is that of producing a derivation tree in reverse order for the starting example *e*. That is, from *e* to the leaves of the tree, that will be clauses from the background knowledge, known (or queried) examples, and clauses from the hypothesis space (i.e. learned clauses).

If the same is done for each positive example given initially, and recursively for each queried positive example, then the first problem of extensionality is solved. If there exists a derivation tree for a positive example in the set {hypothesis space + background knowledge + known examples} such a tree is found, and together with it also the clauses necessary to the derivation. The found clauses belonging to the hypothesis space will be part of the description of the target concept.

However, the second problem of extensionality still remains. How can we avoid a (known) negative example to be derived by a learned description? In fact, the same strategy can be used, but in a complementary way. For each negative example, we test the existence of a derivation tree for it. If such a tree is found, clauses from the hypothesis space involved in the derivation will not be part of any description of the target concept. Let us show this with an example.

Consider the following (wrong) program P for *reverse*:

c1 = reverse(X,Y) :- head(X,H), tail(X,T1), head(Y,H), tail(Y,T2), reverse(T1,T2).
c2 = reverse(X,Y) :- null(X), null(Y).
c3 = reverse([X,Y,Z],[Z,Y,X]).

that can be learned by using this set of examples:

e_1 = reverse+([],[]), e_2 = reverse+([1],[1]), e_3 = reverse+([3,2,1],[1,2,3]).
e_4 = reverse-([3,2,1],[3,2,1]).

Then P $\vdash$ reverse([3,2,1],[3,2,1]), however, P is extensionally consistent with the provided examples. We can avoid learning P by means of a chain of abductive queries as follows: e_4 is matched against the head of c1 generating the query "reverse([2,1],[2,1])?" that is classified as a negative example (let call it e_5) by the user. Then, e_5 is matched against c1 and the clause is found to extensionally cover this example, since "reverse([1],[1])" is a known positive example. As a consequence, c1 extensionally covers a negative example (a queried one, indeed) and is rejected.

On the ground of the above examples, we have devised the following strategy. Every legal, i.e. permitted by the constraints of the learning scheme, clause of type

$$P(X,Y) \ :- \ \alpha(X,Y,Z_1,\ldots,Z_k).$$

is processed with the following Abductive Completion Procedure:

ACP

for every example $P(a,b)$ of P do

(1) ask the user for a set of values $c_1,\ldots,c_k$ for $Z_1,\ldots,Z_k$ such that $\alpha(a,b,c_1,\ldots,c_k)$ is true;

(2) for every $Q(\cdot)$ occurring in α as a conjunct, add the obtained example to the collection of its positive examples.

end Algorithm.

Notice that adding one example may cause the request of others. As a consequence, the procedure must be repeated for every clause, over and over, until no more examples are added.

Consider, for instance, the example about the ancestor relation given above, and the following two clauses:

$$\text{relative}(X,Y) \;:-\; \text{ancestor}(X,Y).$$
$$\text{ancestor}(X,Y) \;:-\; \text{parent}(W,Y), \text{ancestor}(X,W).$$

where

$< b, g >$ is the only positive example of "relative",

$< d, d >$ is the only negative example of "relative", and

$< f, f >$ is the only negative example of "ancestor".

By using the first clause, the user is queried for ancestor(b, g), and this is added to the positive examples of "ancestor". Using the second clause, the user is asked for a value of W such that parent(W, g) and ancestor(b, W) are both true. By inspecting the Parenthood Table we have the answer $W = d$.

The second clause can be used again with $X = b$ and $Y = d$, and we obtain the completed set of positive examples for "ancestor":

$+ < b, g >, < b, d >, < b, r >$

Not all possible examples have been added, only the ones that were useful for those two clauses, given the initial examples. If this is done for all the clauses that are possible a priori, i.e. that satisfy the given constraints, then problems 1 and 2 no longer hold:

THEOREM 7. *Suppose the examples given to an extensional learning system are completed with the ACP. If the learning system successfully exits its main loop and outputs a definition* Π *(for a concept P,) then* $\Pi \vdash P(a, b)$ *implies that* $P(a, b)$ *is extensionally covered by some clause in* Π.

Proof. Suppose by contradiction that (c1) $\Pi \vdash P(a,b)$ but (c2) $P(a,b)$ is not extensionally covered.

Let $P(X,Y) \; :- \; \alpha(X,Y,Z1,\ldots,Z_k)$ be the clause resolved against $P(a,b)$ in a refutation of : $-P(a,b)$. Suppose that $P \vdash \beta(q,r) \wedge \gamma$ and $P \vdash R(r)$, but no such r is a positive example of R. There must be one literal $R(Y)$ having this property, because of assumptions (c1) and (c2). Since the tuples of R must have been completed with the given procedure, the user has been queried for $R(r)$, and this must have been inserted as a negative example. Therefore, it cannot be extensionally covered. We could now repeat the same argument for R. This would produce a non-terminating chain of resolution steps, and a finished proof of $Q(q)$ would never be obtained, contradicting the hypothesis that $P \vdash Q(q)$. ■

As an immediate corollary of the above theorem we have

COROLLARY 8. *If the examples given to an extensional learning system are completed with the ACP, and the learning system successfully exits its main loop and outputs a definition P, then P is consistent.*

It should be noted that this abductive completion is done as a preprocessing step. Then, it guarantees that a solution consistent with the examples is found if it exists. Moreover, it does not require reconsideration of previously generated clauses, as do systems (e.g. MIS [Shapiro, 1983]) that ask for new examples only when they are needed and during the learning process.

7 DEALING WITH FUZZY RELATIONS

It is quite interesting to notice that the relational learning framework described in the previous sections can, to some extent, be extended to the case where fuzzy relations are present. We recall that a fuzzy relation or a fuzzy predicate $P(\cdot)$ no longer has a binary value true or false, but can take on any value, denoted with $\tau(P(\cdot))$ from the interval $[0,1]$. Historically, the semantic of vague (or fuzzy) predicates was introduced in [Zadeh, 1965]. Since it is way beyond our scope to provide here a complete formalization of fuzzy logic, for complete details refer to now classical texts such as [Dubois and Prade, 1980; Klir and Folger, 1988; Zimmerman, 1984] and to [Léa Sombé, 1990] for an interesting and in depth discussion of formalism for uncertain reasoning, including fuzzy logic.

In order to be able to deal with fuzzy predicates or relations we need to appropriately extend all the definitions introduced in the previous sections.

7.1 Fuzzy operators

In order to define a form of fuzzy abduction useful for our discussions we need to recall a few definitions.

- The standard logical operators can be extended in many ways, one possibility is the following

Fuzzy Logic Operators

$\neg P$	$P \vee Q$	$P \wedge Q$	$P \rightarrow Q$
$1-\tau(P)$	$\max(\tau(P),\tau(Q))$	$\min(\tau(P),\tau(Q))$	$\min(\tau(Q)-\tau(P)+1,1)$

- The extension of Modus Ponens in fuzzy logic can be explained as follows

$$\tau(Q) \geq \min(\tau(P \rightarrow Q), \tau(P))$$

We can then propose the following fuzzy abductive inference rule

$$\frac{\alpha \leftarrow \beta \qquad \alpha}{\beta}$$

$$\tau(\beta) \leq \tau(\alpha)$$

7.2 Learning fuzzy relations from examples

It is clear that when we talk about learning fuzzy relations from examples, we are not going to consider positive or negative examples, but instead examples with a"degree of positivity", i.e. the label is a number between zero and one. As a subcase, we can also have binary relations whose definitions depend upon fuzzy predicates as well.

The concept of"extensional definition" of a predicate is not affected by its fuzzyness. As in the binary case, a predicate is extensionally defined if all its values are given. For instance, the unary predicate"old(x)" can be extensively defined over the universe where the parenthood table is defined. The meaning of extensional evaluation, though, needs some new formalization.

DEFINITION 9. Let $\alpha(X_1, X_2, \ldots, X_n)$ be a conjunction of (possibly fuzzy) predicates where there is at least one occurrence of each of the variables (not necessarily distinct) $X_1, \ldots, X_n$. Let $0 \leq \theta \leq 1$. We say that $\alpha(a_1, a_2, \ldots, a_n)$ is *extensionally θ-true* if and only if

- in case $\alpha(X_1, \ldots, X_n) \equiv Q(X_1, \ldots, X_n)$, for some extensionally defined predicate Q, then $Q(a_1, \ldots, a_n)$ is an example of Q with degree of positivity at least θ.

- in case

  $$\alpha \equiv \gamma(X_{i_1}, \ldots, X_{i_h}, T_1, \ldots, T_l), Q(T_1, \ldots, T_l, X_{j_1}, \ldots, X_{j_k}),$$

  for some conjunction γ, predicate Q, $l \geq 0$ and

  $$\{X_{i_1}, \ldots, X_{i_h}, X_{j_1}, \ldots, X_{j_k}\} = \{X_1, \ldots, X_n\},$$

  then there must exist $e_1, \ldots, e_l$ such that both the conjunction $\gamma(a_{i_1}, \ldots, a_{i_h}, e_1, \ldots, e_l)$ and $Q(e_1, \ldots, e_l, a_{j_1}, \ldots, a_{j_k})$ must be extensionally θ-true.

We say that the conjunct $\alpha(X_1, \ldots, X_n)$ *extensionally θ-covers* an example of P, $P(a_1, \ldots, a_n)$ if and only if $\alpha(a_1, \ldots, a_n)$ is extensionally θ-true.

Given the above definition and a collection of examples for P of type $< x, y >, \tau(P(x, y)) >$, we can act as follows:

- Sort in non-decreasing order $0 \leq \theta_1 \leq \cdots \leq \theta_n \leq 1$ the values of the examples;

- for each value $\theta_i > 0$ apply Algorithm 1 (Simplified Foil) by using the θ_i-covering definition. Basically θ_i is used as a threshold value. All examples with value at least θ_i will be considered as "positive" and the others as negative;

- if for some value θ_i the program terminates and outputs a logic program Π_i we check whether Π_i is also a"good" output for any θ_j with $j \neq i$.

- If so the final output will be Π_i otherwise there will be no output.

Basically, we are looking for a program Π such that for any example of type $(< x, y >, \theta)$, with $\theta > 0$ the following two conditions are satisfied

(C1) there exists a threshold value $\theta' \leq \theta$ for which $< x, y >$ is θ'-covered;

(C2) for any threshold value $\theta' > \theta$, $< x, y >$ is not θ'-covered.

Since, by fixing a threshold value, we reduce the fuzzy learning problem to the binary one we can extend definition 3 as follows

DEFINITION 10. Let $0 < \theta \leq 1$ and let $E^{+,\theta}$ and $E^{-,\theta}$ be respectively, the set of examples whose value is at least θ and the set of examples whose value is less than θ.

A definition Π of P is *θ-complete* with respect to the examples $E^{+,\theta}$ of P if and only if

$$(\forall e \in E^{+,\theta})\Pi \vdash P(e).$$

A definition Π of P is *θ-consistent* with respect to the examples $E^{-,\theta}$ if and only if

$$(\forall e \in E^{-,\theta})\Pi \not\vdash P(e).$$

Therefore, we can formalize our learning goal by saying that we are looking for a definition Π of P which is θ-consistent and θ-complete for any θ in the range of values of the given examples.

For instance, suppose we want to learn the target concept TA Tall Ancestor. We use again the parenthood table as example. We also need to provide the truth values of the predicate tall(x) defined over the universe of the parenthood table. For instance suppose such values are

Tall Table

a	.3	b	.5
c	.6	d	.7
f	.8	g	.9
p	.1	r	.6
s	.6	t	.5

Suppose also that the set of examples is the following

> 0 $(< r, g >, .6), (< b, g >, .5), (< b, d >, .5), (< d, g >, .7), (< b, r >, .5), (< r, d >, .6), (< r, p >, .6)$;

$= 0$ $(< d, d >, 0), (< g, p >, 0)(< d, p >, 0)$.

Then by using $\theta_1 = .5$, Foil will output the following program

$$\begin{aligned} \text{TA}(X,Y) \quad &:- \quad \text{parent}(W,Y), \text{TA}(X,W). \\ \text{TA}(X,Y) \quad &:- \quad \text{parent}(X,Y), \text{tall}(X). \end{aligned}$$

It is easy to see that for any $\theta > 0$ such a program covers all the "positive" examples and no"negative" examples. However, differences in the given data may cause our new algorithm to fail producing an output. If for instance, we have as example $(< r, d >, .5)$ then the second clause will θ-cover it as a "positive" example for $\theta = .5$ but it will also cover it as a "negative" example for $\theta = .6$ since $\text{tall}(r) = .6$ and $\text{parent}(r, d)$ has value 1.

7.3 *Using the fuzzy abduction rule*

Obviously, the same problems that we described when dealing with binary relations are also present in the case of fuzzy relations. Moreover, we have some other problems related to the fuzzyness of some of the relations or predicates involved. Indeed, the user may have provided fuzzy truth values for such relations or predicates which are somehow inconsistent with the values attached to the examples. In this case, completing the set of examples using the proposed fuzzy abduction rule will also force the user to change if necessary the values which he/she has provided for the fuzzy predicates or relations. Indeed, suppose the system has to θ-cover the example $P(x, y) = \theta$ for some x, y and that the following candidate clause has been generated:

$$P(X,Y) \; :- \; \alpha(X,Y), Q(Y,Z).$$

Moreover, suppose that $\alpha(x, y)$ is extensionally θ'-true for every $\theta' \geq \theta$. Then, that clause

(case 1) will extensionally θ-cover $P(x, y)$ if there exists some value z such that $Q(y, z)$ is an example (known to the system) of Q with value θ; and

(case 2) it will not θ'-cover $P(x, y)$ for $\theta' > \theta$ only if there is no value z such that $Q(y, z)$ is an example (known to the system) of Q with value $\theta' > \theta$

Three cases are possible

(a) there exists one z such that $Q(z, y)$ has value θ;

(b) there exists w such that $Q(w, z)$ has value less than θ and greater than zero but there is no z such that $Q(z, y)$ has value θ;

(c) for any z $Q(z, y)$ has value greater than θ.

If (a) holds, using our fuzzy abduction rule we can ask for an example z for which $Q(z, y)$ has value θ and will add it to the list of examples. If (b) holds, the clause can never θ-cover $P(x, y)$. Therefore, we would have a definition which is not θ-complete. However, the example might be covered at a lower threshold and for no threshold value a "negative" example will be covered. The user must decide at this point whether there is an inconsistency in the data given to the system or whether to discard the clause

If instead (c) holds, our fuzzy abduction rule will not return any value therefore the clause will be discarded.

8 CLASSIFICATION RULES WITH NON BINARY PREDICATES

Abductive completion of collection of examples during a learning process can be quite useful even when dealing with non-fuzzy classification rules that contain non binary predicates and more generally, fuzzy predicates or sets. A model for learning classification rules in such a setting has been proposed in [Bergadano and Cuttello, 1993; Bergadano and Cutello, 1995; Bergadano and Cutello, 1995]. The main goal of this model is to introduce a theoretical framework for tuning membership functions inside a non-fuzzy classification system so to make the classification system *probably, approximately correct* as defined in [Valiant, 1984]. We recall that the learning process in such a case requires that the learned classification system is approximately correct with high probability and that the induction procedure is polynomial.

Under the requirement that the membership functions of the fuzzy predicates to be learned occur in a structurally given classification system, some interesting results were obtained. Though the general learning problem is computationally hard, under meaningful restrictions either on the underlying probability distribution or on the syntactic structure of the classification systems, fuzzy membership functions can be quickly learned by observing positive and negative examples. The obtained membership functions are probably approximately correct within a particular classification problem. Therefore, with respect to such a classification problem future examples will be correctly classified with high probability.

8.1 *The learning model*

Let $\mathcal{C}$ be a given concept class over a universe $\mathcal{U}$. For the time being we will suppose that $\mathcal{U} \subseteq \Re^k$ for some integer k. Therefore, each element of $\mathcal{U}$ is a k-tuple of real numbers whose intended meaning is to characterize specific subparts of the system under study. Therefore, k is the global dimension of the problem.

Let $C \in \mathcal{C}$ and suppose a system S_m of classification rules for C is given in the following form:

$$\begin{array}{lll} C(x) & :- & Q_1^1(p_{Q_1^1}(x)), \cdots, Q_{n_1}^1(p_{Q_{n_1}^1}(x)) \\ \cdots & \cdots & \cdots \qquad\qquad\qquad \cdots \\ C(x) & :- & Q_1^m(p_{Q_1^m}(x)), \cdots,, Q_{n_m}^m(p_{Q_{n_m}^m}(x)) \end{array}$$

where

- the predicates Q_i^j are taken from a set $\mathcal{P}=\{Q_1, ..., Q_n\}$ of given unary predicates;
- for all Q_j, p_{Q_j} is a projection function that returns the parameter which is of significance for Q_j.
- for every component i, the predicates $Q_1, ..., Q_{m_i}$ corresponding to i (i.e. such that p_{Q_j} returns the value of the i-th component for all $1 \leq j \leq m_i$) define a linguistic order.

Notice that

- it is possible that the same predicate occurs in different rules;
- concerning linguistic orders we have that the membership functions must be ordered in the following sense: if $j > i$ then for every $0 \leq \alpha \leq 1$ we have $\{x|Q_i(x) > \alpha\} \prec \{x|Q_j(x) > \alpha\}$, where $\prec$ is a suitably defined order relation.

The intuitive meaning of the rule system is that a given individual x is classified as a positive example for the concept C if one or more of the rule antecedents are *true* for x.

Therefore we can say that $x \in \mathcal{U}$ is a member of the concept C i.e. $C(x)$ is true if and only if

$$\tau\left(\bigvee_{j=1}^{m}\bigwedge_{i=1}^{n_j} Q_i^j(p_{Q_i^j}(x))\right) = 1$$

where τ is the the truth function. Since we are allowing some predicates in $\mathcal{P}$ to be fuzzy whereas C is not we make the following assumptions:

- A security parameter $0 < \theta < 1$ is given;
- $C(x)$ is θ-true if and only if

  $$\tau\left(\bigvee_{j=1}^{m}\bigwedge_{i=1}^{n_j} Q_i^j(p_{Q_i^j}(x))\right) > \theta$$

  and where the truth value above is computed according to the min-max semantic, i.e.

  $$\tau\left(\bigvee_{j=1}^{m}\bigwedge_{i=1}^{n_j} Q_i^j(p_{Q_i^j}(x))\right) = \max_{j=1,\dots,m}(\min_{i=1,\dots,n_j}(Q_i^j(p_{Q_i^j}(x))).$$

Given a set F=$\{\mu_1, ..., \mu_n\}$ of membership functions associated to the predicates in $\mathcal{P}$, the fuzzy classification system (FCS for short) (1) will be denoted by S_m^F. The notation S_m will then denote the collection of all possible FCS's S_m^F, which in turn can be characterized as the collection of all sets F of membership functions.

We will suppose that the unknown membership functions we are trying to learn are *convex*. In particular, the above implies that membership functions do not have local maxima. Moreover, concerning the linguistic order the convexity hypothesis implies that for every predicate Q the set $\{x|Q(x) > \alpha\}$ is an interval of the real line for every α. So, we can suppose that the order relation $\prec$ is the standard order relation on real intervals:

$$[a, b] \prec [c, d] \text{ iff } a \leq c \text{ and } b \leq d.$$

Such FCS's will be called *convex* fuzzy classification systems.

By definition, S_m is connected to a non fuzzy predicate, i.e. non fuzzy concept C. Our goal is to learn the membership functions from positive and negative examples of the concept C, in the hypothesis that a classification system S_m^F will be used with respect to the fixed threshold value θ. The obtained truth value will be denoted by $S_m^F(x)$. In the end, x is classified as a θ-positive example of C if $S_m^F(x) > \theta$.

8.2 *Learnability of FCS*

As proven in [Bergadano and Cuttello, 1993; Bergadano and Cutello, 1995], the problem of PAC-learning the class of convex FCS can be approached as follows:

- given ϵ and δ draw at least

  $$\max(\frac{4}{\epsilon}\ln\frac{2}{\delta}, \frac{16m}{\epsilon}\ln\frac{13}{\epsilon}),$$

  examples from the distribution D;

- determine the hypercubes of $\Re^k$ where there are θ-positive examples only (the *positive hypercubes*); suppose the hypercubes are $m' \leq m$;

- use one rule of S_m to cover each θ-positive hypercube.

- project the m' positive hypercubes on the i-th component, obtaining $m_i' \leq m_i$ positive intervals $\{[a_1, b_1], ..., [a_{m_i}, b_{m_i}]\}$

- define Q_j^i as a trapezoidal fuzzy set *using* the above defined intervals.

The following result holds (see [Bergadano and Cuttello, 1993; Bergadano and Cutello, 1995]).

THEOREM 11. *The general learnability problem is $\mathcal{NP}$-hard.*

Similarly to any The theorem is proven by using the result proven in [Pitt and Valiant, 1988] that learning a K-term DNF by a K-term DNF is $\mathcal{NP}$-hard. We recall that a K-term DNF is a boolean formula of type:

$$\bigvee_{i=1}^{K} M_i$$

where the M_i are monomials, i.e. for all $i = 1, 2, \ldots, K$

$$M_i = x_{i_1} x_{i_2} \cdots x_{i_a} \cdot \bar{x}_{i_{a+1}} \cdots \bar{x}_{i_{n_i}}$$

In details the problem of learning K-term DNF by K-term DNF is reduced to the problem of learning S_K by S_K.

PAC learnability results are obtained by imposing some restrictions. In particular, if we assume that the underlying distribution is such that every rule *fires* at least once then the learning problem can be quite easily solved. Notice in particular that this assumption is quite reasonable, as we may suppose that the classification rule is given by an expert who will not include a rule in S_m if it is totally useless relatively to the *world* (i.e. the probability distribution of events) he/she lives in.

8.3 Completing the set of examples

All the results above cited were obtained under the following assumptions:

(a1) The teacher (i.e. the experts who provides the structure of the classification system and the labels for the examples) does not need to justify the rules in the system;

(a2) the classification system may not be minimal, i.e. there may be"redundant rules";

(a3) for each component the linguistic order is given.

In order to bypass the hardness result it is obvious that more power must be given to the learner. In particular, in [Bergadano and Cutello, 1995] it was introduced the notion of"equivalence query". The learner was allowed to guess and ask for counter examples. Using this powerful tool and still under assumptions (a1)-(a3), PAC-learnability was achieved.

We now propose a way to extend the results obtained and get rid of conditions (a1)-(a3). In details, the abductive completion procedure (ACP, section 6) in such a framework can be adjusted as follows:

for each rule

$$Q_1(p_{Q_1}(x)) \wedge \cdots \wedge Q_n(p_{Q_n}(x)) \rightarrow C(x)$$

occurring in the system, we ask for a positive example x such that

$$\max_{j=1,\ldots,m}(\min_{i=1,\ldots,n_j}(Q_i^j(p_{Q_i^j}(x))) = \min_{i=1,\ldots,n}(Q_i(p_{Q_i}(x)).$$

If no such a tuple x exists then the rule is dropped from the classification system. It is then clear that

- by forcing the teacher to justify the given rules, the system obtained is"minimized", that is to say rules which have no influence in the classification process (no matter what the value of θ is) are eliminated.
- using a polynomially bounded number of examples, we are able in polynomial time to produce a probably approximately correct classification system, i.e. we achieve PAC-learnability;
- finally, the learner no longer needs to be given the linguistic orders for each component.

9 AUTOMATIC ABDUCTION OF KNOWLEDGE: TESTING

We have seen as abduction can be used in conjunction with induction, in order to alleviate the intrinsic weaknesses of this learning paradigm. However, one may reasonable wonder at which conditions such approach is practical. There are two main issues to consider. First, the abductive procedure terminates only if there is a finite number of answers to each question. This is the case if we are in a finite domain, as in the case of the *ancestor* relationship. In the case of infinite domains, the approach is still acceptable if we work with functional relations. By imposing adequate input/output constraints [Bergadano and Gunetti, 1993], it is possible to work only with positive examples. For each input tuple of the target relation, there is only one positive example matching the tuple, and infinite negative examples. the system can be instructed to work only with the positive examples, assuming the negative examples being all the examples with the same input as the positive ones and different output values. As a consequence, for each query, the user must provide exactly one positive example, and no more.

Second, the number of queries to the user must not only be finite, but also quite small, unless a database of examples is available, and the queries can be handled automatically. This depends mainly on the size of the hypothesis space, i.e. on the number of clauses that can be used to generate queries. In fact, it is becoming more and more clear as learning complex first order theories is quite difficult, and a lot of knowledge must be used. As a consequence, the user must not be a passive agent, just providing the examples of the target relation. She must also be able to design a restricted hypothesis space on the basis of the sought relation [Bergadano and Gunetti, 1996; Van Laer *et al.*, 1994]. Also the number of examples provided initially may influence the number of queries. Experiments based on the presented approach have shown as it is possible to learn complex relations by using very few representative examples of the target concept, and letting the abductive procedure asking for the missing ones. For example, it was possible to learn a program for inserting a key into a balanced tree (rebalancing the tree if necessary) by using an extensional approach and just one well chosen initial positive example. With a hypothesis space of 2^{24} clauses the user was queried for 15 more examples, and the program could be learned in 1831 seconds on a Sun Sparcstation 5. The program was composed of 9 clauses, eight of them recursive [Bergadano and Gunetti, 1996]. In general, it can be shown that an initial positive example is sufficient to learn all the clauses necessary to derive it, if the abductive completion is applied.

In this section we illustrate an interesting application of this combination of abduction and induction, outside the field of Machine Learning. We present a technique to test the correctness of a given program where, moreover, (some of) the test cases are automatically generated by the Abductive Completion Procedure through queries made to the program being tested.

The learning procedure described in Algorithm 1 outputs a set of Horn clauses. As well known, this can be regarded as a logic program. As a consequence, in the rest of this section we will use the term *Inductive Program Learning* (IPL) in place of concept learning, and we will speak of *Logic Programs* instead of concepts described through a set of Horn clauses.

9.1 Induction and Testing

Testing is the field of Computer Science concerned with the problem of detecting the presence of errors in programs by executing them on specific and well-chosen input data. A set of test data allowing us to discover all the errors in the program is called *adequate*. Though it can be formally proved that no algorithm exists to generate an adequate set of test data for an arbitrary program, testing is necessary and widely used in software development. It is practical and (relatively) easy to apply, and can give some (even very precise) information about the correctness of a program.

Testing can be seen as a way of distinguishing a program from all the possible (syntactically correct) alternatives to that program. This is very similar to IPL, where a program in the hypothesis space must be identified from among all the other possible programs on the basis of the given examples. More precisely, testing and IPL are, in some way, symmetric. The latter goes from examples to programs, whereas the former goes from programs to examples (input values).

Most of the time, a program is tested on the basis of the "correct program the user has in mind". It is common in this case to assume that although the correct program is unknown, there is a known *set of programs* that can be seen as alternative implementations and should contain at least one correct solution. This concept of program correctness leads to a practical definition of meaningful test data ([DeMillo and Offutt, 1991]):

DEFINITION 12. ([Howden, 1976]) A test set T for a program P is *reliable* (with respect to a correct program P_c) iff $[\forall x \in T \;\; P(x) = P_c(x)] \rightarrow P \equiv P_c$
If P contains errors, this will be shown by the test cases in T.

DEFINITION 13. Let there be given a test set T for a program P and a set $\mathcal{P}$ of alternative programs. We say that T is *adequate* with respect to $\mathcal{P}$ if and only if it is reliable with respect to every program in $\mathcal{P}$.

As a consequence, if $\mathcal{P}$ contains at least one correct program P_c, then the test set will also be reliable for P. Many different testing techniques are based on a relation between the program P to be tested and a set of possible alternatives P_c. In particular, fault-based methods [DeMillo and Offutt, 1991; King and Offutt, 1991] use a set of typical programming errors defined a priori, and alternative programs are obtained by inserting some of these errors (that is, possible program mutations) in P.

Testing based on the inductive learning of logic programs also relies on definition 13 of adequateness of a test set. As noted above, an intuitive symmetry between induction and testing can almost immediately be noticed [Weyuker, 1983]: induction is the inference from examples to programs; testing is the inference from programs to input values. Given a test set T of input values for a program P, the examples of P for T are defined as $E(P,T)=\{<i,o> \mid i \in T \text{ and } P(i)=o\}$. This notion is formalized by Weyuker as follows [Weyuker, 1983]:

DEFINITION 14. A test set T is *program adequate* for a program P if and only if P_I is inductively inferred from E(P,T) and $P_I \equiv$ P.

The intuitive meaning of the above definition is as follows: If, given a set E(P,T) of input/output examples, we inductively infer a program $P_I \equiv P$, then T is likely to be useful for testing the program P.

In the previous sections we have described a a basic learning method which is terminating and explicitly make use of a finite set of legal programs. More precisely, the hypothesis space is made of Horn clauses, and every possible subset of such Horn clauses is a program that could be learned by the induction procedure, on the basis of the given examples. We will call $\mathcal{P}$ the set of all possible programs. If there are n Horn clauses, there are 2^n possible legal programs in $\mathcal{P}$.

The learned program P_I must belong to $\mathcal{P}$. This is consistent with the fact that all the test case generation procedures that are not specification based refer (sometimes implicitly) to a set $\mathcal{P}$ of alternative programs. As a consequence, definition 14 can then be stated in terms of "sound and complete[2] IPL": T is adequate for P if and only if P is the only program that can be learned from E(T,P) with a sound and complete IPL method. The restriction to a finite set of alternative programs (hypothesis space) $\mathcal{P}$ seems to be acceptable for testing; for instance, all approaches

[2]Formally, a system is defined to be *sound* if it only outputs programs that are consistent w.r.t. the given examples. It is *complete* if it is able to find a consistent program whenever such a program exists in the given hypothesis space. As an instance, the learning procedure of Algorithm 1 (Simplified Foil), is neither sound nor complete, by virtue of Problems 1 and 2.

to fault-based testing such as mutation analysis, are based on this assumption.

The intuitive symmetry between induction and testing can now be made clear: induction is an inference from the pair <E(T,P), $\mathcal{P}$> to a program P in $\mathcal{P}$, whereas testing is an inference from <P, $\mathcal{P}$> to the test set T. In the next subsection this symmetry will be used in actually *generating* the adequate test cases for a (not necessarily logic) program.

9.2 *Abductive Test Case Generation*

Let us call *SFOIL* (Simplified Foil) the learning procedure of Algorithm 1. Let us call $SFOIL_{ACP}$ the *SFOIL* algorithm augmented with the Abductive Completion Procedure and with input/output constraints, as discussed at the beginning of this section. A sequence of inductions of programs from examples is used to generate test cases. Initially, there are no examples and, in the end, the generated set of examples will be adequate in the sense of definition 13.

Let P be the program to be tested. At any given moment, the examples generated so far are used to induce a program P′. New examples that distinguish P from P′ are added, and the process is repeated until no program P′ that is not equivalent to P can be generated. This procedure is described in more detail below.

Test case generation procedure:

Input: a program P to be tested,
a finite set $\mathcal{P}$ of alternative programs
Output: an adequate test set T

T ← ∅;
loop:
<P′,T> ← $SFOIL_{ACP}$(E(T,P),$\mathcal{P}$)
if P′ = "fail" **then return** T
if (∃ i) P′(i) ≠ P(i)
  then T ← T ∪ {i}
  else $\mathcal{P}$ ← $\mathcal{P}$ - P′
goto loop

In order for the above procedure to work, the learning algorithm $SFOIL_{ACP}$ must be sound and complete, as it is proved by theorem 7 and corollary 8. In the test case generation procedure, the test set T is initially empty. The main step in the loop consists of using $SFOIL_{ACP}$ to learn a program P′ that is consistent with the examples generated so far. P′ is then ruled out either by (1) adding an input value to T or by (2) removing it from $\mathcal{P}$. As a consequence, P′ will not be learned again. When $SFOIL_{ACP}$ cannot find a program P′∈ $\mathcal{P}$ that is consistent with the examples, then the only programs with this property are P and those equivalent to it, i.e., the test set T is adequate. This is proved by the following:

THEOREM 15. *Let equivalence be decidable for programs in $\mathcal{P}$. Then the above test case generation procedure outputs an adequate test set T for P.*

Proof. Since equivalence is decidable for programs in $\mathcal{P}$, the procedure terminates, as the condition $(\exists\ i)\ P'(i) \neq P(i)$ always produces an answer. Suppose, by contradiction, that the obtained test set T is not adequate. This means that there is $P' \in \mathcal{P}$ and an input $i \notin T$ such that $P'(i) \neq P(i)$, but $(\forall\ v) \in T,\ P'(v) = P(v)$. However, just before termination, $SFOIL_{ACP}$ failed to induce a program P_I. Moreover, P′ is complete and consistent with respect to E(T,P), because $(\forall\ v) \in T,\ P'(v) = P(v)$. Then, just before termination, and because $SFOIL_{ACP}$ is sound and complete, it must be the case that $P' \notin \mathcal{P}$. As a consequence, P′ must have been removed from $\mathcal{P}$ at some previous iteration. But, in that case, there must have been an input value i such that $P'(i) \neq P(i)$, and this value would have been added to T. This contradicts the assumption that $i \notin T$ and T is not adequate. ∎

If a decision procedure is available for finding an input i such that $P(i) \neq P'(i)$, this is used directly in the above test generation method. As this is not true in general, examples are found by enumerating (in some random or ad hoc order) the possible inputs i, and stopping when an i is found such that $P(i)=o \neq P'(i)=o'$. This enumeration could also be guided by other test case generation techniques, such as path or functional testing. The requirement of decidable equivalence is of course not easily verified or accepted. is a major theoretical and practical issue in program testing. In the implementation of our test case generation method, we approximate it by means of its time-bounded semi-decision procedure. Except for this approximation, the system produces adequate test sets with respect to any finite class of programs $\mathcal{P}$. The reader may find all the details of this testing technique in [Bergadano and Gunetti, 1996b].

9.3 *An example: testing a* merge *program*

Consider the problem of merging two ordered lists, where we allow them to contain repeated elements. The output list must be ordered and should contain every element of the input lists only once. This problem requires a procedure for removing the elements that are repeated. Suppose we are given the following:

```
remove(X,[X|Y],Z) :- !, remove(X,Y,Z).
remove(_,Y,Y).
```

This does just as much as necessary: it removes the initial occurrences of an element in a list, e.g., remove(a,[a,a,b],[b]), but remove(a,[b,a],[b,a]). Let the program to be tested be the following:

P:

```
1) merge(X,Y,Z) :- null(X), Y=Z.
```

2) merge(X,Y,Z) :- null(Y), X=Z.
3) merge(X,Y,Z) :- head(X,X1), tail(X,X2), head(Y,X1), tail(Y,Y2),
merge(X2,Y2,W), remove(X1,W,T), cons(X1,T,Z).
4) merge(X,Y,Z) :- head(X,X1), tail(X,X2), head(Y,Y1), tail(Y,Y2),
X1<Y1, merge(X2,Y,W), remove(X1,W,T), cons(X1,T,Z).
5) merge(X,Y,Z) :- head(X,X1), tail(X,X2), head(Y,Y1), tail(Y,Y2),
X1>Y1, merge(Y2,X,W), remove(Y1,W,T), cons(Y1,T,Z).

In this program the first two clauses are wrong. In fact, any repeated element in the nonempty list will be present in the output. Let $\mathcal{P}$ be defined as any program made of any clause that can be built by using as head *merge(X,Y,Z)* and whose body is made of a subset of the literals occurring in P. As a consequence, $\mathcal{P}$ contains $2^{2^{19}}$ alternative programs, i.e., all possible subsets of the space of clauses. Among the subsets there are versions of the correct implementation of *merge*. A concise representation of $\mathcal{P}$ will be given as input to $SFOIL_{ACP}$. Of course, $\mathcal{P}$ is not generated explicitly. See [Bergadano and Gunetti, 1996b] for the implementative details.

The test case generation procedure starts with an empty test set T_0 of input values, and calls $SFOIL_{ACP}$. As $E(T_0,P)$ contains no examples, the empty program P_0 is an acceptable output of $SFOIL_{ACP}(E(T_0,P),\mathcal{P})$.

Pairs of lists X and Y are then enumerated, so that
$P_0 \vdash \text{merge}(X,Y,Z')$, $P \vdash \text{merge}(X,Y,Z)$ and $Z \neq Z'$
The first such pair that is found is <X,Y>=<[],[]>; for this input, P_0 produces no output and P outputs Z=[]. The new test set is then $T_1 = \{<[],[]>\}$.

$SFOIL_{ACP}(E(T_1,P),\mathcal{P})$ is called again, yielding P_1:
merge(X,Y,Z) :- Y=Z.
This program is an acceptable output of $SFOIL_{ACP}$ because merge([],[],[]) is derived from it, and the output is the same as that of P.

Pairs of lists X and Y are enumerated, so that
$P_1 \vdash \text{merge}(X,Y,Z')$, $P \vdash \text{merge}(X,Y,Z)$ and $Z \neq Z'$. The first such pair that is found is:
<X,Y>=<[1],[]>; for this input, P_1 outputs Z=[] while P outputs Z=[1]. The new test set is then $T_2 = T_1 \cup \{<[1],[]>\}$.
$SFOIL_{ACP}(E(T_2,P),\mathcal{P})$ is called again, yielding P_2:
merge(X,Y,Z) :- X=Z.
This program is an acceptable output of $SFOIL_{ACP}$ because merge([],[],[]) and merge([1],[],[1]) are derived from it.

The test case generation goes on in this way generating new test cases and learning new programs, until no new program can be discovered and $SFOIL_{ACP}$ fails, ending the test generation process. The complete set of learned test cases, with the corresponding outputs, is the following:

1: merge([],[],[])
2: merge([1],[],[1])
3: merge([2],[1],[1,2])
4: merge([2],[],[2]) – abduced by $SFOIL_{ACP}$ from P
5: merge([],[1],[1]) – abduced by $SFOIL_{ACP}$ from P
6: merge([1],[1],[1])
7: merge([1,1,2],[1],[1,2])
8: merge([1,1,2],[],[1,1,2]) – abduced by $SFOIL_{ACP}$ from P, it shows the error
9: merge([1],[2],[1,2])

Only nine examples have been required to locate the error, whereas many more would have been necessary in random testing, if there are many possible element values with respect to the average list length.

We see that some of the examples (one of them showing the error) are added as a consequence of the ACP. Consider, e.g., example 7. When $SFOIL_{ACP}$ is learning a program P_5 consistent with examples 1-7, it has to find a clause covering example 7. During its construction, it comes to the (partial) clause:

merge(X,Y,Z) :- head(X,X1), tail(X,X2), head(Y,X1), tail(Y,Y2), merge(X,Y2,W)

There is no example discovered so far covering literal merge([1,1,2],[],W) and, as consequence, the ACP queries the program P (that is, it runs it) for a value o such that merge([1,1,2],[],o) is true, getting the answer o=[1,1,2] and producing the new testing case represented by example 8 above.

It should also be noted that P does not belong to the hypothesis space $\mathcal{P}$; actually, this is not required by the test case generation system. In fact, the program to be tested could be used as a black box and could, in principle, be written in any programming language.

10 CONCLUSION

The relation between induction and abduction is briefly discussed in this paper. Our main goal was then to show the specific uses of abductive reasoning in Machine Learning. On one hand, it has been used to guide search in a top-down specialization framework related to Explanation-based Learning. On the other hand, we have also shown how abduction can be used to query the user for examples that may be missing. This means that the user must not provide all the needed examples to learn one definition at a time. He/she can forget some examples, and the abductive procedure will ask for them. Observe that only the examples really needed are queried, so it will not waste time trying to cover useless examples. In many extensional systems [Muggleton and Feng, 1990; Quinlan, 1990] the user must provide all the examples at one time, and usually a superset of the examples needed is given, resulting in a lot of time wasted in covering useless examples. We have also highlighted a possible extension of such a learning framework to the case where fuzzy predicates occur. In such a case, the user not only must provide all the missing example but must also provide consistent fuzzy values for all the predicates involved.

ACKNOWLEDGEMENTS

The authors thank esprit project DRUMS II which supported part of this research.

F. Bergadano and D. Gunetti
Computer Science Department, University of Torino, Italy.

V. Cutello,
Mathematics Deptartment, University of Catania, Italy.

REFERENCES

[Bergadano and Besnark, 1994] F. Bergadano and P. Besnard. Abduction and induction by non-monotonic logics, *Workshop on Mathematical and Statistical Methods in Artificial Intelligence*, Udine, Italy, 1994.

[Bergadano and Cuttello, 1993] F. Bergadano and V. Cutello. Learning membership functions. *Proceedings of ECSQARU'93*, 2nd European Conference on Symbolic and Quantitative Approaches to reasoning and uncertainty, Granada, Spain, 1993. *Lecture Notes in Computer Science, Springer-Verlag*, Vol. 747, pp. 25–32, 1993.

[Bergadano and Cutello, 1995] F. Bergadano and V. Cutello. Learning fuzzy sets. In *Proceedings of EUFIT'95*, G. J. Zimmermann, ed. Aachen, Germany, 1995.

[Bergadano and Cutello, 1995] F. Bergadano and V. Cutello. Probably approximate correct (PAC) learning in fuzzy classification systems. *IEEE Transactions on Fuzzy Systems*, **3**, 473–478, 1995.

[Bergadano and Giordana, 1988] F. Bergadano and A. Giordana. A knowledge intensive approach to concept induction. *Proceedings of the Fifth International Conference on Machine Learning*. pp. 305–317, Ann Arbor, MI, 1988.

[Bergadano *et al.*, 1989] F. Bergadano, A. Giordana and S. Ponsero. Deduction in Top-down Inductive Learning, *Proceedings of the Sixth International Conference on Machine Learning*. pp. 23–25, Ithaca, NY, 1989.

[Bergadano and Gunetti, 1993] F. Bergadano and D. Gunetti. An interactive system to learn functional logic programs. In *Proceedings of the International Joint Conference on Artificial Intelligence*, Chambery, France, 1993. Morgan Kaufmann.

[Bergadano and Gunetti, 1996] F. Bergadano and D. Gunetti. *Inductive Logic Programming: from Machine Learning to Software Engineering*. MIT Press, Cambridge, MA, 1996.

[Bergadano and Gunetti, 1996b] F. Bergadano and D. Gunetti. Testing by means of inductive program learning. *ACM Transactions on Software Engineering and Methodology (ACM TOSEM)*, **5**, 1996.

[Birnbaum and Collins, 1991] L. Birnbaum and G. Collins, eds. *Proceedings of the International ML Conference, part VI: Learning Relations*. Morgan Kaufmann, 1991.

[Cohen, 1992] W. Cohen. Abductive explanation-based learning: a solution to the multiple inconsistent explanation problem. *Machine Learning*, **8**, 167–219, 1992.

[Cohen, 1994] W. Cohen. Incremental abductive explanation-based learning. *Machine Learning*, **15**, 5–24, 1994.

[Console and Saitta, 1993] L. Console and L. Saitta. Generalization in learning and abduction. technical report, University of Torino, 1994.

[DeMillo and Offutt, 1991] R. A. DeMillo and A. J. Offutt. Constraint-based automatic test data generation. *IEEE Trans. on Software Engineering*, **17**, 900–910, 1991.

[DeRaedt and Bruynooghe, 1991] L. DeRaedt and M. Bruynooghe. CLINT: A Multistrategy Interactive Concept-Learner and Theory Revision System. In R. S. Michalski and G. Tecuci, eds. *Proceedings of the Workshop on Multistrategy Learning*, pp. 175–190, Harpers Ferry, VA, 1991.

[Dubois and Prade, 1980] D. Dubois and H. Prade *Fuzzy Sets and Systems: Theory and applications*. Academic Press, NY, 1980.

[Ellman, 1989] T. Ellman. Explanation-based learning: a survey of programs and perspectives. *ACM Computing Surveys*, **21**, pp. 163–222, 1989.

[Flach, 1992] P. Flach. Simply Logical, John Wiley and Sons, 1992.

[Howden, 1976] W. E. Howden. Reliability of the path analysis testing strategy. *IEEE Trans. on Software Engineering*, **2**, 208–215, 1976.

[Klir and Folger, 1988] G. J. Klir and T. A. Folger. *Fuzzy Sets uncertainty and Information*. Prentice Hall, Englewood Cliffs, NJ, 1988

[King and Offutt, 1991] K. N. King and A. J. Offutt. A Fortran language system for mutation-based software testing. *Software Practice and Experience*, **21**, 685–718, 1991.

[Léa Sombé, 1990] Léa Sombé *Reasoning Under Incomplete Information in Artificial Intelligence*. John Wiley and Sons Inc., 1990.

[Van Laer *et al.*, 1994] W. Van Laer, L. Dehaspe and L. DeRaedt. Applications of a logical discovery engine. Leuven, Belgium, 1994. Technical Report, Dept. of CS, Univ. of Leuven.

[Mitchell *et al.*, 1986] T. Mitchell, R. M. Keller and S. Kedar-Cabelli. Explanation-based generalization: a unifying view, *Machine Learning*, 47–80, 1986.

[Muggleton and Feng, 1990] S. Muggleton and C. Feng. Efficient induction of logic programs. In *Proceedings of the First Conference on Algorithmic Learning Theory*, Tokyo, Japan, 1990.

[Pazzani *et al.*, 1991] M. Pazzani, C. A. Brunk and G. Silverstein. A Knowledge-intensive approach to learning relational concepts. *Proceedings of the 8th International Conference on Machine Learning*, 1991.

[Pazzani and Kibler, 1992] M. Pazzani and D. Kibler. The utility of knowledge in inductive learning. *Machine Learning*, **9**, 57–94, 1992.

[Pitt and Valiant, 1988] L. Pitt and L. G. Valiant. Computational limitations on learning from examples. *Journal of ACM*, **35**, 965–984, 1988.

[Quinlan, 1990] R. Quinlan. Learning Logical Definitions from Relations. *Machine Learning*, **5**, 239–266, 1990.

[Rouveirol, 1992] C. Rouveirol, ed. *Proceedings of the ECAI Workshop on Logical Approaches to Learning*. ECCAI, Vienna, Austria, 1992.

[Shapiro, 1983] E. Y. Shapiro. *Algorithmic Program Debugging*. MIT Press, 1983.

[Valiant, 1984] L. Valiant. A theory of the learnable. *Communications of the ACM*, **27**, 1134–1142, 1984.

[Weyuker, 1983] E. J. Weyuker. Assessing Test Data Adequacy Through Program Inference. *ACM Trans. on Programming Languages and Systems*, **5**, 641–655, 1983.

[Zadeh, 1965] L. A. Zadeh. Fuzzy sets. *Information and Control*, **8**, 338–353, 1965

[Zimmerman, 1984] H. J. Zimmermann *Fuzzy Set Theory and its applications*. Kluwer-Nijhoff, Boston, 1984.

DIDIER DUBOIS AND HENRI PRADE

AN OVERVIEW OF ORDINAL AND NUMERICAL APPROACHES TO CAUSAL DIAGNOSTIC PROBLEM SOLVING

1 INTRODUCTION

Diagnostic problem solving aims at finding out disorders which may have caused observed symptoms of ill-behaviour. It is often called "abductive reasoning". The pattern which, from the two premises "if disorder is present then symptom is observed" and "symptom is observed", infers the conclusion "disorder is plausible", can be viewed as the simplest pattern of abductive reasoning [Peirce, 1940]. This pattern can be contrasted with deductive inference which, from "if disorder is present then symptom is observed" and "disorder is present" infers that "symptom is observed". Clearly, a key problem in abduction is to provide some status to the abductive conclusion "disorder is plausible" and maybe to assign some degree of plausibility to this disorder, or at least to be able to rank possible causes of an observed state of facts according to their relative plausibilities. The abductive pattern, although considered by many authors (e.g., [Pólya, 1968]), has not been given any logical or numerical formalization until recently, if we except the Bayesian model (which requires more information and where we compute the a posteriori probability of disorders on the basis of observations), and some heuristic, numerically quantified attempts (e.g., [Friedman, 1981; Bandler and Kohout, 1984; Hall, 1987]).

We can distinguish between several directed approaches to abductive reasoning. A first dichotomy is between heuristic methods and causal methods. In heuristic methods, knowledge about the system to be diagnosed is described by means of evidential rules relating a set of symptoms to the expected presence of a disorder. On the contrary, causal methods describe what are the symptoms produced by disorders. Another dichotomy is between classification methods and the abductive methods. In classification methods it is necessary to assume that the disorder explaining the symptom is unique. In abductive methods this assumption is dropped. Typical of the abductive methodology is the relational approach where a (possibly valued) relation encodes the causal knowledge under the form of direct links from disorders to symptoms, and where possible causes are retrieved from observed symptoms by exploiting the topology of the relation. The basic proposal that adopts this point of view is the theory of parsimonious covering of Peng and Reggia [1990]. There are various numerical extensions to that theory according to the underlying semantics of the weights attached to causal links. Fuzzy relational approaches seem to account for the notion of intensity of a symptom (e.g., influenza causes *strong* fever) while other proposals attach uncertainty weights

D. M. Gabbay and R. Kruse (eds.), Abductive Reasoning and Learning, 231–280.

to causal links, expressing how surely a symptom will appear in the presence of a disorder (e.g., influenza *often* causes fever). Such uncertainty weights can be probabilities, degrees of possibility, or degrees of belief in the sense of Shafer [1976].

The simplest relational schemes do not offer to represent prior knowledge about disorders (e.g., with adults, influenza is more likely than chickenpox). Conditional approaches where a causal link is understood in terms of conditional uncertainty measures have the capability of modelling such prior information. The computation of a diagnosis comes down to evaluating a posterior measure for a disorder (a probability or a possibility) through a revision of the priors on the basis of new observations.

Usually, direct links between disorders and symptoms are not available, and represent rather shallow causal knowledge. Deep causal knowledge relate disorders to symptoms via intermediary state variables, thus forming a causal network. Although the non-numerical theory of parsimonious covering can be extended to causal nets, the most widely known method for representing deep causal knowledge is the Bayesian network. Indeed the machinery of conditioning offers a natural way of chaining causal links while the graphical representation takes care of independence assumptions. Recent results suggest that the Bayesian net paradigm can be extended to other uncertainty theories that handle partial ignorance, such as possibility theory.

The diagnostic problem-solving methods surveyed here differ from the logic-based techniques often dubbed "model-based diagnosis" [Hamscher *et al.*, 1992]. These techniques start with a model of normal system behavior, which is not necessarily causal since it is described as a logical theory, while the methods described here handle fault-models and do not explicitly refer to a nominal behaviour. In logical approaches, surveyed in this volume by Flach, diagnosis is sometimes performed via a consistency check. One looks for a minimal set of abnormality assumptions about the system to be diagnosed, such that the observations are consistent with the assumption that the remaining of the system works correctly [Reiter, 1987]. Such a set of abnormality assumptions is called a diagnosis and is consistent with the logical theory formed by the model of the system and the observations. A more refined diagnosis is formed by any minimal set of abnormality assumptions which implies the observations; this is the abductive view of model-based diagnosis (see, e.g., [Console and Torasso, 1992]). Similar modes of diagnosis are already at work in the relational approaches, as indicated in the following. Moreover some logic-based methods have been extended to allow for a quantification of uncertainty.

The paper is organised as follows. In Section 2, the most traditional number-crunching approaches to the diagnosis problem are discussed, namely the statistical Bayesian approach and the pattern-directed rule-based inference school. The former is rigorous but involves very restrictive assumptions, and the latter looks user-friendly but is rather ad hoc. An early diagnosis technique by Smets [1978], that generalizes the early Bayesian scheme to belief functions is also described, as well as a conditional method for possibilistic abduction. These approaches basically come down to a classification procedure. Their limitations are discussed. Section 3 presents the most elementary relational approaches as well as basic issues pertaining to the treatment of uncertainty and intensity of symptoms and disorders in diagnosis problems. First, the parsimonious covering theory developed by Reggia *et al.* [1985], is presented as the underlying framework, and its extension to the handling of incomplete causal knowledge. A distinction is made between fuzzy relational methods on the one hand, first proposed by Sanchez [1977a; 1977b] and Tsukamoto and Terano [1977], which are appropriate when the intensity of the disorders and of the symptoms are a matter of degree; and on the other hand methods where the presence of disorders or symptoms is not a matter of intensity but may be pervaded with uncertainty. In this second case, disorders and symptoms are either present or absent, but we are unsure about the presence of a symptom when a disorder is present or about the observation of a symptom. The lack of confidence due to incomplete information can be either graded in terms of certainty and mere possibility, or in terms of probability. Section 4 presents the probabilistic extension of the parsimonious covering approach that can capture prior information and introduces the notion of causation events. The Bayesian network approach is then briefly presented, its merits and its limitations are discussed. Preliminary investigations on a possibility theory-based version of the conditional approach are reported. Section 5 discusses the links between the relational and the logical approaches and briefly reports on the use of possibilistic logic, belief functions and ATMS for abductive reasoning purposes.

Several sections or subsections of this paper are revised and expanded versions of previous works by the authors, Dubois and Prade [1993b; 1993c].

2 CLASSIFICATION-ORIENTED APPROACHES

Consider a system or device whose current state is observable and described by means of set S of *binary* attributes $\{a_1, \ldots, a_i, \ldots, a_m\}$, viewed as propositional variables. Attribute a_i is supposed to take its value on $\{s_i, \neg s_i\}$. When $a_i = s_i$ we shall say that the symptom s_i is present; when $a_i = \neg s_i$, it means that s_i is absent. Here, negation is denoted by $\neg$. When there is no symptom present, the system is said to be in its normal state and this state is described by the n-tuple $(\neg s_1, \ldots, \neg s_i, \ldots, \neg s_m)$, or equivalently, $\wedge_{i=1,m} \neg s_i$. Let D be a set of possible disorders $\{d_1, \ldots, d_j, \ldots, d_n\}$. A disorder can be present (d_j) or absent $(\neg d_j)$. It is assumed here that the disorders are mutually exclusive so that d_j is equivalent to have $\neg d_k$ for $k \neq j$. This is the unique fault assumption. On the contrary, of course, the symptoms are not mutually exclusive. In the absence of uncertainty, to each d_j we associate the set $M(d_j)$ of symptoms which are entailed, or if we prefer caused, produced, by the presence of d_j *alone*. Let I be the subset of $\{1, 2, \ldots, n\}$ such that the set of present symptoms is $M = \{s_i, i \in I\}$, or equivalently M stands for $\wedge_{i \notin I} \neg s_i \wedge \wedge_{i \in I} s_i$. For short we shall denote this by M_I. Note that M_I

represents complete information about a case.

Diagnostic problem solving has been viewed as a classification task [Clancey, 1985]. The idea behind this view is that disorders are understood just as names for sets of symptoms. Hence one way of modelling the diagnosis task is to describe for each possible disorder a set of supposedly characteristic features, that determines a particular class. Then, given a set of observations, the diagnosis task comes down to a problem of matching the observations against the suitable class description. The main assumption underlying classification approaches is that a unique disorder is present and explains the observations; indeed the "observed object" is supposed to belong to a unique class, since the available knowledge about disorders represents a partition of classes, or at best a hierarchy thereof. However this scheme does not work as such, because uncertainty pervades the links between disorders and their symptoms. These links can be described in two ways, namely, following Pearl [Pearl, 1988a], by means of evidential (explanation-evoking) or causal (expectation-evoking) rules. Causal rules are of the form "if disorder d then symptom s" understood as "if d is true then s is expected". It follows the direction of assumed causality. Evidential rules "if s then d" are understood as "the observation of s suggests d as a possible explanation" (a ready-made abductive step). Historically the Bayesian classification approach to diagnosis relies on causal rules modelled by conditional probabilities, and the expert system approach uses evidential rules weighted by certainty factors. Both exploit a rather shallow form of knowledge. Counterparts of the Bayesian classification approach with belief functions and possibility measures have been devised.

2.1 *Expert Systems Based on Evidential Rules*

Expert systems have been very popular in Artificial intelligence in the late seventies and the early eighties. This popularity was mainly due to the efficiency of two diagnostic problem-solvers in limited application domains, especially MYCIN [Buchanan and Shortliffe, 1984] and PROSPECTOR [Duda *et al.*, 1981], respectively for infectious diseases in medicine, and mineral exploration. Expert systems traditionally consist of three main modules: a factual base F describing a case study, a rule base R containing the domain knowledge under the form of production rules, and an algorithm, called inference engine, that performs deduction. Rules are of the form

$$r\text{: if } \langle \text{COND}_1 \ \& \ \text{COND}_2 \ \& \ldots \& \ \text{COND}_p \rangle \text{ then } \langle \text{CONS} \rangle \ (w_r)$$

where COND_i are Boolean conditions and CONS is the Boolean fact to be deduced and w_r is a certainty factor attached to the rule. Inference in expert systems is of the modus ponens type, whereby CONS is added to F whenever F contains $\text{COND}_1, \ldots, \text{COND}_n$ pertaining to a rule r in R. In the case of diagnostic problem-solving tasks, rules are typically of the explanation evoking type, that is

$$r : \text{ if } \langle s_{i_1} \wedge s_{i_2} \wedge \ldots \wedge s_{i_p} \rangle \text{ then } \langle d_j \rangle \ (w_r).$$

Although the only used inference mode is modus ponens, the strategy for reasoning can be of the forward chaining type or of the backward chaining type (goal-directed). In the forward chaining method, the factual base is enriched until a conclusion of interest is found. In the backward chaining method a network relating the observed facts to the conclusion is built, starting from the disorder d_j towards the actual observations.

Observations can be uncertain and are attached weights w_i that are propagated from the conditions to the conclusions of the rules. A global evaluation w_a of the antecedent part of an enabled rule is obtained as

$$w_a = w_1 *_e w_2 *_3 \ldots *_e w_n$$

where w_i is the weight bearing on COND$_i$ in F, and $*_e$ is a combination operation whose nature is conjunctive. This decomposability may capture an independence assumption between symptoms. w_a is computed either to help select the rule, or, when the rule has been selected for some reason (e.g., in backward chaining), as a preliminary step to weight propagation.

Rule firing requires two major steps in uncertainty management: propagation and fusion. Propagation consists in computing the weight w_c that will be attached to the fact CONS that the consequent of the rule advises to add to the factual base. It is obtained by combining the rule weight and the antecedent evaluation weight, i.e.,

$$w_c = w_r *_p w_a.$$

When adding (CONS, w_c) to the factual base, it may happen that this fact already exists, but with a different weight w. These two weights are then merged into an updated value, by means of some operation $*_f$. This situation mainly occurs in forward chaining, because weight management steps are performed at each cycle of the inference engine. On the contrary, in backward chaining, when some conclusion has to be tested, the inference engine first produces a network of rules that relates the conclusion to the initial facts and uncertainty is propagated through the network in a second step. Then, fusion comes down to aggregate the weights $w_c, i = 1, n$ obtained from the propagation step applied to m parallel rules R_i of the form 'if ANTE$_i$ then CONS'. The weight bearing on CONS is then:

$$w_c = w_{c_1} *_f w_{c_2} *_f \ldots *_f w_{c_n}$$

where w_{c_i} is obtained from R_i. This is the strategy adopted in the MYCIN project where a calculus of certainty factors has been proposed. Many alternative systems or improved certainty factor calculi have been proposed since the early eighties, e.g., ARIES [Appelbaum and Ruspini, 1985], LITO [Lesmo *et al.*, 1985], PROTIS [Soula *et al.*, 1983 ; 1986], SPHINX [Fieschi *et al.*, 1982], SPII [Martin-Clouaire and Prade, 1996], TAIGER [Farreny *et al.*, 1986], FLOPS [Buckley *et al.*, 1986a, b], SLOP [Baldwin, 1986], RUM [Bonissone *et al.*, 1987], MILORD [Godo *et al.*, 1987], etc. However, the certainty factor approach has been much criticized for its *ad hoc* nature (e.g., [Spiegelhalter and Knill-Jones, 1985; Spiegelhalter, 1986; Dubois and Prade, 1989; Peng and Reggia, 1990]).

In the context of diagnostic problem-solving, many difficulties arise with the expert system approach.

First, the type of reasoning performed by diagnostic expert systems looks quite confusing because due to the use of evidential rules, they perform abductive inference in a deductive format. In the PROSPECTOR system, the weight w_r is in fact made of causal conditional probabilities $P(s_{i_1} \wedge s_{i_2} \wedge \dots s_{i_p} \mid d_j)$ and $P(S_{i_1} \wedge s_{i_2} \wedge \dots \wedge s_{i_p} \mid \neg d_j)$, which may be even more confusing.

Evidential rules turn out not to be a very interesting kind of knowledge for diagnosis. Knowledge of system behavior by experts is often causal. A medical doctor knows and can explain the (usual) effects produced by a disease. Even if his usual practice is to recognize characteristic patterns in symptoms, he may have trouble explaining how this pattern recognition proceeds. So evidential rules are not necessarily the type of knowledge an expert would like to supply. However for the purpose of inference, evidential rules are what is needed to perform deductive inference in a diagnosis problem, since abductive patterns of reasoning are out of the scope of regular inference engines.

Evidential rules are typically supplied by an external observer watching a medical doctor who performs diagnosis tasks. For each set of symptoms (corresponding to a patient) the observer notices the resulting diagnosis. As a consequence evidential rules are a very shallow form of knowledge that do not lend itself to explanation capabilities.

Another drawback of the expert system approach is that an evidential rule often means: "if all I know is that symptoms $s_1, s_2, \dots, s_p$ are present then it is plausible (w_R) that d_j is the cause". But the way production rules and certainty factors are handled in expert systems does not comply with such semantics. For instance, there maybe a rule of the form

$$r_1 : s_1 \wedge s_2 \rightarrow d_1 \ (w_{r_1})$$

that will be triggered if three symptoms s1, s2 and s3 are available, even if there exists a specific rule

$$r_2 : s_1 \wedge s_2 \wedge s_3 \rightarrow d_2 \ (w_{r_2})$$

for this case. However maybe situations where r_2 applies are exceptions for rule r_1 so that d_1 would be very implausible in the presence of $s_1 \wedge s_2 \wedge s_3$. Modifying r_1 into $s_1 \wedge s_2 \wedge \neg s_3 \rightarrow d_1$ would create other problems because when only s_1 and s_2 have been observed, neither r_1 nor r_2 will be triggered while any expert would start suspecting d_1 or d_2.

It is also well-known that certainty-factor propagation calculi cannot account for dependencies in the deductive network relating observations to diseases. Some degrees of uncertainty will be used twice in reinforcement patterns, and the weights eventually bearing on disorders will be dubious. The presence of exceptions in rules (that has motivated the introduction of certainty factors) is also a cause of trouble because rules pertaining to opposite conclusions (d_1 and $\neg d_1$) will be triggered and weights will be combined, resulting in non-decisive diagnosis where an expert could see right away what the good conclusion should be (take $d_2 = \neg d_1$ with rules r_1 and r_2 and assume again s_1, s_2 and s_3 are present).

Expert systems for diagnosis are essentially of the classification type because they have problem tackling multiple disorders. Indeed, suppose a set of symptoms M_I that triggers rules in such a way that three disorders d_1, d_2 and d_3 are found. It may turn out that only subsets of two disorders, like $\{d_1, d_2\}$ and $\{d_2, d_3\}$ suffice to explain the manifestations, and the expert system cannot discover this state of fact, unless introducing rules whose conclusions involve combinations of disorders. This solution is only theoretical because it leads to a combinatorial explosion of the number of rules.

It is clear that expert systems using evidential rules and certainty factors are very limited in the diagnosis tasks because they do not represent causal knowledge proper, substitute deduction to abduction, and propagate uncertainty in an ad hoc way. Some attempts have been made to exploit evidential rules, certainty factors, and causal representations simultaneously, such as the CHECK system [Torasso and Console, 1989]. The causal knowledge is used for validating the inferences made by the evidential knowledge. However the treatment of uncertainty in CHECK is again ad hoc. This idea of hybrid diagnostic problem solvers is certainly worth pursuing. However a natural question is whether a rigorous exploitation of causal knowledge might not be enough to solve the diagnosis task, especially, considering that the technical difficulties of expert system with certainty factors exist even for tasks other than diagnosis, namely deductive reasoning tolerant to exceptions, for instance. Much progress has been achieved in that field and if an automated reasoner should be used for handling uncertain evidential rules to-date, the certainty factor technology might not be a good candidate for implementing it.

2.2 The Bayesian Scheme

The alternative traditional approach to diagnosis is Bayesian classification [Duda and Hart, 1973; Charniak, 1983]. In this approach each disorder is viewed as a class of situations that are described by a set of attributes which are the symptoms of the disorder. In the Bayesian approach, it is assumed that some a priori probability assignment on D is available, say $p : \mathtt{D} \to [0, 1]$ with $\Sigma_{j=1,n} p_j = 1$ where p_j is the a priori probability that the unique present disorder is d_j, and such that $\forall j, p_j > 0$. Moreover the description of the criteria for class membership is equated to causal knowledge, of the form $\langle$if disorder d_j then manifestation $M_I\rangle$. This causal knowledge is uncertain and, in the general case, one needs to know the probabilities $P(M_I \mid d_j)$ for all subsets I of $\{1, 2, \ldots, m\}$. They represent the probability that the presence of d_j causes exactly the symptoms s_i for $i \in I$ simultaneously. These $n \times 2^m$ probabilities are necessary to fully describe the uncertain causal knowledge. The conditional probabilities $P(s_i \mid d_j)$ that symptom s_i appears when disorder d_j is present derive from this knowledge, namely:

$$p(s_i \mid d_j) = \Sigma_{I:i \in I} P(M_I \mid d_j).$$

Suppose observations are complete and described by M_I. Then the probability that d_j is *the* disorder explaining these symptoms is computed by means of Bayes theorem as:

$$P(d_j \mid M_I) = \frac{P(M_I \mid d_j) \cdot P(d_j)}{\Sigma_{k=1,n} P(M_I \mid d_k) \cdot P(d_k)}.$$

This approach is very data-demanding. To alleviate the data-gathering burden, the conditional independence of individual symptoms with respect to individual disorders is sometimes assumed, i.e., for any index set I,

$$P(M_I \mid d_j) = \prod_{i \in I} P(s_i \mid d_j) \cdot \prod_{i \notin I} P(\neg s_i \mid d_j).$$

where $\prod_{i \in I}$ denotes the product. Then the knowledge of only n prior probabilities and $m \times n$ conditional probabilities is necessary. And the revised probability $P(d_j \mid M_I)$ based on conjointly observing symptoms in M_I is given by

$$P(d_j \mid M_I) = P(d_j) \cdot \frac{\prod_{i \in I} P(s_i \mid d_j) \cdot \prod_{i \notin I} P(\neg s_i \mid d_j)}{\Sigma_{k=1}^{n} P(M_i \mid d_k) \cdot P(d_k)}.$$

Assume that $\mid \mathtt{D} \mid = 2$ and $\mid \mathtt{S} \mid = 1$, i.e., there are only two mutually exclusive causes $d, \neg d$ and a symptom which is present or not. The causal rule "if d then s" is here understood in terms of the conditional probabilities $P(s \mid d) = 1 - P(\neg s \mid d)$ and $P(s \mid \neg d)$. The a posteriori probability $P(d \mid s)$ of d when s is observed is then:

$$\begin{aligned} P(d \mid s) &= \frac{P(s \mid d) \cdot P(d)}{P(s \mid d) \cdot P(d) + P(s \mid \neg d) \cdot P(\neg d)} \\ &= \frac{1}{\left(1 + \dfrac{P(s \mid \neg d)}{P(s \mid d)} \cdot \dfrac{P(\neg d)}{P(d)}\right)}. \end{aligned}$$

Thus $P(d \mid s)$ is all the larger as the a priori probability of d is larger than the one of $\neg d$ and as symptom s is more probable in the context d than in the context $\neg d$, which is intuitively satisfying. Moreover s will confirm d ($P(d \mid s) > P(d)$) whenever $P(s \mid d) > P(s \mid \neg d)$, if $P(d) \neq 0, 1$.

An additional assumption can be the unconditional independence of observations. Then the posterior probability for disorder d_j has an even simpler expression that presupposes the knowledge of a priori probabilities for symptoms:

$$P(d_j \mid M_I) = P(d_j) \cdot \prod_{i \in I} \frac{P(s_i \mid d_j)}{P(s_i)} \cdot \prod_{i \notin I} \frac{P(\neg s_i \mid d_j)}{P(\neg s_i)}.$$

This result leads to bridging the gap with evidential approaches of rule based classification, by noticing that $P(s_i \mid d_j) = P(d_j \mid s_i) \cdot P(s_i)/P(d_j)$. If genuine evidential rules are given (like in MYCIN), then a combination rule is obtained

$$p(d_j \mid M_I) = P(d_j)^{1-n} \cdot \prod_{i \in I} P(d_j \mid s_i) \cdot \prod_{i \notin I} P(d_j \mid \neg s_i).$$

The Bayesian classification approach is mathematically sound but has severe limitations, that have been well discussed in the literature [Peng and Reggia, 1990]. First at all, the assumption underlying the classical classification paradigm is sometimes irrealistic, viz., that disorders are mutually exclusive. Even with the unique fault assumption, the number of conditional probabilities to be supplied is large, and seem difficult to get when statistical data are not available. Moreover, prior probabilities for disorders may fail to be available. The simplifying assumption, saying that symptoms are conditionally independent when the disorder is fixed, looks quite unrealistic since most of the time, the symptoms produced by a disorder simultaneously will be strongly correlated. Lastly, the classification framework offers no provision for hypothesis generation and testing, because all observations are supposed to be available at once; this is due to the fact that the knowledge is "hidden" in the probabilities, and one has no explicit access to inner causal associations.

2.3 *Moving to a possibilistic framework*

As said earlier, precise prior probabilities for disorders may not exist, and causal knowledge may be imprecise. A natural question that has been addressed is whether it can exist a machinery similar to Bayesian classification in other uncertainty frameworks accepting incomplete information, such as possibility theory. Possibility theory is simpler than probability theory because instead of the additivity rule of probabilitites, it is based on a "maxitive" rule that only requires a totally ordered set for evaluating degrees of uncertainty.

Possibility measures are set-functions governed by the characteristic axiom $\Pi(A \cup B) = \max(\Pi(A), \Pi(B))$ for any events A, B. In this paper an event is viewed as the subset of a set Ω. Complementation is denoted as $\bar{A}$ or $\neg A$, according to whether we emphasize the set-theoretic view or the logical view. The conventions for representing uncertainty in possibility theory differ from those in probability theory, but possibility measures agree with the meaning of upper probabilities. Namely $\Pi(A) = 0$ means that A is impossible, while $\Pi(A) = 1$ only says that A is possible. A possibility measure on D, in the finite case, is entirely specified by means of a possibility distribution π : D $\rightarrow [0,1]$ with $\max_{j=1,n}\pi_j = 1$ where π_j is the a priori possibility that, in our context, the unique present disorder is d_j. When $\pi_j = 1$ means that disorder d_j is unsurprising, plausible. $\Pi(A) = \sup_{u\in A}\pi(u)$ qualifies the degree of possibility of A, defined by assuming that, if it is only known that A occurs, then the most plausible situation compatible with A prevails. For instance the degree of possibility of the event "the light is off" is computed after the degree of possibility of "someone switched off the light" rather than "the bulb is broken", insofar as the latter is very unusual compared to the former. Possibility theory can account for complete ignorance, such that $\Pi(A) = 1$ holds for all events A except the impossible event (empty set). It corresponds to a uniform possibility distribution, set to 1.

A dual degree of necessity $N(A) = 1 - \Pi(\neg A) = \inf_{u\notin A}1 - \pi(u)$ can be defined. $N(A) = \alpha > 0$ means that the most plausible situation where A is false is rather impossible, i.e., not possible to a level greater than $1-\alpha$. Set functions Π and N are simple ordinal models of graded uncertainty [Dubois and Prade, 1988a; Dubois and Prade, 1998]. Their particular character lies in their ordinal nature, i.e., the unit interval is used only to rank-order the various possible situations in U, in terms of their compatibility with the normal course of things as encoded by the possibility distribution π. The above view contrasts with the probabilistic encoding of knowledge, which relies on an additivity assumption. Only ordering is requested in possibility theory. Moreover probability theory cannot account for ignorance in the Bayesian view.

A systematic assumption in possibility theory is that the actual situation is normal, i.e., it is any u such that $\pi(u)$ is maximal given other known constraints. It justifies the evaluation $\Pi(A)$, and contrasts with the probabilistic evaluation of the likelihood of events. Moreover $N(A) > 0$ means that A holds in all the most normal situations. Since the assumption of normality is always made, $N(A) > 0$ means that A is an accepted belief, i.e., one may act as if A were true. This assumption is always a default one and can be revised if further pieces of evidence contradict it.

Possibility theory is an approach to uncertainty where statements of the form "A is possible" and "A is believed" are clearly distinguished via the two set-functions Π and N respectively. This distinction cannot be made in Bayesian probability where the negation of "A is believed" is always interpreted as "not A is believed", instead of "A is not believed" (the latter does not commit into believing "not A"). This feature of possibility theory is shared by alternative settings such as belief functions and upper-lower probabililies. If viewed as genuinely numerical, function N is indeed a particular case of belief function and lower probability [Dubois and Prade, 1998]. However the rules of possibility theory makes it extremely simple with respect to its competitors, which also means that it has a more limited expressive power since it can be construed as purely ordinal.

Conditional possibility can then be defined in a very qualitative way from the relation

$$\Pi(A \cap B) = \Pi(A \mid_* B) * \Pi(B),$$

for some conjunctive operation $*$. It has been first proposed when operation * is minimum by [Hisdal, 1978]. Another possible interpretation of $*$ is the product. For $* =$ minimum one gets:

$$\begin{aligned}\Pi(A \mid_{\min} B) &= 1 \text{ if } \Pi(B) = \Pi(A \cap B) \\ &= \Pi(A \cap B) \text{ if } \Pi(A \cap B) < \Pi(B),\end{aligned}$$

if we obey the minimal specificity principle which allocates the greatest possibility degrees in agreement with the constraint(s), here expressed by relation $\Pi(A \cap B) = \min(\Pi(A \mid B), \Pi(B))$. For $*$=product, one gets a more usual form of conditioning: $\Pi(A \mid B) = \Pi(A \cap B)/\Pi(B)$. The duality relation between possibility and necessity leads to define conditional necessity measures as $N(A \mid B) = 1 - \Pi(\neg A \mid B)$ which expresses that, in the context B, A is all the more certain as $\neg A$ is more impossible. Conditional possibility measures thus obey the following counterpart of Bayes theorem:

$$\Pi(A \mid_* B) * \Pi(B) = \Pi(B \mid_* A) * \Pi(A).$$

It should be pointed out that the behavior of the possibilistic approach greatly differs according to the notion of conditioning that is adopted. Using the ordinal conditioning (based on $* =$min), the updating capabilities of the prior certainty values $(N(A), N(\neg A))$ are limited. Assume for simplicity $N(A) > 0$ (and $N(\neg A) = 0$, as a consequence since $\max(\Pi(A), \Pi(\neg A)) = 1$). Then only the following situations may occur

→ *confirmation*: $N(A \mid B) \geq N(A)$. This situation occurs only if $\Pi(A \cap B) > \Pi(\neg A \cap B)$;

→ *cancellation*: $N(A \mid B) = N(\neg A \mid B) = 0$. The observation of B cancels our belief about A. This happens iff $\Pi(A \cap B) = \Pi(\neg A \cap B)$, but $\Pi(A \cap \neg B) = 1 > \max(\Pi(\neg A \cap \neg B), \Pi(A \cap B), \Pi(\neg A \cap B)) = \Pi(\neg A \cup B)$;

→ *belief revision*: $N(\neg A \mid B) > 0$. The observation of B reverses our belief about A. This is observed when $\Pi(A \cap \neg B) = 1 > \Pi(\neg A \cap B) > \Pi(A \cap B)$ and $\Pi(A \cap \neg B) > \Pi(\neg A \cap \neg B)$; then it is clear that $N(\neg A \mid B) = 1 - \Pi(A \cap B) > N(A)$.

In case $N(A) = N(\neg A) = 0$, we remain in a state of ignorance ($N(A \mid B) = N(\neg A \mid B) = 0$) in the context B iff $\Pi(A \cap B) = \Pi(\neg A \cap B)$, while $N(A \mid B) > 0$ iff $\Pi(A \cap B) > \Pi(\neg A \cap B)$ (this case requires here $\Pi(\neg A \cap \neg B) = 1$ in order to have $\Pi(\neg A) = 1$ and $N(A) = 0$).

The case when $0 < N(A \mid B) < N(A)$ can never happen with the ordinal conditioning (see Figure 1). This is due to a monotonic behavior of $N(A \mid B) = N(\neg B \cup A)$ in that case. The situation $0 < N(\neg A \mid B) < N(A)$ does not occur either for a similar reason as explained above.

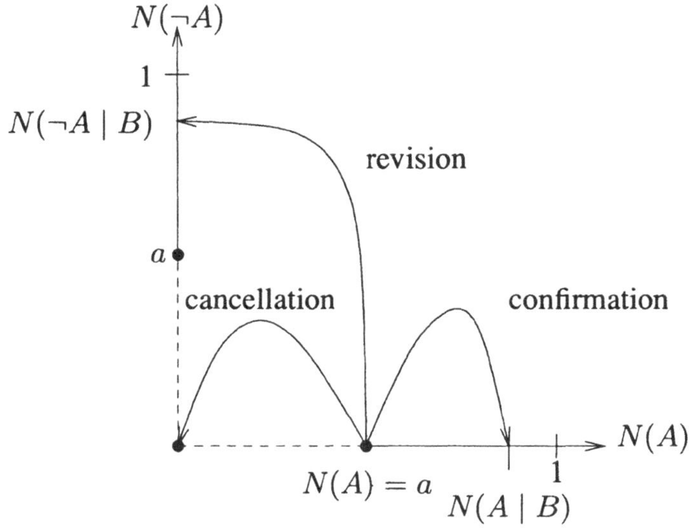

Figure 1. Belief change with ordinal conditioning

On the contrary, if $\Pi(A \mid B) = \frac{\Pi(A \cap B)}{\Pi(B)}$, then it is permitted to observe $0 < N(A \mid B) < N(A)$ (attenuation) or $0 < N(\neg A \mid B) < N(A)$ (soft revision). These remarks precisely describe the limitations of ordinal approaches in terms of expressive power in the conditional approach.

2.4 *A Possibilistic Classification Method*

Assume there is an a priori possibility distribution π on D. If there is no a priori knowledge then $\pi_j = 1$ for all disorders. The causal knowledge is described by a set of conditional possibilities $\Pi(M_I \mid d_j)$ for all subsets I of $\{1, 2, \ldots, m\}$. They represent the possibility that the presence of d_j causes exactly the symptoms s_i for $i \in I$ simultaneously. Again if some of these quantities are lacking, then, let $\Pi(M_I \mid d_j) = 1$. It can be checked that for $* =$ product one gets

$$\pi(d_j \mid M_I) = \frac{\Pi(M_I \mid d_j) \cdot \pi_j}{\max_{k=1,n} \Pi(M_I \mid d_k) \cdot \pi_k}$$

and for $*$ =minimum:

$$\begin{aligned}\pi(d_j \mid_{\min} M_I) &= 1 \text{ if } \min(\Pi(M_I \mid_{\min} d_j), \pi_j) = \\ &\quad = \max_{k=1,n} \min(\Pi(M_I \mid_{\min} d_k), \pi_k) \\ &= \min(\Pi(M_I \mid_{\min} d_j), \pi_j) \text{ otherwise}\end{aligned}$$

If the symptoms are non-interactive in the presence of disorder d_j (i.e., the relation linking the symptoms, if any, is unknown) there holds the following decomposability property (e.g., [Dubois and Prade, 1988b])

$$\Pi(M_I \mid_* d_j) = \min(\min_{i \in I} \Pi(s_i \mid_* d_j),\ \min_{i \notin I} \Pi(\neg s_i \mid_* d_j)).$$

The conditional possibility $\Pi(s_i \mid_* d_j)$ represents to what extent s_i is a possible symptom of the presence of d_j, viewed as a cause. Strictly speaking we may have $\sup_j \Pi(s_i \mid_* d_j) \leq 1$ only. It is natural to assume that $\sup_j \Pi(s_i \mid_* d_j) = 1$ if there is at least one cause d_j that makes the appearance of s_i somewhat certain; in other words s_i is a completely relevant symptom for the set D of disorders. This reads $N(s_i \mid_* d_j) > 0$ for some i and implies $\Pi(s_i \mid_* d_j) = 1$. Indeed, if for instance $\sup_j \Pi(s_i \mid_* d_j) = 0$, it would mean that it is impossible to observe s_i due to a cause in D, that is, this attribute is always in its normal state ($\neg s_i$). In the following we take the assumption that $\sup_j N(s_i \mid_* d_j) > 0$ for granted, and call it the *postulate of observation relevance*. It is worth noticing that the statement $N(s_i \mid_* d_j) > 0$ has all the properties of a non-monotonic inference such as "if cause d_j is present then normally s_i is observed" [Dubois and Prade, 1991].

The non-interactivity assumption supplies the least informative value of $\Pi(M_I \mid_* d_j)$ when only the individual links between a disorder and the presence or absence of a symptom are known. It takes a pessimistic stand that refrains from assuming any reinforcement effect between symptoms. On the contrary if we assume the independence between symptoms created by a disorder, then one must change the minimum into a product in the expression of $\Pi(M_I \mid_* d_j)$.

If there is no a priori knowledge about disorders, the a priori possibility distribution is vacuous, i.e., $\pi_j = 1$, for all j. Then, results very close to Edwards' [1972] notion of likelihood are obtained; especially , for the product-based conditioning, the identity

$$\pi(d_j \mid M_I) = \frac{\Pi(M_I \mid d_j)}{\max_{k=1,n} \Pi(M_I \mid d_k)}.$$

Namely, this identity is used by non-Bayesian statisticians that do not assume any prior probability exists, when they face a classification problem. They make the assumption that the probability of classifying an object d_j in the face of evidence M_I is proportional to the likelihood of d_j. The connection between likelihoods and possibility theory is now well established [Smets, 1982; Dubois *et al.*, 1997]. It is again possible to make the connection with evidential rule-based approaches, if non-interactivity or independence between symptoms is assumed, and in the absence of prior information about disorders. It is then clear that

$\Pi(s_i \mid d_j) = \Pi(d_j \mid s_i)$ using the identity $\pi(s_i \mid d_j) * \pi_j = \Pi(d_j \mid s_i) * \Pi(s_i)$ and noticing that, due to observation relevance, $\Pi(s_i) = 1$. Consequently, the classification possibility for product-based conditioning, can be written in the evidential style, denoting s_i^ε any of s_i or $\neg s_i$ according to whether $i \in I$ or not:

$$\pi(d_j \mid M_I) = \frac{\Pi(d_j \mid s_1^\varepsilon) * \Pi(d_j \mid s_2^\varepsilon) * \ldots * \Pi(d_j \mid s_m^\varepsilon)}{\max_{k=1,n} \Pi(d_k \mid s_1^\varepsilon) * \Pi(d_k \mid s_2^\varepsilon) * \ldots * \Pi(d_k \mid s_m^\varepsilon)}$$

When $*$ is the product (symptom conditional independence), this is equivalent to the combination formula of MYCIN, if one restricts disorders to $\{d_j, \neg d_j\}$ [Dubois and Prade, 1988d]. For $* = \min$ this is the combination rule in possibility theory.

In the following we briefly consider the simple situation where $|\ \mathrm{D}\ | = 2$ and $|\ \mathrm{S}\ | = 1$ and verify conditions under which a revision of prior possibilities occurs. Moreover $\min(\pi(d), \pi(\neg d)) > 0$, i.e., none of d or $\neg d$ is a priori certain; $\pi(d \mid s)$ and $\pi(\neg d \mid s)$ are only related by the normalization condition $\max(\pi(d \mid s), \pi(\neg d \mid s)) = 1$. We also have $\max(\pi(d), \pi(\neg d)) = 1$. These two conditions are immediate consequences of the basic axiom of possibility theory, assuming that the possibility of a tautology should be equal to 1. With the qualitative form of conditioning, the possibilistic version of Bayes theorem then reads:

$$\begin{aligned} \pi(d \mid s) &= \min(\Pi(s \mid d), \pi(d)) \text{ if } \Pi(d \wedge s) < \Pi(\neg d \wedge s) \\ &= 1 \text{ otherwise} \end{aligned}$$

and the same for $\pi(\neg d \mid s)$ changing d into $\neg d$. When $\min(\Pi(s \mid d), \pi(d)) \neq \min(\Pi(s \mid \neg d), \pi(\neg d))$, one of $\pi(d \mid s)$ and $\pi(\neg d \mid s)$ is less than 1 and the other is 1. For instance if $\Pi(d \wedge s) > \Pi(\neg d \wedge s)$ then $N(d \mid s) > 0$, and the plausible conclusion when observing s is that the cause is d. If $\Pi(d \wedge s) = \Pi(\neg d \wedge s)$ then $\pi(d \mid s) = \pi(\neg d \mid s) = 1$ and d and $\neg d$ become equally uncertain, i.e., we are led to ignorance about the causes. Similarly, a uniform probability, i.e., $P(d \mid s) = P(\neg d \mid s) = 1/2$ is obtained in the probabilistic model when $P(s \mid d) \cdot P(d) = P(s \mid \neg d) \cdot P(\neg d)$.

Let us check when the observation of symtom s leads us to conclude that d is plausible, namely, when does $N(d \mid s) > 0$ hold? Equivalently this reads $\pi(d \mid s) = 1$ and $\pi(\neg d \mid s) < 1$. These conditions are again equivalent to $\Pi(d \wedge s) > \Pi(\neg d \wedge s)$. Expressing these degrees of possibility in terms of causal conditional possibilities leads to

$$\Pi(s \mid d) > \min(\Pi(s \mid \neg d), \pi(\neg d))$$

$$\text{and } \pi(d) > \min(\Pi(s \mid \neg d), \pi(\neg d)).$$

Assume $\pi(\neg d) \geq \Pi(s \mid \neg d)$.This is true in particular if the absence of disorder is a priori believed ($N(\neg d) > 0$) or there is no a priori belief ($\pi(d) = \pi(\neg d) = 1$).

Then $N(d \mid s) > 0$ implies $\Pi(s \mid d) > \Pi(s \mid \neg d)$. However if $\pi(\neg d) < \Pi(s \mid \neg d)$ (which means that d is firmly believed a priori), then

$$N(d \mid s) = 1 - \pi(\neg d \mid s) = 1 - \min(\Pi(s \mid \neg d), \pi(\neg d)) = 1 - \pi(\neg d) = N(d)$$

Then no confirmation effect takes place, nor any rebuttal is possible, even if $\Pi(s \mid d) < \Pi(s \mid \neg d)$, that is if symptom s is more plausible when d is absent than when it is present.

Condition $\Pi(s \mid d) > \Pi(s \mid \neg d)$ is similar to the condition $P(s \mid d) > P(s \mid \neg d)$ for probabilistic confirmation. However the possibilistic inequality is not enough to improve or enforce the a posteriori belief in d. The modification of the prior belief more generally requires that $\Pi(s \mid \neg d) < \min(\pi(d), \pi(\neg d))$, i.e. no strong opinion is a priori held in favor or against d. This condition also reads $\max(N(d), N(\neg d)) < N(\neg s \mid \neg d)$, that is, our certainty of both antagonistic hypotheses is lower than our certainty of not observing s whenever d is absent (in other words our certainty that only d can produce s). If our a priori opinion is strong enough, observing a symptom will not modify our beliefs.This behavior of the possibilistic Bayesian like inference in the qualitative case has been first noticed by Garbolino [1989] and further studied by Dubois and Prade [1993b]. See also Zwirn and Zwirn [1996] for a systematic analysis of possibilistic confirmation, and its links with non-monotonic reasoning, in the scope of epistemology.

Hence, the above possibilistic model of abduction, due to its purely qualitative setting, has two limitations if a learning technique is envisaged:

- the impossibility to modify strongly entrenched a priori opinion. This rigidity appears in the probabilistic model only when $P(d) = 1$ or 0;
- the convergence of the learning process in one step when the same evidence s repeatedly appears. This is due to the idempotence of min.

These limitations disappear if we define conditional possibility based on product. Then

$$\pi(d \mid s) = \min\left(1, \frac{\Pi(s \mid d) \cdot \pi(d)}{\Pi(s \mid \neg d) \cdot \pi(\neg d)}\right).$$

Then it is easy to see that $\Pi(s \mid d) > \Pi(s \mid \neg d)$ is now necessary and sufficient for the actual confirmation of d under observing s. Moreover learning becomes progressive when s is repeatedly observed.

2.5 Classification Diagnosis Using Belief Functions

A more general approach than the previous ones consists in representing causal knowledge under the form of a conditional belief function Bel$(\cdot \mid d)$ on the space of attribute values, for each $d \in \mathtt{D}$. This idea has been proposed and formalized by Smets [1978; 1993]. This method works again without the necessity of having prior knowledge on D. A belief function on a finite set Ω is defined by a mass assignment $m : 2^\Omega \to [0, 1]$, such that $m(A)$ can be viewed as the probability of knowing only A. Of course $m(\varnothing) = 0$ and $\Sigma_{A \subseteq \Omega} m(A) = 1$ and the belief function is defined as

$$\forall B \subseteq \Omega, \mathrm{Bel}(B) = \Sigma_{A \subseteq B} m(A).$$

It plays the role of a degree of certainty. When the family of focal sets $\{A, m(A) > 0\}$ is nested then Bel is a necessity measure, and $\mathrm{Pl}(\neg B) = 1 - \mathrm{Bel}(B)$, called the plausibility function, is a possibility measure. On the contrary, if Pl = Bel then the focal sets are singletons and Pl = Bel is a probability function.

Here, $\mathrm{Bel}(\cdot \mid d)$ is a belief function on the Cartesian product $\Omega = \times_{j=1,m}\{s_j, \neg s_j\}$, for all $d \in$ D. Smets actually assumes that for each $d \in$ D, there are m belief functions $\mathrm{Bel}_j(\cdot \mid d)$, one per observable attribute. The problem is then to compute $\mathrm{Bel}(d \mid M_J)$ where M_J is a set of present symptoms, from the knowledge of $\mathrm{Bel}_j(\cdot \mid d)$ only. Smets [1993; 1998] proposes a generalized Bayes theorem that leads to the following expression

$$\mathrm{Pl}(d \mid M_J) = \mathrm{Pl}(M_J \mid d); \quad \mathrm{Pl}(\neg d \mid M_J) = 1 - \prod_{d' \neq d} (1 - \mathrm{P}(M_J \mid d'))$$

where $\mathrm{Pl}(M_J \mid d)$ is obtained by combination of the belief functions pertaining to each attribute. This combination is conjunctive and is done using Dempster's rule [Shafer, 1976]. Since $\mathrm{Bel}_j(\cdot \mid d_i)$ is defined on a simple universe $\{s_j, \neg s_j\}$, it is entirely defined by two numbers $c^+_{ij} = \mathrm{Bel}_j(s_j \mid d_i)$ and $c^-_{ij} = \mathrm{Bel}_j(\neg s_j \mid d_i)$. These coefficients can be interpreted in the terminology of the last section as $c^+_{ij} = 1 - \Pi(\neg s_j \mid d_i)$ and $c^-_{ij} = 1 - \Pi(s_j \mid d_i)$. However the condition $\min(c^-_{ij}, c^+_{ij}) = 0$ is requested by the possibilistic approach but not here. Only $c^+_{ij} + c^-_{ij} \leq 1$ is requested. It can be shown that, using Dempster's rule [Smets, 1993], the conditional plausibility of d given M_J is

$$\mathrm{Pl}(M_J \mid d) = \prod_{j \in J} (1 - c^-_{ij}) \cdot \prod_{j \notin J} (1 - c^+_{ij}).$$

In other words this method comes very close to the numerical version of the possibilistic approach with an independence assumption of symptoms, conditioned on the disorder, typical of classification approaches. However while in possibility theory, the set $\{\mathrm{Pl}(d \mid M_J), d \in \mathrm{D}\}$ is enough to represent the information fully, this is no longer true in the setting of belief functions because combining possibility measures by means of Dempster rule does not preserve the consonance of the result. The information is contained in the mass function. On the set $\{\neg s_j, s_j\}$ the mass function associated to $\mathrm{Bel}_j(\cdot \mid d_i)$ is

$$m_{ij}(s_j) = c^+_{ij}; m_{ij}(\neg s_j) = c^-_{ij}; m_{ij}(T_j) = 1 - c^+_{ij} - c^-_{ij}$$

where T_j is the tautology referring to attribute a_j. Then the mass function associated to any subset $S \subseteq \times_{j=1,m}\{s_j, \neg s_j\}$ is characterized, via Dempster rule as

$$\begin{aligned} m(S) &= \prod_{j=1,n} m_{ij}(s_j^\varepsilon), S = s_1^\varepsilon \wedge s_2^\varepsilon \wedge \ldots \wedge s_n^\varepsilon \\ &= 0 \text{ for any } S \text{ of another form} \end{aligned}$$

where in the above formula the following conventions are adopted:

- S is viewed as a formula on the language generated by $\mathtt{S}$;
- $$\begin{aligned} s_j^\varepsilon &= s_j \text{ if } \varepsilon = 1 \\ &= \neg s_j \text{ if } \varepsilon = -1 \\ &= T_j \text{ if } \varepsilon = 0. \end{aligned}$$

This kind of approach has been studied on its own although applied to case studies (see for instance the GENDIAG system of [Gouvernet *et al.*, 1985], and [Smets, 1998]). It would be interesting to study its agreement with the alternative uncertainty settings. However, Smets [1998] provides an illustrative comparative example and points out that the belief function approach "takes into account only the information really available, whereas too often the Bayesian is forced into filling missing probabilities by pure guesses, which might be called an artificial hyperprobabilization". Note that the probabilistic Bayesian approach (with an assumption of symptoms independence conditioned on disorders) is retrieved when $\mathrm{Bel}_j(\cdot \mid d_i)$ is nothing but a conditional probability, for all $d_i \in \mathtt{D}$, and a uniform probability on disorders is assumed (at least up to a normalization factor).

3 CAUSAL RELATIONAL APPROACHES WITH MULTIPLE FAULTS

3.1 The Parsimomious Covering Theory

Reggia *et al.* [1985] (see also [Peng and Reggia, 1990]) have extensively studied a relation-based formulation of diagnosis problems. In their model they assume the knowledge of a relation R between disorders and symptoms, such that the fact that the pair (d_j, s_i) satisfies this relation "means d_j may directly cause s_i. Note that this does *not* mean that d_j *necessarily* causes s_i, but only that it *might*", as stated in [Peng and Reggia, 1990]. This choice of interpretation for the "causal" association between d_j and s_i can be understood as assigning a non-extreme probability to it. Moreover the set M^+ of symptoms known to be present is supposed to be available. Note that M^+ does not necessarily contain all actually present symptoms: the presence of some symptoms may have not been checked yet.

For each disorder d, let $M(d) = \{s, (d,s) \in R\}$ be the set of symptoms possibly attached to d through R. Peng and Reggia further assume that several disorders may be jointly present and do not make the unique fault assumption. They how-

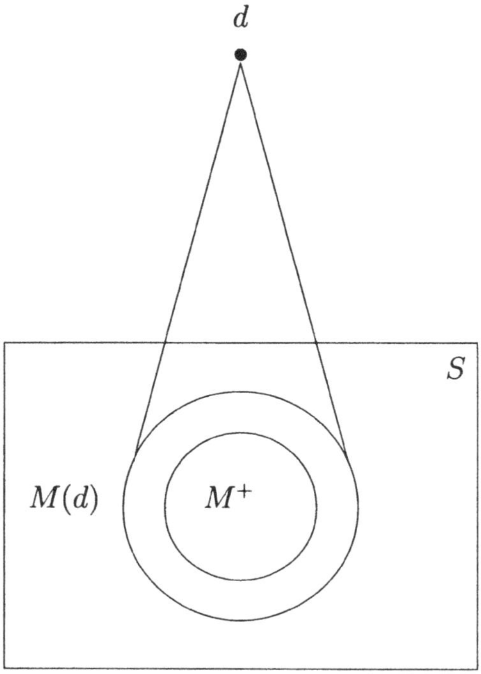

Figure 2. $\{d\}$ is a cover of M^+

ever make the hypothesis that disorders are independent so that their effects can be added and do not interfere, namely

$$M(\{d_i, d_j\}) = M(d_i) \cup M(d_j).$$

The set of potential causes of a symptom s is $D(s) = \{d, (d, s) \in R\}$ and the set of potential causes of a set M^+ of symptoms is $D(M^+) = \{d, \exists s \in M^+, (d, s) \in R\}$. An explanation of M^+ is a subset D of disorders that covers the symptoms, i.e., such that $M^+ \subseteq M(D)$ (D is then called a cover of M^+ and $\widehat{DD}$ denotes the set of covers) and is minimal in some sense (see Figure 2).

Peng and Reggia more particularly look for so-called "*parsimonious*" covers, especially relevant, irredundant and minimal covers of M^+. D is a *relevant* cover if it is a cover and if $\forall d \in D, \exists s \in M^+, (d, s) \in R$. See Figure 3. It means that D is contained in the set $D(M^+)$ of potential causes. An *irredundant* cover is such that none of its proper subset is also a cover of M^+; D is a *minimal* cover if its cardinality is smallest among all covers of M^+. Clearly a one-disorder cover is a minimal cover, a minimal cover is an irredundant cover, an irredundant cover is a relevant cover, a relevant cover is obviously a cover. The set of relevant covers is

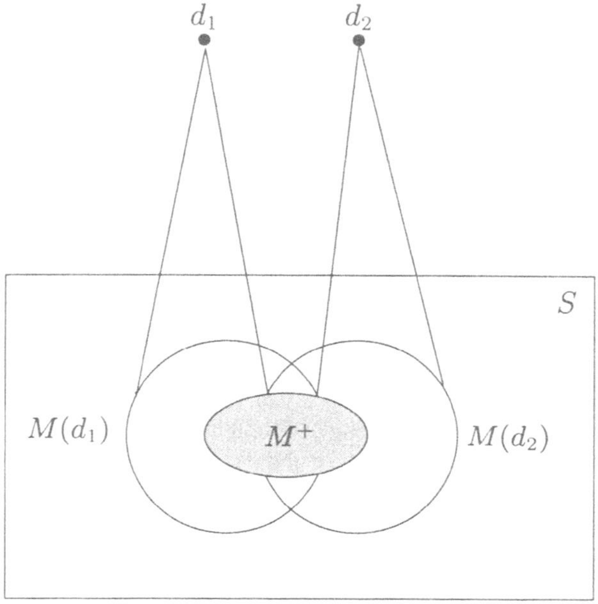

Figure 3. $\{d_1, d_2\}$ is a relevant cover of M^+

also defined by

$$DD* = \{D \in \widehat{DD}, \forall d \in D, M(d) \cap M^+ \neq \varnothing\}.$$

The diagnosis problem then comes down to computing covers, especially irredundant ones. This problem is NP-hard. Algorithms for carrying out this task are given in the book of Peng and Reggia.

Limitations of this model, in the framework of direct causal relationships are twofold:

- Causes may be dependent, in the sense that adding a cause may cancel an already present effect. If this phenomenon happens, a subset $M(D)$ of entailed symptoms should be prescribed for each subset $D \subseteq \mathtt{D}$ of disorders which can be jointly present. In other words, we then work with a relation on $2^{\mathtt{D}} \times \mathtt{S}$, rather than on $\mathtt{D} \times \mathtt{S}$. If some disorders can never be jointly present, 2^D should be replaced by the appropriate set $\mathtt{A} \subseteq 2^{\mathtt{D}}$ of associations of disorders which indeed make sense;

- The model does not distinguishes between effects that are certainly present when a disorder is present and effect that are only possible, or more exactly only cares about the latter.

3.2 A Generalized Parsimonious Covering Theory

In order to cope with the second limitation, an extension of the parsimonious covering theory has been devised by Dubois and Prade [1995]. Instead of one relation expressing causality between disorders and symptoms, the extended model assumes that two relations are available, namely R^+ and R^- that to each disorder d associates the set $M(d)^+$ of symptoms which are certainly produced by d alone and the set $M(d)^-$ of symptoms which certainly cannot be caused by d alone (which cannot be present when d alone is present). The union of the set $M(d)^+$ and the set $M(d)^-$ no longer necessarily covers S, i.e., $M(d)^+ \cup M(d)^- \neq \mathrm{S}$, but, we always have $M(d)^+ \cap M(d)^- = \varnothing$. Denoting $M(d)^0 = \mathrm{S} - (M(d)^+ \cup M(d)^-)$, $s \in M(d)^0$ means that s is only a possible symptom of d. In particular s may be absent or present when d is present. An unknown disorder can always be introduced, i.e., d_0 such that $M(d_0)^0 = \mathrm{S}$, i.e., it may cause any symptom. Hence it is not a closed-world model. In Peng and Reggia [1990] the causal relation R is defined by $(d, s) \in R \Leftrightarrow s \in \overline{M(d)^-}$, using our present notations. The available evidence is composed not only of a set M^+ of certainly present symptoms but also of a set of certainly absent ones M^-. When not all the information is available, the set M^+ and the set M^- again do not form a partition of S; indeed we have $M^+ \cap M^- = \varnothing$ but $M^+ \cup M^- \neq \mathrm{S}$, or if we prefer, there may exist a non-empty set $M^0 = \mathrm{S} - (M^+ \cup M^-)$ of symptoms, the presence or the absence of which are completely unknown. In other words, we may be not aware of the presence of some symptoms, and perhaps, only a subset of the symptoms which are absent is available. In the parsimonious covering model, the set M^- is generally not considered; when it is, the completeness of observations is reached, namely $M^+ \cup M^- = \mathrm{S}$. This is not assumed in the following. Let us assume again that disorders are independent so that

$$\begin{cases} M(\{d_i, d_j\})^+ = M(d_i)^+ \cup M(d_j)^+ \\ \text{and} \\ M(\{d_i, d_j\})^- = M(d_i)^- \cap M(d_j)^-. \end{cases}$$

Note the dissymmetry underlying this assumption: complex combinations of disorders tend to produce many symptoms of abnormality and very few (and sometimes no) indications of normal behaviour ($M(D)^- = \varnothing$). This is due to the fact that $M(d)^+$ gathers the symptoms that d suffices to produce, but that remain if other disorders are present. However $s \in M(d)^-$ does not mean that d suffices to prevent s, but only that d alone is not enough to produce it. Moreover $M(d)^+ \cap M(d)^- = \varnothing$ for all d imply $M(D)^+ \cap M(D)^- = \varnothing$ for all subsets of disorders D too.

The set of potential explanations (covers) is then defined by (see Figure 4)

$$\widehat{DD} = \{D \in \mathrm{A} \subseteq 2^{\mathrm{D}}, M(D)^+ \subseteq \overline{M^-} \text{ and } M(D)^- \subseteq \overline{M^+}\}.$$

This also writes

$$\widehat{DD} = \{D \in \mathrm{A} \subseteq 2^{\mathrm{D}}, M(D)^+ \cap M^- = \varnothing \text{ and } M(D)^- \cap M^+ = \varnothing\}.$$

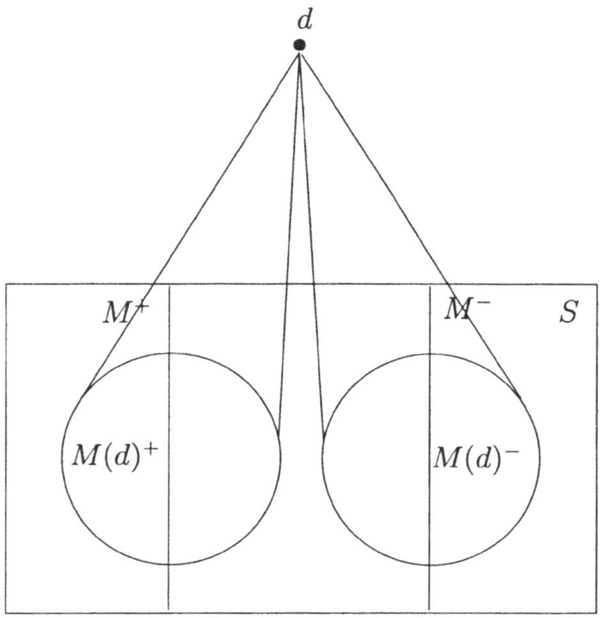

Figure 4. $\{d\}$ is a potential explanation of (M^+, M^-)

Explanations D that do not belong to $\widehat{DD}$ are inconsistent with the available observations, in the sense that such an explanation D is supposed to justify a symptom that has proved to be absent or to rule out a symptom that has been observed. Of course the most interesting explanations are those made of single disorders each of which can alone explain both M^+ and M^-, that is, covers of the form $\{d\}$. In Peng and Reggia's theory, since $M(D)^+ = \varnothing$ and $M^- = \varnothing$, the set $\widehat{DD}$ is just the set of covers in their sense, namely the covering of *surely* present symptoms by the ones *possibly* caused by D. Clearly the knowledge of non-trivial sets $M(D)^+$ and M^- reduces the number of covers.

Another particular case is when $M(d)^+ \cup M(d)^- = \mathsf{S}$, so that $M(d)^+ = \overline{M(d)^-}$, but we have not necessarily a complete observation of the symptoms which are present. Then the cover test becomes

$$\widehat{DD} = \{D \in \mathsf{A} \subseteq 2^{\mathsf{D}}, M^+ \subseteq M(D)^+ \subseteq \overline{M^-}\}$$

since we only know that the set of symptoms which are indeed present is lower bounded by the set M^+ and upper bounded by the set $\overline{M^-}$. Alternatively the causal knowledge may be incomplete but the observations are complete ($M^+ = \overline{M^-}$). Then

$$\widehat{DD} = \{D \in \mathsf{A} \subseteq 2^{\mathsf{D}}, M(D)^+ \subseteq M^+ \text{ and } M(D)^- \subseteq M^-\}.$$

Of course, in the unfrequent ideal case when knowledge and observations are complete, the set $\widehat{DD}$ contains all full explanations of the symptoms, also called "perfect covers".

Explanations in $\widehat{DD}$ are not necessarily relevant. Especially, the unknown disorder d_0 is always a cover. The cover membership test is thus very permissive, i.e., $\widehat{DD}$ can be very large. Namely an explanation D may belong to $\widehat{DD}$ even if $M(D)^+ \cap M^+ = \varnothing$ and $M(D)^- \cap M^- = \varnothing$. Indeed, in this case, $M(D)^+ \cup M(D)^- \subseteq M^0$. This means that $\widehat{DD}$ includes explanations no sure symptoms of which are observed and no forbidden symptoms are for sure absent. Such an explanation D may still be present when some M^+ and M^- are observed, since all the available, certain information we have about D pertains to symptoms in M^0 about which no observation exists (this is true for the unknown disorder d_0). Then, among the disorders in $\widehat{DD}$ we may prefer those, if any, which have at least a weak relevance to the observations, namely the subset $\widehat{DD}*$ of $\widehat{DD}$, defined by Dubois and Prade [1995]:

$$\widehat{DD}* = \{D \in \widehat{DD}, M(D)^+ \cap M^+ \neq \varnothing \text{ or } M(D)^- \cap M^- \neq \varnothing\}.$$

This eliminates irrelevant explanations which are such that $M(D)^+ \cup M(D)^- \subseteq M^0$, i.e., the disorders which are not suggested by the observations without being ruled out by them (and d_0 particularly). It is worth-noticing that the refinement of $\widehat{DD}$ by $\widehat{DD}*$ makes sense only in the general case of incomplete information. Indeed, $\widehat{DD} = \widehat{DD}*$, when $M^0 = \varnothing$ or $M(d)^0 = \varnothing$. Obviously, we may still refine $\widehat{DD}*$ by strengthening the requirements on d, by asking for a conjunction rather than a disjunction in the relevance test or by replacing the conditions by more demanding ones like $M(D)^+ \subseteq M^+$ (all sure symptoms of D are observed) or $M(D)^+ \supseteq M^+$ (all observed symptoms are among the ones which certainly accompany D), and similar conditions for the subsets pertaining to the absence of symptoms. In particular, the subset $\widehat{DD}_{**}$ of $\widehat{DD}$ defined by (see Figure 5)

$$\widehat{DD}_{**} = \{D, M(D)^+ \supseteq M^+ \text{ and } M(D)^- \supseteq M^-\}$$

gathers the explanations, if any, which cover the observations in a much more demanding way. Namely $\widehat{DD}_{**}$ gathers the explanations D such that sure effects of D cover surely observed symptoms and sure non-effects of D cover surely absent symptoms. But some sure effects or sure non-effects of D (i.e., which are in M^0) are maybe unobserved yet. In order to check that $\widehat{DD}_{**} \subseteq \widehat{DD}$, note that if $D \in \widehat{DD}_{**}$ then $M^+ \subseteq M(D)^+ \subseteq \overline{M(D)^-}$, and $M^- \subseteq M(D)^- \subseteq \overline{M(D)^+}$.

Note that Peng and Reggia's notion of relevance does not coincide with the ones defined above for two reasons. First, with Peng and Reggia's definition, a one-disorder cover is necessarily relevant, or in other words their notion of relevance is non-trivial only with covers which are not singletons, and this is not the case with the above introduced notions. Second, the set $\widehat{DD}*$, is ever empty in Peng and Reggia's particular case since $M(D)^+ = \varnothing$ and $M^- = \varnothing$. Peng and Reggia's

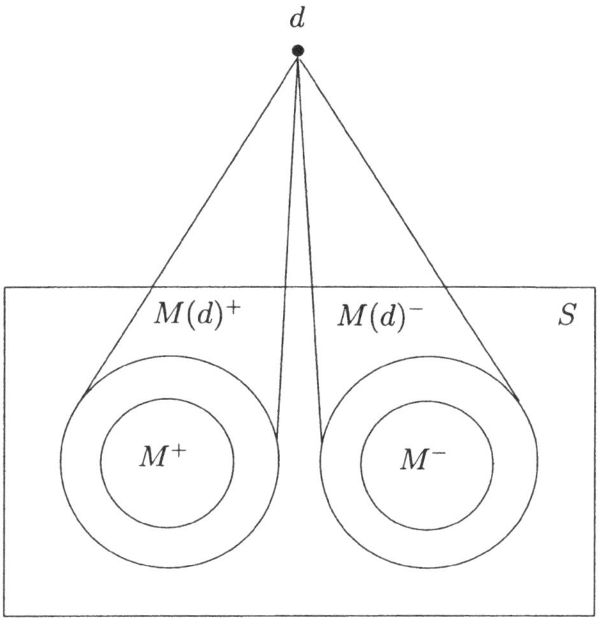

Figure 5. $\{d\}$ is a relevant cover of (M^+, M^-)

definitions could easily be extended to a framework where possible and sure symptoms are told apart. For instance, the set of relevant covers, in their sense, could be defined by the subset of disorders $\{D \in \widehat{DD}, \forall d \in D, \overline{M(d)^-} \cap M^+ \neq \varnothing$ or $\overline{M(d)^+} \cap M^- \neq \varnothing\}$. Consequently, our relational diagnosis model is more general than that of Peng and Reggia, in the incompletely informed case. However this notion of relevance again does not apply to single disorders. More natural definitions may be obtained by requiring non-empty intersection or inclusion relationships between M^+ and $M(d)^+$ or/and M^- and $M(d)^-$ for all d in D. These stronger forms of relevance cannot be expressed in Peng and Reggia's model since again $M(D)^+ = \varnothing$ and $M^- = \varnothing$ are assumed. In the extended incomplete information setting, it makes of course sense to select, among relevant covers, irredundant or minimal ones in accordance with the parsimony principle.

3.3 Graded Relational Methods

It seems desirable in practice to model a more gradual association between symptoms and disorders, and to take into account the uncertainty or vagueness pervading the prediction or the observation of symptoms. It is not right away clear what should be the precise meaning of the degrees of association $\mu_R(d_j, s_i)$, and of the degrees of presence $\mu_{M^+}(s_i)$ when the relation R or the subset M^+ is graded, nor the meaning of the degrees attached to disorders which are then proposed as explanation of M^+. For instance Kitowski and Bargiel [1987] spoke of the partial occurrence of a disorder, or of the uncertain observation of a symptom, but other interpretations can be thought of.

There are basically two types of interpretations which can be contemplated, one pertaining to the level of fulfilment of a gradually appearing symptom, the other to the uncertainty of clearcut statements pertaining to observations. Namely, on the one hand we may be uncertain about the presence of a symptom because its observation is difficult, costly, or even untimely; on the other hand symptoms may be a matter of degree (e.g., in medical diagnosis, the fever of a patient may be more or less high). In the first interpretation the uncertainty pervading our knowledge of actual symptoms stems from the use of easy, or quick observation methods whose results are imprecise. This may occur even if the presence or absence of a symptom is an all or nothing matter: the presence of a binary symptom can be more or less certain ($\mu_{M^+}(s_i) > 0$) or more or less possible ($1 - \mu_{M^-}(s_i) > 0$) or more or less probable ($\mu_{M^+}(s_i) = 1 - \mu_{M^-}(s_i) > 0$) due to unreliable or imprecise measurement.

On the contrary, $\mu_{M^+}(s_i)$ can model the *intensity* of a symptom s_i. This second interpretation is only possible if the underlying attribute a_i is no longer binary. For instance, instead of considering that symptom s_i means "fever" as opposed to "not fever", it looks more natural to distinguish between fevers which are more or less high. The vagueness pertaining to the word "high" reflects the fact that fever is a matter of degree and can be measured on a numerical scale T (ranging from 35 to 42 Celsius degrees, say). "High fever" refers to a fuzzy set F of T (that means "close to 40", say) in a given context. If $x \in T$ is the temperature of the patient, then μ_{M^+}(high fever) $= \mu_F(x) = 1 - \mu_{M^-}$(high fever), since M^- refers to the absent symptoms and μ_{M^-}(high fever) $= 1$ means "not high fever". When grades of membership express intensities of presence of symptoms the complementation is dictated by an equivalence between strong intensity of "high fever" and low intensity of "not-high fever". The intensity of the disorder then simply reflects the level of matching between (numerical) measurements and the most acute forms of the symptoms caused by the disorder (like "high fever").

Similarly the degree of association $\mu_R(d_j, s_i)$ may account for the uncertainty that s_i follows from the presence of d_j, or for the intensity of symptom s_i when d_j is present (which again supposes that the severeness of the effect can be graded); if the disorder d_j may itself be more or less severe $\mu_R(d_j, s_i)$ may even correspond to a gradual rule, like "the more severe the disorder, the stronger the symptom".

On the contrary, if we interpret $\mu_R(d_j, s_i)$ as a degree of uncertainty it may either mean to what extent s_i *necessarily* follows from the presence of d_j or only to what extent it is possible that s_i is present when d_j is present. In particular if the fuzzy set $M(d_j)(\mu_{M(d_j)}(s_i) = \mu_R(d_j, s_i))$ gathers the symptoms which more or less certainly or necessarily follow from d_j, the fact that s_i is such that $\mu_R(d_j, s_i) = 0$ means that either s_i does not follow from d_j, or that we do not know (i.e., we have absolutely no certainty) that s_i is caused by d_j; by contrast if $M(d_j)$ is the fuzzy set of symptoms which more or less possibly follows from $d_j, \mu_R(d_j, s_i) = 0$ means that s_i cannot be caused by d_j, but $\mu_R(d_j, s_i) = 1$ expresses no certainty that s_i should accompany d_j, it only expresses that it is fully possible. If $\mu_R(d_j, s_i)$ does reflect the uncertainty of presence of symptom s_i due to disorder d_j, but the end-points of the unit interval are interpreted as impossible (0) and certain (1), then it is difficult to imagine that $\mu_R(d_j, s_i)$ represents something else but the probability that d_j causes s_i when d_j is present, as used by Peng and Reggia [1990], this probability being subjective and objective. This leads us to Bayesian diagnosis methods.

According to the interpretation we have in mind, graded intensity or graded uncertainty, it will lead to different models with different interpretations of the results. To some extent the literature on fuzzy relational equations for diagnosis suffered from a lack of concern for these representational issues. In the next section we consider the case where the intensity is graded. Further on, the possibilistic extension of Peng and Reggia's model is presented. It offers an ordinal view of uncertainty only requiring the comparison of levels of uncertainty, in order to deal with the incomplete information case when the uncertainty is a matter of degree. Lastly, the probabilistic version of the parsimonious covering is recalled, in the subsequent part of this paper.

3.4 Fuzzy Relational Methods for Handling the Intensity of Symptoms

In the late seventies, Sanchez [1977a; 1977b], Tsukamoto and Terano [1977] had developed diagnosis methods based on a fuzzy relational model, thus before Reggia and his co-workers [Reggia *et al.*, 1985] proposed the relational model as a starting point for discussing diagnosis problems. Several slightly different proposals can be found in the fuzzy set literature (e.g., [Pappis and Sugeno, 1985; Adlassnig *et al.*, 1986; Asse *et al.*, 1987; Kitowski and Bargiel, 1987; Yager, 1985; Yager, 1988; Gmytrasiewicz *et al.*, 1990]. In the following we discuss how a fuzzy relational model can handle the intensity of presence of symptoms or disorders.

The simplest version is recalled here. Now R is a *fuzzy* relation defined on $\mathtt{D} \times \mathtt{S}$, and M^+ is a fuzzy set. R induces for each $d \in \mathtt{D}$, the fuzzy set $M(d)$ of symptoms fuzzily associated with d, i.e., $\forall d \in \mathtt{D}, \forall s \in \mathtt{S}, \mu_{M(d)}(s) = \mu_R(d, s)$ (where μ denotes a membership function). The membership grades $\mu_{M^+}(s)$ and $\mu_{M(d)^+}(s)$ are respectively interpreted as the intensity, estimated on the normalized interval $[0, 1]$, of the observed presence of symtom s, and intensity of the presence of s when disorder d is present with its maximal intensity. As already explained in the previous section, we have, by definition, $\mu_{M^-} = 1 - \mu_{M^+}$ and $\mu_{M(d)^-} = 1 - \mu_{M(d)^+}$. It is important to see that the intensity model based on fuzzy sets extends the crisp completely informed model ($M(d)^+ \cup M(d)^- = \mathtt{S}$ and $M^+ \cup M^- = \mathtt{S}$) over to non-binary symptoms, but is not concerned with the treatment of uncertainty. Given fuzzy M^+ and R, the diagnosis problem is then stated as the determination of some (possibly fuzzy) subset $D \subseteq \mathrm{D}$ of disorders such that the composition of D with R yields M^+, which can be written symbolically $D \circ R = M^+$, and which in the simplest model translates into the fuzzy relation equation

$$\forall i = 1, n,\ \max_{j=1,k} \mu_R(d_j, s_i) * \mu_D(d_j) = \mu_{M^+}(s_i).$$

In this equation, $\mu_R(d_j, s_i) * \mu_D(d_j)$ can be interpreted as the expected intensity of s_i due only to disorder d_j, taking into account that d_j is present with intensity $\mu_D(d_j)$. Possible choices for $*$ are 'min' (the intensity of the effect is just upper-bounded by the intensity of the cause) or the product (the intensity of the effect is proportional to the intensity of the cause). The use of 'max' for aggregating the effects of disorders reflects the assumption that in case of the effective presence of several disorders, the intensity of the resulting symptom is just the greatest of the expected intensities associated with each disorder, which might be debatable (we might think of using a bounded or probabilistic sum instead of 'max' for instance so as to account for a cumulative effect of disorders). The causal equation expresses that the unknown fuzzy set of (non mutually exclusive) disorders D has produced, via R, the fuzzy set of symptoms M^+. The value $\mu_D(d)$, computed from the equation, is the plausible intensity of presence of d induced by the intensity-qualified symptom. Assuming that D is a crisp set, we get a fuzzy set of resulting symptoms $M(D) = \cup_{d \in D} M(d)$ where the union is expressed by the maximum operation. The fuzzy model thus underlies an assumption of "independent" symptoms like Peng and Reggia's model, although the term "independence" in the graded case rather evokes a non-idempotent operation for the combination of causal effect intensities. A better name for the assumption is rather "monotonic causality" in the sense that adding disorders cannot cancel already present symptoms. This assumption can be straightforwardly given up by defining R on $2^{\mathrm{D}} \times \mathrm{S}$ [Sanchez, 1977b].

The fuzzy causal relation equation with an operation $*$ different from minimum, such as the product or the operation $\max(0, \cdot + \cdot - 1)$ has been extensively considered in the fuzzy set literature (see [Di Nola *et al.*, 1989] for instance). In the following we only consider the use of minimum for the sake of simplicity. The fuzzy causal equation is an extension of the causal equation selecting covers (see Subsection 3.1) when one assumes complete information on the symptoms which are present and on those which are for sure associated with a given disorder. Indeed the conditions $M(D)^+ \cap M^- = \varnothing$ and $M(D)^- \cap M^+ = \varnothing$ then reduce to $M(D)^+ = \cup_{d \in D} M(d)^+ = M^+$, which yields a set of perfect covers $\widehat{DD} = \{D, R\hat{o}D = M^+\}$ with $(d, s) \in R \Leftrightarrow s \in M(d)^+ = \overline{M(d)^-}$. Thus, when the presence or the absence of symptoms and disorders is a binary matter, the fuzzy causal equation reduces to the condition giving perfect covers in the completely informed case.

The fuzzy causal equation where M^+ and R are known but D is unknown may have several solutions, or no solution. If several solutions exist, the minimal ones (in the sense of fuzzy set inclusion) defined by $F \subseteq G \Leftrightarrow \mu_F \leq \mu_G$), are worth-considering; see [Pappis and Sugeno, 1985]. They correspond to the idea of irredundant covers in Peng and Reggia's model.

If the equation $D\hat{o}R = M^+$ has no solution, most authors relax it into $D\hat{o}R \subseteq M^+$ or $D\hat{o}R \supseteq M^+$. The fuzzy inclusion $D\hat{o}R \subseteq M^+$ can be interpreted as looking only for partial explanations to the set of symptoms; alternatively R may be interpreted as a lower bound of the expected intensity of symptoms. The other inclusion $D\hat{o}R \supseteq M^+$ mirrors the assumption that M^+ is only a partial observation of the symptoms intensities (they are underestimated), or that R only estimates an upper bound of the expected intensity of symptoms. Peng and Reggia's model corresponds to the first of these last two assumptions, since a solution to $D\hat{o}R \supseteq M^+$ does correspond to a cover of the partial observations M^+ and a minimal solution is an irredundant cover.

Most authors in the literature envisage solving equation $D \circ R \subseteq M^+$ as a substitute to $D \circ R = M^+$ and compute its largest solution $\underline{D}$. It is obtained as is

$$\mu_{\underline{D}}(d_j) = \min_i \mu_R(d_j, s_i) \to \mu_{M^+}(s_i)$$

where $\to$ denotes the residuation of operation $*$, namely $\alpha \to \beta = \sup\{x, \alpha * x \leq \beta\}$. When $* =$ minimum, $\to$ is the so-called Gödel implication, i.e., $\alpha \to \beta = 1$ if $\alpha \leq \beta$ and $\alpha \to \beta = \beta$ if $\alpha > \beta$. Note that the fuzzy set inclusions $M(d_i) \subseteq M^+$ and $M(d_j) \subseteq M^+$ imply $M(\{d_i, d_j\}) = M(d_i) \cup M(d_j) \subseteq M^+$, where the fuzzy union is defined by the max operation on the membership degrees. Hence $\underline{D}$ is unique and is also the largest solution of $D \circ R = M^+$ when a solution exists. In the latter situation, exhibiting $\underline{D}$ as the solution to the diagnosis problem is debatable because it violates the parsimony principle: $\underline{D}$ may accumulate several distinct explanations of a subset of symptoms. The interest of the maximal solution is to discard disorders inconsistent with it (namely d_j such that $\mu_{\underline{D}}(d_j) = 0$) The parsimony principle rather suggests to look for minimal solution(s) D to the equation $D \circ R = M^+$ (more generally to $D \circ R \supseteq M^+$ since minimal solutions to the former are also minimal solutions to the latter); see [Di Nola *et al.*, 1989] for algorithms computing these solutions (there may be several ones which cannot be compared with respect to set inclusion). These minimal solutions correspond to irredundant covers *à la* Peng and Reggia.

When the equation $D \circ R \supseteq M^+$ has no solutions one may compute minimal solutions to $D \circ R = \underline{D} \circ R$. Each such minimal solution is made of a set of disorders which altogether are sufficient to explain the greatest subset of observations in M^+ which is explainable according to R (that is $\underline{D} \circ R$) and which has no effect greater than the one recorded in M^+.

Another way of tackling the problem when the equation $D \circ R \supseteq M^+$ has no solutions is to evaluate the plausibility of a fuzzy covering D of M^+ by a degree of inclusion of M^+ in $M(D) = D \circ R$. There are many ways of defining such inclusion degrees (see [Dubois and Prade, 1980]). One classical measure can be

$$I(M(D) \subseteq M^+) = \frac{| M(D) \cap M^+ |}{| M^+ |}$$

where $| F |$ is the cardinality of the fuzzy set F, that is the sum of the membership grades. This is proposed by Dalton [1993] who considers only crisp covers. Note that selecting the best covers in the sense of this index tends to favour large sets D and this goes against the parsimony principle. The degree of perfect covering of of M^+ by the set of disorders D is then a measure of equality of M^+ and $M(D)$, that can be chosen of the form

$$d(M(D) = M^+) = \min(I(M(D) \subseteq M^+), I(M^+ \subseteq M(D)).$$

See Dubois and Prade [1993b] for further discussions and solution sets to fuzzy diagnosis problems.

3.5 Possibilistic Extension of the Parsimonious Covering Theory under Incomplete Information

The incompletely informed case in Subsection 3.2 has been extended to graded uncertainty (Dubois and Prade [1993a; 1995]). Namely, M^+ and M^- are now independent fuzzy sets of symptoms which are more or less certainly present, and more or less certainly absent respectively. Degrees of membership are not supposed to be numerical but only belong to a finite, totally ordered set of levels ranging between a bottom 0 and a top 1, i.e., $\lambda_1 = 0 < \lambda_2 < \ldots < \lambda_n = 1$ instead of $[0, 1]$, with $\min(\lambda_i, \lambda_k) = \lambda_i$ and $\max(\lambda_i, \lambda_k) = \lambda_k$ if $i \leq k$, and equipped with an order-reversing map $1 - \lambda_i = \lambda_{n+1-i}$. Degrees of membership reflect uncertainty about all-or-nothing symptoms or disorders. Especially the requirement $M^+ \cap M^- = \varnothing$ is maintained (where the intersection is defined by the min operation), i.e., we cannot be somewhat certain both of the presence and of the absence of the same symptom simultaneously. Similarly, $M(d)^+$ (more generally $M(D)^+$) and $M(d)^-$ (more generally $M(D)^-$) will denote the fuzzy sets of symptoms which are respectively more or less certainly present and more or less certainly absent when disorder d alone is present (more generally when the subset D of disorders is present). The independence assumption between causes when they hold, easily extend to the fuzzy case under the form

$$\begin{cases} \mu_{M(\{d_i,d_j\})^+} = \max(\mu_{M(d_i)^+}, \mu_{M(d_j)^+}) \\ \mu_{M(\{d_i,d_j\})^-} = \min(\mu_{M(d_i)^-}, \mu_{M(d_j)^-}) \end{cases}$$

Again this choice of connectives presupposes that if there is certainty of presence of symptom s due to two causes, these local certainties neither cancel nor reinforce each other. Similarly, the more numerous the disorders the less numerous are the symptoms of which we are sure that they will not show up. Obviously, we also assume $\forall d, M(d)^+ \cap M(d)^- = \varnothing$ (so that $\forall D, M(D)^+ \cap M(D)^- = \varnothing$).

By complementation (defined by $\mu_{\bar{F}} = 1 - \mu_F$) of M^- and $M(d)^-$, we obtain the fuzzy sets $\overline{M^-}, \overline{M(d)^-}$, (and more generally $\overline{M(D)^-}$) of symptoms which are more or less possibly present, respectively, in the considered situation, when d is present, or when disorders in D are altogether present. This corresponds to the usual duality between what is (more or less) certain, i.e., necessarily true, and what is (more or less) possibly true. Note that $M^+ \subseteq \overline{M^-}, M(d)^+ \subseteq \overline{M(d)^-}$, and $M(D)^+ \subseteq \overline{M(D)^-}$ in the sense of a fuzzy set inclusion stronger than usual. Since $M^+ \cap M^- = \varnothing$, we have

$$\{s_i \in \mathrm{S}, \mu_{M^+}(s_i) > 0\} \subseteq \{s_i \in \mathrm{S}, \mu_{M^-}(s_i) = 0\} = \{s_i \in \mathrm{S}, \mu_{\overline{M^-}}(s_i) = 1\}$$

i.e., the support of M^+ (i.e., the elements with a positive membership grade) is included in the core of $\overline{M^-}$ (the elements which belong with degree 1); the same holds for $M(d)^+, \overline{M(d)^-}$, or $M(D)^+, \overline{M(D)^-}$. This is in agreement with possibility theory: for crisp events $A, N(A) > 0 \Leftrightarrow \Pi(\neg A) < 1 \Rightarrow \Pi(A) = 1$ since then one of A or $\neg A$, at least, should be completely possible. The pair of fuzzy sets $(M^+, \overline{M^-})$ such that $M^+ \cap M^- = \varnothing$ is called a twofold fuzzy set [Dubois and Prade, 1987a]. Twofold fuzzy sets have been introduced for modelling incompletely known sets, as is the set of present symtoms M (bracketed from below by M^+ and from above by $\overline{M^-}$), as well as the set $M(d)$ of symptoms caused by d in the diagnosis problem. Also the causal independence assumptions lead to equations that compute $M(D)^+, \overline{M(D)^-}$ in terms of $M(d)^+, \overline{M(d)^-}$ in a way coherent with the union of twofold fuzzy sets, where the union of pairs of fuzzy sets is obtained componentwise.

The extension to fuzzy sets of the equations defining covers can be done very simply. It requires that the extent to which two fuzzy sets F and G of S intersect be evaluated. The consistency between F and G is simply defined as [Zadeh, 1979]

$$\mathrm{cons}(F, G) = \sup_{s \in \mathrm{S}} \min(\mu_F(s), \mu_G(s)).$$

It computes the degree of existence of some common element for F and G. The degree of covering of the known symptoms, as specified by the pair (M^+, M^-), by a crisp set D of disorders is:

$$\begin{aligned} \mu_{\widehat{DD}}(D) &= \min(1 - \mathrm{cons}(M(D)^+, M^-), 1 - \mathrm{cons}(M(D)^-, M^+)) \\ &= 1 - \max(\mathrm{cons}(M(D)^+, M^-), \mathrm{cons}(M(D)^-, M^+)) \end{aligned}$$

where the minimum operator expresses the conjunction of the conditions $M(D)^+ \cap M^- = \varnothing$ and $M(D)^- \cap M^+ = \varnothing$. It is easy to check that this equation can be written in terms of inclusion indices under the form

$$\mu_{\widehat{DD}}(D) = \min(I(M(d)^+, \overline{M^-}), I(M(d)^-, \overline{M^+}))$$

where the degree of inclusion of fuzzy set F into fuzzy set G is defined by

$$I(F, G) = 1 - \mathrm{cons}(F, \bar{G}) = \inf_{s \in \mathrm{S}} \max(1 - \mu_F(s), \mu_G(s)).$$

Note that $I(F, G) = 1 \Leftrightarrow \mathrm{support}(F) \subseteq \mathrm{core}(G) \Leftrightarrow F \cap \bar{G} = \varnothing$, i.e., the inclusion index extends the strong inclusion encountered in twofold fuzzy sets and guarantees that D fully belongs to $\widehat{DD}$ (i.e., with degree 1) if and only if $M(d)^+$ and $\overline{M^-}$ do not overlap at all, nor $M(d)^-$ and $\overline{M^+}$. The inclusion index contrasts with the cardinality-based one of the previous section that assumed numerical membership grades. When the latter are only elements of a total order, the scalar cardinality of a fuzzy set is undefined. In contrast, $\mathrm{cons}(M(D)^+, M^-)$ is defined on a totally ordered set and so is the degree of inclusion using the order-reversing map. The degree of covering $\mu_{\widehat{DD}}(D)$ clearly expresses that a disorder d is all the less a candidate explanation as the fuzzy set of its more or less certain effects overlaps the fuzzy set of symptoms more or less certainly absent, or as the fuzzy set of effects which are more or less certainly absent when d is present overlaps the fuzzy set of symptoms which are more or less certainly present.

The idea of relevance expressed by the set of covers $\widehat{DD}*$ in the crisp case can be easily extended using the consistency index. Namely,

$$\mu_{\widehat{DD}*}(D) = \min(\mu_{\widehat{DD}}(D),\ \max(\mathrm{cons}(M(D)^+, M^+),\ \mathrm{cons}(M(D)^-, M^-))).$$

A more demanding relevance notion can be obtained by changing max into min in the above definition. The subset $\widehat{DD}_{**}$ of sets of disorders whose sure symptoms or non-symptoms completely cover of the actual observations (but some sure effects or sure non-effects of which may be unobserved yet) can be also extended by

$$\mu_{\widehat{DD}**}(D) = \min(\mathsf{I}(M^+, M(d)^+), \mathsf{I}(M^-, M(d)^-))$$

where $\mathsf{I}(F, G)$ could be a less drastic inclusion index than $1 - \mathrm{cons}(F, \bar{G})$, since we only require that $\mathsf{I}(F, G) = 1 \Leftrightarrow \mu_F \leq \mu_G$ for expressing the coverage property. An example of such inclusion index is

$$\mathsf{I}(F, G) = \inf_s \mu_F(s) \to \mu_G(s)$$

with $a \to b = 1$ if $a \leq b$ and $a \to b = b$ if $a > b$. Note that the fuzzy inclusion $\widehat{DD}_{**} \subseteq \widehat{DD}$ does not hold, i.e., $\mu_{\widehat{DD}**} \leq \mu_{\widehat{DD}}$ is not always true because of the disparity between inclusion indices I and I. However, if $\mu_{\widehat{DD}**}(D) = 1$ then $\mu_{M^+} \leq \mu_{M(D)^+}$ and since $\mathrm{support}(M(D)^+) \subseteq \mathrm{core}(\overline{M(D)^-})$ then $\mathrm{support}(M^+) \subseteq \mathrm{core}(\overline{M(D)^-})$. Likewise, one gets $\mathrm{support}(M^-) \subseteq \mathrm{core}(\overline{M(D)^+})$. So, $\mathrm{core}(\widehat{DD}_{**}) \subseteq \mathrm{core}(\widehat{DD})$.

Taking into account the incomplete nature of the information about the presence or absence of symptoms decreases the discrimination power when going from the completely informed case to the incomplete information case, since then the number of possible disorders in $\hat{D}$ increases. This is due to the fact that with incomplete information there are symptoms which are neither certain nor impossible and consequences of the presence of a given disorder d which are only possible, while $M(d)^+ = M(d)$ and $M(d)^- = \overline{M(d)}$ when causal information is complete. This suggests that, in order to improve the discrimination power of the model, we have to refine the non-fuzzy model in such a way that consequences (resp. symptoms) previously expressed as certain (resp. certainly present) and impossible (resp. certainly absent) remain classified in the same way and where some consequences considered as possible (resp. possibly present symptoms) are now allowed to be either somewhat certain (resp. somewhat certainly present) or somewhat impossible (resp. somewhat certainly absent). See Figure 6. The introduction of grades of uncertainty enables us to rank-order the possible subsets of covers in $\widehat{DD}$. This counterbalances the increase of candidate explanations due to the incompleteness of the information.

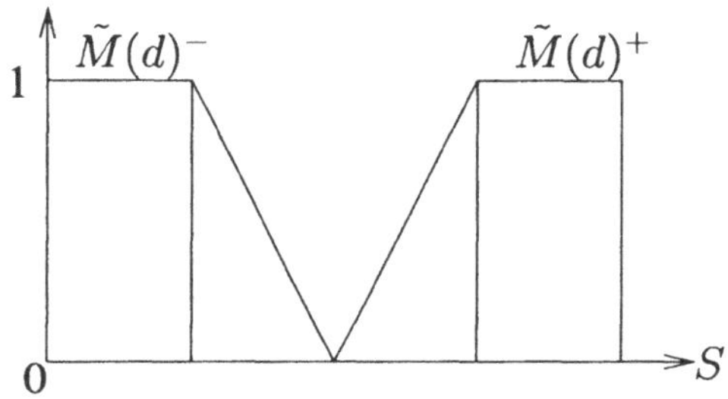

Figure 6. Partially informed situation with preferences

It can be easily shown that adding preference levels on top of incompleteness modeling can at the same time enable the discrimination power of the completely informed situation to be recovered, and also enable the extra disorders obtained on the partially informed situation to be ranked in terms of their plausibility levels. Indeed, consider for simplicity the case of single faults and assume that the dichotomy $M(d), \overline{M}(d)$ represents a first approximation of the fuzzy sets $\tilde{M}(d)^+$ and $\tilde{M}(d)^-$ the sense that (see Figure 6):

$$M(d) = \text{ support}(\tilde{M}(d)^+); \quad \overline{M(d)} = \text{ support}(\tilde{M}(d)^-)$$

and moreover $M(d)^+ = \text{ core}(\tilde{M}(d)^+), M(d)^- = \text{ core}(\tilde{M}(d)^-)$. Then it can be proved that both results of the completely informed and partially informed situations can be recovered by the fuzzy model. Namely, let $\hat{D}_c$, $\hat{D}_p$, and $\hat{D}$ be the sets of single disorder covers obtained in the completely informed, incompletely informed and possibilistic graded cases respectively. We shall assume non-fuzzy incomplete observations, in which case $\hat{D}_c = \{d \in \mathrm{D}, M^+ \subseteq M(d) \subseteq \overline{M}^-\}$. Then we can show the following equalities, $\hat{D}_c = \text{ core}(\hat{D}); \hat{\overline{D}}_p = \text{ support}(\hat{D})$ (see [Dubois and Prade, 1993a] for a proof). This result indicates that our approach, although much more qualitative than one based on probability theory, still possesses the ability to rank-order the set of plausible unique disorders explaining an incomplete set of symptoms.

EXAMPLE 1 (Dubois and Prade, [1995]). Let us show the benefit of grading uncertainty on an example. Let $\mathrm{D} = \{d_1, d_2, d_3, d_4\}, \mathrm{S} = \{s_1, s_2, s_3\}$ and let us first consider the completely informed case

$$\begin{aligned} M(d_1)^+ &= \{s_2\}, & M(d_1)^- &= \{s_1, s_3\} \\ M(d_2)^+ &= \{s_3\}, & M(d_2)^- &= \{s_1, s_2\} \\ M(d_3)^+ &= \{s_1, s_2, s_3\}, & M(d_3)^- &= \varnothing \\ M(d_4)^+ &= \{s_2, s_3\}, & M(d_4)^- &= \{s_1\}, \end{aligned}$$

and

$$M^+ = \{s_2, s_3\}, \qquad M^- = \{s_1\}.$$

Then, we get $\hat{D} = \{d_4\}$. Let us weaken our example into an incompletely informed case

$$\begin{aligned} M(d_1)^+ &= \varnothing, & M(d_1)^- &= \{s_1\} \\ M(d_2)^+ &= \{s_3\}, & M(d_2)^- &= \varnothing \\ M(d_3)^+ &= \{s_3\}, & M(d_3)^- &= \varnothing \\ M(d_4)^+ &= \varnothing, & M(d_4)^- &= \{s_1\}. \end{aligned}$$

Then $\hat{D} = \{d_1, d_2, d_3, d_4\}$, i.e., all causes are possible. Let us now consider the following table where the previous incompletely informed case is completed with some gradual uncertainty information that is in accordance with the complete information case.

Table 1.

	$M(d_1)^+$	$M(d_1)$	$M(d_2)$	$M(d_2)$	$M(d_3)$	$M(d_3)$	$M(d_4)^+$	$M(d_4)^-$	S	
		-	+	-	+	-				
membership s_1	0	1	0	0.9	0.7	0	0	1	0	M^-
grades s_2	0.8	0	0	0.6	0.5	0	0.7	0	1	M^+
s_3	0	0.8	1	0	1	0	0.3	0	1	M^+
cons$(M(d)^+, M^-)$	0		0		0.7		0			
cons$(M(d)^-, M^+)$		0.8		0.6		0		0		
$\mu_{\hat{D}}(d)$	0.2		0.4		0.3		1			

Thus we obtain that d_4 is a better cover than d_2, itself a better cover than d_3, which is a better cover than d_1, i.e., a ranking of the possible single disorders is obtained. This is in accordance with the complete information case as well as the incomplete information case, but significantly improves their findings. Note that all disorders are relevant in the sense of $\widehat{DD}*$, but only d_4 is fully relevant in the sense of $\widehat{DD}_{**}$.

Clearly, when the set of single-fault covers $\hat{D}$ is empty, we have to look for two-element subsets D which may account for M^+ and M^-, and then for three-element subsets if there is no two-element one, and so on until a plausible explanation is found. However, if we look for multiple disorders explaining a given set of symptoms, it is clear that we shall have a trade-off problem between small sets of disorders which are plausible in the sense of the parsimony principle, and bigger sets of disorders which are more plausible because they ensure a better covering of the observed symptoms. This topic, along with the proper representation of independent disorders, requests further investigation.

The graded approach presented in this section has been applied (with a single fault assumption, though) to a fault diagnosis problem in satellites (Cayrac *et al.* [1995; 1996]).

4 HANDLING UNCERTAIN PRIOR KNOWLEDGE IN RELATIONAL METHODS

The approaches described in Section 3 have in common the absence of any assumption regarding a priori knowledge about the presence of disorders. More precisely these methods can be applied when no such knowledge is available, but as such they cannot use prior knowledge when it exists. Another common feature to methods of Section 3 is the assumption of direct causal relationships between disorders and their observable symptoms. This section surveys methods that can handle prior knowledge and/or unobservable states that make the causal relationship between disorders and observable symptoms indirect. Advanced methods of that kind are basically probabilistic. However, for the sake of completeness, some insight is given into possibilistic approaches, for which only partial results exist.

4.1 Probabilistic Extension of the Parsimomious Covering Theory

Peng and Reggia [1990] present a method of diagnosis problem solving that refines their parsimonious covering theory, and generalizes the probabilistic classification method of Section 2.2 as well. The aim of the diagnosis method is to compute the probability $P(D \mid M^+)$ of explanation $D \subseteq \mathtt{D}$, given a set of observed symptoms M^+. In this section it is assumed that observations are complete, i.e., $M^+ \cup M^- = \mathtt{S}$: a symptom not in M^+ is not present. Note that both D and M^+ are conjunctive sets of disorders and symptoms, contrary to Section 2.2 where due to the unique fault assumption, a subset D represented a disjunctive set, i.e., a set of disorders *one* of which is present. Here D is the set of disorders *all* of which are present, and only them. To simplify notations we shall denote $D_I = \{d_i \in \mathtt{D}, i \in I\}$, and M^+ is denoted $M_J = \{s_j \in \mathtt{S}, j \in J\}$. Then D_I represents the event $\wedge_{i \in I} d_i \wedge \wedge_{i \notin I} \neg d_i$, and similarly for M_J. Of course using Bayes theorem, we again find

$$P(D_i \mid M_J) = \frac{P(M_J \mid D_I) \cdot P(D_I)}{P(M_J)}.$$

Ranking the explanations D_I only requires the computation of $P(M_J \mid D_I)$ and $P(D_I)$. This computation can be tricky if no assumption is made. Peng and Reggia have proposed assumptions so as to make the data-gathering task not burdensome and to remain coherent with the parsimonious covering method. However these assumptions are much more realistic than in the Bayesian classification approaches.

The data is supposed to be made of two items: a probabilistic causal relationship R, and the prior probabilities of single disorders. The entries of R are defined as conditional probabilities

$$\mu_R(d_i, s_j) = P(s_j : d_i \mid d_i)$$

where $s_j : d_i$ is called a causation event, and represents the event that d_i actually causes s_j. Occurrence of $s_j : d_i$ means that s_j and d_i are present *and* that d_i actually causes s_j. So $s_j : d_i$ implies $s_j \wedge d_i$ but not conversely. Particularly $P(s_j : d_i \mid d_i)$ generally differs from (and is smaller than) $P(s_j \mid d_i)$. The latter conditional probability also accounts for situations where s_j and d_i are present together but s_j is not due to d_i (but to some other disorder, present as well). It is supposed that the set of disorders contain the cause of any symptom s_j, that is

s_j is equivalent to $\wedge_{d \in \mathrm{D}}\, s_j : d$ (mandatory causation).

This assumption is closely related to the one of observation relevance in Section 2.4. Of course if s_j can only be caused by d then $s_i : d = s_i \wedge d$, so that $P(s_i : d \mid d)$ reduces to $P(s_i \mid d)$. But one may have $P(s_j : d_i \mid d_i) = 0$ while $P(s_j \mid d_i) > 0$, namely if s_j is not caused by d_i, but d_i turns out to be present.

The set of prior probabilities $\{P(d), d \in D\}$ is of course certainly not such that $\Sigma_d P(d) = 1$ since the disorders are not mutually exclusive. Probabilistic assumptions that underly the multiple-fault probabilistic diagnosis method are as follows, where $p_i = P(d_i)$ and $c_{ij} = P(s_j : d_i \mid d_i)$

- disorder independence:

$$\begin{aligned} P(D_I) &= (\textstyle\prod_{i \in I} p_i) \cdot (\prod_{i \notin I} (1 - p_i)) \\ &= k \cdot \textstyle\prod_{i \in I} \frac{p_i}{1 - p_i} \end{aligned}$$

  where $k = \prod_{d \in \mathrm{D}} P(d)$ is a constant. This statement replaces the unique fault assumption;

- causation independence. Let X be the conjunction of any cause or causation events other than $s_j : d_i$ or its negation. Then $P(s_j : d_i \mid d_i \wedge X) = P(s_j : d_i \mid d_i)$. X is called a context of $s_j : d_i$.

  It means that the particular influence of d_i on m_j is not affected by other events. It replaces the assumption of independence of symptoms of a given disorder, that traditional Bayesian classification adopts. On such a basis the following results and properties are obtained:

  - if X is a context for $s_j : d_i$ then

$$\begin{aligned} P(s_j : d_i \wedge X \mid d_i) &= c_{ij} \cdot P(X \mid d_i) \\ P((s_j : d_i) \wedge X) &= P(d_i \wedge X) \cdot c_{ij} \end{aligned}$$

- single disorder: if $D_J = \{d_i\}$ then

  $$P(M_J \mid \{d_i\}) = \Pi_{j\in J} c_{ij} \cdot \Pi_{j\notin J}(1 - c_{ij}).$$

  This is the form of the result obtained in Section II.2 for Bayesian classification.

- multiple disorders:

  $$P(s_j \mid D_I) = 1 - \Pi_{i\in I}(1 - c_{ij}).$$

  This result expresses that effects of several disorders accumulate, as in the parsimonious covering model (Section 3.1). It is an extension of $M(D_i) = \cup_{i\in I} M(d_i)$ to fuzzy sets of symptoms induced by disorders d_i, where membership grades $\mu_{M(d_i)}(s_j) = c_{ij}$. Indeed the fuzzy union operation $a + b - ab$ is used in the above. In contrast, it is changed into the maximum in the fuzzy and possibilistic versions of the parsimonious covering approach (Sections 3.4 and 3.5). More generally

  $$P(M_J \mid D_I) = \Pi_{j\in J} P(s_j \mid D_I) \cdot \Pi_{j\notin J}(1 - P(s_j \mid D_I)).$$

From this expression it is easy to verify that

- if D_I does not cover M_J then $P(M_J \mid D_I) = 0$. Indeed it means that $c_{ij} = 0$ for some symptom s_j and all $d_i \in D_I$;

- if d_i causes s_j for sure ($c_{ij} = 1$) but $s_j \notin M_J$ then $P(M_J \mid D_I) = 0$ for any D_I containing d_i. This is a case where an explanation is rejected because a surely implied symptom is not observed.

This situation is not met in the Boolean version of the parsimonious covering theory but is met in the extension due to Dubois and Prade [1995] (see Sections 3.2 and 3.5). It is interesting to compare $P(M_J \mid D_I)$ and $\mu_{\widehat{DD}}(D_I)$ obtained in the possibilistic approach of Section 3.5, in the case when $M^+ = \overline{M^-}$ is not fuzzy. Then

$$\begin{aligned} \mathrm{cons}(M(D_I)^+, \overline{M_J}) &= \sup_{j\notin J} \max_{i\in I} \mu_{M(d_i)^+}(s_j) \\ \mathrm{cons}(M(D_I)^-, \overline{M_J}) &= \sup_{j\in J} \max_{i\in I} \mu_{M(d_i)^-}(s_j) \end{aligned}$$

and

$$\begin{aligned} \mu_{\widehat{DD}}(D_I) &= \min(\min_{j\notin J} \min_{i\in I} 1 - \mu_{M(d_i)^+}(s_j), \\ &\qquad \min_{j\in J} \max_{i\in I} 1 - \mu_{M(d_i)^-}(s_i)) \\ &= \min(\min_{j\notin J} 1 - \mu_{M(D_I)^+}(s_j), \\ &\qquad \min_{j\in J} 1 - \mu_{M(D_I)^-}(s_j)). \end{aligned}$$

Note that the expressions of $\mu_{\widehat{DD}}(D_I)$ and $P(M_J \mid D_I)$ are similar if we change min into product, and identify $1 - \mu_{M(D_I)^-}(s_i)$ with $P(s_i \mid D_I)$ (since $P(s_i \mid D_I) > 0$ means that s_i is a possible symptom of D_I). Identifying $1 - \mu_{M(D_I)^+}(s_i)$

with $1 - P(s_i \mid D_I)$ does not really make sense unless one admits $\overline{M(D_I)^+} = M(D_I)^-$, which never holds in the possibilistic setting except if these sets are not fuzzy.

Peng and Reggia have studied whether the ranking of covers in the probabilistic parsimonious model is in accordance with the plausible ranking of the non-probabilistic model. Namely let D_1, D_2, D_3, D_4 be simple, relevant, irredundant and minimal covers of M_J in the sense of Section 3.1. Then does it hold that $P(D_1 \mid M_J) \leq P(D_2 \mid M_J) \leq P(D_3 \mid M_J) \leq P(D_4 \mid M_J)$? It turns out that these inequalities do not always hold:

- $P(D_1 \mid M_J) \leq P(D_2 \mid M_J)$ almost almost always hold. Let $D_1 = D_2 \cup \{d_k\}$ where D_1 is a cover not relevant due to the presence of d_k that causes none of $s_j \in M_J$. Then the inequality is violated if d_k is a priori very probable (p_k close to 1), has very few manifestations and all of them are unlikely (c_{kj} small). This is a very uncommon situation;

- $P(D_2 \mid M_J) \leq P(D_3 \mid M_J)$ may fail to hold. Let D_3 be an irredundant cover, $D_2 = D_3 \cup \{d_k\}$ be a redundant, relevant cover. Then the inequality is violated whenever d_k is a priori very probable, and has a symptom which is very frequent and present in M_J. Then D_2 is as interesting an explanation as D_3, even if redundant;

- $P(D_3 \mid M_J) \leq P(D_4 \mid M_J)$ often does not hold. Let D_4 be minimum (in the sense of cardinality) while D_3 is irredundant. Then there is generally no reason why the inequality should hold. The inequality makes sense only when the a priori probabilities of disorders are very small and almost equal and the causal strength values c_{ij} are large.

Lastly, the probabilistic multiple-fault diagnosis method can be shown to involve only mild independence assumptions, namely:

- interdependence between symptoms and disorders
  $P(s_j \wedge d_i) \geq P(s_j) \cdot P(d_i)$, with equality only if d_j does not cause s_i;

- interdependence between manifestations
  $P(s_k \wedge s_j) \geq P(s_k) \cdot P(s_j)$ with equality only if s_k and s_j have no common cause, and
  $P(s_k \wedge s_j \mid d_i) \geq P(s_k \mid d_i) \cdot P(s_j \mid d_i)$ with equality only if d_i is the single common cause of s_k and s_j

- conditional interdependence between disorders
  $P(d_1 \wedge d_2 \mid s_j) \geq P(d_1 \mid s_j) \cdot P(d_2 \mid s_j)$ with equality holds only if one of d_1 or d_2 does not cause s_j.

4.2 From Deep Causal Knowledge to Bayes Nets and Beyond

The parsimonious covering methodology can be extended to the case when the state-variables describing a system are no longer exclusively made of disorders and symptoms. Namely, a system can be described by means of a set $\mathtt{H} = \{h_1, \ldots, h_p\}$ of state variables, supposedly Boolean, containing a subset $\mathtt{D} \subseteq \mathtt{H}$ of disorders and a subset $\mathtt{S} \subseteq \mathtt{H}$ of observable symptoms. Of course $\mathtt{D} \cap \mathtt{S} = \varnothing$. State-variables in $\mathtt{H} - \mathtt{D} \cup \mathtt{S}$ are non-observable and are supposed to be intermediary variables. This means that there exists a causal relation $R \subseteq \mathtt{H} \times \mathtt{H}$ such that $(h, h) \in R$ does not hold, nor does it belong to the transitive closure of R; the domain of R is $\mathtt{H} - \mathtt{S}$ and its range is $\mathtt{H} - \mathtt{D}$. Peng and Reggia [1990] study this kind of generalized causal model which can be termed "deep" since it involves the inner behaviour of the system.

The causal matrix R is often viewed as a directed acyclic graph whose root nodes are the elements of D and the end nodes are the elements of S. However, these latter assumptions are not compulsory, and deep causal knowledge is more simply described by an acyclic graph whose nodes are elements of H. The most popular methodology that adopts this point of view is the Bayesian network approach, whose emergence is due to the works of Pearl [1988b], and Lauritzen and Spiegelhalter [1988]. Given a causal network, it is easy to characterize a joint probability distribution on the set of variables in H, using the following assumptions:

- The probability of any state is positive;
- Each subset of variables $\{h_1, h_2, \ldots, h_k\} \cup \{h\}$ such that $\{h_1, h_2, \ldots, h_k\} = \{h_i, (h_i, h) \in R\}$ is weighted by a conditional probability distribution $P(h \mid h_1 \wedge h_2 \wedge \ldots \wedge h_k)$;
- Each state variable h such that $\{h_i, (hi, h) \in R\} = \varnothing$ is weighted by an a priori probability $P(h)$.

Then, if $C(h) = \{h_i, (h_i, h) \in R\}$ denote the set of state variables causing h directly, a joint probability $P(h_1, h_2, \ldots, h_p)$ is obtained as

$$P(h_1, h_2, \ldots, h_p) = \prod_{h \in \mathtt{H}} (h \mid C(h)).$$

It comes down to assuming, as in the probabilistic parsimonious covering problem, that

- first causes $D = \{h, \nexists h_i, (h_i, h) \in R\}$ are independent;
- $\forall h$ in H, h depends only on $C(h)$, and is independent of any $h' \in \mathtt{H} - C(h) = \{h\}$, conditionally on $C(h)$.

However, as pointed out by Peng and Reggia, $P(h \mid C(h))$ has no proper causal meaning, although it is convenient to use a causal graph that is acyclic, as a knowledge representation device, and to read independence assumptions on it (see [Dubois and Prade, 1990] for a discussion on causality in belief nets). As arguments against the intrinsic presence of causality in Bayesian nets, it must be pointed out that

- In the framework of statistics, Bayes nets appears just like one way of encoding a big joint probability distribution function. The causal ordering of variables is then arbitrary, and the best one is chosen as the one (if any) that yields the simplest directed acyclic graph;
- Alternative representations involving non-directed Markov graphs have been used with equal success by Lauritzen and Spiegelhalter [1988]. Causality is just a convenient modeling assumption.

The advantage of Bayesian nets is their expressive power and their data compression power, as well as their computational efficiency when the causal graph is tree-like, more specifically when any pair (h_i, h_j) is related via a single path. A lot of literature exists on Bayesian nets and it is beyond the scope of this paper to scan this literature.

Bayes nets are used for question-answering purposes, and can address the following queries, especially:

- Given a set $M^+ \cup M^-$ of observations, where M^+ is made of variables h_i assigned to TRUE, and M^- are other state-variables assigned to FALSE, what is the posterior probability $P(h \mid M^+ \cup M^-)$ for some unobservable (or unobserved yet) state variables (including disorders);
- What are the most likely states of the unobserved variables? This is the quest for the most likely explanations of a set of observations [Seroussi and Golmard, 1994].

Bayesian nets are certainly a useful tool for handling statistical data involving numerous variables. They even give provision for learning probabilities that remain unknown. However, in the scope of handling expert knowledge, they have several limitations regarding expressiveness:

1. The results of the approach heavily rely on the independence assumptions as well as the numerical values put in the network. These values must be precise numbers which in practice, and when no statistical knowledge is available, may be out of reach (for instance Prob(other symptom | other disease) when exhaustive lists of observable symptoms and diseases are not available)
2. The network building method never produces inconsistencies. The "expert" is asked exactly the amount of data required for ensuring the unicity of the underlying distribution. Hence this distribution depends upon the set of questions asked to the expert.

The second objection is by far the most significant problem. When representing knowledge, inconsistency is almost unavoidable at the acquisition level and tools for detection of such inconsistencies must be used in order to improve the available pieces of information. Inconsistency handling is simply ignored by the Bayesian network approach. The acyclicity assumption is instrumental in preserving consistency, using the causality interpretation to prevent loops in the data. But as said before, $P(h \mid C(h))$ is not the probability that $C(h)$ alone causes h when present, and only refers to joint occurrence of h and $C(h)$, when $C(h)$ is present. This may be very confusing for an "expert".

The first objection can be tackled by using sensitivity analysis techniques and more generally by admitting that knowledge can be represented by a set of conditional probability bounds that constrain a set of admissible probability distributions. For instance we know that $P(B \mid A) \in [0.7, 0.8]$, or that "most As are B", where the linguistic quantifier "most" is modelled by an interval restricting the possible values of the value of the proportion $\mid A \cap B \mid / \mid A \mid$ (where $\mid \cdot \mid$ denotes the cardinality).

See Ramoni and Riva [1994] for the treatment of interval-valued probabilities in Bayesian nets. There are other frameworks that can account for incomplete information, such as belief function networks [Shenoy and Shafer, 1986; Saffiotti, 1990], or upper and lower probabilities [de Campos *et al.*, 1994], or possibility theory [Kruse and Schwecke, 1990]. However these methods rely on constraints relating subsets of variables and use hypergraph structures as a basis for uncertainty propagation. They are not handling concepts of causality explicitly and thus are beyond the scope of this paper. It does not mean that they cannot be applied to diagnosis problems, and indeed they can, especially when the set of uncertain constraints is learned from data. This is the topic of the following chapters of this book, by Borgelt and Kruse, by Gebhardt, and by de Campos and Moral. However, uncertain constraint-propagation tools are also suitable for data fusion which is a problem that noticeably differs from diagnosis.

4.3 *Towards a Possibilistic Conditional Approach to Causal Reasoning*

It is rather clear that the probability-based approach to causal reasoning using causation-events as in Section 4.1 could be envisaged beyond the probabilistic setting. The mandatory causation assumption makes sense regardless of the uncertainty theory under concern. Consider the setting of possibility theory. Then the set of available data would consist in the following:

- Prior knowledge: if disorder d_i is a priori considered as present then $d_i \in D^+$ and $N(d_i) = \alpha_i^+ > 0$. If disorder d_i is a priori considered as absent then $d_i \in D^-$ and $N(\neg d_i) = \alpha_i^- > 0$. D^+ and D^- are disjoint subsets of

D. This means

$$\begin{array}{ll} \forall d_i \in D^+ : & \Pi(d_i) = 1, \Pi(\neg d_i) = 1 - \alpha_i^+, \\ \forall d_i \in D^- : & \Pi(d_i) = 1 - \alpha_i^-, \Pi(\neg d_i) = 1, \\ \forall d_i \notin D^- \cup D^+ : & \Pi(d_i) = \Pi(\neg d_i) = 1 \text{ (ignorance);} \end{array}$$

- Conditional knowledge: for $s_j \in M(d_i)^+$, define

$$\mu_{M(d_i)^+}(s_j) = c_{ij}^+ = N((s_j : d_i) \mid_* d_i).$$

This is the degree of certainty that d_i causes s_j. Similarly for $s_j \in M(d_i)^-$,

$$\mu_{M(d_i)^-}(s_j) = c_{ij}^- = N(\neg(s_j : d_i) \mid_* d_i).$$

This is the degree of certainty that d_i does not cause s_j. These conditional degrees of necessity are based on some operation $*$ such that $\Pi(A \cap B) = \Pi(A \mid_* B) * \Pi(B)$, and $N(A \mid_* B) = 1 - \Pi(\neg A \mid_* B)$.

It is then easy to state the disorder independence assumption as follows: let D_I be the event $\wedge_{i \in I} d_i \wedge \bigwedge_{i \notin I} \neg d_i$, and $\bar{I}$ is the complement of I. Then

$$\begin{array}{rcl} \Pi(D_I) & = & (*_{i \in I}(1 - \alpha_i^-)) * (*_{i \notin I}(1 - \alpha_i^+)) \\ & = & (*_{d_i \in D_I \cap D^-}(1 - \alpha_i^-)) * (*_{d_i \in D_{\bar{I}} \cap D^+}(1 - \alpha_i^+)). \end{array}$$

Note that $\Pi(D_I) = 1$ as soon as $D_I \cap D^- = \varnothing$ and $D_{\bar{I}} \cap D^+ = \varnothing$, that is as soon as no disorder in D_I is a priori rejected and no disorder in $D_{\bar{I}}$ is *a priori* accepted. Moreover

$$\begin{array}{rcl} N(D_I) & = & \min(\min_{i \in I} \alpha_i^+, \min_{i \in \bar{I}} \alpha_i^-) \\ & = & 0 \text{ except if } D_I = D^+ \text{ and } D_{\bar{I}} = D^-. \end{array}$$

In other words $N(D_I) = 0$ as soon as $\exists i, \Pi(d_i) = \Pi(\neg d_i) = 1$, i.e. $D^+ \cup D^-$ does not cover D.

The causation independence assumption then writes

$$\begin{array}{c} N(s_j : d_i) \mid d_i \wedge X) = N((s_j : d_i) \mid d_i) \\ N(\neg(s_j : d_i) \mid d_i \wedge X) = N(\neg(s_j : d_i) \mid d_i) \end{array}$$

where X is a context not involving d_i nor s_j. Computing posterior degrees of possibility for disorders is a matter of further research. Section 2.4 has already proposed results along this line for a very simple case (see [Dubois and Prade, 1993b]).

A further step of research along this line is to work with possibilistic networks whose structure expresses conditional (logical) independence assumptions, and to develop local computation methods just as Pearl [1988b] did in the probabilistic framework; see Fonck [1994], Fonck and Straszecka [1991] for a preliminary attempt along this line. The basic issue is to find a proper definition of independence in the possibilistic setting, that ensures good properties for uncertainty propagation algorithms. Such definitions of independence in possibilistic-like settings are given by Fonck [1994], Studeny [1995]. See also De Campos, Gebhardt and Kruse [1995]. There is no definite agreement on the notion of possibilistic independence yet [Dubois *et al.*, 1994; Dubois *et al.*, 1999a]. As it turns out, possibility distributions being akin both to probability distributions and to sets, have decomposability properties similar to Bayesian nets as well as relations whose decomposition is studied in the theory of databases. A graphical representation, counterpart of a possibilistic logic representation (as the one used in the next Section 5), has been recently proposed (see Benferhat et al., 1998).

Causal reasoning phenomena such as explaining away may be discussed in the possibilistic setting: namely if a symptom has two independent causes, it is often acknowledged that observing this symptom, and further getting certainty for the presence of one of its cause does reduce the plausibility of the other cause. This is in accordance with the parsimony principle, since having selected an explanation tends to discard other explanations. This phenomenon has been discussed at length in the probabilistic setting (e.g., [Wellman and Henrion, 1991]) and has received preliminary investigation in the possibilistic setting [Benferhat *et al.*, 1996].

The above results suggests that to some extent the Bayesian diagnosis method can be reconducted in the possibilistic setting. To achieve this program would bridge the gap between the simple relational models that cannot account for prior knowledge of causes and the probabilistic models that must account for priors, namely a conditional possibilistic framework offers an approach where prior knowledge is welcome but not compulsory (this remark also applies to techniques based on belief functions and probability bounds). Clearly, achieving this research program is a long range project in which important notions such as independence and causality must be carefully addressed.

5 WEIGHTED LOGICAL DIAGNOSIS METHODS

This section is devoted to the links between model-based logical approaches to diagnosis and the above numerical relational methods. In the model-based paradigm [Hamscher *et al.*, 1992], two methodologies emerge: consistency-based diagnosis and abductive diagnosis. In consistency-based diagnosis, a model of the behavior of the system is supposed to be available under the form of a logical theory $\mathtt{T}$. This model may be restricted to the good, normal behavior of the system; in such a case one can only detect the presence of disorders (inconsistency of the expected normal behavior with what is observed), without having any means for trying to identify them (if no description of fault modes is provided). A system is made of components connected by links. Components are represented by propositional variables and links are expressed by non-atomic formulas. A set of observations is also available as another set of formulas $M = M^+ \cup M^-$. Let $\mathtt{D}$ be a set of atomic symbols representing disorders. An explanation is then a conjunction of formulas from D, such that positive literals indicate fault assumptions while negative ones are non-fault assumptions. In the consistency-based paradigm, an explanation is a conjunction D of assumptions such that $D \cup M \cup \mathrm{T}$ is consistent. This technique is good to eliminate fault assumptions d such that $\mathrm{T} \cup M^+ \cup M^- \models \neg d$.

In the abductive approach the model T should include a description of fault modes. A so-called abductive explanation of a formula M (representing the observations) with respect to a theory T is usually defined as a conjunction D of literals, such that $\mathrm{T} \cup D \models M$, where D is also supposed to be consistent with T. In the general case, the literals do not necessarily correspond to symptoms or to disorders only ("intermediary entities" are allowed), there is no restriction on the formulas in T, or on M, and the literals are not sorted. See for instance [el Ayeb *et al.*, 1993] for an extension of this definition to first order logic. The problem is then to compute (a) conjunction(s) D which is/are minimal (in the sense of subsumption) and non- trivial (i.e., not such that $M \models D$).

The parsimonious covering theory in its weighted possibilistic version can be easily related to the model-based approaches, if the theory T is taken as representing the relational causal incomplete knowledge. Cayrac *et al.*, [1997] have envisaged the following logical translation of the possibilistic relational framework: the fuzzy set $M(d_i)^+$ of sure symptoms of d_i is modelled by a collection of weighted clauses $(\neg d_i \vee s_j, \alpha_{ij})$ expressing $N(\neg d_i \vee s_j) \geq \alpha_{ij}$. Similarly the set $M(d_i)^-$ of symptoms not caused by d_i alone is modelled by a collection of weighted clauses $(\neg d_i \vee \neg s_j, \beta_{ij})$ expressing $N(\neg d_i \vee \neg s_j) \geq \beta_{ij}$.

In possibilistic logic [Dubois and Prade, 1987b], classical logic formulas are weighted by lower bounds of the necessity degree with which the formula is held for certainly true. In the framework of possibilistic logic, the extended resolution principle reads

$$N(S) \geq \alpha, N(t) \geq \beta \models N(\text{Resolvent}(s,t)) \geq \min(\alpha, \beta).$$

The causal influence of a disorder d as considered in Section 3.5 translates into the following pattern of inference expressed in terms of a necessity measure N (using the possibilistic logic resolution rule): $N(d_i) \geq \gamma_i$ and $N(\neg d_i \vee s_j) \geq \alpha_{ij}$ entail $N(s_j) \geq \min(\gamma_i, \alpha_{ij})$.

Consider the following knowledge base $\mathrm{T} = \{N(\neg d_i \vee s_j) \geq \alpha_{ij}\} \cup \{N(\neg d_i \vee \neg s_j) \geq \lambda_{ij}\}$ made of uncertain implications between disorders and symptoms (expressing both positive and negative information) and of uncertain symptoms $M = M^+ \cup M^-$ where $M^+ = \{s_j, N(s_j) \geq \beta_j\}$ and $M^- = \{s_j, N(\neg s_j) \geq \rho_j\}$. Then using the extended resolution rule it holds:

$$N(\neg d_i) \geq \max[\max_{s_j \in M^-} \min(\lambda_{ij}, \beta_j), \max_{s_j \in M^+} \min(\alpha_{ij}, \rho_j)].$$

This corresponds to the consistency-based approach where disorders inconsistent with the observations are excluded; it defines the fuzzy set of potential explanations formed of single disorders (of section 3.5) which have not been ruled out, since $\mu_{\widehat{DD}}(\{d_i\}) = 1 - N(\neg d_i)$ (with $\mu_{M(d_i)^+}(s_j) = \alpha_{ij}, \mu_{M(d_i)^-}(s_j) =$

$\lambda_{ij}, \mu_{M^+}(s_j) = \beta_j, \mu_{M^-}(s_j) = \rho_j$). Thus the logical approach does capture the disorders which are somewhat *possible*, i.e., which are not more or less certainly ruled out by the positive or negative observations. In the case where $M^- = \varnothing, M(d)^- = \varnothing$, the available information cannot lead to a selection of covers since all single disorder assumptions in D become potential explanations. In other words the consistency-based approach no longer applies.

The abductive approach consists in finding those disorders d such that $\mathrm{T} \cup \{d\} \models M^+ \cup M^-$. Basically, d is an abductive explanation of $N(s) \geq \alpha$ if $N(s) \geq \alpha$ can be derived from T, now made of weighted clauses, together with the weighted clause encoding $N(d) = 1$ (provided that these clauses are consistent). In the possibilistic context, from $N(d_i) = 1$ and $N(\neg d_i \vee s_j) \geq \alpha_{ij}$ one derives $N(s_j) \geq \alpha_{ij}$, and, in possibilistic logic, this in turn implies $N(s_j) \geq \beta_j$ only if $\beta_j \leq \alpha_{ij}$. In this case d is an abductive explanation of s. More generally, d is an abductive explanation of s to level $\beta_j \to \alpha_{ij}$ where the arrow denotes a residuated implication such that $\beta \to \alpha = 1$ whenever $\beta \leq \alpha$. If $\beta_j > \alpha_{ij}$ then the plausibility of d can be taken as α_{ij} using Gödel implication. On the whole one could compute

$$\Delta(d) = \min(\min_{s_j \in M^+}(\beta_j \to \alpha_{ij}), \min_{s_j \in M^-}(\rho_j \to \lambda_{ij})) = \mu_{\widehat{DD}_{**}}(\{d_i\}).$$

Namely the fuzzy set of relevant covers made of single faults $\hat{D}_{**}$ of Section 3.5 is recovered. In practice only the abductive explanations of anomalous symptoms is possible and only $\Delta(d) = \min_j(\beta_j \to \alpha_{ij})$ is computed (see [Cayrac *et al.*, 1995]). If $\forall i, \Delta(d_i) = 0$, we shall consider to what extent a combination of causes $d_{i_1}, \ldots, d_{i_r}$ can cause all the more or less certain observations, by means of the following expression $\mu_{\hat{D}_{**}}(\{d_{i_1}, \ldots, d_{i_r}) = \min_j[\beta_j \to \max(a_{i_1 j}, \ldots, \alpha_{i_r j})]$, starting with $r = 2$, i.e., looking for the possibility that the conjunction of two causes explains M^+. This can be achieved, using an ATMS language, by computing the environments of the set of weighted formulas $N(s_j) \geq \beta_j, j = 1, k$, with r assumptions. Obviously, in the general case we should directly work with a possibilistic theory T allowing for weighted justifications of the form $N(\neg d_{i_1} \vee \ldots \vee \neg d_{i_{s(k)}} \vee s_j) \geq \alpha_{kj}$, if the disorders interact in the symptoms they cause.

However the modelling of causal information by means of logical clauses has some difficulties when multiple faults are looked for and negative information is used. Actually, a clause $\neg d \vee s$ means that s occurs when d is present regardless of other causes, while, in the method of Peng and Reggia, $s : d$ presupposes that d is alone. The logical representation of the possibilistic approach to the parsimonious covering approach is not totally satisfactory, though, since the clause $\neg d \vee \neg s$ means that the presence of d *prevents* the occurence of s, and not that d alone does not cause d (the latter would be better expressed by the negation of a causation event $\neg(s : d)$ while the logical clause expresses a negative causation event "$\neg s : d$"). Similarly the causation event $s_j : d_i$ as described in Section 4.1 is not accurately described by $\neg d_i \vee s_j$, since the causation event is more restricted than the conjunction $d_i \wedge s_j$.

Note that despite this gap between the logical model and the possibilistic version of the parsimonious covering approach, the results of both frameworks agree on single faults. Unfortunately this is no longer true for multiple faults. It is easy to see that

$$\left.\begin{array}{l} N(\neg d_i \vee s_j) \geq \alpha_{ij} \\ N(\neg d_k \vee s_j) \geq \alpha_{kj} \end{array}\right\} \text{ imply } N(\neg(d_i \wedge d_k) \vee s_j) \geq \max(\alpha_{ij}, \alpha_{kj}).$$

This is in accordance with the disorder superposition assumption $\mu_{M^+(D)} = \max_{d \in D} \mu_{M^+(d)}$ of Section 3.2. Indeed due to superposition, effect s_j of d_i cannot be suppressed by adding a new disorder, so that $\neg d_i \vee s_j$ and $s_j : d_i$ are in agreement even if $\neg d_i \vee s_j$ does not presuppose that d_i is alone: the logical modelling implies superposition since $\neg d_i \vee s_j$ implies $\neg d_i \vee \neg d_k \vee s_j$. Unfortunately the above inference also holds if s_j is changed into $\neg s_j$, which leads to $\mu_{M^-(D)} = \max_{d \in D} \mu_{M^-(d)}$. On the contrary the disorder superposition assumption of Section 3.4 postulates that $\mu_{M^-(D)} = \min_{d \in D} \mu_{M^-(d)}$. The latter can be explained by the use of causation events. Since $s : D = \wedge_{d \in D} s : d$ and using superposition of causal effects, it is clear that the "non-effects" do not accumulate: $\neg(s : D) = \wedge_{d \in D} \neg(s : d)$.

This brief discussion suggests a partial agreement between the relational approach (in the incomplete information case with graded uncertainty) and the model-based approach cast in possibilistic logic, when the single fault assumption is made. A difficulty in the logical approach with respect to the parsimonious covering approach is that it is hard, in the former, to describe explicitly that a disorder does not cause a symptom without preventing it to occur if other disorders are present. One way out could be the explicit handling of causation events as a special class of propositional variables.

The possibilistic diagnosis approach could be also related to works aiming at developing explanation capabilities of expert systems handling uncertainty (e.g., [Farreny and Prade, 1992]) and more generally to assumption-based truth maintenance systems (ATMS). Model-based diagnosis techniques often rely on the use of ATMS algorithms in order to find out disorder configurations. Abductive methods use prime implicants. Methods based on the search of prime implicates or prime implicants have been proposed in the literature [Reiter, 1987; Reiter and de Kleer, 1987; Jackson, 1989; Jackson, 1991a]. Jackson [1991b] has proposed to extend his prime implicate-based approach to possibilistic logic. We are then very close to possibilistic ATMS introduced in [Dubois *et al.*, 1991]. In this approach pieces of knowledge are represented by necessity-weighted formulae, just as in the above logical weighted causal methods. Special propositional variables represent disorder assumptions. Labels of formulas representing observations M^+ (labels are are weakly consistent, sound, complete and minimal collections of assumptions) can be computed ; such a label is made of a collection of weighted sets of assumptions, that form the set of abductive explanations. See [Benferhat *et al.*, 1994] for applications to diagnosis problems.

On the contrary, the belief function approach to truth-maintenance starts from a set of belief functions (one per assumption) that quantify a priori knowledge about disorders. Then the degree of belief of an abductive explanation can be computed applying Dempster rule to combine in a conjunctive way the degrees of belief of assumptions involved in the corresponding explanation. The main issue is to cope with explanations that are inconsistent with the knowledge and the observations. These explanations are called "no-goods" in the ATMS terminology and have zero belief (and plausibility). The degree of belief of an abductive explanation must then be conditioned by the set of consistent explanations (potential ones). Algorithms for computing such degrees of belief are proposed by Laskey and Lehner [1988], Pearl [1988b] and Provan [1989]. These works have been systematized by Kohlas and Monney [1995] and Smets [1991] who have considered the assumption-based truth maintenance paradigm as a possible foundation of belief functions (after the suggestion made by Pearl).

6 CONCLUDING REMARKS

The paper has provided an overview of graded relational methods for diagnosis problems.The relational model can be viewed as the result of a compilation of the logical model where we produce the clauses relating literals pertaining to disorders to literals pertaining to symptoms. Probabilistic methods have been drastically improved by getting rid of the single-fault assumption, and introducing powerful tools for capturing and exploiting independence assumptions. As for non-probabilistic methods, although those based on fuzzy sets and fuzzy relations date back to the late 70s, it is only recently that the difference between graded uncertainty on the satisfaction of a binary property and graded satisfaction of a non-binary property (called here "graded intensity") has been made clear in this framework. The handling of uncertainty on non-binary attributes in diagnosis problems is still largely unexplored, if we except the Bayesian net approach (but see [Dubois *et al.*, 1999b]).

The possibilistic relational model with graded uncertainty has been shown to be in agreement with the logical approach where uncertainty is represented in the possibility/necessity model. Clearly, the relational model presented here is not capable of capturing the most general kind of incomplete information. For instance, we cannot express that we are certain that symptoms s_i or s_j are present (but perhaps one of them is absent), or a similar information stating that when disorder d is present it is certain that s_i or s_j are present (and that s_k or s_e are absent) for instance. See Dubois and Prade [1988a] for some hints on the the modelling of such pieces of knowledge in the framework of possibility theory and information systems. The treatment of the most general kind of incomplete information would require to work with a (fuzzy) relation R defined on $2^{\mathrm{D}} \times 2^{\mathrm{S}}$.

Lastly, it must be noticed that neither non-probabilistic approaches such as fuzzy relational ones, nor logical methods handle a priori information as the probabilistic methods, based on a revision process, do. We have suggested some ideas to incorporate prior knowledge in possibilistic methods whose handling of uncertainty is more ordinal. Indeed the mathematical equations that govern conditioning and lead to calculate posterior degrees of plausibility of disorders are very similar to the probabilistic equations. It may open the road to less informed but simpler methods where a priori knowledge can be accounted for if present, but where the lack of a priori knowledge can be coped with.

Institut de Recherche en Informatique de Toulouse, C.N.R.S.
Université Paul Sabatier,
118 route de Narbonne,
31062 TOULOUSE Cedex 4, France.

REFERENCES

[Adlassnig *et al.*, 1986] K. P. Adlassnig, W. Scheithauer and G. Kolarz. Fuzzy medical diagnosis in a hospital. In : *Fuzzy Logic in Knowledge Engineering*, H. Prade and C. V. Negoita, eds. pp. 275–294. Verlag TüV Rheinland, Kln, 1986.

[Appelbaum and Ruspini, 1985] L. Appelbaum and E. H. Ruspini. ARIES: An approximate reasoning inference engine. In *Approximate Reasoning in Expert Systems*, M. M. Gupta, A. Kandel, W. Bandler and J.B. Kiszka, eds. pp. 745–765. North-Holland, Amsterdam, 1985.

[Asse *et al.*, 1987] A. Asse, A. Maizener, A. Moreau and D. Willaeys. Diagnosis based on subjective information in a solar energy plant. In *Approximate Reasoning in Intelligent Systems, Decision and Control*, E. Sanchez and L. A. Zadeh, eds. pp. 159–173. Pergamon Press, New York, 1987.

[Baldwin, 1986] J. F. Baldwin. Support logic programming. *Int. J. of Intelligent Systems*, **1**, 73–104, 1986.

[Bandler and Kohout, 1984] W. Bandler and L. J. Kohout. The four modes of inference in fuzzy expert systems. In *Cybernetics and Systems Research 2*, R. Trappl, ed. pp. 581–586. North-Holland, Amsterdam, 1984.

[Benferhat *et al.*, 1994] S. Benferhat, D. Dubois, J. Lang and H. Prade. Hypothetical reasoning in possibilistic logic: basic notions and implementation issues. In *Between Mind and Computer, Fuzzy Science and Engineering, Vol. I*,P. Z. Wang and K. F. Loe, eds. pp. 1–29. World Scientific Publ., Singapore, 1994

[Benferhat *et al.*, 1996] S. Benferhat, D. Cayrac, D. Dubois and H. Prade. Explaining away in a possibilistic setting. *Proc. of the 6th Int. Conf. on Information Processing and Management of Uncertainty in Knowledge-Based Systems (IPMU-96)*, Granada, Spain, July 1-5, Vol. II, pp. 929–934, 1996.

[Benferhat *et al.*, 1998] S. Benferhat, D. Dubois, L. Garcia and H. Prade. Directed possibilistic graphs and possibilistic logic. *Proc. of the 7th Int. Conf. on Information Processing and Management of Uncertainty in Knowledge-Based Systems (IPMU-98)*, Paris, July 6-10, pp. 1470–1477, 1998.

[Bonissone *et al.*, 1987] P. Bonissone, S. Gans and K. Decker. RUM: A layered architecture for reasoning with uncertainty. *Proc. of the 10th Inter. Joint Conf. on Artificial Intelligence (IJCAI'87)*, Milano, Italy, pp. 891–898, 1987.

[Buchanan and Shortliffe, 1984] B. G. Buchanan and E. H. Shortliffe. *Rule-Based Expert Systems The MYCIN Experiments of the Stanford Heuristic Programming Project*. Addison-Wesley, Reading, MA, 1984.

[Buckley *et al.*, 1986] J. J. Buckley, W. Siler and D. Tucker. A fuzzy expert system. *Fuzzy Sets and Systems*, **20**, 1–16, 1986

[Cayrac *et al.*, 1995] D. Cayrac, D. Dubois and H. Prade. Practical model-based diagnosis with qualitative possibilistic uncertainty. *Proc. of the 11th Conf. on Uncertainty in Artificial Intelligence*, P. Besnard and S. Hanks, eds. pp. 68–76, 1995.

[Cayrac *et al.*, 1996] D. Cayrac, D. Dubois and H. Prade. Handling uncertainty with possibility theory and fuzzy sets in a satellite fault diagnosis application. *IEEE Trans. on Fuzzy Systems*, **4**, 251–269, 1996.

[Cayrac *et al.*, 1997] D. Cayrac, D. Dubois and H. Prade. Possibilistic handling of uncertainty in practical fault isolation. In *Fuzzy Information Engineering*, d. Dubois, H. Prade and R. Yager, eds. pp. 437–456, Wiley, New York, 1997.

[Charniak, 1983] E. Charniak. The Bayesian basis of common sense medical diagnosis. *Proc. of the National Conf. on Artificial Intelligence (AAAI'83)*, pp. 70–73, 1983.

[Clancey, 1985] W. J. Clancey. Heuristic classification. *Artificial Intelligence*, **25**, 289–350, 1985.

[Console and Torasso, 1992] L. Console and P. Torasso. (1992) A spectrum of logical definitions of model-based diagnosis. In *Readings in Model-Based Diagnosis*, W. Hamscher, L. Console and J. De Kleer, eds. pp. 78–88. Morgan & Kaufmann, San Mateo, CA, 1992.

[Dalton, 1993] R. E. Dalton. Towards a theory of fuzzy causality for diagnosis. *Information Sciences*, **80**, 195–211, 1993.

[de Campos *et al.*, 1995] L. de Campos, J. Gebhardt and R. Kruse. Axiomatic treatment of possibilistic independence. In *Symbolic and Quantitative Approaches to Reasoning and Uncertainty*, C. Froidevaux and J. Kohlas, eds. pp. 77–88. LNAI 946, Springer Verlag, 1995.

[de Campos *et al.*, 1994] L. de Campos, J. F. Huete and S. Moral. Probability intervals: A tool for uncertain reasoning. *Int. J. of Uncertainty, Fuzziness and Knowledge-Based Systems*, **2**, 167–196, 1994.

[Di Nola *et al.*, 1989] A. di Nola, S. Sessa, W. Pedrycz and E. Sanchez. *Fuzzy Relation Equations and their Applications to Knowledge Engineering*. Kluwer Academic Publ., Dordrecht, 1989.

[Dubois and Prade, 1980] D. Dubois and H. Prade. *Fuzzy Sets and Systems: Theory and Applications*. Academic Press, New York, 1980.

[Dubois and Prade, 1987a] D. Dubois and H. Prade. Twofold fuzzy sets and rough sets — Some issues in knowledge representation. *Fuzzy Sets and Systems*, **23**, 3–18, 1987.

[Dubois and Prade, 1987b] D. Dubois and H. Prade. Necessity measures and the resolution principle. *IEEE Trans. on Systems, Man and Cybernetics*, **17**, 474–478, 1987.

[Dubois and Prade, 1988a] D. Dubois and H. Prade. Incomplete conjunctive information. *Computers and Mathematics with Applications*, **15**, 797–810, 1988.

[Dubois and Prade, 1988b] D. Dubois and H. Prade. (with the collaboration of H. Farreny, R. Martin-Clouaire and C. Testemale) *Possibility Theory — An Approach to Computerized Processing of Uncertainty*. Plenum Press, New York, 1988. French editions, Masson, 1985 and 1987.

[Dubois and Prade, 1988c] D. Dubois and H. Prade. Modelling uncertainty and inductive inference. *Acta Psychologica*, **68**, 53–78, 1988.

[Dubois and Prade, 1988d] D. Dubois and H. Prade. Representation and combination of uncertainty with belief functions and possibility measures. *Computational Intelligence (Canada)*, **4**, 244–264, 1988.

[Dubois and Prade, 1989] D. Dubois and H. Prade. Handling uncertainty in expert systems: Pitfalls, difficulties, remedies. In *The Reliability of Expert Systems*, E. Hollnagel, ed. pp. 64–118. Ellis Horwood, Chichester, UK, 1989.

[Dubois and Prade, 1990] D. Dubois and H. Prade. Probability theory in artificial intelligence: Book Review of 'Probabilistic Reasoning in Intelligent Systems: Networks of Plausible Inference' by J. Pearl. *J. Mathematical Psychology*, **34**, 472–482, 1990.

[Dubois and Prade, 1991] D. Dubois and H. Prade. Possibilistic logic, preference models, non-monotonicity and related issues. In *Proc. of the Inter. Joint Conf. on Artificial Intelligence (IJCAI-91)*, Sydney, Australia, Aug. 24-30, pp. 419–424, 1991.

[Dubois and Prade, 1993a] D. Dubois and H. Prade. A fuzzy relation-based extension of Reggia's relational model for diagnosis handling uncertain and incomplete information. In *Proc. of the 9th Conf. on Uncertainty in Artificial Intelligence*, D. Heckerman and A. Mamdani, eds. pp. 106–113, 1993.

[Dubois and Prade, 1993b] D. Dubois and H. Prade. Possibilistic abduction. In *IPMU'92 — Advanced Methods in Artificial Intelligence (Proc. of the 4th Inter. Conf. on IPMU'92)*, Palma de Mallorca, Spain, July 1992, pp. 3–12. Lecture Notes in Computer Science, Vol. 682, Springer Verlag, Berlin, 1993.

[Dubois and Prade, 1993c] D. Dubois and H. Prade. Fuzzy set and possibility theory-based models for diagnostic problem solving. Tech. Rep. IRIT/93-28-R, October 1993.

[Dubois and Prade, 1995] D. Dubois and H. Prade. Fuzzy relation equations and abductive reasoning. Special Issue on "Equations and Relations on Ordered Structures: Mathematical Aspects and Applications, A. Di Nola, W. Pedrycz and S. Sessa, eds. *Fuzzy Sets and Systems*, **75**, 119–134, 1995.

[Dubois and Prade, 1998] D. Dubois and H. Prade. Possibility theory: quanitative and quantitative aspects. In *Handbook on Defeasible Reasoning and Uncertainty Management Systems—Volume 1: Quantified Representation of ncertainty and Imprecision*, P. Smets, ed. pp. 169–226. Kluwer, 1998.

[Dubois *et al.*, 1991] D. Dubois, J. Lang and H. Prade. A possibilistic assumption-based truth maintenance system with uncertain justifications, and its application to belief revision. In *Truth Maintenance Systems (Proc. of the ECAI'90 Workshop)*, Stockholm, August 1990. J. P. Martins and M. Reinfrank, eds. pp. 87–106. Lecture Notes in Computer Sciences, Vol. 515, Springer Verlag, Berlin, 1991

[Dubois *et al.*, 1994] D. Dubois, L. Fariñas del Cerro, A. Herzig and H. Prade. An ordinalview of independence with application to plausible reasoning. In *Proc. of the 10th Conf. on Uncertainty in Artificial Intelligence (UAI-94)*, R. Lopez de Mantaras and D. Poole, eds. Seattle, July 29- 31, pp. 195–203. Morgan Kaufmann Publ., San Francisco, 1994.

[Dubois *et al.*, 1997] D. Dubois, S. Moral and H. Prade. A semantics for possibility theory based on likelihoods. *Journal of Mathematical Analysis and Applications*, **205**, 359–380, 997.

[Dubois *et al.*, 1999a] D. Dubois, L. Fariñas del Cerro, A. Herzig and H. Prade. A roadmap of quanitative independence. In *Fuzzy Sets, Logics and Reasoning about Knowledge*, d. Dubois, H. Prade and E. P. Klement, eds. pp. 325–350. Vol. 15 Applied Logic Series, Kluwer, 1999.

[Dubois *et al.*, 1999b] D. Dubois, M. Grabisch and H. Prade. A general approach to diagnosis in a fuzzy setting. In *Proceedings of the 8th International Fuzzy Sets Association (IFSA'99)*, Taipei, Taiwan, pp. 680–684, 1999.

[Duda *et al.*, 1981] R. O. Duda, J. Gaschnig and P. E. Hart. Model design in the PROSPECTOR consultant system for mineral exploration. In *Expert Systems in the Micro-Electronic Age*, D. Michie, ed. Edinburgh University Press, Edinburgh, 1981.

[Duda and Hart, 1973] R. O. Duda and P. E. Hart. *Pattern Classification and Scene Analysis*. Wiley, New York, 1973.

[Edwards, 1972] A. W. F. Edwards. *Likelihood* . Cambridge University Press, Cambridge, UK, 1972.

[el Ayeb *et al.*, 1993] B. el Ayeb, P. Marquis and M. Rusinowitch. Preferring diagnoses by abduction. IEEE Trans. on Systems, *Man and Cybernetics*, **23**, 792–808, 1993.

[Farreny and Prade, 1992] H. Farreny and H. Prade. Positive and negative explanations of uncertain reasoning in the framework of possibility theory. In *Fuzzy Logic for the Management of Uncertainty*, L. A. Zadeh and J. Kacprzyk, eds. pp. 319–33. Wiley, New York, 1992.

[Farreny *et al.*, 1986] H. Farreny, H. Prade and E. Wyss. Approximate reasoning in a rule-based expert system using possibility theory: a case study. In *Proc. of the 10th World IFIP Congress on Information Processing '86*, Dublin, 1-5 Sept, H. J. Kugler, ed. pp. 407–413. North-Holland, Amsterdam, 1986.

[Fieschi *et al.*, 1982] M. Fieschi, M. Joubert, D. Fieschi, G. Soula and M. Roux. SPHINX: An interactive system for medical diagnosis aids. In *Approximate Reasoning in Decision Analysis*, M. M. Gupta and E. Sanchez, eds. pp. 269–275. North-Holland, Amsterdam, 1982.

[Fonck, 1994] P. Fonck. Conditional independence in possibility theory. In *Proc. of the 10th Conf. on Uncertainty in Artificial Intelligence*, Seattle, WA, July 29-31, R. López de Mátaras and D. Poole, eds. pp. 221–229, 1994.

[Fonck and Straszecka, 1991] P. Fonck and E. Straszecka. Building influence networks in the framework of possibility theory. *Ann. Univ. Sci. Budap., Sect. Comp.*, **12**, 101–106, 1991.

[Friedman, 1981] L. Friedman. Extended plausible inference. In *Proc. of the 7th Inter. Joint Conf. on Artificial Intelligence (IJCAI'81)*, Vancouver, pp. 487–495, 1981.

[Garbolino, 1989] P. Garbolino. On the Plausibility of Inverse Possibility. Annali dell'Università di Ferrara, Nuova Serie, Sezione III, Filosofia, Discussion Paper n. 8, Univ. degli Studi di Ferrara, Italy, 1989.

[Gmytrasiewicz *et al.*, 1990] P. Gmytrasiewicz, J. A. Hassberger and J. C. Lee. Fault tree based diagnostics using fuzzy logic. In *IEEE Trans. on Pattern Analysis & Machine Intelligence*, **12**, 1115–1119, 1990.

[Godo *et al.*, 1987] L. Godo, R. López de Mántaras, C. Sierra and A. Verdaguer. MILORD: The architecture and the management of linguistically expressed uncertainty. *Int. J. of Intelligent Systems*, **4**, 471–504, 1987.

[Gouvernet *et al.*, 1985] J. Gouvernet, M. Caraboeuf and S. Ayme. GENDIAG: A computer-assisted facility in medical genetics based on belief functions. *Methods of Information in Medicine*, **24**, 177–180, 1985.

[Hall, 1987] L. O. Hall. On the fuzzy logic modes of inference: Confirmation and denial. In *Proc. of the 2nd Inter. Fuzzy Systems Assoc. (IFSA'87) Congress*, Tokyo, Japan, July 20-25, pp. 24–26, 1987.

[Hamscher *et al.*, 1992] W. Hamscher, L. Console and J. de Kleer, eds. *Readings in Model-Based Diagnosis*. Morgan & Kaufmann, San Mateo, CA, 1992.

[Hisdal, 1978] E. Hisdal. Conditional possibilities — Independence and non-interactivity. *Fuzzy Sets and Systems*, **1**, 283–297, 1978.

[Jackson, 1989] P. Jackson. Propositional abductive logic. In *Proc. of the 7th Conf. of the Society for the Study of Artificial Intelligence and Simulation of Behaviour*, A. G. Cohn, ed. pp. 89–94. Pitman & Morgan-Kaufmann, 1989.

[Jackson, 1991a] P. Jackson. Possibilistic prime implicates and their use in abduction. Research Note, McDonnell Douglas Research Lab., St Louis, MO, 1991.

[Jackson, 1991b] P. Jackson. Prime implicates: their computation and use. In *Proc. of the 11th Inter. Conf. on Expert Systems & Applications*, Avignon, May 27-31, Vol. 1, pp. 53–64, 1991.

[Kitowski and Bargiel, 1987] J. Kitowski and M. Bargiel. Diagnosis of faulty states in complex physical systems using fuzzy relational equations. In *Approximate Reasoning in Intelligent Systems, Decision and Control*, E. Sanchez and L. A. Zadeh, eds. pp. 175–194. Pergamon Press, New York, 1987.

[Kohlas and Monney, 1995] J. Kohlas and P. A. Monney. *A Mathematical Theory of Hints*. Lecture Notes in Economics and Mathematical Systems, Vol. 425, Springer Verlag, Berlin, 1995.

[Kruse and Schwecke, 1990] R. Kruse and E. Schwecke. Fuzzy reasoning in a multidimensional space of hypothesis. *Int. J. of Approximate Reasoning*, **4**, 47–68, 1990.

[Laskey and Lehner, 1988] K. B. Laskey and P. E. Lehner. Belief maintenance: An integrated approach to uncertainty management. In *Proc. of the 7th National Conference on Artificial Intelligence*, Saint Paul, Aug. 21-26, pp. 210–214. 1988.

[Lauritzen and Spiegelhalter, 1988] S. L. Lauritzen and D. J. Spiegelhalter. Local computations with probabilities on graphical structures and their application to expert systems. *J. of the Royal Statistical Society*, **B-50(2)**, 157–224, 1988.

[Lesmo *et al.*, 1985] L. Lesmo, L. Saitta and P. Torasso. Evidence combination in expert systems. *Int. J. of Man-Machine Studies*, **22**, 307–326, 1985.

[Martin-Clouaire and Prade, 1996] R. Martin-Clouaire and H. Prade. SPII-1 — A simple inference engine capable of accommodating imprecision and uncertainty. In *Computer-Assisted Decision-Making*, G. Mitra, ed. pp. 117–131. North-Holland, Amsterdam, 1996.

[Pappis and Sugeno, 1985] C. P. Pappis and M. Sugeno. Fuzzy relational equations and the inverse problem. *Fuzzy Sets and Systems*, **15**, 79–90, 1985.

[Pearl, 1988a] J. Pearl. Embracing causality in default reasoning. *Artificial Intelligence*, **35**, 259–271, 1988.

[Pearl, 1988b] J. Pearl. *Probabilistic Reasoning in Intelligent Systems: Networks of Plausible Inference*. Morgan Kaufmann, San Mateo, CA, 1988.

[Peirce, 1940] C. S. Peirce. Abduction and induction. In *Selected Philosophical Writings*, J. Buchler, ed. Republished by Dover, 1955.

[Peng and Reggia, 1990] Y. Peng and J. A. Reggia. *Abductive Inference Models for Diagnostic Problem-Solving*. Springer Verlag, New York, 1990.

[Pólya, 1968] G. Pólya. *Mathematics and Plausible Reasoning — Vol. II: Pattern of Plausible Inference*. First edition, 1954. Second edition, Princeton University Press, 1968.

[Provan, 1989] G. Provan. An analysis of ATMS-based techniques for Computing Dempster–Shafer belief functions. In *Proc. 11th Int. Joint Conf. on Artificial Intelligence*, Detroit, Aug. 20-25, pp. 1115–1120, 1989.

[Ramoni and Riva, 1994] M. Ramoni and A. Riva. Belief maintenance in Bayesian networks. In *Proc. of the 10th Conf. on Uncertainty in Artificial Intelligence*, Seattle, WA, R. Lopez de Mantaras, D. Poole, eds. pp. 498–505, 1994.

[Reggia *et al.*, 1985] J. A. Reggia, D. S. Nau, P. Y. Wang and H. Peng. A formal model of diagnostic inference. *Information Sciences*, **37**, 227–285, 1985.

[Reiter, 1987] R. Reiter. A theory of diagnosis from first principles. *Artificial Intelligence*, **32**, 57–96, 1987.

[Reiter and de Kleer, 1987] R. Reiter and J. de Kleer. Foundations of assumption-based truth maintenance systems: Preliminary report. In *Proc. of the 6th National Conf. on Artificial Intelligence (AAAI'87)*, pp. 183–188, 1987.

[Saffiotti, 1990] A. Saffiotti. A hybrid framework for representing uncertain knowledge. In *Proc. of the National Conf. on Artificial Intelligence (AAAI'90)*, Boston, MA., pp. 653–658, 1990.

[Sanchez, 1977a] E. Sanchez. Solutions in composite fuzzy relation equations: Application to medical diagnosis in Brouwerian logic. In *Fuzzy Automata and Decision Processes*, M. M. Gupta, G. N. Saridis and B. R. Gaines, eds. pp. 221–234. North-Holland, Amsterdam, 1977.

[Sanchez, 1977b] E. Sanchez. Inverses of fuzzy relations, application to possibility distributions and medical diagnosis. *Fuzzy Sets and Systems*, **2**, 75–86. Preliminary version in *Proc. of the 1977 IEEE Conf. on Decision and Control*, New Orleans, Louisiana, Dec. 7-9, , pp. 1384–1389, 1977.

[Seroussi and Golmard, 1994] B.Seroussi and J. L. Golmard. An algoirthm directly finding the k most probable configurations in Bayesian networks. *International Journal of Approximate Reasoning*, **11**, 205–233, 1994.

[Shafer, 1976] G. Shafer. *A Mathematical Theory of Evidence*, Princeton University Press, Princeton, NJ, 1976.

[Shenoy and Shafer, 1986] P. P. Shenoy and G. Shafer. Propagating belief functions with local computations. *IEEE Expert*, Fall, 43–52, 1986.

[Smets, 1978] P. Smets. *Un modèle mathématico-statistique simulant le processus du diagnostic médical.* Doctoral dissertation, Université Libre de Bruxelles, Belgium, 1978.

[Smets, 1982] P. Smets. Possibilistic inference from statistical data. In *Proc. of the 2nd World Conf. on Mathematics at the Service of Man*, Universidad Politecnica de la Palmas, Spain, A. Ballester, D. Cardus and E. Trillas, eds. pp. 611–613, 1982.

[Smets, 1991] P. Smets. Probability of provability and belief functions. *Logique & Analyse*, **133/134**, 177–195, 1991.

[Smets, 1993] P. Smets. Belief functions: The disjunctive rule of combination and the generalized Bayesian theorem. *Int. J. of Approximate Reasoning*, **9**, 1–35, 1993.

[Smets, 1998] P. Smets. Application of the transferable belief model to diagnostic problems. *Int. J. of Intelligent Systems*, **13**, 127–157, 1998.

[Soula *et al.*, 1986] G. Soula, B. Vialettes, J. L. San Marco, X. Thirion and M. Roux. PROTIS: A fuzzy expert system with medical applications. In *Fuzzy Logic in Knowledge Engineering*, H. Prade and C. V. Negoita, eds. pp. 295–310. Verlag TüV Rheinland, Köln, 1986.

[Spiegelhalter, 1986] D. Spiegelhalter. A statistical view of uncertainty in expert systems. In *Artificial Intelligence and Statistics*, W. A. Gale, ed. Addison-Wesley, Reading, MA, 1986.

[Spiegelhalter and Knill-Jones, 1985] D. Spiegelhalter and R. P. Knill-Jones. Statistical and knowledge based approaches to clinical decision support systems. *J. of the Royal Statistical Society*, A-147, 35–177, 1985.

[Studeny, 1995] M. Studeny. Conditional independence and natural conditional functions. *Int. J. of Approximate Reasoning*, **12**, 43–68, 1995.

[Torasso and Console, 1989] P. Torasso and L. Console. *Diagnostic Problem Solving — Combining Heuristic, Approximate and Causal Reasoning*. North Oxford Academic, London, 1989.

[Tsukamoto and Terano, 1977] Y. Tsukamoto and T. Terano. Failure diagnosis by using fuzzy logic. In *Proc. of the IEEE Conf. on Decision and Control*, New Orleans, Louisiana, pp. 1390–1395, 1977.

[Wellman and Henrion, 1991] M. P. Wellman and M. Henrion. Qualitative intercausal relations or explaining "explaining away". In *Proc. 2nd Int Conf on Principles of Knowledge Representation and Reasoning (KR'91)*, pp. 535–546. Morgan Kaufmann, San Francisco, 1991.

[Yager, 1988] R. R. Yager. Measures of accountability and credibility in knowledge based diagnosis systems. In *Engineering Risk and Hazard Assessment Vol. III*, A. Kandel and E. Avni, eds. pp. 1–15 CRC Press, Boca Raton, FL, 1988.

[Yager, 1985] R. R. Yager. Explanatory models in expert systems, *Int. J. of Man-Machine Studies*, **23**, 539–550, 1985.

[Zadeh, 1978] L. A. Zadeh. Fuzzy sets as a basis for a theory of possibility. *Fuzzy Sets and Systems*, **1**, 3–28, 1978.

[Zadeh, 1979] L. A. Zadeh. A theory of approximate reasoning. In *Machine Intelligence*, Vol. 9, J. E. Hayes, D. Michies and L. I. Mikulich, eds. pp. 149–194, Elsevier, New York, 1979.

[Zwirn and Zwirn, 1996] D. Zwirn and H. Zwirn. Metaconfiramation, *Theory and Decision*, **41**, 195–228, 1995.

CHRISTIAN BORGELT AND RUDOLF KRUSE

ABDUCTIVE INFERENCE WITH PROBABILISTIC NETWORKS

1 INTRODUCTION

Abduction is a form of non-deductive logical inference. Examples given by [Peirce, 1958], who is said to have coined the term "abduction", include the following:

> I once landed at a seaport in a Turkish province; and as I was walking up to the house which I was to visit, I met a man upon horseback, surrounded by four horsemen holding a canopy over his head. As the governour of the province was the only personage I could think of who would be so greatly honoured, I inferred that this was he. This was a hypothesis.
>
> Fossils are found; say remains like those of fishes, but far in the interior of the country. To explain the phenomenon, we suppose the sea once washed over this land. This is another hypothesis.
>
> Numberless documents and monuments refer to a conqueror called Napoleon Bonaparte. Though we have not seen him, what we have seen, namely all those documents and monuments, cannot be explained without supposing that he really existed. Hypothesis again.

On the other hand, probabilistic networks [Pearl, 1992] are a method to structure a multivariate probability distribution and to compute efficiently (conditioned) marginal distributions on subspaces. Hence, at first glance, there seems to be little connection between abduction and probabilistic networks. Therefore we start this chapter by showing how the two notions are connected through hypothesis assessment and statistical explanations. In addition, we use these starting sections to argue for a certain way of defining abductive inferences in contrast to inductive ones (although, unfortunately, this involves repeating some parts of what was already discussed in the introductory chapter). We believe that this view could help to avoid a lot of confusion that seems to prevail w.r.t. the term "abduction".

Next we discuss a general model of abductive inference. However, this model is not suited for implementation, because it needs too much storage space. Direct approaches to simplify the model render it manageable, but require strong independence assumptions that are hardly acceptable in applications. Therefore a modeling technique is desired, by which we can take into account dependences between the involved variables, but which nevertheless lets us exploit (conditional) independences to simplify the model. One such technique, which has become very popular nowadays, are probabilistic networks. We review this modeling technique and discuss how probabilistic networks can be used for abductive inference.

D. M. Gabbay and R. Kruse (eds.), Abductive Reasoning and Learning, 281–314.

2 OUR CATEGORIZATION OF LOGICAL INFERENCES

Logic, in the most general sense, describes the structure of languages in which one can argue. That is, logic is the (formal) theory of arguments, where an *argument* is a group of statements that are related to each other. An argument consists of one statement representing the *conclusion* and one or more statements that give reasons supporting it. The latter statements are called *premisses* [Salmon, 1973]. The process of deriving the conclusion from the premisses (using an argument) is called an *inference*. Arguments are studied by analyzing the *inference rule* used to draw the inference. Such rules are usually stated in the form of *argument schemes*. For example, the well-known inference rule of *modus ponens* is characterized by the following argument scheme:

$$\begin{array}{l} A \rightarrow B \\ A \\ \hline B \end{array}$$

If a given argument can be constructed from this scheme by replacing A and B with suitable statements, then the *modus ponens* was used to draw the inference.

As already said in the introduction, abduction is a form of logical inference. Therefore, in this section we briefly study the specific properties of *abductive* inferences in contrast to *deductive* and *inductive* inferences (as we see them) by discussing their characteristic features. We do so, because our interpretation of the term "abduction" is a specific one that differs (although only slightly) from explications given by some other authors. Thus, the main aim of this section is to avoid confusion by introducing precise notions. For a more detailed discussion of other categorizations of logical inferences and especially of other criteria to distinguish between induction and abduction than those we use, see the introductory chapter.

2.1 Deduction and Reduction

Łukasiewicz showed (according to [Bochenski, 1954]), that all logical inferences can be divided into two classes, which he called *deduction* and *reduction*. The idea underlying this division is the following: By exploiting logical equivalences we can modify the premisses of all arguments in such a way that arguments with only two premisses result. One of these premisses is a *conditional* or an *implication* (an if-then-statement), the other is equivalent either to the antecedent or to the consequent of this conditional. Written as argument schemes these two cases look like this:

$$\text{Deduction:} \quad \begin{array}{l} A \rightarrow B \\ A \\ \hline B \end{array} \qquad\qquad \text{Reduction:} \quad \begin{array}{l} A \rightarrow B \\ B \\ \hline A \end{array}$$

(To avoid some technical problems, we implicitly assume throughout this chapter that conditionals may or may not be (multiply) universally quantified. This saves the (always possible) derivation of e.g. $A(c) \rightarrow B(c)$ from $\forall x : A(x) \rightarrow B(x)$ for some appropriate constant c.)

Obviously, both of these inference rules are based on the logical tautology $((A \rightarrow B) \land A) \rightarrow B$. However, they use it in different ways. The first scheme is the *modus ponens* in its usual form. It corresponds exactly to the tautology, since the inference is drawn in the direction of the implication from A to B. The second scheme looks like some kind of reversal of the first, since the inference is drawn in the opposite direction. In the following we briefly study the properties of these two kinds of inferences.

Deduction serves the purpose to make explicit all truths that are determined by a set of statements. We can find these truths by constructing appropriate deductive arguments. Deduction is the basis of the hypothetico-deductive method used in science [Hempel, 1966]: A set of statements, the so-called *axioms*, is fixed. Then one tries to find all consequences of the axioms using deductive inferences (in which conclusions derived earlier can also be used as premisses). In the natural sciences these consequences are eventually compared to experimental findings to check the validity of the axioms (which, if the predicted experimental results do not show up, are refuted by applying the *modus tollens*). (A more detailed discussion can be found in the section on Peirce's inferential theory in the introductory chapter.) For this method to be acceptable, it is necessary that deductive inferences yield only true statements provided that the premisses are true. This can be guaranteed only, if no information is added to the information already present in the premisses. (If information was added, we could not guarantee the truth of the conclusion, simply because we would not know whether the additional information is correct.) Obviously deduction fulfills these requirements. Thus, the basic properties of deduction are that it is infallible, but it does not tell us anything new (w.r.t. the premisses). These properties are a consequence of the fact that deductive inferences corresponds exactly to tautologies.

Reduction serves the purpose to find explanations for statements that describe e.g. observations made. Obviously, the second premiss of a reductive argument can be obtained from the first premiss (the conditional) and the conclusion by a deductive inference. This is the rationale underlying reductive arguments: The premiss B becomes a logical consequence and is thus "explained" by the conclusion (provided the other premiss, i.e., the conditional, is given). However, there is a serious drawback: Reductive inferences are not truth preserving. Indeed, this type of inference is well-known in (deductive) logic as the fallacy of confirming the consequent [Salmon, 1973]. The conclusion may be false, even if both premisses are true. This is not surprising, since information about the statement A is added, namely, that it is not possible that A is false if B is true (although the conditional would be true in this case). Obviously, this additional information could be false. Thus, the basic properties of reduction are that it is fallible, but as a compensation it tells us something new (w.r.t. the premisses). These properties are a consequence of the fact that there is no tautology to which reductive inferences correspond.

2.2 *Induction and Abduction*

Often abduction is defined as what we (following Łukasiewicz) called reduction. All explanatory inferences, that is, all inferences yielding conclusions from which one of the premisses can be derived deductively (given the other premiss) are then called abductive. This approach is closely connected to the tradition of calling all non-deductive arguments inductive arguments [Salmon, 1973], which seems to go back to Aristotle [Losee, 1993] (see section 2.1 of the introductory chapter for a more detailed discussion of this view).

Another approach, which traces back to [Peirce, 1958], contrasts induction and abduction by associating them with different argument schemes, namely:

Induction: $\dfrac{\begin{array}{l}A\\B\end{array}}{A \rightarrow B}$ Abduction: $\dfrac{\begin{array}{l}A \rightarrow B\\B\end{array}}{A}$

(Obviously it can be argued that both schemes are based on the aforementioned tautology $((A \rightarrow B) \wedge A) \rightarrow B$.) More details on this view can be found in section 2.2 of the introductory chapter of this book.

Although we are more sympathetic to the second approach, we consider both to be unacceptable. The first we reject, because it does not provide grounds on which to distinguish between induction and abduction (see also below). However, the two notions are not used interchangeably. In practice a distinction is made (although it is often rather vague).

The second approach we reject because of the strange form given to the argument scheme of inductive inference, since it leads to all kinds of problems. In the first place, where does the conclusion come from? It is not part of the premisses as in the other two argument schemes. Why is "$A \rightarrow B$" the conclusion and not "$B \rightarrow A$"? (Obviously this depends on the order of the premisses, i.e., on something that is irrelevant to the other two schemes.) If the resulting conditional is (universally) quantified[1], but the premisses are not, how are the constants chosen that are to be replaced by variables? However, a more important objection is the following: Łukasiewicz showed (according to [Bochenski, 1954]) that induction is only a special case of reduction. A simple example (taken from [Bochenski, 1954]) will make this clear: Let us assume that we experimented with three pieces of phosphorus, a, b, and c, and found that they caught fire below 60^oC. We infer that all pieces of phosphorus behave in this fashion. What is the argument scheme of this inference? Obviously it is the following:

If all pieces of phosphorus catch fire below 60^oC,
then a, b, and c will.
a, b, and c catch fire below 60^oC.
____
All pieces of phosphorus catch fire below 60^oC.

[1] As already said above, we implicitly assume that all conditionals may or may not be (multiply) universally quantified.

Clearly, this is a reduction. Thus, induction and abduction, the latter of which is characterized by the reductive argument scheme in the second approach, would be indistinguishable. Or, to put it differently: It sounds unreasonable to call the same inference once an induction and once an abduction depending on the form of the argument scheme, if the schemes can be transformed into each other. Note also, that seeing induction as a special case of reduction removes the problem how we arrive at the special form of the conclusion: It is a part of the first premiss. Of course, we now face the problem where the first premiss comes from. However, it is not a problem of logic where the premisses come from. In logic they are always assumed to be given and it is merely studied how the conclusion is derived from them. We see it as an advantage that the problem of how to generate the hypotheses that enter the inferences is removed from the realm of logic. (Describing induction as an inference leading from A and B to $A \rightarrow B$ is a trial to incorporate the hypothesis generation process into logic.)[2]

In contrast to the approaches discussed above, our own distinction between induction and abduction refers to [Popper, 1934][3]:

> An inference is usually called an inductive inference or an induction, if it goes from *particular statements*, which describe e.g. observations, experiments, etc., to *general statements*, i.e., to hypotheses and theories.

That is, we do not base our distinction on the argument scheme used, since for both induction and abduction it is the reductive scheme, but on the type of the statement inferred, which indeed differs for these two types of inferences.

Note that this distinction is not contained in the second approach discussed above. If we conclude from A and B that $A \rightarrow B$, we will not necessarily have inferred a general statement. If A and B are particular statements, then $A \rightarrow B$ is also a particular statement (provided no quantifier is added). That it is often seen as a general statement nevertheless is due to the fact that general laws often come in the form of conditional statements. However, that a statement is a *law*, i.e., that it is valid in general, is expressed by variables and quantifiers. Of course, we could require the conclusion to be universally quantified, but, as already indicated, which constants are to be replaced by variables is completely arbitrary and not determined by logical rules.

With induction being an inference leading to a general statement, it is natural to define abduction as its counterpart [Borgelt, 1992]:

> Abduction is a reductive inference in which the conclusion is a particular statement.

Using a notion of formal logic, we may also say that abduction is a reductive inference with the conclusion being a *ground formula*. That is, the conclusion must

[2]Thus we place ourselves in the tradition that distinguishes the *context of discovery* and the *context of justification* of (scientific) hypotheses [Losee, 1993] and declare that logic belongs to the latter.

[3]Our translation from the German, italics in the original.

not contain variables, neither bound nor free. Using a term introduced in the introductory chapter, we can say that we define abduction by requiring that the set of *abducibles* must contain only ground formula of a given formal language. We do not admit existentially quantified formulae (as some other authors do), since $\exists x : A(x)$ is equivalent to $\neg\forall x : \neg A(x)$ and thus tantamount to a general statement. By the restriction to ground formulae we want to make explicit that specific facts are inferred and *not* general laws (for which we need variables and quantifiers). In contrast to this, induction infers general laws and *not* specific facts.[4] Hence, with induction one generalizes observations, with abduction one learns about unobserved or even unobservable (particular) facts.

One may ask why an analogous distinction is not made for deductive inferences. To this we reply that such a distinction *is* made in the theory of the hypothetico-deductive method of science. This distinction is necessary, because only particular statements can be confronted with experimental findings. We can never observe directly a general law [Popper, 1934]. Thus we need to distinguish inferences which, for example, derive statements of lower generality from statements of higher generality (in order to find more specific theories) from inferences which yield particular statements (which can be tested in experiments). However, as far as we know, there are no special names for these two types of deductive inferences and thus this distinction may slip attention. Unfortunately a closer inspection of this astonishingly far reaching analogy is beyond the scope of this paper.

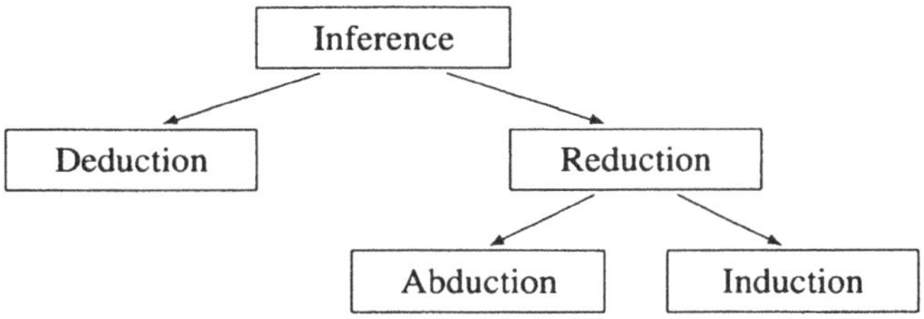

Figure 1. Categories of logical inferences.

The classification scheme for logical inferences we arrived at is shown in figure 1. The distinction between abductive and reductive inferences is based on the form of the argument scheme, the distinction between inductive and abductive inferences is based on the form of the conclusion. This scheme looks very much like the one discussed in the section on Peirce's syllogistic theory in the introductory chapter. However, it should be noted that the distinction in the non-deductive branch is based on a different criterion and hence the two schemes are not identical.

[4]It can be seen from the examples given by [Peirce, 1958] that he must have had something like this in mind, since in his examples of abductive arguments the result is a particular statement, see e.g. the examples cited in the introduction. However, he may have permitted existentially quantified formulae.

3 HYPOTHESIS ASSESSMENT

As already mentioned above, a reductive—and thus an abductive—inference can yield a false conclusion, even if the premisses are true. This is usually made explicit by calling the conclusion of a reductive argument a *hypothesis* or a *conjecture*. Of course, we do not want our conclusions to be wrong. Therefore we look for criteria to assess the hypotheses obtained by reductive arguments in order to minimize the chances that they are wrong. This is a perfectly reasonable intention. However, it has lead to fairly unpleasant situation: One of the main problems why there is such confusion about what the term "abduction" means is the fact that nearly nobody distinguishes clearly between the logical inference and the assessment of its result. This can already be seen from a frequently used characterization of abduction, namely, that it is an "inference to the best(!) explanation" (cf., for example, [Josephson and Josephson, 1996]). That is, not all explanatory inferences (even if their conclusions are only ground formulae, see above) qualify as abductive inferences. To qualify, they must yield a "better" explanation than all other explanatory inferences (or at least an equally good one). However, it is clear that depending on the domain of application different criteria may be used to specify what is meant by "best". Thus it is not surprising that there are as many different interpretations of the term "abduction" as there are criteria to assess hypotheses. In the preceding section we tried to give a purely logical definition of the term *abduction*, which enables us to study now in a systematic way different criteria to assess hypotheses without introducing any ambiguity w.r.t. the term "abduction".

We do not claim our list of criteria to be complete, but only give some examples in order to make clear the main difficulties involved. (Note that for the most part of this section we do not distinguish between abduction and induction, since the criteria studied apply to all reductive arguments.) More details on hypothesis assessment can be found in the introductory chapter.

Relation between antecedent and consequent. If we require only that a reductive argument explains its second premise by yielding a hypothesis from which it can be derived deductively (given the conditional), a lot of useless or even senseless "explanations" are possible. For example, consider the conditional "If sparrows can fly, then snow is white." (which, from a purely logical point of view, is true due to the "material" implication). However, we are not satisfied with the explanation "sparrows can fly", if we try to figure out why snow is white, because we cannot see any connection between the two facts. The situation gets even worse, if we replace the antecedent by a contradiction, since: *ex contradictio quodlibet* (from a contradiction follows whatever you like). Therefore, from a logical point of view, a contradiction is a universal explanation.

The problem is that we want an explanation to give a reason, or even better, a cause for the fact to be explained.[5] However, "reason" and "cause" are semantical

[5]Here the difference between the if-then in logic and the if-then in natural language is revealed most clearly. In natural language most often a causal connection or a inherence relation is assumed between

notions, referring to the *meaning* of statements. Yet the meaning of statements cannot be captured by formal logic, which handles only the "truth functionality" of compound statements.

Relation to other statements. Usually drawing an inference is not an isolated process, but takes place in an "environment" of other statements, which are known or at least believed to be true. These other statements can (but need not) have a bearing on the conclusion of a reductive argument. They may refute or support it. For example, the fact that clouds are white may be explained by the hypothesis that they consist of cotton wool. However, we also know that cotton wool is heavier than air. Thus we can refute the hypothesis (although indirectly, since we need a (deductive) argument to make the contradiction explicit). On the other hand, if we observe white animals, we may conjecture that they are swans. This conjecture is supported, if we learn also that the animals can fly and have orange bills. That is, a hypothesis gets more plausible, if several reductive arguments lead to it.

Parsimony (Ockham's razor). Not only do we want explanations to give reasons or causes and to be compatible with our background knowledge, we also want them to be as simple as possible. For example, suppose that your car does not start and that the headlights do not come on. Considering as explanations that both headlights are burned out *and* that the starter is broken or that simply the battery is empty, you would surely prefer the latter. The rationale of this preference is expressed in Ockham's razor[6]: *pluralitas non est ponenda sine necessitate*, that is, multiple entities should not be assumed without necessity. (Note that this is also a semantic criterion, although it may be turned into a syntactical one by fixing the formal language and introducing a measure of complexity for the formulae.)

Probability. Consider the following two abductive arguments:

Water is (1) a liquid, (2) transparent, and (3) colorless.
This substance is (1) a liquid, (2) transparent, and (3) colorless.
———————————————
This substance is water.

Tetrachloromethane is (1) a liquid, (2) transparent, and (3) colorless.
This substance is (1) a liquid, (2) transparent, and (3) colorless.
———————————————
This substance is tetrachloromethane.

W.r.t. the criteria mentioned above they are equivalent. Nevertheless we prefer the first, since in daily life we deal with water much more often than with tetrachloromethane. Due to this difference in frequency we judge the first conclusion to be much more *probable* than the second. This is one way in which the notion of *probability* enters our discussion, namely as a means to assess hypotheses.

---

the antecedent and the consequent of a conditional. In formal logic, no such assumption is made.

[6]William of Ockham, 1280–1349.

4 PROBABILISTIC INFERENCES

Up to now we assumed implicitly that the conditional that appears in deductive as well as in reductive arguments is known to be absolutely correct. However, in real world applications we rarely find ourselves in such a favorable position. To quote a well-known example: Even the statement "If an animal is a bird, then it can fly." is not absolutely correct, since there are exceptions like penguins, ostriches etc. To deal with such cases—obviously, confining ourselves to absolutely correct conditionals would not be very helpful—we have to consider statistical syllogisms, statistical generalizations, and (particular) statistical explanations.

4.1 Statistical Syllogisms

By *statistical syllogism*, a term which was already mentioned in the introductory chapter, we mean a deductively shaped argument, in which the conditional is a statistical statement. For example, consider the following argument:

80% of the beans in box B are white.
This bean is from box B.

This bean is white.

(To make explicit the conditional form of the first premise, we may rewrite it as: "If x is a bean from box B, then x is white (with probability 0.8).")

The most important thing to note here is that with a statistical conditional a deductively shaped argument loses its distinctive mark, namely, it is no longer infallible. Since 20% of the beans in box B are not white, the conclusion may be false, even though the premisses are true. It would be a genuine deductive argument only, if the probability in the conditional were 1 (this is why we only called it a *deductively shaped* argument).

Nevertheless we are rather confident that the conclusion of the argument is true, since the probability of the bean being white is fairly high (provided the bean was picked at random). We may express this confidence by assigning to the conclusion a *degree of belief* (or a *degree of confidence*) equal to the probability stated in the conditional. We may even call this degree of belief a probability, if we switch from an *empirical* (or sometimes called *frequentistic*) interpretation of probabilities to a *subjective* (or *personalistic*) interpretation [Savage, 1954], or if we interpret the argument as a rule for decisions, which is to be applied several times. In the latter case the degree of belief measures the relative frequency of those cases in which a decision made according to this rule is correct (and thus can be seen as an empirical probability).[7]

[7]Note that we cannot interpret the number 0.8 assigned to the conclusion of the argument as an empirical probability for a single instance of the argument, since in a single instance the bean is either white or it is not white. There is no random element involved and thus the probability of the bean being white is either 0 or 1, even if we do not know whether it is 0 or 1.

4.2 *Statistical Generalizations*

A *statistical generalization*, which is called an inductive generalization in the introductory chapter, is an argument like, for instance:

If 50% of all children born are girls,
then 50% in this sample will be girls.
50% of the children in this sample are girls.

50% of all children born are girls.

Here the sample may be determined, for instance, as all children born in a specific hospital. (Note that often the first premise (the conditional) is missing, although the argument would not be logically complete without it.) Obviously, a statistical generalization is the probabilistic analog of an inductive argument. It would be a genuine inductive argument, if the percentages appearing in it were 100% (or the probabilities were 1). Statistical generalizations are one of the main topics of statistics, where they are used, for instance, to predict the outcome of an election from polls. In statistics it is studied what it takes to make reliable the conditional of the argument above (e.g. the sample has to be *representative* of the whole population), what is the best estimate of parameters for the whole population (like the relative frequency of 50% in the argument above—since it is, of course, logically possible that the relative frequency of girls in the whole population is 60%, but in the sample it is only 50%), and how to compute these best estimates.

4.3 *(Particular) Statistical Explanations*

We use *(particular) statistical explanation*[8] as a name for the probabilistic analog of an abductive inference. That is, a statistical explanation is an argument like

70% of all patients having caught a cold develop a cough.
Mrs Jones has a cough.

Mrs Jones has caught a cold.

Again, the above would be a genuine abductive argument, if the relative frequency in the conditional were 100% (the conditional probability were 1).

Although this argument seems plausible enough, there is a problem hidden in it, namely that we cannot assign a *degree of belief* to the conclusion as easily as we could for statistical syllogisms. The reason is that in a statistical syllogism the inference is drawn in the direction of the conditional probability—just as in a deductive argument the inference is drawn in the direction of the implication—whereas in a statistical explanation it is drawn in the opposite direction—just as in a reductive argument the inference is drawn opposite to the direction of the

[8] The word *particular* is added only to emphasize that the explanation must be a particular statement, since statistical generalizations—due to their reductive structure—also yield statistical explanations. In the following we drop it, since there is no danger of confusion.

implication. Thus, to assign a degree of belief to the conclusion of a statistical explanation, we have to invert the conditional probability using *Bayes' rule*

$$P(A|B) = \frac{P(B|A)P(A)}{P(B)},$$

where $P(B|A)$ is the conditional probability appearing in the first premise of the argument (with $A \hat{=}$ "The patient P has caught a cold." and $B \hat{=}$ "The patient P has a cough.") and $P(A)$ and $P(B)$ are the prior probabilities of the events A and B, respectively.[9]

It is obvious that using Bayes' rule we can always invert a conditional probability, provided the prior probabilities are known. That is, with Bayes' rule we can always transform a statistical explanation into a statistical syllogism, simply by turning the conditional around. Thus, in probabilistic reasoning the difference between abductively and deductively shaped arguments vanishes—which is not surprising, since the distinctive mark of deductive inferences, their infallibility, is lost (see above).

However, the distinction between arguments the conclusion of which is a general statement and those the conclusion of which is a particular statement remains valid, since it is next to impossible to know or even define the prior or posterior probability of a general statement. This is especially true, if the conclusions to be inferred are whole theories and consequently [Popper, 1934] has argued (convincingly, in our opinion) that there is no way to prefer, for instance, Einstein's theory of gravitation over Newton's based on their respective probabilities. Therefore in the following we confine ourselves to inferences of particular statements, which is no real restriction, since our topic is abductive inference anyway.

5 A GENERAL MODEL OF ABDUCTIVE INFERENCE

In this section we introduce a general model of abductive inference incorporating hypothesis assessment that is based on [Bylander *et al.*, 1991], but also closely related to [Peng and Reggia, 1989]. Since this model cannot be implemented directly (it would require too much storage space) we have to look for simplifications, which finally lead us to probabilistic networks.

5.1 Formal Definition

We start by giving a formal definition of an abductive problem. This definition is intended to describe the framework in which the abductive reasoning takes place by fixing the statements that may be used in abductive arguments and a mechanisms to assess the hypotheses resulting from such arguments.

[9]Note that Bayes rule is of little value for genuine abductive arguments, since for these prior and posterior probability must be the same (since they must be either 0 or 1) and thus $P(A|B) = P(A)$. That is, in order to apply Bayes' rule in this case, we already have to know what we want to infer.

DEFINITION 1. An **abductive problem** is a tuple $\mathcal{AP} = \langle D_{\text{all}}, H_{\text{all}}, e, pl, D_{\text{obs}} \rangle$, where

- D_{all} is a finite set of possible atomic data,
- H_{all} is a finite set of possible atomic hypotheses,
- e is a relation of $2^{D_{\text{all}}}$ and $2^{H_{\text{all}}}$, i.e., $e \subseteq 2^{D_{\text{all}}} \times 2^{H_{\text{all}}}$,
- pl is a mapping from $2^{D_{\text{all}}} \times 2^{H_{\text{all}}}$ to a partially ordered set Q, and
- $D_{\text{obs}} \subseteq D_{\text{all}}$ is the set of observed data.

The sets D_{all} and H_{all} contain the statements that can be used in inferences, the former the possible observations (or *data*, which is why this set is denoted by the letter D), the latter the possible *hypotheses*. Since we deal with abductive inferences we require all statements in D_{all} and H_{all} to be particular, i.e., expressible as ground formulae. Note that we do not require the two sets to be disjoint, although this requirement is useful for most applications.

The relation e (for *explanation*) connects sets of observations with sets of hypotheses which explain them. That is, it describes the set of conditionals that can be used in abductive arguments (if $(H, D) \in e$, then $H \rightarrow D$). It thus contains information about what is accepted as an explanation (cf. section 3).

The mapping *pl* (for *plausibility*) assesses an inference by assigning an element of the set Q, which states the *quality* of the inferred (compound) hypothesis w.r.t. the given data. It may be a partial mapping, since we need values only for the elements of the relation e. In the following the set Q will always be the interval $[0, 1]$, as we deal only with probabilities and degrees of belief. However, this restriction is not necessary and therefore we chose to make the definition more general. Note that w.r.t. the mapping *pl* we may choose to add to D_{all} statements that do not need an explanation, but may have a bearing on the assessment of possible hypotheses. For example, in medical diagnosis we register the sex of a patient, although it is clearly not a fact to be explained by the diagnosis, simply because the likelihood of certain diseases differs considerably for the two sexes.

D_{obs} is the set of *observed* data, i.e., the set for which an explanation is desired. Of course, this set is not actually a part of the framework for abductive reasoning, which is fixed by the first four elements of the five-tuple. This framework may be used with several different sets D_{obs} and therefore it is reasonable to consider a general representation that is independent of D_{obs}. However, it has to be admitted that without a set of observations to be explained, there is no *problem* to be solved, and therefore, in contrast to [Bylander *et al.*, 1991], we chose to add it to the definition.

Before we can proceed further, we have to say a few words about the interpretation of subsets of the sets D_{all} and H_{all}, for instance, the interpretation of D_{obs}. Given a set $D \subseteq D_{\text{all}}$, we assume that *all* statements contained in D must hold, i.e., D is interpreted as a conjunction of its elements. This is straightforward, but what about the statements in $D_{\text{all}} \backslash D$? Obviously it would be impractical to require that they are all false, since this is equivalent to requiring perfect knowledge about all possible observations, which in applications is rarely to be had. Unfortunately, we can neither assume that nothing is known about their truth value, because there may be semantical relations between possible observations. For example, D_{all} may contain the statements "This bean is white.", "This bean is red." and "This bean is green." As soon as we observe that the given bean is white, we know that the other two statements are false. (Obviously the same problem arises, if we consider whether a given set D of observations is satisfiable or whether it contains mutually exclusive statements.) Since this is a problem on the semantical level, it cannot be taken care of automatically in the formal model. Rather we have to assume that the relation e is defined in such a way that it connects only consistent sets of observations with consistent sets of hypotheses. In addition we have to assume that the set D_{obs} is consistent (which is no real restriction, since the observed reality should be consistent).

We can support that these requirements are met, though, by introducing some structure on the sets D_{all} and H_{all}. The simplest way to achieve this is to group atomic observations and hypotheses in such a manner that the statements in each group are mutually exclusive and exhaustive (to achieve the latter it may be necessary to introduce additional atomic statements that cover the remaining situations). After we did this, we can reconstruct the sets D_{all} and H_{all} as the Cartesian product of the domains of certain *variables*, each of which represents a group of mutually exclusive and exhaustive statements. That a set of observations and hypotheses is (formally) consistent can now easily be ensured by requiring that it must not assign more than one value to a variable, i.e., must not select more than one statement from the corresponding set of mutually exclusive statements. It should be noted, though, that this approach only excludes formal (or syntactical) inconsistencies, whereas factual inconsistencies (for instance, a nine year old girl with two children or a car that weighs 10 grams) still have to be taken care of by the explanation relation e.

An abductive problem is solved by finding the best explanation(s) for the data observed. This motivates the next two definitions.

DEFINITION 2. In an abductive problem $\mathcal{AP} = \langle D_{\text{all}}, H_{\text{all}}, e, pl, D_{\text{obs}} \rangle$ a set $H \subseteq H_{\text{all}}$ is called an **explanation** (of the data D_{obs}), iff $(H, D_{\text{obs}}) \in e$.

Often an explanation is required to be parsimonious (see above). We may add this requirement by defining

$$H \text{ is an explanation, iff } (H, D_{\text{obs}}) \in e \ \wedge \ \neg\exists H' \subset H : (H', D_{\text{obs}}) \in e,$$

that is, H is an explanation only, if no proper subset of H is also an explanation.

DEFINITION 3. In an abductive problem $\mathcal{AP} = \langle D_{\text{all}}, H_{\text{all}}, e, pl, D_{\text{obs}} \rangle$ an explanation H is called a **best explanation** (of the data D_{obs}), iff there is no explanation H' that is better than H w.r.t. the mapping pl, i.e., iff

$$\neg\exists H' : (H', D_{\text{obs}}) \in e \ \wedge \ pl(H, D_{\text{obs}}) < pl(H', D_{\text{obs}}).$$

Clearly, which explanation is selected depends on how the hypothesis assessment function pl ranks the possible explanations. Therefore this function should be chosen with special care. Fortunately, we can identify the ideal choice, which may serve as a guideline.

DEFINITION 4. The **optimal hypothesis assessment function** is

$$pl_{\mathrm{opt}}(H, D) = P(H|D).$$

The best explanation under pl_{opt} is called the **most probable explanation.**

Strictly speaking, the notation $P(H|D)$ used in this definition is formally incorrect, because the sets H and D are no events. However, this notation is intended as an abbreviation for $P\left(\{\omega \in \Omega \mid \bigwedge_{h \in H} h(\omega)\} \mid \{\omega \in \Omega \mid \bigwedge_{d \in D} d(\omega)\}\right)$, where Ω is the underlying sample space. Here we view the statements h and d as random variables that assume the value *true*, if they hold for the elementary event ω, and the value *false* otherwise.

The probability of the hypothesis given the data is the optimal hypothesis assessment function, because it is easy to show that deciding on the hypothesis it advocates is, in the long run, superior to any other decision strategy—at least w.r.t. the relative number of times the correct decision is made. If the alternatives carry different costs in case of a wrong decision, an different function may be better. Nevertheless, the probability of the hypothesis given the data is still very important in this case, because it is needed to compute the hypothesis assessment function that optimizes the expected benefit.

Note that regarding a probabilistic assessment function as the best one possible does not exclude other uncertainty calculi. For a probabilistic approach to be feasible, specific conditions have to hold, which may not be satisfied in a given application. For example, if the available information about the modeled domain is not precise enough to compute a probability distribution, other approaches have to be considered. This justifies using, for example, possibility distributions [Dubois and Prade, 1988] or mass assignments [Baldwin *et al.*, 1995] for hypothesis assessment. In this chapter, however, we assume that we can treat abductive problems probabilistically and therefore drop the subscript *opt* in the following. An approach based on possibility distributions, which is closely related to the one presented here, can be found in the next chapter.

The relation e and the mapping pl of an abductive problem can easily be represented as a table in which the lines correspond to the possible sets of hypotheses and the columns correspond to the possible sets of observations (or vice versa, if you like). A sketch is shown in table 1. In this table each entry states the value assigned by the mapping pl to the pair (H, D) corresponding to the table field, provided that this pair (H, D) is contained in the relation e. For pairs (H, D) not contained in the relation e the value of the mapping pl is replaced by a 0, which serves as a kind of marker for unacceptable explanations. This way of marking explanations presupposes that any acceptable explanation has a non-vanishing probability, which is a reasonable assumption.

Table 1. Sketch of a possible representation of the relation e and the mapping pl of an abductive problem. The zeros indicate unacceptable explanations, i.e., pairs (H, D) not contained in e. The p_i are the conditional probabilities $P(H|D)$.

pl	$\emptyset$	$\{d_1\}$	$\{d_2\}$	$\{d_3\}$	$\cdots$	$\{d_1, d_2\}$	$\{d_1, d_3\}$	$\cdots$
$\emptyset$	1	0	0	0	$\cdots$	0	0	$\cdots$
$\{h_1\}$	0	p_1	p_6	p_{10}		p_{13}	p_{17}	
$\{h_2\}$	0	p_2	p_7	0		p_{14}	p_{18}	
$\{h_3\}$	0	p_3	0	p_{11}		0	p_{19}	
$\vdots$	$\vdots$				$\ddots$		$\vdots$	
$\{h_1, h_2\}$	0	p_4	p_8	0		p_{15}	0	
$\{h_1, h_3\}$	0	p_5	p_9	p_{12}	$\cdots$	p_{16}	p_{20}	$\cdots$
$\vdots$	$\vdots$						$\vdots$	$\ddots$

Note that inconsistent sets of observations or hypotheses correspond to all zero columns or lines and thus may be removed from the table. Note also that we actually need to mark the acceptable explanations, because we *cannot* define the relation e by

$$e = \{(H, D) \in 2^{H_{\text{all}}} \times 2^{D_{\text{all}}} \mid P(H|D) > 0\}.$$

We *cannot* even define the relation e as

$$e = \{(H, D) \in 2^{H_{\text{all}}} \times 2^{D_{\text{all}}} \mid P(H|D) > P(H)\}.$$

If we did so, we would accept unreasonable explanations. For example, if we learn that one or several bathing accidents have occurred, there is a much higher probability that the ice-cream sales have been high recently than without the bathing accidents. However, the high ice-cream sales are obviously not an acceptable explanation for the bathing accidents.[10] Generalizing, we cannot define the relation e as shown above, because what makes an explanation acceptable is a semantical relation between the hypotheses and the data. However, probability theory, just as logic, cannot capture semantical relations. Therefore we have to indicate separately which explanations are acceptable.

Solving an abductive problem with the representation just discussed is especially simple. All one has to do is to visit the table column that corresponds to the observed data and to find the line of this column that holds the highest probability. The set of hypotheses corresponding to this line is the most probable explanation for the observed data. However, it is clear that for any real world problem worth considering we cannot set up the table described above, since it would have too many lines and columns. Therefore we have to look for simplifications.

[10]The reason for this correlation is the simple fact that most bathing accidents occur in summer, because more people go bathing when it is warm. They also buy more ice-cream when it is warm.

5.2 *Simplifications*

In the following we consider, in two steps, simplifications of the general model introduced in the preceding section. The first simplification is based on the idea to replace the explanation relation e by a function mapping from $2^{H_{\text{all}}}$ to $2^{D_{\text{all}}}$, which assigns to a set H of hypotheses the union of all sets D of observations that H can explain. In this case we may just check for set inclusion to determine whether H explains D_{obs}. Of course, this simplification is not always possible, if we want the result to be equivalent to the original abductive problem. Specific conditions have to hold, which are given in the following definition.

DEFINITION 5. An abductive problem $\mathcal{AP} = \langle D_{\text{all}}, H_{\text{all}}, e, pl, D_{\text{obs}}\rangle$ is called **functional**, iff

1. $\forall H \subseteq H_{\text{all}}: \quad \forall D_1, D_2 \subseteq D_{\text{all}}:$
   $((H, D_1) \in e \wedge D_2 \subseteq D_1) \Rightarrow (H, D_2) \in e$

2. $\forall H \subseteq H_{\text{all}}: \quad \forall D_1, D_2 \subseteq D_{\text{all}}:$
   $((H, D_1) \in e \wedge (H, D_2) \in e \wedge D_1 \cup D_2$ is consistent$) \Rightarrow (H, D_1 \cup D_2) \in e$

Note that the first condition is no real restriction, since we interpret sets of observations as conjunctions. Consequently, if a set $D \subseteq D_{\text{all}}$ can be explained by a set $H \subseteq H_{\text{all}}$, all subsets of D should also be explainable by H. The second condition, however, is a restriction, since counterexamples can easily be found. For instance, due to the sexual dimorphism in mallards (*Anas platyrhynchos*), we can explain the observation of a female bird with webbings as well as the observation of a bird with webbings and a shining green-black head by the hypothesis that the bird is a mallard. However, the conjunction of the two observations cannot be explained by this hypothesis, since only male mallards have a shining green-black head [Thiele, 1997].[11]

The relation e of a functional abductive problem can be replaced, as the term "functional" already indicates, by a function e_f defined as follows

$$\forall H \subseteq H_{\text{all}}: \quad e_f(H) = \{d \in D_{\text{all}} \mid \exists D \subseteq D_{\text{all}} : d \in D \wedge (H, D) \in e\}.$$

With this function an explanation (for a consistent set D_{obs}) can be defined—as already indicated—as a set $H \subseteq H_{\text{all}}$, such that $D_{\text{obs}} \subseteq e_f(H)$.

The simplification that can be achieved for a functional abductive problem becomes most obvious, if the function e_f is represented as a relation e_1 of $2^{H_{\text{all}}}$ and D_{all}, which can be defined by

$$\forall H \subseteq H_{\text{all}}: \quad (H, d) \in e_1 \quad \Leftrightarrow \quad d \in e_f(H).$$

This relation can be represented as a table with one column for each $d \in D_{\text{all}}$ and one line for each $H \subseteq H_{\text{all}}$. Of course, lines corresponding to inconsistent sets

[11] The conjunction of the observations is not inconsistent, though, since in another kind of ducks, *tadorna tadorna*, which exhibits no sexual dimorphism w.r.t. plumage, male *and* female have a shining green-black head [Thiele, 1997].

Table 2. Sketch of a possible representation of the relation e and the mapping pl of a D-independent abductive problem. The zeros indicate possible observations that cannot be explained by the corresponding set of hypotheses.

pl	d_1	d_2	d_3	$\cdots$	d_n
$\emptyset$	p_1	p_2	p_3	$\cdots$	p_4
$\{h_1\}$	p_5	0	p_6		p_7
$\{h_2\}$	p_8	p_9	0		0
$\{h_3\}$	0	0	p_{10}		p_{11}
$\vdots$	$\vdots$				$\vdots$
$\{h_1, h_2\}$	p_{12}	p_{13}	p_{14}		p_{15}
$\{h_1, h_3\}$	p_{16}	0	p_{17}		p_{18}
$\vdots$	$\vdots$				$\vdots$

of hypotheses can be removed, just as for the table used to represent the original relation e. However, this table only represents the relation e in a more compact form. To state the hypothesis assessment function pl we still need a table with $2^{|H_{\text{all}}|}$ lines and $2^{|D_{\text{all}}|}$ columns. In order to represent pl with a table similar to the one needed to represent e_1, we need much stronger assumptions.

DEFINITION 6. A functional abductive problem $\mathcal{AP} = \langle D_{\text{all}}, H_{\text{all}}, e, pl, D_{\text{obs}} \rangle$ is called **D-independent**, iff

$$\begin{aligned} &\forall H \subseteq H_{\text{all}} : \forall D \subseteq D_{\text{all}} : \\ &\qquad D \text{ is consistent} \quad \Rightarrow \quad P(D|H) = \prod_{d \in D} P(d|H). \end{aligned}$$

Intuitively, D-independence means that the probability of a possible observation d is independent of any other observations that may be present given any set of hypotheses H. Note that the set H may be empty. That is, for all $D \subseteq D_{\text{all}}$ we also have $P(D) = \prod_{d \in D} P(d)$.

The relation e and the mapping pl of a D-independent abductive problem can easily be represented as a table in which the lines correspond to possible sets of hypotheses and the columns correspond to possible (atomic) observations. A sketch is shown in table 2. Each field of this table holds a conditional probability $P(d|H)$. Note that, as for table 1, we can delete from this table all lines that correspond to inconsistent sets of hypotheses. From this table we can then compute the assessment of any consistent set of hypotheses using Bayes' rule, provided we also know the prior probability $P(H)$ of the set of hypotheses:

$$P(H|D) = \frac{P(D|H)P(H)}{P(D)} = P(H) \prod_{d \in D} \frac{P(d|H)}{P(d)}.$$

Thus the best explanation is found by computing $P(H|D_{\text{obs}})$ for all (consistent) sets H of hypotheses using the above formula and then selecting that set H for which this probability is highest. This approach, although restricted to one element sets $\{h\}$, was suggested in [Charniak and McDermott, 1985].

As we have shown, the representation of the explanation relation e and the hypothesis assessment function pl of an abductive problem can be considerably simplified, if it is D-independent. However, the table of conditional probabilities required is still too large to be stored for any reasonably sized real world problem, simply because it still has too many lines. Therefore, in a second simplification step, we consider independent abductive problems. In such problems the possible hypotheses independently explain observations.

DEFINITION 7. A functional abductive problem $\mathcal{AP} = \langle D_{\text{all}}, H_{\text{all}}, e, pl, D_{\text{obs}} \rangle$ is called **independent**, iff

$$\forall H \subseteq H_{\text{all}}: \quad e_f(H) = \bigcup_{h \in H} e_f(\{h\}),$$

where e_f is the function by which the relation e can be represented in a functional abductive problem (see above).

Of course, this is a very strong restriction. It requires that there is no interaction of hypotheses in the sense that no combination of atomic hypotheses can to explain an observation, neither of the atomic hypotheses can explain on its own. In addition, the explanatory powers of atomic hypothesis must not cancel each other (there must not be "destructive interference"). This requirement excludes such commonplace situations as the following: Suppose we have a computing center that is equipped with its own emergency power supply. If we observe that the computing center is not working, we can explain this neither by the assumption that the electric power station is down nor by the assumption that the emergency power supply is broken. Only a conjunction of both is an acceptable hypothesis. On the other hand, cancellation must not occur, that is, one hypothesis must not have a subtractive effect on another. This is common in medicine, though. For example, in the domain of acid-base disorders, one disease might explain an increased blood pH, and another might explain a decreased pH, but together the result might be a normal pH ([Patil *et al.*, 1982] as cited in [Bylander *et al.*, 1991]).

Although excluding interactions between hypotheses is a strong restriction, we may choose to do so, because it enables us to simplify the representation of the framework for abductive reasoning. In an independent abductive problem the relation e can be replaced by a relation e_2, which is defined as follows:

$$\forall h \in H_{\text{all}}: \quad (h, d) \in e_2 \quad \Leftrightarrow \quad d \in e_f(h).$$

With this relation, an explanation can be defined as a consistent set H of hypotheses, such that

$$D_{\text{obs}} \subseteq \{d \in D_{\text{all}} \mid \exists h \in H : (h, d) \in e_2\}.$$

The relation e_2 can be represented as a simple table with $|H_{\text{all}}|$ lines and $|D_{\text{all}}|$ columns. In this table an entry corresponding to a pair (h, d) is marked, if the atomic hypothesis h can explain the atomic datum d. From this table all possible explanations H of the observed data D_{obs} can easily be constructed, if it is taken care that only consistent sets H are selected. However, as in the first simplification step, only the relation e can be represented in a more compact form. To store the hypothesis assessment function pl, we still need a table with $2^{|H_{\text{all}}|}$ lines and $|D_{\text{all}}|$ columns. In order to represent pl with a table similar to the one needed to represent e_2, we need much stronger assumptions [Borgelt, 1992]:

DEFINITION 8. A D-independent abductive problem $\mathcal{AP} = \langle D_{\text{all}}, H_{\text{all}}, e, pl, D_{\text{obs}}\rangle$ is called **HD-independent**, iff

$$\begin{aligned}&\forall H \subseteq H_{\text{all}} : \forall d \in D_{\text{all}} :\\ &H \text{ is consistent } \Rightarrow P(H|d) = \prod_{h \in H} P(h|d) \ \wedge \ P(H) = \prod_{h \in H} P(h).\end{aligned}$$

The idea underlying this definition is the following: To compute the assessment of a set H of hypotheses in a D-independent abductive problem we needed $P(H)$ and $P(d|H)$. These probabilities force us to store a table with $2^{|H_{\text{all}}|}$ lines. However, with the above relations, we can compute these probabilities from a much smaller set of probabilities. For $P(H)$ this is obvious as the computation formula is given in the definition. For $P(d|H)$ we have to apply Bayes' rule:

$$P(d|H) = \frac{P(H|d)P(d)}{P(H)} = P(d) \prod_{h \in H} \frac{P(h|d)}{P(h)}.$$

Another application of Bayes' rule to the factors of the product of the above formula yields

$$P(d|H) = P(d) \prod_{h \in H} \frac{P(d|h)P(h)}{P(d)P(h)} = P(d) \prod_{h \in H} \frac{P(d|h)}{P(d)}.$$

Combining this formula and the one derived for D-independent abductive problems we arrive at the following formula to compute the conditional probability of a set H of hypotheses given a set D of possible observations [Borgelt, 1992]:

$$P(H|D) = \prod_{h' \in H} P(h') \prod_{d \in D} \left(\prod_{h \in H} \frac{P(d|h)}{P(d)} \right).$$

As a consequence we only need to store the probabilities $P(h)$, $P(d)$, and $P(d|h)$ for all $h \in H_{\text{all}}$ and all $d \in D_{\text{all}}$, that is, only $(|H_{\text{all}}| + 1) \cdot (|D_{\text{all}}| + 1) - 1$ probabilities. These probabilities can be organized in a simple table as sketched in table 3. In addition to this table we need the table that represents the relation e_2, because we cannot incorporate the relation e_2 in this table by marking entries with a zero (since we need all probabilities). Nevertheless, this representation is easily manageable even for large sets H_{all} and D_{all}.

Table 3. Sketch of a possible representation of the mapping pl of an HD-independent abductive problem.

pl	prior	d_1	d_2	$\cdots$	d_n
prior		p_{01}	p_{02}	$\cdots$	p_{0n}
h_1	p_{10}	p_{11}	p_{12}		p_{1n}
h_2	p_{20}	p_{21}	p_{22}		p_{2n}
$\vdots$	$\vdots$			$\ddots$	$\vdots$
h_m	p_{m0}	p_{m1}	p_{m2}	$\cdots$	p_{mn}

6 PROBABILISTIC NETWORKS

The simplifications introduced in the preceding section finally led to a manageable model of probabilistic abductive inference. However, the assumptions needed to arrive at this model are hardly acceptable. In practice we rarely encounter a problem in which all independence assumptions are satisfied. To cope with this problem, the idea suggests itself to search for a model, in which we can take dependences into account, but nevertheless can exploit all existing independences to reduce the amount of storage needed and to make inferences in complex domains tractable. Probabilistic networks are such a model.

6.1 Decomposition and Abductive Reasoning

Stated as concisely as possible, the basic ideas underlying probabilistic networks are these: Under certain conditions a probability distribution P on a multi-dimensional domain, which encodes *prior* or *generic knowledge* about this domain, can be decomposed into a set $\{P_1, \ldots, P_n\}$ of probability distributions on lower-dimensional subspaces. This decomposition is based on dependence and independence relations between the attributes of the domain. If such a decomposition is possible, it is sufficient to know the distributions on the subspaces to compute all probabilities that can be computed using the original distribution P. Since such a decomposition is usually represented as a network (or graph), it is commonly called a *probabilistic network* or a *probabilistic graphical model.* Reasoning in such a network consists in conditioning the represented probability distribution on the observed values of some attributes.

To elaborate a little: By *multi-dimensional domain* we mean that each state of the section of the world to be modeled can be described by stating the values of a finite set $V = \{A_1, \ldots, A_m\}$ of attributes. Each such attribute—or, more precisely, the set $\mathrm{dom}(A_k) = \{a_{k,1}, \ldots, a_{k,n_k}\}$ of its possible values—forms a dimension of the domain. Of course, to form a dimension, the possible values have to be *exhaustive* and *mutually exclusive.* With these restrictions the *universe of discourse* or *frame of discernment* is the multi-dimensional domain

$\Omega = \mathrm{dom}(A_1) \times \cdots \times \mathrm{dom}(A_m)$. Each possible state of this domain is described by a tuple $\omega = (a_{1,i_1}, \ldots, a_{m,i_m})$ containing the values which the attributes in V assume for this state.

The *probability distribution* P on Ω assigns to each point $\omega \in \Omega$ the (prior) probability that the modeled section of the world is in a state corresponding to that point. These probabilities are usually estimated by human domain experts or computed from a statistical analysis of available data.

By *decomposition* we mean that the probability distribution P on the domain as a whole can be reconstructed from the distributions $\{P_1, \ldots, P_r\}$ on lower-dimensional subspaces, where a subspace $\Omega_W \subseteq \Omega$ is the joint domain of a subset $W \subseteq V$ of attributes, i.e., $\Omega_W = \times_{A \in W} \mathrm{dom}(A)$. The distributions $\{P_1, \ldots, P_r\}$ on the subspaces assign conditional or marginal probabilities (depending on the network type—see below) to projections of the tuples $\omega \in \Omega$ to the corresponding subspaces. From these probabilities the original probability distribution P on Ω can be recomputed.

A decomposition of a probability distribution has several advantages, the most important being that it can usually be stored much more efficiently and with less redundancy than the original distribution. However, just being able to store a probability distribution more efficiently would not be of much use for reasoning tasks, were it not for the possibility to draw inferences in the underlying domain using only the distributions $\{P_1, \ldots, P_r\}$ on the subspaces without having to reconstruct the original distribution P. If we have obtained *evidential knowledge* about the current state ω_0 of the domain under consideration, which consists in observations of the values of some of the attributes, we can *condition* the represented probability distribution on the observed values by passing the conditioning information from subspace distribution to subspace distribution until all have been updated. This process is usually called *evidence propagation*.

Obviously, mapping our general model of abductive inference to probabilistic networks is—for the most part—very simple. When considering the interpretation of subsets H and D of hypotheses and possible observations above, we already mentioned that it is useful to form groups of mutually exclusive and exhaustive statements to support checking for the requirement that these subsets are consistent. These groups of statements correspond to the attributes considered in probabilistic networks. We only have to join the sets H_{all} and D_{all} to form one universe of discourse. (If we need the information where the attributes came from, we can keep it in attribute markers.) The hypothesis assessment function pl corresponds directly to the probability distribution P on the domain, since from this probability distribution we can compute the probability $P(H|D)$ for all sets H and D. Thus the decomposition can be used to simplify the representation of the hypothesis assessment function pl. The observed data D_{obs} corresponds to the evidential knowledge. The only element of an abductive problem for which there is no direct analog is the explanation relation e, which identifies the acceptable explanations.

In the following sections we first review the theory of probabilistic networks in more detail and later we turn to the problem of hypothesis selection.

6.2 Conditional Independence

Whether and how a given probability distribution P can be be decomposed into a set $\{P_1, \ldots, P_r\}$ of distributions on subspaces is determined by the dependence structure of the attributes of the domain Ω underlying P. It is clear that attributes that directly depend on each other must appear together at least once in a distribution of the decomposition. A separation is possible only, if they are *conditionally independent* [Dawid, 1979; Pearl, 1992]. That is, if two attributes get independent, if certain other attributes are fixed, their dependence is not genuine, but mediated through other attributes. In this case these two attributes need not appear together in a distribution of the decomposition, but it suffices that there is a "path" connecting them through the mediating attributes. (This already suggests that a decomposition may be representable as a network or graph.)

Conditional probabilistic independence is defined in accordance with the usual notion of stochastic independence as follows:

DEFINITION 9. Let X, Y, and Z be three disjoint subsets of attributes in V. X is called **conditionally independent** of Y given Z w.r.t. P, written $X \perp\!\!\!\perp Y \mid Z$, iff

$$\forall \omega \in \Omega: \quad P(\omega_{X \cup Y} \mid \omega_Z) = P(\omega_X \mid \omega_Z) \cdot P(\omega_Y \mid \omega_Z)$$

whenever $P(\omega_Z) > 0$.

Here a conditional probability distribution is defined in the usual way, i.e., as

$$P(\omega_X \mid \omega_Y) = \frac{P(\omega_{X \cup Y})}{P(\omega_Y)}.$$

The notion of conditional independence provides, as already mentioned, the connection to a graph representation. It has been shown in general that a notion of conditional independence satisfying certain axioms, which are known as the *semi-graphoid axioms* [Dawid, 1979; Spohn, 1980; Pearl and Paz, 1987; Smith, 1989], can be used to define a graph structure on the set of attributes. These axioms are

symmetry: $(X \perp\!\!\!\perp Y \mid Z) \Longrightarrow (Y \perp\!\!\!\perp X \mid Z)$

decomposition: $(W \cup X \perp\!\!\!\perp Y \mid Z) \Longrightarrow (W \perp\!\!\!\perp Y \mid Z) \wedge (X \perp\!\!\!\perp Y \mid Z)$

weak union: $(W \cup X \perp\!\!\!\perp Y \mid Z) \Longrightarrow (X \perp\!\!\!\perp Y \mid Z \cup W)$

contraction: $(W \perp\!\!\!\perp Y \mid Z) \wedge (X \perp\!\!\!\perp Y \mid Z \cup W) \Longrightarrow (W \cup X \perp\!\!\!\perp Y \mid Z)$

The *symmetry* axiom states that in any state of knowledge Z, if Y tells us nothing new about X, then X tells us nothing new about Y. The *decomposition* axiom asserts that if two combined items of information are judged irrelevant to X, then each separate item is irrelevant as well. The *weak union* axiom states that learning irrelevant information W cannot help the irrelevant information Y become relevant to X. The *contraction* axiom states that if we judge W irrelevant to X after learning some irrelevant information Y, then W must have been irrelevant before we learned Y. Together the weak union and contraction properties mean that irrelevant information should not alter the relevance of other propositions in the system; what was relevant remains relevant, and what was irrelevant remains irrelevant [Pearl, 1992]. It is plausible that any reasonable notion of conditional independence should satisfy these axioms and, indeed, probabilistic conditional independence does.

6.3 Graph Representation

Given an appropriate notion of conditional independence, i.e., one that satisfies the semi-graphoid axioms, a *conditional independence graph* $G = (V, E)$ can be defined. Each node of this graph corresponds to an attribute of the underlying domain. The topology of the graph, i.e., which edges are present and which are missing, is used to represent an independence model, i.e., a set of conditional independence statements, of the domain under consideration [Pearl, 1992; Spirtes *et al.*, 1993]: The conditional independence of two sets of attributes given a third is expressed by *separation* of the corresponding node sets by the nodes that correspond to the conditioning attributes.

What is to be understood by "separation" depends on whether the graph is chosen to be directed or undirected. If it is undirected, separation is defined as follows: If X, Y, and Z are three disjoint subsets of nodes in an undirected graph, then Z separates X from Y, iff after removing the nodes in Z and their associated edges from the graph there is no path, i.e., no sequence of consecutive edges, from a node in X to a node in Y. Or, in other words, Z separates X from Y, iff all paths from a node in X to a node in Y contain a node in Z.

For directed graphs, which have to be acyclic, the so-called *d-separation criterion* is used [Pearl, 1992; Verma and Pearl, 1990]: If X, Y, and Z are three disjoint subsets of nodes in a directed acyclic graph (DAG), then Z is said to *d-separate* X from Y, iff there is no path, i.e., no sequence of consecutive edges (of any directionality), from a node in X to a node in Y along which the following two conditions hold:

1. every node with converging edges either is in Z or has a descendant in Z,
2. every other node is not in Z.

With the described notions of separation, we can define the so-called *Markov properties* of graphs [Whittaker, 1990]:

pairwise: Attributes, whose nodes are non-adjacent in the graph, are conditionally independent given all remaining attributes.

local: Given the attributes of the adjacent nodes (the neighbors), an attribute is conditionally independent of all remaining attributes.

global: Any two subsets of attributes, whose corresponding node sets are separated by a third node set, are conditionally independent given the attributes corresponding to the nodes in the third set.

Note that the local Markov property is contained in the global, and the pairwise Markov property in the local.

Since the pairwise Markov property refers to the independence of only two attributes, it would be most natural (at least for undirected graphs) to use it to define an independence graph: If two attributes are dependent given all other attributes, there is an edge between their corresponding nodes, otherwise there is no edge [Whittaker, 1990]. But, unfortunately, the three types of Markov properties are not equivalent in general, and it is obvious that we need the *global* Markov property for inferences from multiple observations. However, the above definition can be used, if—in addition to the semi-graphoid axioms—the following axiom holds:

intersection: $(W \perp\!\!\!\perp Y \mid Z \cup X) \wedge (X \perp\!\!\!\perp Y \mid Z \cup W) \Longrightarrow (W \cup X \perp\!\!\!\perp Y \mid Z)$

The semi-graphoid axioms together with this one are called the *graphoid axioms*. If they hold for a given notion of conditional independence, an independence graph can be defined via the pairwise Markov condition, since the intersection axiom allows us to infer the global Markov property from the pairwise. If the intersection axiom does not hold, the global Markov property has to be used to define an independence graph. If can be shown that a strictly positive probability distribution satisfies the intersection axiom [Pearl, 1992] and therefore the probability distribution on the domain to be modeled is often required to be strictly positive.

It is clear that an independence graph for a given domain is easy to find. For example, the complete undirected graph, i.e., the graph in which every node is connected directly to every other, always is an independence graph. Similarly, a directed acyclic graph which becomes a complete undirected graph if the directions of the edges are neglected is a trivial independence graph. However, using such graphs would not reduce the amount of data that needs to be stored (see below). Therefore we have to add the condition that the independence graph has to be *sparse* or even *minimal*, i.e., should contain as few edges as possible.

Whether directed acyclic graphs or undirected graphs are used is, to a certain degree, a matter of taste. However, it should be noted that the two types of graphs represent conditional independence relations in fundamentally different ways. There are undirected graphs that represent a set of conditional independence statements that cannot be represented by a single directed acyclic graph, and vice versa.

6.4 *Factorization*

The conditional independence graph is also called the *qualitative* part of a probabilistic network, since it specifies which attributes are dependent and which are independent, but not the details of the dependences. These are represented in the *quantitative* part of a probabilistic network which consists of a set of probability distributions. W.r.t. the quantitative part of a probabilistic network the conditional independence graph describes a *factorization* of the joint probability distribution P on Ω. The exact representation of the quantitative information and the factorization formula depends on the type of the conditional independence graph.

Bayesian networks. The most popular kind of probabilistic networks in artificial intelligence is the *Bayesian network*, also called *belief network* [Pearl, 1986; Pearl, 1992]. A Bayesian network consists of a directed acyclic graph and a set of conditional probability distributions $P(\omega_A \mid \omega_{\text{parents}(A)})$, $A \in V$, where parents(A) is the set of attributes corresponding to the parents of the node that corresponds to attribute A. That is, there is one probability distribution for each attribute and each distinct instantiation of its parent attributes in the graph. If an attribute does not have any parents, its associated distribution is simply an unconditional distribution.

A Bayesian network describes a factorization of a strictly positive joint probability distribution P on Ω w.r.t. a directed acyclic graph into a set of conditional probability distributions according to

$$\forall \omega \in \Omega: \quad P(\omega) = \prod_{A \in V} P(\omega_A \mid \omega_{\text{parents}(A)}).$$

These equations can easily be derived from the (generally valid) *chain rule of probability*,

$$\forall \omega \in \Omega: \quad P(\omega) = \prod_{i=1}^{m} P(\omega_{A_i} \mid \omega_{\{A_1,\ldots,A_{i-1}\}}),$$

and a set of conditional independence statements. The reason is that a conditional independence statement $X \perp\!\!\!\perp Y | Z$ implies

$$P(\omega_X \mid \omega_{Y \cup Z}) = P(\omega_X \mid \omega_Z).$$

Therefore we can cancel from the conditions all attributes of which the attribute A_i is independent given the remaining attributes and thus arrive at the factorization formula shown above. It should be noted, though, that the achievable simplification may depend on the order of the attributes.

Markov networks. An alternative type of probabilistic networks uses undirected graphs and is called a *Markov network* [Lauritzen and Spiegelhalter, 1988; Pearl, 1992]. It represents so-called Markov random fields. Similar to a Bayesian network a Markov network describes a factorization of the joint probability distribution P on Ω, but it uses a *potential representation*: A strictly positive probability distribution P on Ω factorizes w.r.t. an undirected graph $G = (V, E)$, iff

$$\forall X \in \text{cliques}(G): \exists \phi_X: \forall \omega \in \Omega: \quad P(\omega) = \prod_{X \in \text{cliques}(G)} \phi_X(\omega_X),$$

where cliques(G) is the set of all maximal cliques of G, each of which is represented by the set of attributes whose corresponding nodes are contained in it. The *factor potentials* ϕ_X are strictly positive functions defined on Ω_X, $X \subseteq V$.

Whether a Bayesian network or a Markov network is chosen to model a given domain is, as already indicated above, to a certain degree a matter of taste. However,

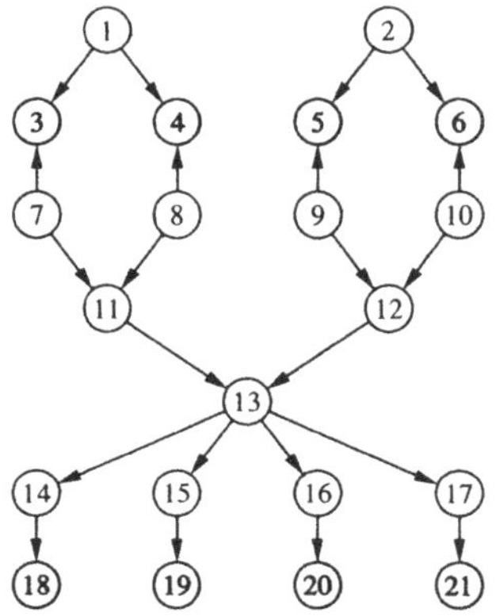

21 attributes:
1 – dam correct?
2 – sire correct?
3 – stated dam ph.gr. 1
4 – stated dam ph.gr. 2
5 – stated sire ph.gr. 1
6 – stated sire ph.gr. 2
7 – true dam ph.gr. 1
8 – true dam ph.gr. 2
9 – true sire ph.gr. 1
10 – true sire ph.gr. 2
11 – offspring ph.gr. 1
12 – offspring ph.gr. 2
13 – offspring genotype
14 – factor 40
15 – factor 41
16 – factor 42
17 – factor 43
18 – lysis 40
19 – lysis 41
20 – lysis 42
21 – lysis 43

The grey nodes correspond to observable attributes.

Figure 2. Conditional independence graph of a graphical model for genotype determination and parentage verification of Danish Jersey cattle in the F-blood group system [Rasmussen, 1992].

since a Bayesian network is a directed graph, it is well-suited to represent direct causal dependencies between variables: Often we can choose the directions of the edges (and thus the "directions" of the conditional probabilities) in such a way that they coincide with the direction of the causal influence. This is quite natural for knowledge representation, especially in applications for diagnostic reasoning, i.e., abductive inference, and thus one may prefer Bayesian network models. However, the causal interpretation of Bayesian networks should be handled with care, since it involves strong assumptions about the statistical manifestation of causal dependences [Borgelt and Kruse, 1999].

6.5 An Example Network

As an example of a probabilistic network we consider an application for blood group determination of Danish Jersey cattle in the F blood group system, the primary purpose of which is parentage verification for pedigree registration [Rasmussen, 1992]. The underlying domain is described by 21 attributes, eight of which are observable. The size of the domains of these attributes ranges from two to eight possible values and the total frame of discernment has $2^6 \cdot 3^{10} \cdot 6 \cdot 8^4 =$ 92 876 046 336 possible states. This space is clearly much too large to be handled as a whole and therefore a decomposition of the expert knowledge about this domain is necessary to make reasoning feasible. Figure 2 lists the attributes and shows the conditional independence graph of this graphical model, which was designed by human domain experts (the graphical model is a Bayesian network and thus the conditional independence graph is a directed acyclic graph). The grey nodes correspond to the observable attributes.

Table 4. Conditional probability distributions for a subgraph of the conditional independence graph shown in Figure 2.

sire correct	phenogroup true sire	phenogroup stated sire F1	V1	V2
yes	F1	1	0	0
yes	V1	0	1	0
yes	V2	0	0	1
no	F1	0.58	0.10	0.32
no	V1	0.58	0.10	0.32
no	V2	0.58	0.10	0.32

In this example, a decomposition of the joint probability distribution according to the conditional independence graph shown in figure 2 leads to a considerable simplification. Instead of having to determine the probability of each of the 92 876 046 336 elements of the 21-dimensional frame of discernment Ω, only 306 conditional probabilities in subspaces of at most three dimensions need to be specified. An example of a conditional probability table is shown in table 4, which is for the phenogroup 1 of the stated sire of a given calf conditioned on the phenogroup of the true sire of the calf and whether the sire was correctly identified. The numbers in this table are derived from statistical data and the experience of domain experts. The family of all 21 conditional probability tables forms the quantitative part of the graphical model for the Danish Jersey cattle example.

6.6 *Evidence Propagation*

After a probabilistic network has been constructed, it can be used to do reasoning. In the Danish Jersey cattle example, for instance, the phenogroups of the stated dam and the stated sire can be determined and the lysis values of the calf can be measured. With this information it becomes possible to assess whether the stated parents of the calf are the true parents or whether the breeder has made a mistake (or has attempted to deceive the registration office).

However, reasoning in a probabilistic network is not always completely straightforward. Considerations of efficiency make it often advisable to transform a graphical model into a form that is better suited for propagating the evidential knowledge and computing the resulting marginal distributions for the unobserved attributes. We briefly sketch here a popular efficient reasoning method known as *clique tree propagation* (CTP) [Lauritzen and Spiegelhalter, 1988; Castillo *et al.*, 1997], which involves transforming the conditional independence graph into a clique tree. A well-known interactive software tool for probabilistic reasoning in clique trees is HUGIN [Andersen *et al.*, 1989].

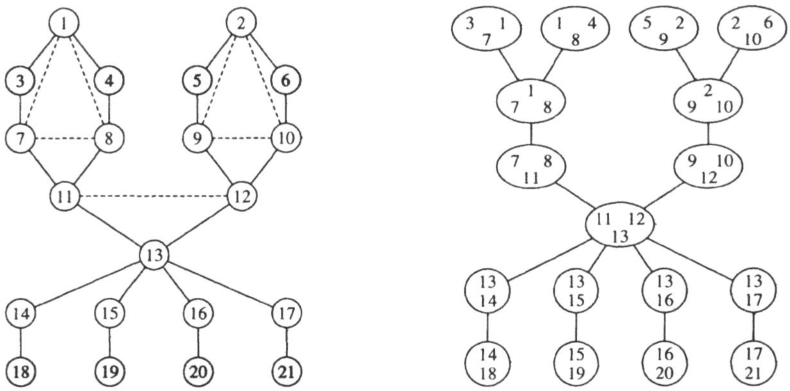

Figure 3. Triangulated moral graph (left) and clique tree (right) for the graphical model shown in Figure 2. The dotted lines are the edges added when parents were "married".

The transformation into a clique tree is carried out as follows: If the conditional independence graph is a directed acyclic graph, it is first turned into an undirected graph by constructing its associated *moral graph* [Lauritzen and Spiegelhalter, 1988]. A moral graph is constructed from a directed acyclic graph by "marrying" the parent nodes of all nodes (hence the name "moral graph"). This is done by simply adding undirected edges between the parents. The directions of all other edges are discarded. In general the moral graph satisfies only a subset of the independence relations of the underlying directed acyclic graph, so that this transformation may result in a loss of independence information. The moral graph for the Danish Jersey Cattle example is shown on the left in figure 3. The edges that were added when parents were "married" are indicated by dotted lines.

In a second step, the undirected graph is triangulated. (If the conditional independence graph is an undirected graph right from the start, this is the first step to be carried out.) An undirected graph is called *triangulated*, if all cycles containing at least four nodes have a chord, where a chord is an edge that connects two non-adjacent nodes of the cycle. To achieve triangulation, it may be necessary to add edges, which may result in a (further) loss of independence information. In the Danish Jersey cattle example, however, the moral graph shown on the left in figure 3 is already triangulated, so no new edges need to be introduced.

Finally, the triangulated graph is turned into a clique tree by finding the maximal cliques, where a clique (see above) is a fully connected subgraph, and it is maximal, if it is not contained in another clique. In the clique tree there is one node for each maximal clique of the triangulated graph and its edges connect nodes that represent cliques having attributes in common. It should be noted that in general the clique tree is not unique, because often different sets of edges can be chosen.

The clique tree for the Danish Jersey cattle example is shown on the right in figure 3. Detailed information on triangulation, clique tree construction and other related graph-theoretical problems can be found in [Castillo *et al.*, 1997].

The quantitative part of a graphical model, of course, has to be transformed, too. From the quantitative information of the original graphical model one has to compute a marginal distribution for each of the subspaces represented by the nodes of the clique tree. For the Danish Jersey cattle example, we have to compute a marginal distribution for the subspace formed by the attributes 1, 3, 7, one for the subspace formed by the attributes 1, 4, 8, and so on.

Having constructed a clique tree, which is merely a preliminary operation to make evidence propagation more efficient, we can finally turn to evidence propagation itself. Evidence propagation in clique trees is basically an iterative extension and projection process. When evidence about the value of an attribute becomes available, it is first extended to a clique tree node the attribute is contained in. This is done by conditioning the associated marginal distribution. We call this an extension, since by this conditioning process we go from restrictions on the values of a single attribute to restrictions on tuples of attribute values. Hence the information is extended from a single attribute to a subspace formed by several attributes. Then the conditioned distribution is projected to all intersections of the clique tree node with other nodes. Via these projections the information can be transferred to other nodes, where the process repeats: First it is extended to the subspace represented by the node, then it is projected to the intersections connecting it to other nodes. The process stops when all nodes have been updated.

The propagation scheme outlined above and the subsequent computation of posterior marginal distributions for the unobserved attributes can easily be implemented by locally communicating node- and edge-processors. These processor also serve the task to let pieces of information "pass" each other without interaction. Such bypassing is necessary, if the propagation operations in the underlying uncertainty calculus are not idempotent, that is, if incorporating the same information twice can invalidate the results. This is the case in probabilistic reasoning. This problem is also the reason why the clique graph is usually required to be a *tree*: If there were loops, information could travel on two or more different paths to the same destination and thus be incorporated twice.

Of course, clique tree propagation is not the only known evidence propagation scheme. Others include bucket elimination [Dechter, 1996; Zhang and Poole, 1996] and iterative proportional fitting [Whittaker, 1990]. Commonly used evidence propagation algorithms differ from each other w.r.t. the network structures they support. For example, both bucket elimination and iterative proportional fitting can also be used with networks that contain cycles.

6.7 *Hypothesis Selection*

As already indicated at the beginning of this section, probabilistic networks mainly provide means to represent the hypothesis assessment function pl. However, in a probabilistic network there is no direct analog to the explanation relation e of our general model of an abductive problem. That is, with a probabilistic network, we can easily compute the plausibility of a given hypothesis, but we cannot check whether the hypothesis is semantically acceptable.

In applications tailored for a specific application this is often irrelevant, because from the application it can be clear what attributes we are interested in and therefore we only have to compute the most probable tuple of values for the subspace formed by these attributes. As an example consider again the Danish Jersey Cattle example discussed above. Its primary purpose is, as already mentioned, parentage verification. That is, we are interested in the values of two variables, namely "Dam correct?" and "Sire correct?". All other unobservable variables are only there to structure the available expert knowledge about the domain. For the two attributes, however, which indicate whether the stated dam is the true dam and whether the stated sire is the true sire, we compute the probability of all possible combinations and select the most probable one. Actually, in the original network there is an additional node "Parent correct?" which combines these hypotheses in a single attribute. In the discussion above, we discarded this node, because it does not carry real information, but only makes the clique tree unnecessarily complex. In cases like these, where a set of attributes can be fixed in advance, a probabilistic network approach to abductive reasoning is most powerful.

To identify the most probable tuple in a subspace formed by a set of attributes, may not always be appropriate, though. If, for instance, not all acceptable (compound) hypotheses consist of the same number of atomic hypotheses, we cannot use it, because it may result in hypotheses too specific for a given problem. However, even in this case the probabilistic network alone may contain enough information to select the best acceptable hypothesis. For example, the structure of the network can provide information how to restrict the set of attributes we have to take into account to form a (compound) hypothesis. Obviously, it is sufficient to select a set of explanatory attributes (i.e., attributes derived from H_{all}) in such a way that the observed attributes and the remaining attributes are conditionally independent given the selected attributes. A (compound) hypothesis formed from these attributes has to be considered complete because due to the interpretation of the semi-graphoid axioms (see above) the remaining attributes are irrelevant for the observations. Of course, such a restriction does not guarantee that the selected (compound) hypothesis is semantically acceptable, but it may help to restrict the set of hypotheses one has to consider. The idea can be enhanced by the requirement that a reasonable hypothesis should make the observed data more likely than it is without it or that the observed data should make an acceptable hypothesis more likely. However, as we showed with a simple example above, a higher likelihood is not sufficient for a semantically acceptable hypothesis.

If these approaches, which try to do without additional information, are not feasible, we have to add some structure to represent (a simplification of) the explanation relation e. A very manageable structure results, if we have an independent abductive problem and thus can represent the relation e as a simple table with

H_{all} lines and D_{all} columns. In this case the probabilistic network helps to avoid the strong probabilistic independence assumptions underlying D-independent and HD-independent abductive problems. We only need the logical independence assumptions needed to simplify the representation of the relation e. If even these logical assumptions are too strong, we can enhance the table of the relation e for an independent abductive problem by an explicit list of (compound) hypotheses, for which "constructive" or "destructive inference" occurs, i.e., those (compound) hypotheses which can explain more than the sum of their elements and those, which can explain less (examples of such situation we gave above). Provided this list is of moderate size, the problem remains tractable. A more detailed discussion can be found in [Bylander *et al.*, 1991; Borgelt, 1992].

6.8 *Learning from Data*

A probabilistic network is a powerful tool to support reasoning—as soon as it is constructed. Its construction by human experts, however, can be tedious and time consuming. Therefore recent research in probabilistic graphical models focused on learning them from a database of sample cases. In accordance with the two components of graphical models, one distinguishes between *quantitative* and *qualitative* (or *structural*) *network induction.*

Quantitative network induction for a given network structure consists in estimating the joint probability distribution P, where P is selected from a family of parameterized probability distributions. A lot of approaches have been developed in this field, using methods such as maximum likelihood, maximum penalized likelihood, or fully Bayesian approaches, which involve different computational techniques of probabilistic inference such as the expectation maximization (EM) algorithm, Gibbs sampling, Laplace approximation, and Monte Carlo methods. For an overview, see [Buntine, 1994; Spiegelhalter *et al.*, 1993].

Qualitative network induction consists in learning a network structure from a database of sample cases. In principle one could use the factorization property of a probabilistic network to evaluate its quality by comparing for each $\omega \in \Omega$ the probability computed from the network with the relative frequency found in the database to learn from. However, this approach is usually much too costly.

Other methods are based on linearity and normality assumptions [Pearl and Wermuth, 1993], rely on the extensive testing of conditional independences (CI tests) [Verma and Pearl, 1992], or use a Bayesian approach [Cooper and Herskovits, 1992; Lauritzen *et al.*, 1993]. Unfortunately, the first group is fairly restrictive, CI tests tend to be unreliable unless the volume of data is enormous, and with an increasing number of vertices they soon become computationally intractable. Bayesian learning requires debatable prior assumptions (for example, default uniform priors on distributions, uniform priors on the possible graphs) and also tends to be inefficient unless greedy search methods are used.

Nevertheless, several network induction algorithms have successfully been applied. The oldest example is an algorithm to decompose a multi-variate probability distribution into a tree of two-dimensional distributions [Chow and Liu, 1968]. It uses mutual information as an evaluation measure and optimum weight spanning tree construction as a search method.

Another example is the $K2$ algorithm [Cooper and Herskovits, 1992], which uses a greedy parent search and a Bayesian evaluation measure. Its drawback, which consists in the fact that it needs a topological order of the attributes, can be overcome by a hybrid algorithm [Singh and Valtorta, 1993], which combines CI tests (to find a topological order) and $K2$ (to construct the Bayesian network with respect to this topological order). Unfortunately, $K2$ can deal only with complete and precise data. The treatment of missing values and hidden variables is clear only from a theoretical point of view [Cooper and Herskovits, 1992].

A third algorithm, which uses a backward search strategy, has been described in [Højsgaard and Thiesson, 1994].

Several evaluation measures, which can be used with optimum weight spanning tree construction and greedy parent search as well as other search methods, are surveyed in [Borgelt and Kruse, 1997].

7 SUMMARY

In this paper we considered how probabilistic networks can support abductive reasoning. Starting from a definition of an abductive inference as a reductive, i.e., explanatory inference whose conclusion is a particular statement, we showed how probability theory enters ours consideration due to two reasons: In the first place, if we want to handle real world problems, we have to take into account statistical conditionals. Secondly, in order to reduce the chances of an incorrect result, we have to assess and compare the conclusions of abductive inferences. Based on a general model of abductive inference we showed that a direct approach to represent a hypothesis assessment function is not feasible and thus simplifications are required. Although straightforward simplifications lead to a manageable model, they involve strong presuppositions which cannot reasonably be expected to hold in applications. As a solution probabilistic networks suggest themselves as well-established technique to decompose a multi-variate probability distribution in order to make reasoning in high-dimensional domains possible. They are very well-suited to represent the hypothesis assessment function of abductive problem solving. However, it may be necessary to enhance them by a method to identify the acceptable hypotheses, because the raw probabilistic information they represent is often not sufficient for this task.

Department of Knowledge Processing and Language Engineering, Otto-von-Guericke-University of Magdeburg, Germany.

REFERENCES

[Andersen *et al.*, 1989] S. K. Andersen, K. G. Olesen, F. V. Jensen and F. Jensen. HUGIN — a shell for building Bayesian belief universes for expert systems. *Proceedings of the 11th International Joint Conference on Artificial Intelligence (IJCAI'89, Detroit, MI)*, 1080–1085. Morgan Kaufman, San Mateo, CA, 1989.

[Baldwin *et al.*, 1995] J. F. Baldwin, T. P. Martin and B.W. Pilsworth. *FRIL — Fuzzy and Evidential Reasoning in Artificial Intelligence.* Research Studies Press/J. Wiley & Sons, Taunton/Chichester, 1995.

[Bochenski, 1954] I. M. Bocheński. *Die zeitgenössischen Denkmethoden.* Franke-Verlag, Tübingen, 1954.

[Borgelt, 1992] C. Borgelt. *Konzeptioneller Vergleich verschiedener numerischer und logischer Ansätze abduktiver Inferenz.* Diplomarbeit, TU Braunschweig, 1992.

[Borgelt and Kruse, 1997] C. Borgelt and R. Kruse. Evaluation measures for learning probabilistic and possibilistic networks. *Proceedings of the 6th IEEE International Conference on Fuzzy Systems (FUZZ-IEEE'97, Barcelona)*, Vol. 2:1034–1038. IEEE Press, Piscataway, NJ, 1997.

[Borgelt and Kruse, 1999] C. Borgelt and R. Kruse. A Critique of Inductive Causation. *Proc. 5th European Conf. on Symbolic and Quantitative Approaches to Reasoning and Uncertainty (ECSQARU'99, London), LNAI 1638*, 68–79. Springer, Heidelberg, 1999.

[Buntine, 1994] W. Buntine. Operations for learning with graphical models. *Journal of Artificial Intelligence Research* 2:159–225, 1994.

[Bylander *et al.*, 1991] T. Bylander, D. Allemang, M. C. Tanner and J. R. Josephson. The computational complexity of abduction. *Artificial Intelligence* 49: 25–60. North-Holland, Amsterdam, 1991.

[Castillo *et al.*, 1997] E. Castillo, J. M. Gutierrez and A. S. Hadi. *Expert Systems and Probabilistic Network Models.* Springer, New York, NY, 1997.

[Charniak and McDermott, 1985] E. Charniak and D. McDermott. *Introduction to Artificial Intelligence.* Addison-Wesley, Reading, MA, 1985.

[Chow and Liu, 1968] C. K. Chow and C. N. Liu. Approximating discrete probability distributions with dependence trees. *IEEE Trans. on Information Theory* 14(3):462–467. IEEE Press, Piscataway, NJ, 1968.

[Cooper and Herskovits, 1992] G. F. Cooper and E. Herskovits. A Bayesian method for the induction of probabilistic networks from data. *Machine Learning* 9:309–347. Kluwer, Dordrecht, 1992.

[Dawid, 1979] A. Dawid. Conditional independence in statistical theory. *SIAM Journal on Computing* 41:1–31, 1979.

[Dechter, 1996] R. Dechter. Bucket elimination: a unifying framework for probabilistic inference. *Proceedings of the 12th Conference on Uncertainty in Artificial Intelligence (UAI'96, Portland, OR)*, 211–219. Morgan Kaufman, San Mateo, CA, 1996.

[Dubois and Prade, 1988] D. Dubois and H. Prade. *Possibility Theory.* Plenum Press, New York, NY, 1988.

[Hempel, 1966] C. G. Hempel. *Philosophy of Natural Science.* Prentice-Hall, Englewood Cliffs, NJ, 1966.

[Højsgaard and Thiesson, 1994] S. Højsgaard and B. Thiesson. BIFROST — block recursive models induced from relevant knowledge, observations, and statistical techniques. *Computational Statistics and Data Analysis*, 1994.

[Josephson and Josephson, 1996] J. R. Josephson and S. G. Josephson. *Abductive Inference — Computation, Philosophy, Technology.* Cambridge University Press, Cambridge, MA, 1996.

[Kruse *et al.*, 1991] R. Kruse, E. Schwecke, and J. Heinsohn. *Uncertainty and Vagueness in Knowledge-based Systems: Numerical Methods (Series: Artificial Intelligence).* Springer, Berlin, Germany 1991

[Lauritzen and Spiegelhalter, 1988] S. L. Lauritzen and D. J. Spiegelhalter. Local computations with probabilities on graphical structures and their application to expert systems. *Journal of the Royal Statistical Society, Series B*, 2(50):157–224. Blackwell, Oxford, 1988.

[Lauritzen *et al.*, 1993] S. L. Lauritzen, B. Thiesson and D. Spiegelhalter. Diagnostic systems created by model selection methods — a case study. *Proceedings of the 4th International Workshop on Artificial Intelligence and Statistics (Fort Lauderdale, FL)*, pp. 93–105, 1993.

[Losee, 1993] J. Losee. *A Historical Introduction to the Philosophy of Science (3rd edition).* Oxford University Press, Oxford, 1993.

[Patil *et al.*, 1982] R. S. Patil, P. Szolovits and W. B. Schwartz. Modeling knowledge of the patient in acid-base and electrolyte disorders. In: P. Szolovits, ed. *Artifical Intelligence in Medicine*, pp. 191–226. Westview Press, Boulder, CO, 1982.

[Pearl, 1986] J. Pearl. Fusion, Propagation, and Structuring in Belief Networks. *Artificial Intelligence* 29:241–288, 1986.

[Pearl, 1992] J. Pearl. *Probabilistic Reasoning in Intelligent Systems: Networks of Plausible Inference (2nd edition)*. Morgan Kaufman, San Mateo, CA, 1992.

[Pearl and Paz, 1987] J. Pearl and A. Paz. Graphoids: A Graph Based Logic for Reasoning about Relevance Relations. In: B. D. Boulay et al., eds. *Advances in Artificial Intelligence 2*, pp. 357–363. North Holland, Amsterdam, 1987.

[Pearl and Wermuth, 1993] J. Pearl and N. Wermuth. When can association graphs admit a causal interpretation? *Proceedings of the 4th International Workshop on Artificial Intelligence and Statistics (Fort Lauderdale, FL)*, pp. 141–150, 1993.

[Peirce, 1958] C. S. Peirce. (C. Hartshorne, P. Weiss and A. Burks, eds.) *Collected Papers of Charles Sanders Peirce*. Havard University Press, Cambridge, MA, 1958.

[Peng and Reggia, 1989] Y. Peng and J. A. Reggia. *Abductive Inference Models for Diagnostic Problem Solving*. Springer, New York, NY, 1989.

[Popper, 1934] K. R. Popper. *Logik der Forschung*. 1st edition: Julius Springer, Vienna, 1934. 9th edition: J. C. B. Mohr, Tübingen, 1989. English edition: *The Logic of Scientific Discovery*, Hutchinson, London, 1959.

[Rasmussen, 1992] L. K. Rasmussen. *Blood Group Determination of Danish Jersey Cattle in the F-blood Group System (Dina Research Report 8)*. Dina Foulum, Tjele, 1992.

[Salmon, 1973] W. C. Salmon. *Logic (2nd edition)*. Prentice-Hall, Englewood Cliffs, NJ, 1973.

[Savage, 1954] L. J. Savage. *The Foundations of Statistics*. J. Wiley & Sons, New York, NY, 1954.

[Singh and Valtorta, 1993] M. Singh and M. Valtorta. An algorithm for the construction of Bayesian network structures from data. *Proceedings of the 9th Conference on Uncertainty in Artificial Intelligence (UAI'93)*, pp. 259–265. Morgan Kaufman, San Mateo, CA, 1993.

[Smith, 1989] J. Q. Smith. Influence Diagrams for Statistical Modeling. *Annals of Statistics* 17(2):654–672, 1989.

[Spiegelhalter *et al.*, 1993] D. Spiegelhalter, A. Dawid, S. Lauritzen and R. Cowell. Bayesian Analysis in Expert Systems. *Statistical Science* 8(3):219–283, 1993.

[Spirtes *et al.*, 1993] P. Spirtes, C. Glymour and R. Scheines. *Causation, Prediction, and Search (Lecture Notes in Statistics 81)*. Springer, New York, NY, 1993.

[Spohn, 1980] W. Spohn. Stochastic independence, causal independence, and shieldability. *Journal of Philosophical Logic* 9:73–99, 1980.

[Thiele, 1997] W. Thiele. *Wasservögel und Strandvögel: Arten der Küsten und Feuchtgebiete*. BLV Naturführer, Munich, 1997.

[Verma and Pearl, 1990] T. S. Verma and J. Pearl. Causal networks: semantics and expressiveness. In: R. D. Shachter, T. S. Levitt L.N. Kanal, and J.F. Lemmer, eds. *Uncertainty in Artificial Intelligence 4*, pp. 69–76. North Holland, Amsterdam, 1990.

[Verma and Pearl, 1992] T. S. Verma and J. Pearl 1992. An algorithm for deciding if a set of observed independencies has a causal explanation. *Proceedings of the 8th Conference on Uncertainty in Artificial Intelligence (UAI'92, Stanford, CA)*, pp. 323–330. Morgan Kaufman, San Mateo, CA, 1992.

[Whittaker, 1990] J. Whittaker. *Graphical Models in Applied Multivariate Statistics*. J. Wiley & Sons, Chichester, 1990.

[Zhang and Poole, 1996] N. L. Zhang and D. Poole. Exploiting causal independence in Bayesian network inference.*Journal of Artificial Intelligence Research* 5:301–328, 1996.

JÖRG GEBHARDT

LEARNING FROM DATA: POSSIBILISTIC GRAPHICAL MODELS

1 INTRODUCTION

One of the major problems of handling imperfect information in knowledge-based systems is finding a computationally appealing description of the available data, which is both economical in using storage and supports efficient reasoning techniques. The existence of such a description strongly rests on whether dependencies among the data items are decomposable into local, more basic dependencies. The relational database model [Maier, 1983; Ullman, 1988; Ullman, 1989] has tackled this problem by storing a database in form of a lossless join decomposition, namely a collection of projections, from which the original relation can be reconstructed. Some recent work in this field concerns structure identification in relational data [Dechter and Pearl, 1992] and the clarification of cross-references to the solution of constraint-satisfaction problems [Gyssens *et al.*, 1994]. Although many problems in artificial intelligence, database theory, graph theory, and operations research are connected with decomposition problems, very few general results have been established so far. In the field of graphical modeling and its widespread application areas in diagnostics, expert systems, KDD systems, planning systems, data analysis, and control, the most advanced deliverables refer to probabilistic graphical models [Whittaker, 1990; Castillo *et al.*, 1997; Lauritzen, 1997]. In addition to efficient evidence propagation techniques [Pearl, 1986; Lauritzen and Spiegelhalter, 1988], there are many deliverables that refer to the induction of Bayesian networks or Markov networks from statistical data. For an overview, see [Buntine, 1994; Madigan and Raftery, 1994; Fisher and Lenz, 1996]. The corresponding algorithms are based on linearity and normality assumptions [Pearl and Wermuth, 1993], the extensive testing of conditional independence relations [Spirtes and Glymour, 1991; Verma and Pearl, 1992], or Bayesian approaches [Cooper and Herskovits, 1992; Lauritzen *et al.*, 1993]. Some crucial problems of these methods concern their computational complexity, their limited reliability unless the amount of data is enormous, and strong presuppositions like the requirement of an ordering of the nodes and a priori distribution assumptions. In order to overcome the complexity problems some heuristic approaches like, for example, the K2 algorithm [Cooper and Herskovits, 1992] with respect to a appropriate quality measure [Geiger and Heckerman, 1995] or the fusion of conditional independence tests with Bayesian learning [Singh and Valtorta, 1993] have turned out to be unavoidable.

Probabilistic graphical models provide a well-founded normative framework for knowledge representation and belief updating in the presence of uncertainty in precise data, but extending pure probabilistic settings to the treatment of imprecise

D. M. Gabbay and R. Kruse (eds.), *Abductive Reasoning and Learning*, 315–389.

(multi-valued) information, for obvious reasons by using families of probability distributions, restricts the computational tractability of inference mechanisms and structure induction algorithms. It therefore seems to be convenient to discuss appropriate strategies of information compression and to investigate non-standard uncertainty calculi in order to support efficient propagation without essentially affecting the expressive power and correctness of decision making procedures. Information compression is acceptable for systems that perform approximate reasoning, justified by a weak sensitivity with respect to an imperfect modeling of the system's behavior.

A quite general approach to uncertain reasoning under imprecision was presented by G. Shafer and P. Shenoy who proposed the concept of a valuation network [Shenoy, 1989; Shenoy and Shafer, 1990; Shenoy, 1993]. It covers upper and lower probabilities [Walley and Fine, 1982; Kyburg, 1987; Walley, 1991], Dempster-Shafer belief functions [Dempster, 1967; Dempster, 1968; Shafer, 1976; Kohlas and Monney, 1995; Shafer and Pearl, 1990; Smets and Kennes, 1994], and possibility theory [Zadeh, 1978; Dubois and Prade, 1988a; Dubois and Prade, 1991]. Nevertheless, quantitative learning of distribution functions as well as qualitative learning of dependency structures in these calculi has not been studied in much detail yet.

This chapter examines possibilistic graphical models. It therefore combines the benefits of the knowledge structuring properties of graphical models with possibility theory as the most promising candidate among the mentioned calculi to deal with the necessity of reasonable information compression in order to reduce complexity problems when treating uncertain and imprecise information together.

Section 2 is devoted to the theoretical foundations of possibility theory. As an alternative to standard approaches, where the concept of a possibility distribution is often introduced by possibility measures [Dubois and Prade, 1988a] or in terms of possibilistic logics [Dubois and Prade, 1992], we prefer a semantic background that is more on the line of interpreting a possibility distribution as an information-compressed representation of a database of (multi-valued) sample cases. The chosen underlying concept of an imperfect specification supports the comparison with other well-known calculi of uncertainty modeling such as the probabilistic and the belief function setting, so that special properties, benefits, and limits of each setting can be clarified. Additionally, the provided semantic underpinnings help in justifying how to operate on possibility distributions, with the result that Zadeh's extension principle, well-known for this purpose, can be verified as an extension theorem instead of only being a principle of generalizing set operations to the possibilistic framework.

Based on the provided theoretical background, section 3 deals with fundamental topics of possibilistic graphical modeling, which are possibilistic conditioning, possibilistic conditional independence, and the concept of a possibilistic conditional independence graph. Different from other views of possibility theory, where conditioning is sometimes interpreted as a special case of Dempster's rule, possibilistic conditional independence turns out to be strongly related to conditional possibilistic non-interactivity. Similar to the probabilistic setting, this concept leads to a factorization criterion for possibilistic conditional independence. On the other hand, a difference is that one has to deal with the problem that pairwise Markov property, local Markov property, and global Markov property are not equivalent when referred to possibility distributions, so that possibilistic conditional independence graphs are defined with respect to the global Markov property as the most restrictive one.

Section 4 presents a decomposition theorem for possibility distributions that can be regarded as a possibilistic counterpart of a reformulation of the Hammersley-Clifford Theorem for Markov Networks. Another main result of this section is the development of an algorithm for approximating tightest hypertree-structured decompositions of possibility distributions. Tightness is referred here to U-Uncertainty being a measure of nonspecificity for possibility distributions that can be justified by information-theoretical considerations. The algorithm is useful in order to learn possibilistic graphical models with hypertree structure from databases of (multi-valued) sample cases. For an algorithm that supports the induction of possibilistic networks with a directed acyclic conditional independence graph, the reader is referred to [Gebhardt and Kruse, 1996b].

Section 5 documents the applicability of the presented research results with respect to an example of blood-group determination and parentage verification for pedigree registration in modern cattle breeding programs. Results are obtained with the aid of the software tool POSSINFER, which is an implementation of the concepts and results of this contribution for learning and evidence propagation in possibilistic graphical models.

Note that in the following, we examine the concept of a possibility distribution in a way that is suitable for the understanding of possibilistic graphical models in a self-contained manner. On the other hand, the contribution is not intended to serve as an introduction to possibility theory, neither with respect to the aims of possibilistic logics in artificial intelligence, nor with respect to other interpretations such as, for example, the relationships to fuzzy set theory or Dempster-Shafer theory. For more details on possibility theory in this wider sense the reader is, for example, referred to [Dubois and Prade, 1988a; Kruse *et al.*, 1994].

2 THEORETICAL FOUNDATIONS OF POSSIBILITY THEORY

This section is devoted to some foundational issues on possibility theory as far as they are helpful for later investigations in the field of possibilistic graphical modeling. Similar to the variety of interpretations of subjective probabilities, there are different views that concern the meaning of a degree of possibility and a possibility distribution.

Referred to reasoning with possibilistic graphical models, especially learning such models from database of (multi-valued) sample case, it is near at hand to interpret possibility distributions as information-compressed specifications of uncertain object states of interest within a refined random set framework. In this connection, we profit from some earlier research results in the field of uncertainty modeling, namely the **context model** [Gebhardt and Kruse, 1993a; Gebhardt, 1992; Gebhardt and Kruse, 1997a] as a uniform framework for knowledge representation in the probabilistic setting, the belief function setting, and the possibilistic setting, respectively. Following this line, section 2.3 briefly introduces the basic concepts of the context model and its support to a comparative discussion of the mentioned uncertainty calculi.

Section 2.3 focuses in more detail on the semantic aspects of possibility theory. Section 2.4 prepares the consideration of reasoning with possibilistic graphical models by presenting some general results that refer to operating on possibility distributions. One of the essential deliverables is to confirm Zadeh's extension principle as the only way of generalizing set operations to operations on possibility distributions that is consistent with the context model and its desiderata.

2.1 On the Modeling of Imperfect Information

The last ten years of research in automated reasoning, oriented toward the implementation of common sense reasoning in artificial intelligence, made it apparent that probability theory, although the obvious candidate for a normative modeling of imperfect knowledge, is too limited to cover all faces of the representation and efficient handling of uncertain, imprecise and inconsistent data, as they occur in various application fields of industrial practice. New models have been developed, based either on **numerical methods** (e.g. the theory of evidence and possibility theory) or **symbolic methods** (e.g. non-monotonic logics, default logics, and autoepistemic logics). We will restrict ourselves to the numerical methods, since they are of major importance for graphical models being the main topic of the following sections.

Before using any numerical model of imperfect information, it is necessary to **explain the meaning of the numbers** that we are operating on, saying that we have to provide an interpretation of the corresponding quantification of uncertainty, for example (subjective) probabilities, degrees of belief, plausibility, necessity, possibility, and so forth. Furthermore, one has to **justify the choice of the semantic underpinnings** by natural requirements, for instance an underlying system of reasonable axioms. This also includes to check the validity and appropriateness of the derived model with respect to applicational needs, referred, for example, to the purpose of adequate knowledge representation, knowledge acquisition, inference mechanisms, and decision making.

The importance of such considerations has been shown impressively by the history of the so-called **certainty factor (CF) model** [Shortliffe and Buchanan, 1975]. It was one of the first approaches in the field of artificial intelligence that tackled the problem of representing and propagating uncertain expert knowledge by assigning numerical weights to logical formulas (rules and facts). In the CF model, the treatment of uncertain pieces of evidence during the reasoning process is quite simple and heuristic in nature. Especially the expert system MYCIN for medical diagnosis [Shortliffe, 1976; Buchanan and Shortliffe, 1984], but also some of its extensions and improvements, incorporated in the inference systems FLOPS [Buckley *et al.*, 1986], RUM [Bonissone *et al.*, 1987], and MILORD [Godo *et al.*, 1989], gave hints about the appropriateness of the CF model for particular domains of industrial applications. Unfortunately, by representing the whole operational behavior of certainty factors within a simulating probabilistic framework, it is possible to prove that the original definition of certainty factors is inconsistent with the method of parallel combination and the associated desiderata proposed in the CF model [Heckerman, 1986]. Additionally, if certainty factors are redefined in accordance with these desiderata, the treated evidence must satisfy some very restrictive assumptions on conditional independence, and the considered inference network has to be tree-structured. Such implicit assumptions hidden in the CF model are rarely true in practice, so that the success of MYCIN-like implementations can be explained by their simplicity and the fact that there seem to be many applications where the correctness of decisions is not very sensitive to violations of the required restrictions. It is quite easy to construct examples such that the inference mechanism used in the CF model produces completely wrong results. However, Howard's **influence diagrams** [Howard and Matheson, 1981] and Pearl's well-known **Bayesian networks** [Pearl, 1986; Pearl, 1992a] can be regarded as consistent generalizations of the CF model.

The tendency of tolerating intuitively appealing ad-hoc methods for the handling of complex inference problems was one of the major obstacles concerning the establishment of new approaches to the modeling of imperfect knowledge. For this reason, the different types of numerical calculi dealing with imperfection met with only little attention first, accompanied by controversial opinions on the utility of any non-probabilistic approach in this area [Lindley, 1982; Cheeseman, 1986]. As an impression, we refer to the polemic discussions on belief functions [Pearl, 1990; Pearl, 1992b; Smets, 1992] and fuzzy sets [Lindley, 1987; Elkan, 1994], respectively.

Meanwhile, due to the enormous industrial success of some of the new approaches (for instance, the world-wide marketing of concepts of **fuzzy control**, cf. [Kruse *et al.*, 1994]) and the improvement of theoretical issues (for example, the justification of fuzzy control as interpolation in vague environments on the base of similarity relations [Klawonn *et al.*, 1995]), several of the non-standard models for the treatment of imperfect information can nowadays be regarded as verified applicable tools, being cooperative with probabilistic approaches rather than competitive to them.

One of these non-standard tools is the **Dempster–Shafer theory of evidence** [Dempster, 1967; Dempster, 1968; Matheron, 1975; Shafer, 1976; Shafer and Pearl, 1990] and related approaches like **random set theory** [Nguyen, 1978], the **probability of provability** approach [Smets, 1993; Pearl, 1992a], and the **theory of hints** [Kohlas and Monney, 1995]. Others to be mentioned are **upper and lower probabilities** [Walley and Fine, 1982; Kyburg, 1987; Walley, 1991], the **trans-**

ferable belief model [Smets and Kennes, 1994], and **possibility theory** [Zadeh, 1978; Dubois and Prade, 1988a] with its relationship to **fuzzy set theory** [Zadeh, 1975; Kruse *et al.*, 1994].

As indicated in the introductory section, our investigations will be focused on the particular uncertainty calculus of possibility theory. In order to support a more formal discussion of semantical underpinnings, we will take advantage of the **context model** [Gebhardt and Kruse, 1993a; Gebhardt, 1992]. It provides suitable tools for a uniform consideration of concepts of the probabilistic setting, the belief function setting, and the possibilistic setting.

As an example, the context model has been applied to clarify the semantic basis for the correct or the inconsistent behavior of particular rules of parallel combination of pieces of uncertain evidence with respect to various numerical settings. An extensive study of this topic, which is quite important in the modern research fields of **data fusion** (cf. [Luo and Kay, 1995]) and **data mining** (cf. [Fayyad *et al.*, 1996]), can be found in [Gebhardt and Kruse, 1997a]. For the needs of graphical modeling the relevance of combination techniques is obvious when constructing a network with the aid of imperfect knowledge issued by different information sources such as sensors, experts, and databases of sample cases. We will not consider the specific problems of combination in more detail here, but restrict ourselves to those parts of the context model that help in understanding the basics of possibility theory and possibilistic graphical models.

Our view on the common purpose of the probabilistic, the belief function, and the possibilistic setting is to deal with uncertainty about the **current state** ω_0 of an object *obj* under consideration. As an example referred to database theory, suppose that *obj* is fully characterized by a finite set of attributes $A_1, \ldots, A_n$ with attached domains $\Omega_1, \ldots, \Omega_n$. In this case, the product space $\Omega = \Omega_1 \times \ldots \times \Omega_n$ occurs as the **set of alternatives** (the **universe of discourse**) that consists of all possible states of *obj*.

Accepting the **closed world assumption** (cf. [Smets, 1988]) the current state ω_0 of *obj*, which in artificial intelligence is also called the **actual state of affairs** or the **actual world**, can therefore always be represented as an element of Ω.

As a matter of fact, any imperfect information about ω_0 is always **source-dependent**. Since typical **uncertainty measures** (probability measures, plausibility measures, possibility measures, and so forth) are not capable of holding any information about an underlying information source, they play the role of **induced concepts** rather than basic elements of uncertainty modeling. For this reason the context model uses the more general concept of an **imperfect specification** of ω_0, which supplies a formalism in order to present imperfect data, and where these data come from.

As we are content with the needs of the above mentioned numerical settings, especially possibility theory, it is sufficient to study only two special types of imperfect information, which are **imprecision** and **uncertainty**, respectively. We refer to imprecision as crisp sets of possible alternatives, but do not pay any attention to fuzziness in this work. For a survey on the many other aspects of imperfect information see [Smithson, 1989].

The following subsection will particularize the context model's view of imprecision and uncertainty.

Imprecision (Multi-valued Data)

A very basic type of imperfect knowledge arises from **imprecision** in the available data about what is the value of ω_0. In our terms it means that an information source S provides a *set-valued specification* $A \subseteq \Omega$ of ω_0 rather than to determine the current object state ω_0 precisely. The power set $\mathfrak{P}(\Omega) = 2^\Omega = \{A \mid A \subseteq \Omega\}$ is therefore the set of all possible set-valued specifications of ω_0.

Let $A, B \subseteq \Omega$. A is called **precise**, iff $|A| = 1$, imprecise, iff $|A| > 1$, and **contradictory** (to the closed world assumption), iff $A = \emptyset$. A is called **correct** w.r.t. ω_0, iff $\omega_0 \in A$; otherwise A is **incorrect** w.r.t. ω_0. The contradictory specification is always incorrect w.r.t. any object state. A is **at least as specific** as B, iff $A \subseteq B$. If A is a proper subset of B, then A is **more specific** than B, whereas B is **less specific** than A.

Presupposing the **total reliability** of S says that a set-valued specification A of ω_0, issued by S, is correct and of maximum specificity w.r.t. ω_0. Correctness implies that $\omega_0 \in A$ is true. **Maximum specificity** means that the information provided by S does not guarantee the correctness of any proper subset of A w.r.t. ω_0.

EXAMPLE 1. Let E be an expert who serves as an information source for the diagnosis of a particular person's disease $\omega_0 \in \Omega$, where $\Omega = \{d_1, d_2, d_3, d_4, d_5\}$ is the set of all possible alternative diseases under consideration.

Suppose that E provides the imprecise specification $A = \{d_3, d_4\}$ of ω_0. In case of total reliability of E, we know that either $\omega_0 = d_3$ or $\omega_0 = d_4$ is true, but there is no evidence about which of these two diseases is the more likely one within the specification given by E.

W.r.t. the expert's information the set $A_1 = \{d_3, d_4, d_5\}$ is less specific than A, and it therefore does not satisfy the maximum specificity criterion. $A_2 = \{d_3\}$ is a precise specification of ω_0, which is more specific than A, but is not guaranteed to be correct w.r.t. ω_0, since in the light of additional evidence it may turn out that $\omega_0 = d_4$. □

Uncertainty (Competing Specification Contexts)

A set-valued specification $A \subseteq \Omega$ of ω_0 does not provide any information about preferences among the elements in A, which would justify to say that one element in A is more likely to be the current object state ω_0 than another element in A. Since such preferences are supplied in many applications, we need a richer structure in order to model imperfect knowledge about ω_0.

Whenever preferences as mentioned above are given, they induce **uncertainty** about a (set-valued) specification of ω_0 in the way that there occur competing and conflicting **specification contexts** for ω_0, where it is not clear with certainty, which one of these at least two contexts should be selected for a set-valued specification of ω_0.

On a quite intuitive and abstract level, a specification context c is just an element of a set C of distinguished alternative viewpoints for a set-valued specification of ω_0. It describes a possible state of the whole world, reduced to those physical and observation-related frame conditions that are relevant for ω_0.

A context might be one of the possible outcomes of an experiment (as an example for a **physical frame condition**), an expert opinion or sensor measurement (which are examples of **source-dependent observation conditions**), or—as in most cases of practical relevance—a description of a mixture of physical and observation-related frame conditions. An appropriate formalization of a context can be given, for instance, with the aid of a logical proposition.

In the following, for the sake of practical relevance and simplicity, we confine to the special case of **finite sets** C of specification contexts, which turns out to be sufficient for many practical problems. Each context $c \in C$ is supposed to be free of any quantifiable uncertainty about ω_0. We assume that c supplies a unique set-valued specification $\gamma(c) \subseteq \Omega$ which is correct and of maximum specificity w.r.t. ω_0, given that the frame conditions described by c are present when observing and specifying ω_0. In this case, we want to call c a **true context** w.r.t. ω_0, and all we know about ω_0 in c is that $\omega_0 \in \gamma(c)$.

Note that in the general case of an imperfect specification, a context c that is not true for specifying ω_0 (it does not describe the adequate frame conditions for the current object state), may nevertheless yield a correct specification $\gamma(c)$ of ω_0. Hence, the truth of a context for specifying ω_0 does normally not coincide with the correctness of a specification of ω_0. This is not surprising, since set-valued specifications $\gamma(c)$ are context-dependent: Assuming the total reliability of contexts, the truth of a context w.r.t. ω_0 always implies the correctness $\gamma(c)$ of w.r.t. ω_0, but not vice versa.

Referred to the above considerations it is adequate and sufficient for our purpose to formalize an **imperfect specification** of ω_0 as a 4-tuple $\Gamma = (\Omega, C, \gamma, P_C)$, where Ω denotes the chosen universe of discourse, C is a finite set of specification contexts, and γ is a mapping $\gamma : C \to 2^{\Omega}$ such that $\gamma(c)$ is a set-valued specification of ω_0 relative to the context c.

P_C is a probability measure on C. $P_C(\{c\})$ can be interpreted as the **probability of selecting c as a true context for specifying** ω_0. One could think of more general approaches that allow alternative uncertainty measures on C, for example belief functions or possibility measures. We restrict ourselves to probability measures, since they appear as the simplest way of quantifying our confidence in contexts with the aid of additive weighting.

Playing the role of a probability of context selection, $P_C(\{c\})$ is either an **objective probability** (due to physical frame conditions) or a **subjective probability** (due to additional individual observation conditions).

We assume that $\gamma(c)$ is correct and of maximum specificity w.r.t. ω_0, given that the context c is true w.r.t. ω_0. In case that c is not true w.r.t. ω_0, it can happen that $\gamma(c) = \emptyset$. From a knowledge representation point of view, it is nevertheless reasonable to refer to this type of contexts, since operating on imperfect specifications, for example expansion (see section 3.1), may produce them in a posterior state of reasoning. On the other hand, within the final step of decision making based on imperfect specifications, these contexts have to be excluded from consideration by conditioning P_C on the subset

(1) $C_\Gamma \overset{\text{Df}}{=} \{c \in C \mid \gamma(c) \neq \emptyset\}$

of those contexts that are possible candidates for being true w.r.t. ω_0 (see section 2.2).

EXAMPLE 2. Reconsider the expert E who specifies his imperfect observation of a particular person's disease $\omega_0 \in \Omega$, where $\Omega = \{d_1, d_2, d_3, d_4, d_5\}$ is the set of all possible alternative diseases of interest. Suppose that E now provides a probability distribution on Ω, defined by

$$P(\{d_1\}) = P(\{d_2\}) = P(\{d_4\}) = P(\{d_5\}) = 0.1, \text{ and}$$
$$P(\{d_3\}) = 0.6.$$

The available imperfect knowledge about ω_0 can be represented as follows:

The expert E refers to the set $C = \{c_1, c_2, c_3, c_4, c_5\}$ of specification contexts, where $c_j, j = 1, \ldots, 5$, describes the physical frame condition "E is observer and d_j is the unknown disease". The underlying context measure space is $(C, 2^C, P_C)$ with $P_C(\{c_j\}) \overset{\text{Df}}{=} P(\{d_j\})$. Assuming the total reliability of E, we obtain the imperfect specification $\Gamma = (\Omega, C, \gamma, P_C)$ of ω_0, where $\gamma : C \to 2^\Omega$, $\gamma(c_j) \overset{\text{Df}}{=} \{d_j\}$.

This is an example for an ad-hoc generalization of the probability measure P to an imperfect specification Γ of ω_0, involving the expert's trivial specification contexts, namely to distinguish between the five possible diseases.

To apply Γ as an imperfect specification of ω_0 means that context c_j is selected with the probability $P_C(\{c_j\})$ as a true context for specifying ω_0. Since $\gamma(c_j) = \{d_j\}$ is assumed to be correct and of maximum specificity w.r.t. ω_0, we obtain the precise information that $\omega_0 = d_j$ holds in context c_j, so that the selection probability $P_C(\{c_j\})$ of this context equals the occurrence probability $P(\{d_j\})$ of disease d_j. We thus have an adequate representation of P in the framework of imperfect specifications.

2.2 *Semantics of Imperfect Specifications*

The idea of using set-valued mappings on probability spaces in order to treat uncertain and imprecise data refers to random set approaches that are well-known for a long time [Strassen, 1964; Dempster, 1968; Nguyen, 1978; Kampé de Fériet, 1982]. On the other hand, for operating on imperfect specifications in the field of belief change in knowledge-based systems, it is important to provide a modified semantics that goes beyond a pure random set approach. This is the idea, for example, of the **theory of hints** [Kohlas and Monney, 1995], which extends the semantical underpinnings of Shafer's theory of evidence [Shafer, 1976]. The context approach [Gebhardt and Kruse, 1993a; Gebhardt and Kruse, 1997a; Gebhardt, 1992] that we introduced in the previous subsection is on a similar line, but has a different purpose in the way that it involves the structure of information sources of imperfect specifications.

Viewing $\Gamma = (\Omega, C, \gamma, P_C)$ as a hint means that the set C of contexts corresponds to a finite set of **possible interpretations** of the imprecise information given by Γ. One of these interpretations is assumed to be the "correct" interpretation, but it is unknown which one. Since the theory of hints is not intended to further specify possible interpretations and therefore does not consider any logical dependency structure between the elements in C, the probability $P_C(\{c\})$ of an interpretation c coincides with the **probability of selecting** c **as a true context for specifying** ω_0 in the context approach.

The additional semantic background of contexts allows a closer look at the meaning of the quantity $P_C(\{c\})$: It is important to recognize that in the general case this probability does **not** equal the **probability of occurrence of the frame conditions described by** c. This is easy to see when we consider, for example, a context c_0 with no restricting frame conditions, so that c_0 is true for any ω_0. In logical terms, c_0 could be represented by a tautology. If $P_C(\{c_0\})$ is interpreted as the probability of occurrence of the frame conditions described by c_0, then it is $\gamma(c_0) = \Omega$ and $P_C(\{c_0\}) = 1$. It follows that $P_C(\{c_0\}) = 0$ for all other contexts $c \in C$, which is counterintuitive, since any useful context should have a positive occurrence probability. In the general case of an imperfect specification it is therefore not always feasible to define an appropriate probability measure P_C on C that quantifies context occurrence. On the other hand, interpreting P_C as a measure of context selection, it might be criticized that finding justified quantities for the probabilities $P_C(\{c\})$ tends to be difficult in practical applications.

We want to deal with this problem at least from a semantic point of view: We will show that any imperfect specification $\Gamma = (\Omega, C, \gamma, P_C)$ with P_C being a probability measure of context selection can be regarded as a representation of a **class** of imperfect specifications $\Gamma^* = (\Omega, C^*, \gamma^*, P_{C^*})$, where each C^* occurs as a modification of C such that $P_{C^*}(\{c^*\})$ equals the occurrence probability of the event $\omega_0 \in \gamma^*(c^*)$.

Sample Spaces of Contexts We start our considerations with the special case of an imperfect specification such that $\{\gamma(c) \mid c \in C\}$ is a **partitioning** of a subset Ω' of Ω. This means that C can be interpreted as a **sample space** of contexts: For any $\omega_0 \in \Omega'$, there is a uniquely determined context c such that $\gamma(c)$ is correct w.r.t. ω_0. $P_C(\{c\})$ can be viewed as the **probability of occurrence** of context c. It equals the probability of being the (uniquely determined) true context for specifying ω_0.

EXAMPLE 3. (example 2 continued) The defined set of specification contexts $C = \{c_1, c_2, c_3, c_4, c_5\}$ satisfies the sample space criterion. The chosen contexts stand for a finite number of elementary events, each of which is reflecting one of the physical frame conditions that are distinguished for a set-valued specification of ω_0. These outcomes form a sample space of specification contexts, and one and only one of them is true w.r.t. ω_0.

The General Case of Context Sets Things become a little more difficult, if C is not supposed to be a sample space. A typical example for this situation is to define C as a set of experts or sensors that supply their individual specifications of ω_0. The corresponding specification contexts may depend from one another, and may share common frame conditions for specifying ω_0. In this case, $\{\gamma(c) \mid c \in C\}$ normally is not a partitioning of a subset of Ω, i.e., the context-dependent specifications $\gamma(c)$ of ω_0 are not necessarily pairwise disjoint. As a consequence, we may find more than a single true context for specifying ω_0, so that $P_C(\{c\})$ is better regarded as the (subjective) probability with which c should be chosen for a set-valued specification of ω_0. Such a **probability of context selection** is different from a probability of context occurrence, but it is strongly related to it in the following sense:

Any imperfect specification $\Gamma = (\Omega, C, \gamma, P_C)$ can be transformed (using context refinement and coarsening of contexts with the same attached context-dependent set-valued specifications of ω_0) into an imperfect specification $\Gamma^* = (\Omega, C^*, \Gamma^*, P_{C^*})$ which satisfies the partitioning criterion, so that C^* can be regarded as a sample space.

Γ^* is called a **context refinement** of Γ, if there exists a **coarsening mapping** $\rho : C^* \to C$ such that for all $c \in C$, the following three conditions hold:

(i) $\rho(C^*) = C$

(ii) $\gamma(c) = \bigcup_{\substack{c^* \in C^* \\ \rho(c^*)=c}} \gamma^*(c^*)$

(iii) $P_C(\{c\}) = \sum_{\substack{c^* \in C^* \\ \rho(c^*)=c}} P_{C^*}(\{c^*\})$

Note that context refinement adds information, so that Γ^* is more informative than Γ. From a conceptual point of view, $\Gamma = (\Omega, C, \gamma, P_C)$ can thus be interpreted as a representation of the class of all imperfect specifications $\Gamma^* = (\Omega, C^*, \gamma^*, P_{C^*})$ of ω_0, relative to any refined set C^* of specification context, such that Γ^* satisfies the partitioning criterion. In other words: Γ occurs as an incomplete description of an imperfect specification Γ^* that is referred to a sample space of contexts. The consequence of this incomplete description is the fact that the coarsening process from Γ^* to Γ transforms the probability measure P_{C^*}, which quantifies probabilities of **context occurrence**, to a probability measure P_C, which only quantifies probabilities of **context selection**.

Relationships to Belief Functions From an application-oriented point of view, the main purpose of using an imperfect specification Γ of ω_0 is to calculate the probability $P_\Gamma(A)$ that $\omega_0 \in A$ is true, referred to any event $A \in S$ relative to an appropriate measurable space $(\Omega, \mathfrak{S})$. It is

(2) $P_{\Gamma^*}(\gamma^*(c^*)) = P_{C^*}(\{c^*\})$ for all $c^* \in C^*$.

Calculating the lower and the upper bound for $P_{\Gamma^*}(A)$ yields

(3) $\mathrm{Bel}_\Gamma(A) \leq P_{\Gamma^*}(A) \leq \mathrm{Pl}_\Gamma(A)$,

where

(4) $\mathrm{Bel}_\Gamma(A) = \sum_{\substack{c \in C \\ \emptyset \neq \gamma(c) \subseteq A}} P_C(\{c\})$ and $\mathrm{Pl}_\Gamma(A) = \sum_{\substack{c \in C \\ \gamma(c) \cap A \neq \emptyset}} P_C(\{c\})$.

Bel_Γ is the **belief function,** Pl_Γ the **plausibility function** induced by Γ [Shafer, 1976].

(3) holds for all context refinements Γ^* of Γ, and the bounds are in fact assumed by particular imperfect specifications Γ^*. This emphasizes the special role of belief functions and plausibility functions with respect to imperfect specifications: $\mathrm{Bel}_\Gamma(A)$ quantifies the **minimum** of the possible probabilities of occurrence of the event A, relative to any context refinement of Γ, and $\mathrm{Pl}_\Gamma(A)$ quantifies the corresponding **maximum**.

EXAMPLE 4. As a slight modification of example 2, suppose that expert E does not supply a distribution of subjective probabilities on Ω, but provides an imperfect specification of ω_0, which is based on two medical tests t_1, t_2. We assume that these tests are not completely reliable, so that they result in imprecise specifications of ω_0. Suppose that t_1 delivers $\omega_0 \in \{d_1, d_2, d_3\}$, whereas t_2 delivers $\omega_0 \in \{d_3, d_4, d_5\}$.

The set of specification contexts considered by E is $C = \{c_1, c_2\}$, where c_j denotes the context "select the result of t_j in order to specify ω_0". Suppose that the subjective context selection probabilities of expert E are quantified by $P_C(\{c_1\}) = 0.7$ and $P_C(\{c_2\}) = 0.3$, respectively.

Collecting all available information, the expert E supplies the imperfect specification $\Gamma = (\Omega, C, \gamma, P_C)$ of ω_0, where $\gamma(c_1) = \{d_1, d_2, d_3\}$ and $\gamma(c_2) = \{d_3, d_4, d_5\}$. Γ may be refined as follows into an imperfect specification $\Gamma^* = (\Omega, C^*, \gamma^*, P_{C^*})$ with a sample space C^* of contexts:

$$\begin{aligned} \gamma^*(c_1^*) &= \{d_1, d_2\}, & c_1^* &= c_1 \wedge \neg c_2, \\ \gamma^*(c_2^*) &= \{d_4, d_5\}, & c_2^* &= \neg c_1 \wedge c_2, \\ \gamma^*(c_3^*) &= \{d_3\}, & c_3^* &= c_1 \wedge c_2. \end{aligned}$$

Note that $\gamma(c_1) \cup \gamma(c_2) = \Omega$ implies that at least one of the specification contexts can be considered as a true context for specifying ω_0. Hence, $C^* = \{c_1^*, c_2^*, c_3^*\}$ distinguishes the three contexts, where either only c_1 is true (c_1^*), or only c_2 is true (c_2^*), or c_1 and c_2 are both true (c_3^*) for specifying ω_0.

The subjective selection probability P_C induces a set of possible occurrence probabilities P_{C^*} which satisfy the conditions

$$P_{C^*}(\{c_1^*\}) \leq P_C(\{c_1\}) = 0.7,$$
$$P_{C^*}(\{c_2^*\}) \leq P_C(\{c_2\}) = 0.3 \quad \text{and}$$
$$P_{C^*}(\{c_1^*\}) + P_{C^*}(\{c_2^*\}) + P_{C^*}(\{c_3^*\}) = 1.$$

From our previous considerations (see eq. (3)), we know that

$$\mathrm{Bel}_\Gamma(A) \leq P_{\Gamma^*}(A) \leq \mathrm{Pl}_\Gamma(A), \quad A \subseteq \Omega.$$

Choosing, for example, $A = \{d_1, d_2, d_3\}$, we obtain

$$0.7 \leq P_{\Gamma^*}(\{d_1, d_2, d_3\}) \leq 1,$$

so that the occurrence probability of event $\omega_0 \in \{d_1, d_2, d_3\}$ is at least 0.7.

Decision Making

The task of decision making based on imperfect specifications Γ of ω_0 concerns the formalization of someone's betting behavior with respect to the truth of the proposition "$\omega_0 \in A$", where $A \in \mathcal{S}$ is taken from an appropriate measurable space $(\Omega, \mathfrak{S})$ of events.

Let B_0, B_1 be two betters and S_0, S_1 their stakes, so that $S_0 + S_1 = 1$. B_0 bets on "$\omega_0 \in A$" versus "$\omega_0 \in \Omega \backslash A$", B_1 vice versa. The **bet function** $\mathrm{bet}_\Gamma : \mathfrak{S} \to [0,1]$ quantifies the stakes for B_0. B_0 accepts the bet, iff $S_0 \leq \mathrm{bet}_\Gamma(A)$ and therefore $S_1 \geq 1 - \mathrm{bet}_\Gamma(A)$. On the other hand, B_1 accepts the bet, iff $S_1 \leq 1 - \mathrm{bet}_\Gamma(A)$.

We minimize the betting risk, exhaust the whole knowledge about ω_0 that is provided by Γ, and make use of a **generalized insufficient reason principle** on the pignistic level of uncertain inference, similar to the **transferable belief model** (see [Smets and Kennes, 1994]). It presupposes uniform distributions on the context-dependent set-valued specifications $\gamma(c)$, $c \in C$, justified by the assumption that there are no preferences among object states contained in $\gamma(c)$. Furthermore, turning over from knowledge representation to decision making, we condition on the set C_Γ of contexts that are possibly true w.r.t. ω_0 (see (1)). Under these conditions, we obtain the following definition of the bet function of Γ:

Let P_Ω denote the uniform distribution on a finite universe of discourse Ω, and let $P_C(C_\Gamma) > 0$.

$$\text{(5)} \quad \mathrm{bet}_\Gamma : \mathfrak{S} \to [0,1], \quad \mathrm{bet}_\Gamma(A) \stackrel{\mathrm{Df}}{=} \frac{1}{P_C(C_\Gamma)} \sum_{\substack{c \in C \\ \gamma(c) \cap A \neq \emptyset}} P_C(\{c\}) \cdot P_\Omega(A | \gamma(c))$$

is called the **bet function** of Γ.

EXAMPLE 5. (example 4 continued) For the event $A = \{d_1, d_2, d_3\}$, we calculate $\mathrm{bet}_\Gamma(A) = 0.8$, which means that for expert E, with a probability of 80%, the person's disease is d_1, d_2 or d_3, respectively.

Selected Numerical Settings

In this subsection, we apply the concept of an imperfect specification and its induced bet function for a closer look at uncertainty modeling in two of the most important types of numerical settings, which are pure probabilistic settings, and the belief function setting, respectively. This is helpful in order to see relationships and differences with respect to the possibilistic setting, which we will examine in much detail in section 2.3.

Probabilistic Settings. In pure probabilistic settings, imperfect knowledge about ω_0 is specified in terms of a probability space $(\Omega, \mathfrak{S}, P)$. The purpose is to calculate the probability $P(A)$ of selected events $A \in \mathfrak{S}$, quantifying the uncertainty about the truth of the proposition "$\omega_0 \in A$". For further semantic issues, see [Cox, 1946; Savage, 1954; Fine, 1973; de Finetti, 1974; Fishburn, 1996].

If there is not only uncertainty about the truth of this proposition, but also imprecision regarding the choice of an appropriate probability measure, knowledge representation changes to a family $\mathcal{P}$ of probability spaces $(\Omega, \mathcal{S}, P_i)$, $i \in I$, referred to an underlying index set I. In this case, bounds of probabilities are $\overline{P}(A) = \sup\{P(A) \mid P \in \mathfrak{P}\}$ and $\underline{P}(A) = \inf\{P(A) \mid P \in \mathfrak{P}\}$, where $\overline{P} : \mathfrak{S} \to [0,1]$ and $\underline{P} : \mathfrak{S} \to [0,1]$ denote the **upper and lower probability function** of $\mathcal{P}$, respectively [Walley and Fine, 1982; Walley, 1991]. Restricting ourselves to a finite set Ω of alternatives, the specification of imperfect knowledge on ω_0 in terms of a probability space $\mathfrak{M} = (\Omega, 2^\Omega, P)$ on $\Omega = \{\omega_1, \ldots, \omega_m\}$ can also be done with the aid of an imperfect specification $\Gamma = (\Omega, C, \gamma, P_C)$. In particular, define $C \overset{\text{Df}}{=} \{c_1, \ldots, c_m\}$, $\gamma : C \to 2^\Omega$, $\gamma(c_j) \overset{\text{Df}}{=} \{\omega_j\}$ and $P_C(\{c_j\}) \overset{\text{Df}}{=} P(\{\omega_j\})$, $j = 1, \ldots, m$, where c_j denotes the trivial context "ω_j *is the current object state*". Hence, there is a one-to-one correspondence between the contexts in C and the possible object states in Ω. It follows that the consideration of a single probability space and especially decision making based on probability measures (as one kind of uncertainty measures) are covered by the concept of an imperfect specification, and it is $P(A) = \mathrm{bet}_\Gamma(A)$ for all $A \subseteq 2^\Omega$.

$\mathfrak{M}$ is useful as an **uncertain**, but **precise** specification of ω_0: Uncertainty is due to the probability measure P; precision occurs since there is a probability on each of the alternatives in Ω. As a consequence, an equivalent modeling based on the context approach results in an imperfect specification Γ with **precise** context-dependent specifications of ω_0 (i.e.: $|\gamma(c_j)| = 1$).

EXAMPLE 6. (example 2 continued) We have already presented the transformation of the probability distribution issued by the expert E into an imperfect specification Γ of the person's disease. For any $A \subseteq \Omega$, we calculate $\mathrm{bet}_\Gamma(A) = P(A)$, for example, $\mathrm{bet}_\Gamma(\{d_1, d_2, d_3\}) = 0.8$.

The Belief Function Setting. In the belief function setting [Dempster, 1967; Shafer, 1976], the interpretation of Dempster-Shafer theory by the transferable belief model [Smets and Kennes, 1994], the theory of hints [Kohlas and Monney, 1995], or related random set approaches [Hestir *et al.*, 1991], imperfect information about ω_0 occurs in terms of a belief function or its dual plausibility function, respectively. On the other hand, for practical purposes, when an expert supplies this imperfect information, it seems to be more convenient to assume that a **mass distribution** (basic probability assignment) $m : 2^\Omega \to [0,1]$ is available, such that

$$(6) \quad m(\emptyset) = 0 \quad \text{and} \quad \sum_{A:A\subseteq\Omega} m(A) = 1,$$

where $m(A) > 0$ holds for only a finite number of events A.

If m represents our imperfect knowledge about ω_0, then the quantity $m(A)$ stands for the probability with which A is selected as a most specific correct set-valued specification of ω_0.

From this we obtain the **belief function** Bel_m and the **plausibility function** Pl_m as induced concepts:

$$(7) \quad \text{Bel}_m(A) = \sum_{B:B\subseteq A} m(B), \qquad \text{Pl}_m(A) = \sum_{B:B\cap A\neq\emptyset} m(B).$$

It is straight-forward to transform a mass distribution into an imperfect specification of ω_0: Let $\mathcal{E} = \{A_1, \ldots, A_k\} = \{A \mid A \subseteq \Omega, m(A) > 0\}$ denote the finite set of events with positive selection probabilities for specifying ω_0.

$\mathcal{E}_m$ induces the set $C_m = \{c_1, \ldots, c_k\}$ of specification contexts, where c_j, $j = 1, \ldots, k$, describes the frame condition that "A_j is the most specific correct specification of ω".

With respect to this interpretation, the mass distribution m corresponds to the imperfect specification $\Gamma_m = (\Omega, C_m, \gamma_m, P_{C_m})$ with $\gamma_m : C \to 2^\Omega$ such that

$$(8) \quad \gamma_m(c_j) \stackrel{\text{Df}}{=} A_j \quad \text{and} \quad P_{C_m}(\{c_j\}) \stackrel{\text{Df}}{=} m(A_j), \qquad j = 1, \ldots, k.$$

$P_{C_m}(\{c_j\})$ is the selection probability as well as the occurrence probability of c_j, since the description of the contexts implies that the true context for specifying ω_0 is uniquely determined.

On the reverse way, if an imperfect specification $\Gamma = (\Omega, C, \gamma, P_C)$ is non-contradictory, which means that $\gamma(c) \neq 0$ is satisfied for all $c \in C$, then it has an induced information-compressed representation in form of the mass distribution

$$(9) \quad m_\Gamma : 2^\Omega \to [0,1], \qquad m_\Gamma(A) \stackrel{\text{Df}}{=} \sum_{\substack{c\in C \\ \gamma(c)=A}} P_C(\{c\}).$$

EXAMPLE 7. (example 4 continued) The information issued by the two medical tests can be represented with the aid of the mass distribution $m : 2^\Omega \to [0,1]$ that is determined by

$$m(\{d_1, d_2, d_3\}) = 0.7 \quad \text{and} \quad m(\{d_3, d_4, d_5\}) = 0.3.$$

The resulting imperfect specification Γ_m of ω_0 corresponds to the imperfect specification Γ that we have considered in this example, but note that the set C of contexts refers to the medical tests and is therefore in a way more informative than the set C_m of contexts which is simply induced by the mass distribution m without any relationship to the underlying information source.

Nevertheless, due to a correct modeling of our imperfect knowledge about ω_0 in both approaches, it is easy to verify that $m_\Gamma = m$. Given the event $A = \{d_1, d_2, d_3\}$, we therefore calculate $\mathrm{bet}_{\Gamma_m}(A) = \mathrm{bet}_\Gamma(A) = 0.8$.

2.3 Semantic Aspects of Possibility Theory

Similar to the various alternative approaches to the semantics of (subjective) probabilities and belief functions, respectively, there are different interpretations of possibility theory. Some of them refer to possibility distributions as epistemic interpretation of fuzzy sets [Zadeh, 1978], the axiomatic view of possibility theory with the aid of the concept of a possibility measure as a particular sub-additive measure of uncertainty [Dubois and Prade, 1988a], the identification of possibility measures as upper envelopes of sets of probabilities [Dubois and Prade, 1988b], and possibility theory based on likelihoods [Dubois *et al.*, 1993]. Spohn's theory of epistemic states [Spohn, 1988; Spohn, 1990] has provided one more semantics of possibility theory [Dubois *et al.*, 1993]. Furthermore, possibility distributions are also considered as contour functions of consonant belief functions [Shafer, 1976], as one-point coverages of random sets [Nguyen, 1978; Goodman, 1982; Hestir *et al.*, 1991], or as falling shadows within the framework of set-valued statistics [Wang, 1983].

In the role of precursors of possibilistic reasoning, the economist [Shackle, 1961] who was interested in a theory of "degrees of potential surprise" has to be mentioned, and the philosophers [Cohen, 1973] and [Rescher, 1976] who studied "grades of inductive support" and "plausibility degrees", respectively.

From the viewpoint of the context model, we want to consider two approaches to possibility theory in some more detail. The first is a standard one that interprets the possibilistic setting as a special case of the belief function setting: If we are given a **consonant** imperfect specification $\Gamma = (\Omega, C, \gamma, P_C)$ of ω_0, which means that Γ has nested non-contradictory context-dependent set-valued specifications $\gamma(c)$ of ω_0, then Pl_Γ specializes to a **possibility measure**, and Bel_Γ to a **necessity measure** in terms of Shafer's evidence theory [Shafer, 1976]. The investigation of these two uncertainty measures can therefore directly be referred to the belief function setting. Stating the consonance property allows to recover the set-valued specifications $\gamma(c)$ in Γ from Bel_Γ and Pl_Γ, but the same time it reduces the scope of possibility theory as the theory of possibility measures and their dual necessity measure to such kind of uncertain knowledge that in essence rests on qualitative preferences only.

For this reason, the second approach to possibility theory that we want to study is more general in the sense that we use imperfect specifications without insisting on the consonance property. The main idea is to introduce possibility distributions as **information-compressed** representations of imperfect specifications $\Gamma = (\Omega, C, \gamma, P_C)$, based on the context-dependent specifications $\gamma(c)$ and some appropriate assumptions on the potential dependency structure of the involved contexts. The acceptability of a certain kind of information compression is justified by the fact that in many applications we are not in the position or, for reasons of efficiency and necessary complexity reduction, not willing to describe sets of specification contexts in all detail.

Under this restriction, the purpose is that we do not want to operate on Γ directly, but only refer to its **degree of α-correctness** w.r.t. ω_0, defined as the total mass of all contexts in C that yield a correct context-dependent specification of ω_0 [Gebhardt and Kruse, 1993b]. More particularly, assuming α-correctness of Γ w.r.t. ω_0 says that the conjunctive combination of any subset C' of contexts in C with a total mass of at least α, carried out by calculating the intersection of the associated context-dependent specifications $\gamma(c')$, $c' \in C'$, is supposed to result in a correct set-valued specification of ω_0.

DEFINITION 8. (α-correctness) Making use of the concepts and symbolics introduced so far, and confining to finite sets of specification contexts, let $\Gamma = (\Omega, C, \gamma, P_C)$ be an imperfect specification of an object state $\omega_0 \in \Omega$.

Γ is called **α-correct** w.r.t. ω_0, iff

$$P_C(\{c \in C \mid \omega_0 \in \gamma(c)\}) \geq \alpha, \qquad 0 \leq \alpha \leq 1.$$

Based on the assumption that Γ is α-correct w.r.t. ω_0, we are searching for the most specific set-valued specification A_α of ω_0, namely the largest subset of Ω such that α-correctness of Γ w.r.t. ω is satisfied for all $\omega \in A_\alpha$. The set A_α therefore consists of all potential candidates ω for the truth of the proposition "$\omega_0 = \omega$". It is easy to prove that the family $(A_\alpha)_{\alpha \in [0,1]}$ consists of all **α-cuts** $[\pi_\Gamma]_\alpha$ of the **possibility distribution π_Γ induced by** Γ, where these two concepts are defined as follows:

DEFINITION 9. (Possibility distribution) Let $\Gamma = (\Omega, C, \gamma, P_C)$ be an imperfect specification of $\omega_0 \in \Omega$. Then,

$$\pi_\Gamma : \Omega \to [0,1], \qquad \pi_\Gamma(\omega) \stackrel{\text{Df}}{=} P_C(\{c \in C \mid \omega \in \gamma(c)\})$$

is called the (unnormalized) **possibility distribution** induced by Γ.

POSS(Ω)= $\{\pi \mid \pi : \Omega \to [0,1] \text{ and } |\pi(\Omega)| \in \mathbb{N}\}$ denotes the set of all possibility distributions that can be induced by imperfect specifications referred to Ω and any finite set of contexts.

The quantity $\pi_\Gamma(\omega)$ in fact may be interpreted as a **degree of possibility** for the truth of the proposition "$\omega_0 = \omega$", relative to the set C of contexts:

- If $\pi_\Gamma(\omega) = 1$, then $\omega \in \gamma(c)$ holds for all $c \in C$, which means that $\omega_0 = \omega$ is possible without any restrictions.

- If $\pi_\Gamma(\omega) = 0$, then $\omega \notin \gamma(c)$ holds for all $c \in C$, which means that $\omega_0 = \omega$ is impossible.
- If $0 < \pi_\Gamma(\omega) < 1$, then there is at least one context that supports "$\omega_0 = \omega$" as well as at least one context that contradicts "$\omega_0 = \omega$".

The quantity $\pi_\Gamma(\omega)$ stands for the total mass of contexts that support "$\omega_0 = \omega$" and that can therefore be conjunctively combined in order to get a correct specification of ω_0.

Note that ignoring the interpretation of specification contexts, π_Γ formally coincides with the **one-point coverage** of Γ, when it is seen as a (not necessarily nested) random set [Nguyen, 1978; Hestir *et al.*, 1991]. On the other hand, from a semantics point of view, operating on possibility distributions may better be strongly oriented at the concept of α-correctness. We will discuss this topic extensively in the next section.

DEFINITION 10. (α-cuts of possibility distributions) Let $\pi \in \text{POSS}(\Omega)$ be a possibility distribution. For any $\alpha \in [0,1]$, the α-**cut** $[\pi]_\alpha$ of α is defined by

$$[\pi]_\alpha \overset{\text{Df}}{=} \{\omega \in \Omega \mid \pi(\omega) \geq \alpha\}.$$

EXAMPLE 11. (ex. 4 continued) The imperfect specification $\Gamma = (\Omega, C, \gamma, P_C)$ of the person's disease ω_0 induces the possibility distribution $\pi_\Gamma : \Omega \to [0,1]$, defined by

$$\begin{array}{lllllllll} \pi_\Gamma(d_1) & = & 0.7, & \pi_\Gamma(d_2) & = & 0.7, & \pi_\Gamma(d_3) & = & 1, \\ \pi_\Gamma(d_4) & = & 0.3, & \pi_\Gamma(d_5) & = & 0.3. \end{array}$$

Assuming that 0.8-correctness of Γ w.r.t. ω_0 is satisfied, then $[\pi_\Gamma]_{0.8} = \{d_3\}$ is correct and of maximum specificity w.r.t. ω_0, and hence we obtain $\omega_0 = d_3$.

Note that 0.8-correctness of Γ w.r.t. ω_0 requires the truth of at least 80% of the total mass of contexts for the specification of ω_0. In our case, this means that the two medical tests both have provided a correct set-valued specification of ω_0, so that $\omega_0 \in \gamma(c_1) \cap \gamma(c_2) = \{d_3\}$.

PROPOSITION 12. *(Correctness and maximum specificity of possibility distributions) Let $\Gamma = (\Omega, C, \gamma, P_C)$ be an imperfect specification of $\omega_0 \in \Omega$. For any $\alpha \in [0,1]$, Γ is α-correct w.r.t. ω_0, if and only if $[\pi_\Gamma]_\alpha$ is correct w.r.t. ω_0.*

Proof. Γ is α-correct w.r.t. ω_0

$$\begin{array}{cl} \overset{\text{Def 8}}{\Leftrightarrow} & P_C(\{c \in C \mid \omega_0 \in \gamma(c)\}) \geq \alpha \\ \Leftrightarrow & \omega_0 \in \{\omega \in \Omega \mid P_C(\{c \in C \mid \omega \in \gamma(c)\}) \geq \alpha\} \\ \overset{\text{Def 9}}{\Leftrightarrow} & \omega_0 \in \{\omega \in \Omega \mid \pi_\Gamma(\omega) \geq \alpha\} \\ \overset{\text{Def 10}}{\Leftrightarrow} & \omega_0 \in [\pi_\Gamma]_\alpha \\ \Leftrightarrow & [\pi_\Gamma]_\alpha \text{ is correct w.r.t. } \omega_0 \end{array}$$

■

Note that due to the maximum specificity assumption that we made with respect to the context-dependent specifications $\gamma(c)$ of ω_0, a trivial corollary of proposition 12 says that if Γ is α-correct w.r.t. ω_0, then $[\pi_\Gamma]_\alpha$ is the most specific correct set-valued specification of ω_0, relative to Γ.

We finish this section with the definition and justification of an uncertainty measure of possibility which is in accordance with the context model:

Suppose that we are given a possibility distribution $\pi \in \mathrm{POSS}(\Omega)$ of ω_0. Based on this imperfect information on ω_0, we do not find a unique imperfect specification Γ of ω_0 such that $\pi \equiv \pi_\Gamma$, even if restrictions on the potential underlying sets of contexts are made. Selecting a particular candidate Γ, note that $[\pi_\Gamma]_\alpha$ is the degree of α-correctness of Γ w.r.t. ω_0, interpreted as the degree of possibility of the truth of the proposition "$\omega_0 = \omega$" relative to Γ and π_Γ.

An appropriate possibility measure $\mathrm{Poss}_{\pi_\Gamma}$, associated with π_Γ, should be chosen in the way that for any event $A \subseteq \Omega$, the quantity $\mathrm{Poss}_{\pi_\Gamma}(A)$ is the degree of possibility that $\omega_0 = \omega$ holds for any element $\omega \in A$.

From this we obtain

$$\max_{\omega \in A} \pi_\Gamma(\omega) \leq \mathrm{Poss}_{\pi_\Gamma}(A) \leq \min\left\{\sum_{\omega \in A} \pi_\Gamma(\omega), 1\right\},$$

where we can always construct imperfect specifications Γ of ω_0 such that $\pi_\Gamma \equiv \pi$ is satisfied, and such that $\mathrm{Poss}_{\pi_\Gamma(A)}$ assumes the two limiting bounds. Preferring the lower bound as the most pessimistic case, the resulting definition of a possibility measure coincides with the standard definition used in possibility theory [Dubois and Prade, 1988a].

DEFINITION 13. (Possibility measure) Let $\pi \in \mathrm{POSS}(\Omega)$ be a possibility distribution. Then, the set function

$$\mathrm{Poss}_\pi : \mathfrak{P}(\Omega) \to [0, 1],$$
$$\mathrm{Poss}_\pi(A) \stackrel{\mathrm{Df}}{=} \max\{\pi(\omega) \mid \omega \in \Omega\},$$

where $\max \emptyset \stackrel{\mathrm{Df}}{=} 0$, is called the **possibility measure** of π.

Note that using the sup-operator is not necessary in the definition, since in our framework, $\pi(\Omega)$ is always a finite set of possibility degrees. Extensions are straight-forward, if adequate measurability criteria are considered. Also note that the possibility distributions π addressed by definition 9 in general do **not** satisfy the normality condition that there exists an element $\omega \in \Omega$ such that $\pi(\omega) = 1$. For this reason, the standard definition $\mathrm{Nec}_\pi(A) = 1 - \mathrm{Poss}_\pi(\Omega \backslash A)$ of a dual **necessity measure** as it is used in other interpretations of possibility theory is not suitable and feasible in our approach. However, except that we do not insist on the normality property, the above concept of a possibility measure meets the view of this uncertainty measure in standard possibility theory.

2.4 *Operating on Possibility Distributions*

Making use of the semantic background of possibility theory introduced in the previous section, we now investigate how to operate on possibility distributions in agreement with the context model. We will need the corresponding results in order to justify appropriate reasoning techniques for possibilistic graphical models.

From a sufficiently general point of view, the problem to be solved can be formulated as follows:

Let $\Gamma_i = (\Omega_i, C_i, \gamma_i, P_{C_i})$, $i = 1, \ldots, n$, denote imperfect specifications of object states $\omega_0^{(i)}$. Assuming α-correctness of Γ_i w.r.t. $\omega_0^{(i)}$ for all $i \in \{1, \ldots, n\}$, we know from proposition 12 that the α-cut $[\pi_{\Gamma_i}]_\alpha$ is the most specific correct set-valued specification of $\omega_0^{(i)}$ relative to Γ_i. Hence, if there is an operation $\Phi : \Omega_1 \times \ldots \times \Omega_n \to \Omega$ that maps the $\omega_0^{(i)}$-values to a new object state $\omega_0 = \Phi(\omega_0^{(1)}, \ldots, \omega_0^{(n)})$, then the task is to calculate the most specific correct set-valued specification of ω_0, given the possibility distributions π_{Γ_i} and the stated α-correctness assumptions.

More generally, we may also consider operations $\varphi : \mathfrak{P}(\Omega_1) \times \ldots \times \mathfrak{P}(\Omega_n) \to \mathfrak{P}(\Omega)$ on set-valued specifications with respect to the chosen universes of discourse. In this case, if we are given non-empty subsets $A_i \subseteq \Omega_i$ of object states, a set-valued specification $B_i \subseteq \Omega_i$ is said to be **correct** w.r.t. A_i, if $A_i \subseteq B_i$.

For any $\alpha \in [0, 1]$, the imperfect specification Γ_i is called α-**correct** w.r.t. A_i, if $P_C(\{c \in C_i \mid A_i \subseteq \gamma_i(c)\}) \geq \alpha$ holds, so that context-dependent specifications $\gamma_i(c)$ of at least a total mass α are correct w.r.t. A_i.

It is easy to verify that this generalization from correctness with respect to a single object state to correctness with respect to a set of object states is in conformity with proposition 12, at least in the way that α-correctness of Γ_i w.r.t. A_i implies the correctness (and maximum specificity) of $[\pi_{\Gamma_i}]_\alpha$ w.r.t. A_i. The inverse direction holds only if $|A_i| = 1$, which is in agreement with the assumptions made in order to show the equivalence in proposition 12.

Referred to the above generalization, the task is to calculate the most specific correct set-valued specification of $\varphi(A_1, \ldots, A_n)$, given the possibility distribution $\pi_{\Gamma_1}, \ldots, \pi_{\Gamma_i}$ and the stated α-correctness assumptions. Unfortunately, it turns out that $\varphi([\pi_{\Gamma_1}]_\alpha, \ldots, [\pi_{\Gamma_n}]_\alpha)$ is correct w.r.t. $\varphi(A_1, \ldots, A_n)$, but it does not necessarily satisfy the maximum specificity criterion.

For this reason it is near at hand to search for necessary and sufficient conditions on φ that ensure the maximum specificity preservation. This kind of structural investigation concerning operations on possibility distributions has been considered in [Gebhardt, 1992]. We will profit here from the corresponding results, but renounce the reproduction of the quite extensive proofs behind them.

Transferred to the presentation of the context model in section 2.1, correctness and maximum specificity of $\varphi([\pi_{\Gamma_1}]_\alpha, \ldots, [\pi_{\Gamma_n}]_\alpha)$ w.r.t. $\varphi(A_1, \ldots, A_n)$ is given, if and only if φ has a property that we want to call **specificity preservation**:

DEFINITION 14. (Preservation properties) Let $\Omega_1, \ldots, \Omega_n$, and Ω denote non-empty universes of discourse, and let $\varphi : \mathfrak{P}(\Omega_1) \times \ldots \times \mathfrak{P}(\Omega_n) \to \mathfrak{P}(\Omega)$ be an operator on set-valued specifications.

(a) φ is called **contradiction-preserving**, iff

$$\forall (A_1, \ldots, A_n) \in \bigtimes_{i=1}^{n} \mathfrak{P}(\Omega_i) : \quad \exists i \in \{1, \ldots, n\} :$$
$$(A_i = \emptyset) \Rightarrow (\varphi(A_1, \ldots, A_n) = \emptyset).$$

(b) φ is **correctness-preserving**, iff for all $A_i \subseteq B_i \subseteq \Omega_i$, $i = 1, \ldots, n$,

$$\varphi(A_1, \ldots, A_n) \subseteq \varphi(B_1, \ldots, B_n).$$

(c) φ is **specificity-preserving**, iff φ is contradiction-preserving and

$$\varphi(A_1 \cup B_1, \ldots, A_n \cup B_n)$$
$$= \bigcup \{\varphi(D_1, \ldots, D_n) \mid \forall j \in \{1, \ldots, n\} : D_j = A_j \vee D_j = B_j\}$$

for all subsets A_i, B_i of Ω_i, $i = 1, \ldots, n$.

Contradiction preservation is the property which ensures that a contradictory specification cannot be combined with other set-valued specifications to a non-contradictory specification.

Correctness preservation is a second natural property one may require: If set-valued specifications B_i, $i = 1, \ldots, n$, are correct w.r.t. A_i, then any reasonable operation φ on these set-valued specifications should lead to a specification $\varphi(B_1, \ldots, B_n)$ that is correct w.r.t. $\varphi(A_1, \ldots, A_n)$. The third property is a technical specialization of correctness and contradiction preservation, which we need in order to characterize specificity preservation for operations on possibility distributions.

THEOREM 15. *(Specificity preservation theorem for possibility distributions) Let $\varphi : \mathfrak{P}(\Omega_1) \times \ldots \times \mathfrak{P}(\Omega_n) \to \mathfrak{P}(\Omega)$ denote a correctness- and contradiction-preserving operator.*

φ is specificity-preserving, if and only if for any non-empty sets $A_i \in \Omega_i$, any $\alpha \in (0, 1]$, and any α-correct imperfect specifications $\Gamma_i = (\Omega_i, C_i, \gamma_i, P_{C_i})$ of A_i, $|C_i| \geq 2$, $i = 1, \ldots, n$, the set-valued specification $\varphi([\pi_{\Gamma_1}]_\alpha, \ldots, [\pi_{\Gamma_n}]_\alpha)$ is correct and of maximum specificity w.r.t. $\varphi(A_1, \ldots, A_n)$.

The above theorem describes how to operate on possibility distributions in agreement with the context model such that correctness and maximum specificity are preserved. The next definition and the subsequent theorem help in getting a closed form of the resulting possibility distribution.

DEFINITION 16. (Possibilistic extensions of operators) Let $\Omega_1, \ldots, \Omega_n$, and Ω be non-empty universes of discourse, and let $\varphi : \mathfrak{P}(\Omega_1) \times \ldots \times \mathfrak{P}(\Omega_n) \to \mathfrak{P}(\Omega)$ be an operator on set-valued specifications.

Furthermore, define $\hat{\varphi} : \mathrm{POSS}(\Omega_1) \times \ldots \times \mathrm{POSS}(\Omega_n) \to \mathrm{POSS}(\Omega)$ such that $\hat{\varphi}(\pi_1, \ldots, \pi_n)$ is the possibility distribution with the following α-cuts:

$$[\hat{\varphi}(\pi_1, \ldots, \pi_n)]_\alpha \stackrel{\mathrm{Df}}{=} \varphi([\pi_1]_\alpha, \ldots, [\pi_n]_\alpha), \quad \alpha \in (0, 1].$$

$\hat{\varphi}$ is called the **possibilistic extension** of φ.

THEOREM 17. *(Extension theorem for possibility distributions) Referring to definition 16, let φ be a specificity-preserving operator. Then, defining* $\sup \emptyset \stackrel{\mathrm{Df}}{=} 0$, *for all $\omega \in \Omega$, it is*

$$\hat{\varphi}(\pi_1, \ldots, \pi_n)(\omega) = \sup\{\min\{\pi_1(\omega_1), \ldots, \pi_n(\omega_n)\} \\ \mid (\omega_1, \ldots, \omega_n) \in \bigtimes_{i=1}^{n} \Omega_i \wedge \omega \in \varphi(\{\omega_1\}, \ldots, \{\omega_n\})\}$$

Proof. Let $\omega \in \Omega$ and $\alpha \in (0, 1]$. It is

$$\omega \in [\hat{\varphi}(\pi_1, \ldots, \pi_n)]_\alpha$$

$$\stackrel{\mathrm{Def\ 16}}{\Leftrightarrow} \quad \omega \in \varphi([\pi_1]_\alpha, \ldots, [\pi_n]_\alpha)$$

$$\stackrel{\mathrm{Def\ 14}}{\Leftrightarrow} \quad \exists(\omega_1, \ldots, \omega_n) \in \times_{i=1}^{n} \Omega_i : \quad \forall i \in \{1, \ldots, n\} : \\ \omega_i \in [\pi_i]_\alpha) \wedge \omega \in \varphi(\{\omega_1\}, \ldots, \{\omega_n\})$$

("$\Rightarrow$" holds because of the supposed specificity preservation of φ, and "$\Leftarrow$" follows from the fact that, in this case, φ is also correctness-preserving)

$$\stackrel{\mathrm{Def\ 10}}{\Leftrightarrow} \quad \exists(\omega_1, \ldots, \omega_n) \in \bigtimes_{i=1}^{n} \Omega_i : \\ \min\{\pi_1(\omega_1), \ldots, \pi_n(\omega_n)\} \geq \alpha \ \wedge \ \omega \in \varphi(\{\omega_1\}, \ldots, \{\omega_n\})$$

$$\Leftrightarrow \quad \sup\{\min\{\pi_1(\omega_1), \ldots, \pi_n(\omega_n)\} \mid \\ (\omega_1, \ldots, \omega_n) \in \bigtimes_{i=1}^{n} \Omega_i \ \wedge \ \omega \in \varphi(\{\omega_1\}, \ldots, \{\omega_n\})\} \geq \alpha$$

("$\Rightarrow$" is trivial, and "$\Leftarrow$" is true, since $|\pi_i(\Omega_i)| \in \mathbb{N}$, so that sup is assumed as a maximum)

With the above consideration, we obtain

$$\hat{\varphi}(\pi_1, \ldots, \pi_n)(\omega)$$

$$= \quad \sup\{\alpha \mid \hat{\varphi}(\pi_1, \ldots, \pi_n)(\omega) \geq \alpha\}$$

$$\stackrel{\mathrm{Def\ 10}}{\Leftrightarrow} \quad \sup\{\alpha \mid \omega \in [\hat{\varphi}(\pi_1, \ldots, \pi_n)]_\alpha\}$$

$$= \quad \sup\{\min\{\pi_1(\omega_1),\ldots,\pi_n(\omega_n)\} \mid (\omega_1,\ldots,\omega_n) \in \bigtimes_{i=1}^{n} \Omega_i$$
$$\wedge\, \omega \in \varphi(\{\omega_1\},\ldots,\{\omega_n\})\},$$

which completes the proof. ■

Theorem 17 is the context model's justification of a slightly generalized version of **Zadeh's extension principle** for fuzzy sets and possibility distributions [Zadeh, 1975]. Note that with the semantics provided by the context model, we have verified the extension principle as the only consistent way of operating on possibility distributions that preserves correctness and maximum specificity. The standard form of the extension principle (cf. [Kruse *et al.*, 1994]) is confirmed as follows:

Starting with a function $\Phi : \Omega_1 \times \ldots \times \Omega_n \to \Omega$, and then defining

$$\varphi : \mathfrak{P}(\Omega_1) \times \ldots \times \mathfrak{P}(\Omega_n) \to \mathfrak{P}(\Omega),$$
$$\varphi(A_1,\ldots,A_n) \stackrel{\text{Df}}{=} \Phi(A_1 \times \ldots \times A_n).$$

it is easy to see that φ is always contradiction-, correctness-, and even specificity-preserving. Applying the extension theorem 17, we calculate

$$(10) \quad \hat{\varphi}(\pi_1,\ldots,\pi_n)(\omega) = \sup\{\min\{\pi_1(\omega_1),\ldots,\pi_n(\omega_n)\}\}$$
$$\mid (\omega_1,\ldots,\omega_n) \in \bigtimes_{i=1}^{n} \Omega_i \wedge \omega = \Phi(\omega_1,\ldots,\omega_n)\}.$$

For reasoning in possibilistic graphical models, we will need the possibilistic extensions of the set-theoretic operators of intersection, projection, and cylindrical extension.

We formalize the latter two of them as follows:

DEFINITION 18. (Projection, cylindrical extension) Let $V = \{X_1,\ldots,X_n\}$ be a finite set of variables with associated domains $\Omega_i = \mathrm{Dom}(X_i)$, $i = 1,\ldots,n$, and let $\Omega \stackrel{\text{Df}}{=} \Omega_1 \times \ldots \times \Omega_n$ denote our universe of discourse. Furthermore, for any non-empty subset $I \subseteq V$ of variables, let Ω^I denote the common domain (the product of the domains) of the variables in I. For any $\omega \in \Omega$, let $\omega(I)$ denote the restriction of ω to the components addressed by I. Finally, let $\emptyset \neq S \subseteq T \subseteq V$.

(a) $\mathrm{proj}_S^T : \mathfrak{P}(\Omega^T) \to \mathfrak{P}(\Omega^S)$,
$\mathrm{proj}_S^T(A) \stackrel{\text{Df}}{=} \{\omega^S \in \Omega^S \mid \exists \omega^T \in A : \omega^T(S) = \omega^S\}$
is called the **projection** of Ω^T onto Ω^S.

(b) $\mathrm{ext}_S^T : \mathfrak{P}(\Omega^S) \to \mathfrak{P}(\Omega^T)$,
$\mathrm{ext}_S^T(B) \stackrel{\text{Df}}{=} \{\omega^T \in \Omega^S \mid \omega^T(S) \in B\}$
is called the **cylindrical extension** of Ω^S to Ω^T.

It is simple to prove that intersection, projection, and cylindrical extension are specificity-preserving and thus correctness- and contradiction-preserving operators. The same is also valid for a special composition of these operations, which is **conditioning** as one of the basic operations for reasoning in graphical models. Section 3.1 will deal with this topic in more detail.

3 BACKGROUND CONCEPTS OF POSSIBILISTIC GRAPHICAL MODELS

In this contribution we refer to a possibilistic graphical model as a pair $\mathcal{N} = (D, \Pi)$, consisting of a qualitative part, which is a possibilistic conditional independence graph $G = (V, E)$ of a possibility distribution $\pi \in \text{POSS}(\Omega^V)$, specifying a set of independence statements on the variables in V, and a quantitative part, which is the family Π of marginal distributions $\pi(C)$ of π, referred to all cliques C in G, that constitutes a lossless join decomposition of π. This means that the possibility distribution π, used as a possibilistic specification of a current object state $\omega_0 \in \Omega^V$ of interest, is representable by the possibilistic graphical model $\mathcal{N}$. Evidential reasoning with $\mathcal{N}$ corresponds to conditioning π on some available factual knowledge about instantiations on the variables in V, with the aim of obtaining more specific possibility distributions $\pi^*(X)$ for the component values $\omega_0(X)$, $X \in V$. When calculating these distributions, efficient propagation algorithms can profit from the decomposability of π, whenever the dimensionalities of the distributions $\pi(C)$ are significantly lower than the dimensionality of π.

This section provides the theoretical background for reasoning with possibilistic graphical models of the considered type. Section 3.1 investigates possibilistic conditioning, where the concept of a conditional possibility distribution (definition 19) is introduced in terms of conditioning as an expansion operation that is in accordance with the semantic background proposed in section 2. Section 3.2 deals with the entailed concept of possibilistic conditional independence (definition 20) which coincides with the conditioned version of possibilistic non-interactivity. Furthermore, the derived factorization criterion for possibilistic conditional independence (theorem 21) as the basic tool for decomposition techniques is presented.

Finally, based on the global Markov property satisfied by a possibility distribution, which turns out neither to be equivalent to the local nor to the pairwise Markov property, section 3.3 provides an adequate formulation of the concept of a possibilistic conditional independence graph (definition 24).

3.1 Possibilistic Conditioning

Conditioning as a concept of belief change in ordinal and numerical uncertainty theories has several semantic faces that give rise to a variety of conditioning rules in order to formalize the dynamic behavior of cognitive states in the light of new information.

In a purely probabilistic framework conditioning is achieved by means of Bayes rule and leads to a conditional probability measure whose unicity under reasonable assumptions can be shown with the aid of different types of axiomatic justifications. As three examples we mention [Gärdenfors, 1988] with respect to a numerical justification, [Heckerman, 1988] for a Coxian-like algebraic justification, and [Williams, 1980] for an information-theoretic justification.

As a consequence of the unicity property, Bayesian conditioning does not provide a clear distinction between the three basic forms of belief change proposed by [Gärdenfors, 1988], which are expansion, revision, and contraction, respectively. Actually, the rule of conditioning in probability theory is controversial: Most researchers in Bayes theory regard conditioning as a **focusing** operation that incorporates new pieces of evidence (factual knowledge) about an object state ω_0 of interest without modifying the generic knowledge embodied in an a priori given probability distribution that uncertainly specifies ω_0. This means that the epistemic change only pertains to the belief held about the particular application in which ω_0 is imperfectly observed.

On the contrary, there are also some philosophers, especially in the field of probability kinematics, who point out that conditioning can be advocated as an operation of **belief revision** that respects a principle of minimal change such that consistency with respect to the (possibly partly contradicting) new information is guaranteed (cf. [Domotor, 1980]).

From the viewpoint of probabilistic graphical modeling as it is used, for example, in Bayesian networks and Markov networks, evidence propagation corresponds to conditioning of the focusing type, namely to an expansion of the state of belief that is represented by a prior probability distribution. This property has to be taken into account when graphical modeling is transferred to non-standard numerical formalisms such as possibility theory and belief function theory.

In possibility theory, expansion as a simple addition of new information means that a possibility distribution $\pi \in \mathrm{POSS}(\Omega)$, representing our generic knowledge about ω_0, receives a non-empty input $A \subseteq \Omega$, considered as the sure information $\omega_0 \in A$ that reflects the available factual knowledge about ω_0. With this background, the expansion of the possibility distribution π is defined as

$$(11)\quad \pi_A(\omega) \stackrel{\mathrm{Df}}{=} \begin{cases} \pi(\omega), & \text{iff } \omega \in A \\ 0, & \text{otherwise.} \end{cases}$$

$\pi_A \in \mathrm{POSS}(\Omega)$ is the most specific possibilistic specification of ω_0 that can be constructed from the conjunction of the prior generic knowledge (represented by π) with the provided application-dependent evidence (represented by A).

More particularly, if π is α-correct and of maximum specificity w.r.t. ω_0, then the additional information that ω_0 holds for sure, leads to $[\pi_A]_\alpha = [\pi]_\alpha \cap A$ as the most specific set-valued correct specification of ω_0. This justifies formula (11).

A problem that may occur is the disjointness of $[\pi]_\alpha$ and A, which means that the assumption of α-correctness of π w.r.t. ω_0 entails a contradiction and therefore a subnormalized possibility distribution π_A, even if π represents a **consistent** cognitive state in the way that $\pi(\omega) = 1$ holds for some $\omega \in \Omega$. Partial inconsistency can be avoided by the additional requirement that A has to be chosen such that $\text{Poss}_\pi(A) = 1$. In this case, (11) is in agreement with the Gärdenfors postulates of expansion, expressed in terms of possibility distributions [Dubois and Prade, 1992]. On the other hand, it is also reasonable to adopt (11) in all cases where $\text{Poss}_\pi(A) > 0$, and to allow subnormalized possibility distributions that reflect partial inconsistency and thus upper bounds of possible α-correctness degrees lower than 1. We will follow this line, since inducing π form a database of (multi-valued) sample cases, formalized as an imperfect specification $\Gamma = (\Omega, C, \gamma, P_C)$ of ω_0, with C being the set of specification contexts for these sample cases, usually leads to a subnormalized possibility distribution $\pi \equiv \pi_\Gamma$.

Note that the expansion π_A of π can be viewed as a **data conditioning** of the underlying imperfect specification $\Gamma = (\Omega, C, \gamma, P_C)$ in the way that $\Gamma_A = (\Omega, C, \gamma_A, P_C)$ is defined as $\gamma_A(c) \overset{\text{Df}}{=} \gamma(c) \cap A$ without any modification of the prior probability $P_C(\{c\})$ of selecting context $c \in C$. In this case, we obtain $\pi_A \equiv \pi_{\Gamma_A}$.

An alternative approach to conditioning is that of belief revision, where the idea is to get a consistent cognitive state even if the input information contradicts the a priori cognitive state. Starting with a normalized possibility distribution π, its revision π^*_A can be understood as an additional **context conditioning** of Γ, which is conditioning P_C on all contexts that remain possible in order to get a non-contradictory set-valued specification of ω_0, once the sure information $\omega_0 \in A$ is given. On the level of belief change, context conditioning can be regarded as a counterpart of the normalization step that we addressed when calculating the bet function for decision making (see formula (5)).

It leads to $\Gamma^*_A = (\Omega, C, \gamma_A, P^*_C)$, where

$$(12)\quad P^*_C(\{c\}) \overset{\text{Df}}{=} \begin{cases} \frac{P_C(\{c\})}{k}, & \text{iff } \gamma_A(c) \neq \emptyset \\ 0, & \text{otherwise,} \end{cases}$$

using the normalization factor

$$(13)\quad k \overset{\text{Df}}{=} P_C(\{c \in C \mid \gamma_A(c) \neq \emptyset\})$$

If Γ is a consonant imperfect specification of ω_0, then $k = \text{Poss}_\pi(A)$ holds. In this case, the induced revised possibility distribution $\pi_d(\cdot|A) \equiv \pi_{\Gamma^*_A}$ writes

$$(14)\quad \pi_d(\omega|A) = \begin{cases} \frac{\pi(\omega)}{\text{Poss}_\pi(A)}, & \text{iff } \omega \in A \\ 0, & \text{otherwise} \end{cases}$$

and its respective possibility measure is defined by

$$(15)\quad \Pi_d(B|A) \overset{\text{Df}}{=} \frac{\text{Poss}_\pi(A \cap B)}{\text{Poss}_\pi(A)},$$

provided that $\mathrm{Poss}_\pi(A) > 0$.

This is **Dempster's rule of conditioning**, specialized to possibility measures [Shafer, 1976]. Besides the disadvantage that this rule requires the consonance of the inducing imperfect specification (which corresponds to the unrealistic assumption of nested sample cases when applied to databases), the possibilistic version of Dempster's conditioning seems to be inappropriate for our view of conditioning in graphical modeling, since it reflects revision rather than expansion.

The same is true for **Hisdal's rule of conditioning** [Hisdal, 1978]. Provided that $\mathrm{Poss}_\pi(A) > 0$, the respective revised possibility distribution is given by

$$(16)\quad \pi_h(\omega|A) \stackrel{\mathrm{Df}}{=} \begin{cases} 1, & \text{iff } \pi(\omega) = \mathrm{Poss}_\pi(A), \omega \in A, \\ \pi(\omega), & \text{iff } \pi(\omega) < \mathrm{Poss}_\pi(A), \omega \in A, \\ 0, & \text{iff } \omega \notin A, \end{cases}$$

and its corresponding possibility measure writes

$$(17)\quad \Pi_h(B|A) = \begin{cases} 1, & \text{iff } \mathrm{Poss}_\pi(A \cap B) = \mathrm{Poss}_\pi(A) > 0 \\ \mathrm{Poss}_\pi(A \cap B), & \text{otherwise.} \end{cases}$$

From a formal point of view, Π_d and Π_h also appear as solutions of an equation of the form

$$(18)\quad \Pi(A \cap B) = \Pi(B|A) \otimes \Pi(A),$$

which is used as the possibilistic counterpart of a conditioning device for revision that is similar to the probabilistic one. Hisdal's rule turns out to be the unique solution of equation (18) for $\otimes = \min$ that satisfies the maximum specificity condition, whereas Dempster's rule is the only solution when choosing the product for $\otimes$, and using a Coxian approach [Dubois and Prade, 1990].

In order to establish conditioning as an expansion operation in the framework of possibilistic graphical models, we refer to (11) and introduce the following concept:

DEFINITION 19. (Conditional possibility distribution) Let $V = \{X_1, \ldots, X_n\}$ be a finite set of variables taking their values in attached domains $\Omega_i = \mathrm{Dom}(X_i)$, $i = 1, \ldots, n$. Furthermore, let $\pi \in \mathrm{POSS}(\Omega)$ denote a possibility distribution on the set $\Omega \stackrel{\mathrm{Df}}{=} \Omega_1 \times \ldots \times \Omega_n$, interpreted as the universe of discourse of a current object state ω_0 under consideration. Finally, let X and Z be two non-empty subsets of V.

For any instantiation $z \in \Omega^Z$ of the variables in Z (saying that $\omega_0(Z) = z$ is assumed to be true), abbreviated by $Z = z$, given that $\widehat{\mathrm{proj}_Z^V}(\pi)(z) > 0$, we call $\pi(X|z) \in \mathrm{POSS}(\Omega^X)$, defined by

$$(19)\quad [\pi(X|z)]_\alpha \stackrel{\mathrm{Df}}{=} [\widehat{\mathrm{proj}_X^V}(\widehat{\mathrm{intersect}}(\pi, \widehat{\mathrm{ext}_Z^V}(\{\mathbb{1}_{\{z\}}\})))]_\alpha,$$

the **conditional possibility distribution of X given $Z = z$.**

This definition is explained and justified as follows:

Suppose that we are given a possibility distribution $\pi \in \mathrm{POSS}(\Omega)$, interpreted as an information-compressed representation of an underlying imperfect specification Γ of ω_0 such that $\pi_\Gamma \equiv \pi$. We assume that π is the specification of our generic knowledge about ω_0.

Now suppose that additional evidential knowledge or factual information on the variables in Z becomes available, formalized by $\omega_0(Z) = z$ for a particular tuple $z \in \Omega^Z$. This information can be represented by the possibility distribution $\mathbb{1}_{\{z\}} \in \mathrm{POSS}(\Omega^Z)$, where $\mathbb{1}_{\{z\}}$ denotes the characteristic function of $\{z\}$ with respect to the referential set Ω^Z.

We are interested in the posterior possibility distribution for $\omega_0(X)$ that results from conditioning our general knowledge on the basis of the given factual information. We apply formula (11) for the conditioning process, and then project the calculated possibility distribution on Ω^X. The composition of the involved operations leads to $\pi(X|z)$ in the proposed form, where $\widehat{\mathrm{intersect}}$ stands for the possibilistic extension of set-theoretic intersection.

Note that ext_Z^V, intersect, and proj_X^V are specificity-preserving, so that assuming α-correctness (and anyhow maximum specificity) of Γ w.r.t. ω_0, theorem 15 confirms that $[\pi(X|z)]_\alpha$ is correct and of maximum specificity w.r.t. $\omega_0(X)$, relative to the imperfect information given by π and $Z = z$.

The presupposition $\widehat{\mathrm{proj}_X^V}(\pi)(Z) > 0$ avoids $\pi(X|z) = 0$ as the result of conditioning based on an instantiation of Z that totally contradicts the prior possibility distribution π.

For the following investigations, we will refer to

$$(20)\quad \pi(X) \overset{\mathrm{Df}}{=} \widehat{\mathrm{proj}_X^V}(\pi)$$

as the **projection of** π **onto** Ω^X. From a formal point of view, permitting $Z \neq \emptyset$, and introducing $\Omega^\emptyset$ as the 0-dimensional universe of discourse with ϵ denoting the empty tuple as the only element of $\Omega^\emptyset$, then $\pi(X) = \pi(X|\epsilon)$ can be regarded as a special case of a conditional possibility distribution.

3.2 Possibilistic Conditional Independence

One of the fundamental concepts of graphical modeling is (conditional) independence of variables. This topic has been studied in-depth for the probabilistic setting (cf. [Dawid, 1979; Lauritzen *et al.*, 1990]), but there are also some results with respect to other theories of uncertain information (cf. [Shenoy, 1992]). Related to possibilistic independence, recent work can be found in [Studeny, 1993; Benferhat *et al.*, 1994; Cooman and Kerre, 1994; Farinas del Cerro and Herzig, 1994; Fonck, 1994; de Campos *et al.*, 1995].

Due to the different semantical underpinnings of the conditioning concepts that we have mentioned in the previous section, it is clear that there are various ways of establishing concepts of possibilistic conditional independence. Based on our formalization of conditional possibility distributions (see definition 19), we can particularize as follows:

Let X, Y, and Z denote disjoint subsets of variables, where $X \neq \emptyset$ and $Y \neq \emptyset$. Furthermore, let $\pi \in \text{POSS}(\Omega)$ be the most specific correct possibilistic specification that represents our imperfect knowledge about ω_0. Additionally, provided that $\pi(Z)(z) > 0$, let $Z = z$ be an instantiation, saying that $\omega_0(Z) = z$ holds for sure. In this case, expansion of π leads to $\pi(V|z)$ as the most specific correct possibilistic specification of ω_0.

X is independent of Y given Z w.r.t. π, if and only if any additional knowledge about Y, represented by an admissible instantiation $Y = y$, does not further specify our knowledge about X. This means that $[\pi(X|y,z)]_\alpha = [\pi(X|z)]_\alpha$ occurs as the most specific correct set-valued specification of $\omega_0(X)$, if α-correctness of $\pi(V|z)$ w.r.t. ω_0 is assumed.

Note that the instantiation of Y is not arbitrary, but restricted as a consequence of the chosen α-correctness assumption: If $\pi(V|z)$ is α-correct w.r.t. ω_0, then we know that $y \in [\pi(Y|z)]_\alpha$ is true, so that admissible instantiations of Y have to be taken from this α-cut in order to avoid inconsistencies.

Summarizing the above considerations, we obtain the following concept of possibilistic conditional independence, conform to the theoretical background of definition 19:

DEFINITION 20. (Possibilistic conditional independence) Let X, Y and Z be disjoint subsets of variables of V, where $X \neq \emptyset$ and $Y \neq \emptyset$ and let $\pi \in \text{POSS}(\Omega)$ denote a possibility distribution.

X is called **independent of** Y **given** Z **w.r.t.** π, for short $I_\pi(X, Y|Z)$, if and only if

(21) $[\pi(X|y,z)]_\alpha = [\pi(X|z)]_\alpha$

holds for all $y \in [\pi(Y|z)]_\alpha$, $\alpha \in (0,1]$, and all $z \in \Omega^Z$ such that $\pi(Z)(z) > 0$.

It is easy to see that this definition does not coincide with the alternative approach that

(22) $\pi(X|y,z) = \pi(X|z)$

should be satisfied for all $y \in \Omega^Y$ and all $z \in \Omega^Z$. This condition is equivalent to

(23) $\text{Poss}_{\pi(X|y,z)}(A) = \text{Poss}_{\pi(X|z)}(A), A \subseteq \Omega^X$,

which at first sight may be regarded as the most natural way of defining possibilistic independence with respect to an appropriate underlying conditioning rule. It looks similar to the corresponding definition in the probabilistic framework, except for the substitution of a probability measure by a possibility measure. Equation (23) simply says that our belief in X, quantified by a possibility measure, is not influenced by Y, once an instantiation of Z is given.

In order to clarify the different background of (21) and (22), we make use of the two-level structure of uncertain reasoning that first was proposed by P. Smets for the transferable belief model (TBM) [Smets and Kennes, 1994], but can also be adopted to other numerical settings such as possibility theory.

One basic idea of the TBM is to distinguish between a **credal level** of knowledge representation and belief change, and a **pignistic level** that only appears for the final task of decision making. In the TBM, beliefs at the credal level are quantified by belief functions, whereas beliefs at the pignistic level are quantified by probability distributions. When a decision must be made, the resulting belief function that has been calculated at the credal level is transformed to a probability distribution at the pignistic level. This transformation is called **pignistic transformation**. It consists of the two subsequent steps of normalization and application of a **generalized insufficient reason principle** [Smets and Kennes, 1994; Gebhardt and Kruse, 1993a]. After the normalization step, the resulting bet function is the same as the one we introduced in formula (4) of section 2.2.

Referred to the expansion type of conditioning a belief function at the credal level of the TBM, **Dempster's rule of conditioning** [Shafer, 1976], except for the normalization factor, can be verified as the adequate choice. Turning over to the possibilistic framework, beliefs at the credal level are quantified by possibility measures Poss_π or their underlying possibility distributions π. The pignistic transformation consists of normalization and application of the insufficient reason principle for both, the selection of an α-correctness level, and the choice of an element from the α-cut $[\pi]_\alpha$. Provided that $\mathrm{Poss}_\pi(\Omega) > 0$ and $|\Omega| \in \mathbb{N}$, the respective possibility distribution π that is calculated at the credal level will be normalized to

$$(24)\quad \pi^* = \frac{\pi}{\mathrm{Poss}_\pi(\Omega)},$$

and then transformed to the probability measure

$$(25)\quad P_\pi(A) = \frac{\sum_{\omega\in A} \pi^*(\omega)}{\sum_{\omega\in\Omega} \pi^*(\omega)}.$$

Note that possibilistic conditioning, taken as an expansion operation at the credal level, leads to definition 20, and normalization is delayed to the pignistic level.

An alternative view is to consider the pignistic level at the borderline between normalization and pignistic transformation. In this case, referred to the expansion type of possibilistic conditioning, we have to relate equation (22) to the normalization π^* of π, so that (22) writes

$$(26)\quad \pi^*(X|y,z) = \pi^*(X|z),$$

where

$$(27)\quad \pi^*(X|z) = \frac{\pi(X|z)}{\pi(V|z)} \quad \text{and}$$

$$(28)\quad \pi^*(X|y,z) = \frac{\pi(X|y,z)}{\pi(V|y,z)},$$

provided that $\pi(V|z) > 0$ and $\pi(V|y,z) > 0$, and that conditioning is as in definition 19.

Formulas (27) and (28) coincide with Dempster's rule of conditioning, specialized to possibility measures and their underlying possibility distributions.

Alternative approaches that are established by revision as a type of possibilistic conditioning, lead to other concepts of possibilistic independence. For more details on an axiomatic treatment of this topic, see [de Campos *et al.*, 1995].

For the purpose of this contribution, namely the investigation of possibilistic graphical models, the concept of possibilistic conditional independence as proposed in definition 20 (which is the unnormalized version of Dempsters's rule in the possibilistic framework) arises as the best choice: It is based on the expansion type of conditioning and referred to the credal level of uncertain reasoning, which in fact is the level where the conditioning induced by instantiation of variables and therefore evidence propagation with all respective operations on possibility distributions take place. Note that Zadeh's extension principle [Zadeh, 1975] (see section 2.4) is settled at the credal level of uncertain reasoning, without any normalization.

The following theorem shows that definition 20 provides a **factorization criterion** which is quite similar to what is known from probability theory. The only modification is the change of a probability distribution into a possibility distribution, and the change of the product into the min-operator.

THEOREM 21. *(Factorization theorem for possibilistic conditional independence) Let X, Y and Z be disjoint subsets of variables of V, where $X \neq \emptyset$ and $Y \neq \emptyset$. Let $\pi \in \mathrm{POSS}(\Omega)$ be a possibility distribution. $I_\pi(X, Y|Z)$ holds, if and only if*

(29) $\pi(X \cup Y|z) = \min\{\pi(X|z), \pi(Y|z)\}$

for all $z \in \Omega^Z$.

Proof. We have to verify that

(30) $[\pi(X \cup Y|z)]_\alpha = \mathrm{ext}_X^{X\cup Y}([\pi(X|z)]_\alpha) \cap \mathrm{ext}_Y^{X\cup Y}([\pi(Y|z)]_\alpha)$

for all $z \in \Omega^Z$ and all $\alpha \in (0,1]$.

Let $\omega \in \Omega$ be an object state with $\omega(X) = x$, $\omega(Y) = y$ and $\omega(Z) = z$, so that $\omega(X \cup Y) \in [\pi(X \cup Y|z)]_\alpha$.

It is $x \in [\pi(X|z)]_\alpha$ and therefore $\omega(X \cup Y) \in \mathrm{ext}_X^{X\cup Y}([\pi(X|z)]_\alpha)$ as well as $y \in [\pi(Y|z)]_\alpha$ and therefore $\omega(X \cup Y) \in \mathrm{ext}_Y^{X\cup Y}([\pi(Y|z)]_\alpha)$ which proves the inclusion $\subseteq$.

On the other hand, let $\omega \in \Omega$ such that $\omega(X) = x$, $\omega(Y) = y$, $\omega(Z) = z$ and $\omega(R) \in \Omega^R$ where $R \overset{\mathrm{Df}}{=} V \setminus (X \cup Y \cup Z)$.

If $\omega(X \cup Y) \in \mathrm{ext}_X^{X\cup Y}([\pi(X|z)]_\alpha) \cap \mathrm{ext}_Y^{X\cup Y}([\pi(Y|z)]_\alpha)$,
then $x \in [\pi(X|z)]_\alpha$ and $y \in [\pi(Y|z)]_\alpha$.

The independence relation $I_\pi(X,Y|Z)$ says that $[\pi(X|z)]_\alpha = [\pi(X|y^*,z)]_\alpha$ for all $y^* \in [\pi(Y|z)]_\alpha$. For this reason, $x \in [\pi(X|z)]_\alpha$ leads to $\omega \in [\pi]_\alpha$ for an appropriate choice $\omega(R) \in \Omega^R$ and thus $\omega(x \cup Y) \in [\pi(X \cup Y|z)]_\alpha$. Hence, the inverse inclusion $\supseteq$ holds, too. This completes the proof. ■

The factorization criterion shows that possibilistic conditional independence in the sense of definition 20 corresponds to a conditional version of **possibilistic non-interactivity** [Hisdal, 1978].

3.3 Possibilistic Conditional Independence Graphs

Graphical models are uncertainty models for multivariate observations whose independence structure can be characterized in terms of a so-called conditional independence graph. Such a graph $G = (V, E)$ is chosen to reflect all independence relations that exist between a set V of considered variables, relative to a given distribution function on their common domain Ω^V. In case of a possibility distribution $\pi : \Omega^V \to [0,1]$ that specifies a current object state ω_0 of interest, the basic aim is to find a one-to-one correspondence between the **dependency model** M_π of π, that is the set of all independence relations $I_\pi(X,Y|Z)$ induced by π, and the **dependency model** M_G of G, that is the set of all independence relations $I_G(X,Y|Z)$ induced by G with respect to its appropriate interpretation as a conditional independence graph.

Transferring some general notions of graphical modeling [Pearl, 1992a] to the possibilistic setting, an undirected graph $G = (V, E)$ is called a **dependency map** of M_π if there is a one-to-one correspondence between M_π and M_G such that for all disjoint subsets X, Y, Z of V, it is

(31) $I_\pi(X,Y|Z) \Rightarrow I_G(X,Y|Z)$.

Conversely, G is called an **independency map** of M_π, if

(32) $I_G(X,Y|Z) \Rightarrow I_\pi(X,Y|Z)$.

Since independence can be viewed as a limiting case (the weakest form) of dependence, the independency map property of G with respect to M_π appears as the more important one: It ensures that M_G covers at least all relevant dependencies that follow from M_π, whereas some suitable independencies that may support a complexity-reducing decomposition of π, perhaps fail to be noticed, which is a loss of information in comparison to M_π.

On the other hand, if G is a dependency map of M_π, then it guarantees that all independence relations in M_π are covered by G, but it is not sure that all relevant dependencies are considered.

G is said to be a **perfect map** (or, equivalently, a **graph-isomorph**) of M_π, if it is both a dependency map and an independency map, and hence a unique graphical representation of the dependency model M_π. In the following we will investigate how to come to a reasonable definition of the concept of a possibilistic conditional independence graph G with its induced set of independence relations $I_G(X, Y|Z)$.

As a first starting point we refer to what is known from probabilistic graphical models (cf. [Whittaker, 1990]): In the probabilistic setting, an undirected graph $G = (V, E)$ is called the conditional independence graph of a probability distribution p on Ω, if it satisfies the **pairwise Markov property**, saying that there is no edge between two vertices in G, if and only if the pair of the corresponding variables is independent, given any instantiations of the remaining variables. Formally spoken,

(33) $E = V \times V \setminus \{(X_i, X_j) \mid I_p(X_i, X_j | V \setminus (X_i, X_j))\}$,

where I_p stands for probabilistic independence with respect to p.

A remarkable benefit of the probabilistic setting is that it provides two alternative formulations of the pairwise Markov property: The first one is the **local Markov property,** which expresses the independence relations in the graph in terms of a specified vertex and its local neighbors in the boundary set.

More particularly, it is required that

(34) $I_p(X_i, \text{rest}(X_i) | \, \text{boundary}(X_i))$ for all $X_i \in V$,

where

$$\begin{aligned} \text{boundary}(X_i) &\stackrel{\text{Df}}{=} \{X_j \in V \mid (X_i, X_j) \in E\} \quad \text{and} \\ \text{rest}(X_i) &\stackrel{\text{Df}}{=} V \setminus (\{X_i\} \cup \text{boundary}(X_i)). \end{aligned}$$

A second equivalent formulation in case of strictly positive probability distributions p is the **global Markov property**, which addresses the independence relations in G in terms of the so-called **separation criterion**. It says that for all disjoint subsets X, Y and Z of variables in V, whenever X and Y are separated by Z in G (which means that every path joining two vertices in X and Y, respectively, contains at least one vertex from the separating subset Z), then $I_p(X, Y|Z)$ holds. This is a "global" property in the sense that X, Y and Z are arbitrary disjoint subsets of vertices, void of their location in the graph.

For the purpose of graphical modeling it is helpful when the equivalence of the mentioned Markov properties is given. Unfortunately, the possibilistic setting does not provide this equivalence, where an in-depth understanding of the structural reasons for this fact is best reached when directly referred to a general system of axioms for independence relations in a variety of uncertainty calculi. Such a system was suggested by J. Pearl and A. Paz in order to capture the intuitive notion of independence of variables, void of any reference to numerical forms [Pearl and Paz, 1985]. The corresponding axioms resulted from a study of the concept of conditional independence in probability theory, knowledge-based systems, relational database theory, and graph theory. The addressed properties are similar to those stated first by [Dawid, 1979] for probabilistic conditional independence, those stated by [Spohn, 1990] for causal independence, and those stated by [Smith, 1989] for generalized conditional independence.

Given four disjoint subsets X, Y, Z and W of variables from V, the axioms read as follows:

Symmetry:

(A1) $I(X, Y|Z) \Rightarrow I(Y, X|Z)$

Decomposition:

(A2) $I(X, Y \cup W|Z) \Rightarrow I(X, Y|Z)$

Weak Union:

(A3) $I(X, Y \cup W|Z) \Rightarrow I(X, W|Z \cup Y)$

Contraction:

(A4) $(X, Y|Z)$ and $I(X, W|Z \cup Y) \Rightarrow I(X, Y \cup W|Z)$

Intersection:

(A5) $I(X, Y|Z \cup W)$ and $I(X, W|Z \cup Y) \Rightarrow I(X, Y \cup W|Z)$

Symmetry says that in any state of prior knowledge Z, if Y tells us nothing new about X, then X tells us nothing new about Y. **Decomposition** asserts that if two combined pieces of knowledge Y and W are judged to be irrelevant to X, then each separate item is irrelevant as well. **Weak Union** states that learning additional irrelevant information W does not help the irrelevant information Y to get relevant to X. **Contraction** means that if we judge W to be irrelevant to X after learning some irrelevant piece of information Y, then the combined items of information are irrelevant to X. Finally, **intersection** states that if Y is irrelevant to X when W is instantiated, and W is irrelevant to X when Y is instantiated, neither W nor Y nor their combination can be relevant to X.

Any three-place relation $I(\cdot, \cdot|\cdot)$ that satisfies (A1) through (A4) is called a **semi-graphoid**, and if it also obeys (A5), it is called a **graphoid** [Pearl and Paz, 1985].

(A1) through (A4) are provided by many well-known dependency models, for example by conditional independence in probability theory, but also by vertex separation in undirected graphs (hence the term graphoids), by d-separation in directed acyclic graphs [Verma and Pearl, 1992], by embedded multi-valued dependencies in relational databases [Fagin, 1977], by conditional independence in Spohn's theory of epistemic states [Spohn, 1988; Hunter, 1991], and by qualitative conditional independence [Shafer *et al.*, 1987]. Some of these dependency models even satisfy (A5) and hence the full set of graphoid axioms. As a quite general example those models should be mentioned that fit the axiomatic framework of **valuation-based systems** (VBS) [Shenoy, 1989; Shenoy, 1992]. VBS are able to represent concepts of various uncertainty calculi, for instance probability theory, Dempster-Shafer theory [Shafer, 1976], but also Zadeh's possibility theory [Zadeh, 1978], as far as possibility measures associated with possibility distributions are interpreted in terms of plausibility measures of consonant belief functions. On the other hand, if we consider the pure relational setting and the more general possibilistic setting when approached by our favored view of possibility distributions being (non-normalized) one-point coverages of random sets, only the semi-graphoid axioms are satisfied [de Campos *et al.*, 1995].

The fact that the intersection axiom (A5) does not hold in the two latter settings is by no means surprising: (A5) can be satisfied for strictly positive probability distributions. This corresponds to the treatment of the relation $R = \Omega^V$ in the relational setting, and therefore to the treatment of the possibility distribution $\pi \equiv \mathbb{1}_{\Omega^V}$ in the possibilistic setting. For these particular cases (A5) is satisfied, but not in the general case when we have $R \subseteq \omega$ or $\pi \leq \mathbb{1}_{\Omega^V}$, respectively.

The problem that $I_\pi(\cdot, \cdot|\cdot)$ is only in agreement with the semi-graphoid axioms has slightly inconvenient consequences for the aims of possibilistic graphical modeling:

Concerning the construction of possibilistic dependency models, we have to consider a result by J. Pearl and A. Paz which says that a necessary and sufficient condition for any dependency model to be a graph-isomorph is that it satisfies the five independent axioms of symmetry, decomposition, intersection, strong union, and transitivity [Pearl and Paz, 1985]. As $I_\pi(\cdot, \cdot|\cdot)$ does not agree with the intersection axiom, we have to accept that in the general case there are no graph-isomorphs (perfect maps) of M_π.

As a way out of this dilemma, it is near at hand to drop the dependency model assumption, and to confine to **Markov networks** (minimal independency maps) [Isham, 1981; Lauritzen, 1982]:

An undirected graph G is a Markov network of a dependency model, if deleting any edge of G would cancel its independency map property.

Unfortunately it has been shown that symmetry, decomposition, and intersection are essential in order to construct a Markov network, in which every cutset corresponds to a genuine independence condition [Pearl, 1992a]. As a consequence, even Markov networks are not useful as the basic structure for a general definition of a possibilistic conditional independence graph. For this reason, we restrict ourselves to simply define a possibilistic conditional independence graph as a (not uniquely determined) independency map of the underlying dependency model M_π. In this connection, we have to think about the appropriate type of Markov property that should be the basis for defining the independence relations $I_G(\cdot, \cdot|\cdot)$ induced by such an independency map. To find an answer to this question is the topic of the following considerations.

A closer look at the proof that pairwise, local, and global Markov property are equivalent with respect to probabilistic conditional independence graphs, or more general, to graphoids, is quite helpful for getting hints to what one can (or should not) expect from conditional independence graphs when referred to possibility distributions:

It is straight forward to see that the global Markov property always implies the local Markov property, because the boundary set of a vertex in the graph is always

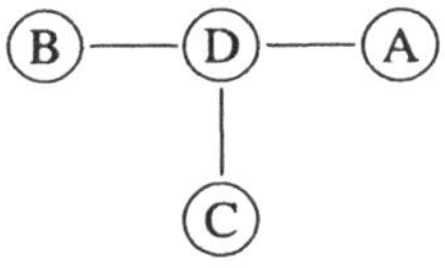

Figure 1. The undirected graph G_1

a separating subset. In order to verify that the local Markov property implies the pairwise Markov property, the weak union axiom (A3) is applied. For proving that the pairwise Markov property implies the global Markov property (in order to complete the proof of equivalence of the three properties), one needs the decomposition axiom (A2), the weak union axiom (A3), and also the intersection axiom (A5). Validity of (A2) coincides with the **reduction lemma**, and validity of (A3) and (A5) coincides with the **block independence lemma** for graphical models in multivariate statistics (cf. [Whittaker, 1990]). We recognize that it is out of reach to prove the equivalence of the Markov properties for pure semi-graphoids like relational conditional independence and possibilistic conditional independence in the sense of definition 20. It is easy to see that

global Markov property $\Rightarrow$ local Markov property
$\Rightarrow$ pairwise Markov property,

but the inverse direction of this chain of implications is not true. We will demonstrate this fact with the aid of two very simple examples taken from the relational setting, which means that they are also applicable in our more general possibilistic framework:

EXAMPLE 22. Let $V_1 \stackrel{\text{Df}}{=} \{A, B, C, D\}$ and
$\Omega \stackrel{\text{Df}}{=} \{a_1, a_2\} \times \{b_1, b_2\} \times \{c_1, c_2, c_3, c_4\} \times \{d_1, d_2, d_3\}$.
With respect to the relation

$$R_1 = \{(a_1, b_1, c_1, d_1), (a_1, b_2, c_2, d_2), (a_2, b_2, c_3, d_3), (a_1, b_1, c_4, d_3)\}$$

and the corresponding possibility distribution $\pi_1 \stackrel{\text{Df}}{=} \mathbb{1}_{R_1}$, we consider the undirected graph $G_1 = (V_1, E_1)$ given in figure 1. It is

$$I_{\pi_1}(A, B|\{C, D\}), I_{\pi_1}(A, C|\{B, D\}), I_{\pi_1}(B, C|\{A, D\})$$

which means that the pairwise Markov property is satisfied. But note that none of the independence statements

(35) $I_{\pi_1}(A, B|D), \quad I_{\pi_1}(A, C|D), \quad I_{\pi_1}(B, C|D)$

and

(36) $I_{\pi_1}(A, \{B, C\}|D), \quad I_{\pi_1}(B, \{A, C\}|D), \quad I_{\pi_1}(C, \{A, B\}|D)$

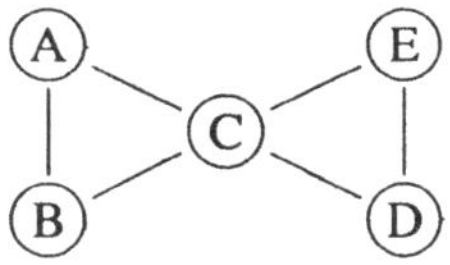

Figure 2. The undirected graph G_2

is true, so that neither the global Markov property (see (35)) nor the local Markov property (see (35)) is satisfied. We conclude that

pairwise Markov property $\not\Rightarrow$ local Markov property.

EXAMPLE 23. Let $V_2 \stackrel{\text{Df}}{=} \{A, B, C, D, E\}$ and $\Omega \stackrel{\text{Df}}{=} \{a_1, a_2\} \times \{b_1, b_2\} \times \{c\} \times \{d_1, d_2\} \times \{e_1, e_2\}$. Using the relation

$$R_2 = \{(a_1, b_1, c, d_1, e_1), (a_2, b_2, c, d_2, e_2)\}$$

and the induced possibility function $\pi_2 \stackrel{\text{Df}}{=} 1\!\!1_{R_2}$, we now refer to the undirected graph $G_2 = (V_2, E_2)$ shown in figure 2. It is

$$I_{\pi_2}(A, \{D, E\}|\{B, C\}), I_{\pi_2}(B, \{D, E\}|\{A, C\}),$$
$$I_{\pi_2}(D, \{A, B\}|\{C, E\}), I_{\pi_2}(E, \{A, B\}|\{C, D\})$$

but none of the independence statements

$$I_{\pi_2}(A, D|C), I_{\pi_2}(A, E|C), I_{\pi_2}(B, D|C), I_{\pi_2}(B, E|C)$$

is true. From this it follows that

local Markov property $\not\Rightarrow$ global Markov property.

The previous two examples show that in order to be general enough, neither the pairwise Markov property (satisfied by π_1) nor the local Markov property (satisfied by π_2) is adequate as a basis for the interpretation of a possibilistic conditional independence graph. We thus refer to the global Markov property and come to the following definition:

DEFINITION 24. (Possibilistic conditional independence graph) An undirected graph $G = (V, E)$ is called a **conditional independence graph** of a possibility distribution $\pi \in \text{POSS}(\Omega^V)$, if for all disjoint subsets X, Y and Z of variables in V, for which each vertex in X is separated from each vertex in Y by the subset Z (in symbols: $I_G(X, Y|Z)$), it is $I_\pi(X, Y|Z)$.

Note that this definition does not specify a uniquely determined conditional independence graph G of a possibility distribution π. It only requires that all independence statements that result from the topological property of separation in G are valid for π. On the other hand, it is not necessary that all independence relations $I_\pi(X, Y|Z)$ are covered by G. This means that G is an independency map of M_π, but it need not to be a perfect map of M_π. For example, the complete undirected graph $G = (V, V \times V)$ is a conditional independence graph of every possibility distribution on Ω^V, but it does not provide any useful independence statement.

We call a conditional independence graph $G = (V, E)$ of $\pi \in \text{POSS}(\Omega^V)$ **more informative** than a conditional independence graph $G' = (V, E')$ of the same possibility distribution if E is a proper subset of E'. Markov networks (minimal independence maps) then occur as most informative conditional independence graphs.

4 LEARNING POSSIBILISTIC GRAPHICAL MODELS FROM DATA

Graphical models provide a qualitative and quantitative framework for reasoning with uncertainty and independencies. Since we restrict ourselves to deal with finite universes of discourse, flexible software systems support efficient inference techniques [Andersen *et al.*, 1989; Heckerman, 1991; Cowell, 1992]. Some of the well-known evidence propagation algorithms [Lauritzen and Spiegelhalter, 1988; Shafer and Shenoy, 1988] are general in the way that they do not insist on a particular uncertainty calculus, but only require a certain class of conditional independence graphs (for example, directed acyclic graphs or hypertrees), and a definition how to operate on distribution functions, especially referred to the standard operations of extension, projection (marginalization), and conditioning. We do not go into further theoretical details on propagation techniques in possibilistic graphical models here, since they are straight-forward due to the mentioned propagation algorithms in connection with theorem 17, definition 18, definition 19, and theorem 26.

The main interest of this section therefore concerns a different important topic, which is the automated generation of possibilistic graphical models: If expert knowledge is not sufficient to fully specify the qualitative as well as the quantitative part of a possibilistic graphical model, then the question arises whether this specification can be induced with the aid of a database of (imprecise, multi-valued) sample cases.

More particularly, let $\mathcal{D} = (D_j)_{j=1}^m$ denote a database with m sample cases, where $D_j \subseteq \Omega$, $D_j \neq \emptyset$, is interpreted as a set-valued specification of a sample object state $\omega_j \in \Omega$. Reconsidering the semantic underpinnings that we proposed in section 2, the set D_j is assumed to be correct and of maximum specificity w.r.t. ω_j, relative to an underlying specification context c_j of ω_j. This means that $\mathcal{D}$ can be represented and interpreted in terms of an imperfect specification $\Gamma = (\gamma, P_C)$, where $C \stackrel{\text{Df}}{=} \{c_1, \dots, c_m\}$, $\gamma : C \to 2^\Omega$, $\gamma(c) \stackrel{\text{Df}}{=} D_j$ and $P_C(\{c\}) \stackrel{\text{Df}}{=} \frac{1}{m}$, $c \in C$, provided that the probability of selecting context c in order to specify the current object state ω_0 is $\frac{1}{m}$ for all of the involved specification contexts. Assuming the uniform distribution on C is natural as long as there is no preference information about context selection.

Reconsidering definition 9, we know that Γ induces the possibility distribution π_Γ as an information-compressed representation of $\mathcal{D}$. The learning problem consists in finding a best approximating decomposition of π_Γ, relative to a chosen class of possibilistic conditional independence graphs. As already indicated, for practical reasons we will focus our attention to independence graphs $G = (V, E)$ with hypertree structure $H(G) = (V, \mathrm{Cliques}(G))$. Concerning structure induction in the completely different environment of directed acyclic graphs, the reader is referred to [Gebhardt and Kruse, 1996a; Gebhardt and Kruse, 1996b].

In order to solve our specific learning problem, this section is organized as follows: In section 4.1, we investigate the decomposability of possibility distributions. It turns out that for possibilistic graphical models with hypertree structure, there is an analogy to the Hammersley-Clifford-Theorem for Markov networks: Theorem 26 characterizes possibilistic conditional independence graphs $G = (V, E)$ as those that are associated with a factorization (lossless join decomposition) of the underlying possibility distribution $\pi \in \mathrm{POSS}(\Omega^V)$ with respect to the cliques of G. In case that a graph G is not a conditional independence graph of π, the decomposition of π constitutes a new possibility distribution $\pi_{H(G)}$ which is less specific than π. This loss of information can be quantified with help of U-uncertainty being the adequate measure of nonspecificity for possibility distributions [Higashi and Klir, 1983; Klir and Folger, 1988]. In section 4.2, we show how to calculate the nonspecificity $\mathrm{Nonspec}(\pi_{H(G)})$ of $\pi_{H(G)}$ (algorithm 1). The learning problem is to minimize $\mathrm{Nonspec}(\pi_{H(G)}) - \mathrm{Nonspec}(\pi)$ relative to the class of all hypertrees with V as their set of nodes. This problem is identified as being NP-hard, so that similar to learning probabilistic graphical models, heuristics are not avoidable. An additional main result of this section consists in the presentation of a method for calculating an approximation of the tightest hypertree decomposition of π, referred to the U-uncertainty measure of nonspecificity. The corresponding efficient learning algorithm has successfully been applied in several real world examples, one of which is chosen to be discussed in section 5.

4.1 Factorization and Decomposition

In applications of reasonable complexity, multidimensional joint domains Ω^V are intractable as a whole, so that efficient knowledge representation and reasoning techniques should take advantage of decomposition operations that reduce considerations to low-dimensional subspaces of Ω^V.

In database theory such decompositions are often guided by functional dependencies (cf. [Ullman, 1988; Ullman, 1989]). Decomposition based on independence relations between variables has extensively been studied in the field of probabilistic graphical models (cf. [Lauritzen and Spiegelhalter, 1988]). A very important result is the reformulation of the **Hammersley-Clifford-Theorem** [Hammersley and Clifford, 1971; Isham, 1981]. For Markov networks it reads as follows [Besag *et al.*, 1991]:

Any probability distribution p on Ω^V, if strictly positive on Ω^V, satisfies the local Markov property (and thus the pairwise as well as the global Markov property), if and only if $p(V)$ has the functional representation

$$(37)\quad p(V) = \prod_{C \in \mathrm{Cliques}(G)} f_C(C)$$

for some functions $f_C > 0$.

G is supposed to be the conditional independence graph of p. Cliques(G) denotes the set of **maximal cliques** of G, that is the set of all subsets W of V such that W consists of the vertices of a complete subgraph of G, and there is no proper subset of W with the same property.

In this section, we will investigate how to get the possibilistic counterpart of the factorization shown in formula (37).

It turns out that there is in fact an analogous theorem for possibility distributions π on Ω, except from the main difference that the product has to be changed into the min-operator, and that the semi-graphoid nature of conditional possibilistic independence forces us to confine to those cases where the hypergraph $H(G) = (V, \mathrm{Cliques}(G))$ induced by the considered possibilistic conditional independence graph G of π has hypertree structure.

Introducing the possibilistic setting as a generalization of the relational setting, it is convincing to handle the factorization of π in terms of lossless join decomposability with respect to G. This means that we do not tolerate any loss of information when representing π by the family $(\pi(C))_{C \in \mathrm{Cliques}(G)}$ of marginal possibility distributions.

Note that the join of these possibility distributions is given by

$$\begin{aligned}(38)\quad \pi_{H(G)} &\overset{\mathrm{Df}}{=} \widehat{\mathrm{intersect}}_{C \in \mathrm{Cliques}(G)}(\widehat{\mathrm{ext}}_C^V(\widehat{\mathrm{proj}}_C^V(\pi))) \\ &= \widehat{\mathrm{intersect}}_{C \in \mathrm{Cliques}(G)}(\widehat{\mathrm{ext}}_C^V(\pi(C))).\end{aligned}$$

which is the extension of the composition of projection and join from the set-theoretical (relational) to the possibilistic setting, using theorem 17.

It is easy to verify that

$$(39)\quad \pi_{H(G)} \geq \pi,$$

so that $\pi_{H(G)}$ is at most as specific as π, reflecting the potential loss of information accompanied with the decomposition of π to $(\pi(C))_{C \in \mathrm{Cliques}(G)}$.

In case that $\pi_{H(G)} = \pi$ holds, we say that the conditional independence graph G and thus the hypergraph $H(G)$ constitutes a **lossless join decomposition** of π.

With the aid of formula (38), the lossless join decomposability condition leads to the possibilistic counterpart of the probabilistic factorization formula (37):

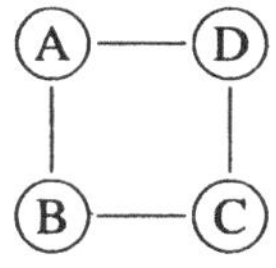

Figure 3. The undirected graph G_3

(40) $\pi(V) = \min\limits_{C \in \text{Cliques}(G)} \pi(C).$

If we are given a possibility distribution $\pi \in \text{POSS}(\Omega^V)$ and a possibilistic conditional independence graph G of π, we could try to derive equality (40) by subsequent application of theorem 21.

The next example clarifies that this is not feasible for arbitrarily chosen topologies of G.

EXAMPLE 25. Let $V \stackrel{\text{Df}}{=} (A, B, C, D)$ and
$\Omega \stackrel{\text{Df}}{=} \{a_1, a_2\} \times \{b_1, b_2, b_3\} \times \{c_1, c_2\} \times \{d_1, d_2\}$.
Referred to the relation

$$R \stackrel{\text{Df}}{=} \{(a_1, b_1, c_1, d_2), (a_2, b_2, c_1, d_1), (a_1, b_3, c_2, d_1)\}$$

and the corresponding possibility distribution $\pi \equiv \mathbb{1}_R$, we consider the following undirected graph $G_3 = (V, E)$: It is easy to see that this 4-cycle is a conditional independence graph for π, since

$$I_\pi(A, C|\{B, D\}) \text{ and } I_\pi(B, C|\{A, C\}).$$

Trying to get the equality

(41) $\pi(V) = \min\{\pi(\{A, B\}), \pi(\{B, C\}), \pi(\{C, D\}), \pi(\{D, A\})\},$

we observe that

$$\pi(a_1, b_1, c_1, d_2) = \pi(a_2, b_2, c_1, d_1) = \pi(a_1, b_3, c_2, d_1) = 1$$

requires

$$\pi(a_1, b_1) > 0, \quad \pi(b_1, c_1) > 0, \quad \pi(c_1, d_1) > 0, \quad \pi(d_1, a_1) > 0$$

and thus

$$\pi(a_1, b_1, c_1, d_1) > 0$$

This is a contradiction to the presupposition that $\pi(a_1, b_1, c_1, d_1) = 0$. We conclude that equality (41) does not hold, and that π is not factorizable in the way expected from formula (40). □

G_3 of example 25 is the trivial case of a non-triangulated graph. Recall that an undirected graph G is **triangulated**, if and only if every simple cycle in G of length strictly greater than 3 possesses a **chord**, that is an edge between two non-subsequent vertices within this cycle.

Since the triangulation property of a graph G implies that its induced hypergraph $H(G)$ is a hypertree (cf. [Kruse *et al.*, 1991]), a possibilistic counterpart of the Hammersley-Clifford-Theorem may at least hold for possibilistic conditional independence graphs G with hypertree structure $H(G)$. In fact, theorem 26 will confirm this presumption.

As a preparation of the proof of this theorem, let us remark that a hypergraph $H = (V, \mathcal{E})$ is called a **hypertree (acyclic)**, if and only if no hyperedge in $\mathcal{E}$ is a subset of another hyperedge in $\mathcal{E}$, and if the hyperedges in $\mathcal{E}$ can be ordered such that the **running intersection property** (RIP) is satisfied [Lauritzen *et al.*, 1984]. This means that there is an ordering $E_1 < E_2 < \ldots < E_k$ on $\mathcal{E}$ such that

$$(42)\quad \forall i \in \{2, \ldots, k\} : \exists n(i) \in \{1, \ldots, i-1\} :$$
$$E_1 \cap (E_1 \cup \ldots \cup E_{i-1}) \subseteq E_{n(i)}.$$

The RIP corresponds to the existence of a **construction sequence** $(E_1, \ldots, E_k)$ for H, so that E_i is a **twig** in the hypertree $H_i \overset{\text{Df}}{=} (V_i, \mathcal{E}_i)$, where $V_i \overset{\text{Df}}{=} E_1 \cup \ldots \cup E_i$ and $\mathcal{E}_i \overset{\text{Df}}{=} \{E_j \mid j \in \{1, \ldots, i\}\}$. The notion of a twig coincides with that of a leaf in the case of ordinary graphs, since hyperedge $E_{n(i)}$ can be regarded as screening the twig E_i against the rest of the hypertree H_i.

THEOREM 26. *(Decomposition Theorem for possibility distributions) Let* $\pi \in \text{POSS}(\Omega^V)$, *and let* $G = (V, E)$ *be an undirected graph with hypertree structure* $H(G) = (V, \text{Cliques}(G))$. G *is a conditional independence graph of* π, *if and only if* π *factorizes into*

$$(43)\quad \pi = \min_{C \in \text{Cliques}(G)} \pi(C)$$

Proof. "$\Rightarrow$": Let G be a conditional independence graph of π. We will show that an appropriate subset of all independence relations $I_G(X, Y|Z)$ that are represented by G suffices in order to satisfy the stated factorization. These independence relations are strongly referable to the intersecting hyperedges in any construction sequence of $H(G)$, and to the special properties of $H(G)$ that result from its hypertree structure.

More particularly, since $H(G)$ is assumed to be a hypertree, there is an ordering $C_1 < C_2 < \ldots < C_k$ on the set $\text{Cliques}(G) = \{C_1, \ldots, C_k\}$ of its hyperedges, such that the running intersection property is fulfilled, and $(C_1, \ldots, C_k)$ is a construction sequence of $H(G)$. Reconsidering the symbolics we used in (42), let $S_i \overset{\text{Df}}{=} C_i \cap C_{n(i)} = C_i \cap V_{i-1}$ denote the set of all vertices in C_i that are already contained in H_{i-1}, and let $R_i \overset{\text{Df}}{=} C_i \setminus S_i$ be the set of new vertices that are added to H_{i-1} when the next hyperedge C_i of the construction sequence is attached to H_{i-1} in order to obtain H_i.

Referring to definition 24 and thus the interpretation of G in terms of the global Markov property, we know that

(44) $I_\pi(R_i, V_{i-1} \setminus S_i | S_i), i = 2, \ldots, k.$

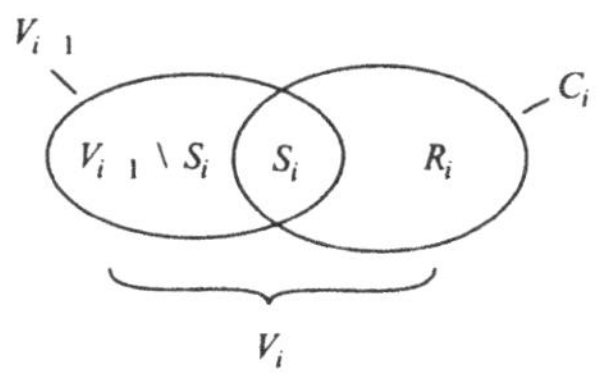

Application of the factorization criterion for possibilistic conditional independence (theorem 21) yields that the above independence relations hold if and only if

$$\pi(R_i \cup (V_i \setminus S_i)) = \min\{\pi(R_i | S_i), \pi(V_{i-1} \setminus S_i | S_i)\}, \quad i = 2, \ldots, k,$$

for arbitrary instantiations of the variables in S_i

Since $\pi(X|Z) = \pi(X \cup Z)$, and observing that $R_i \cup V_{i-1} = V_i$, $R_i \cup S_i = C_i$ and $(V_{i-1} \setminus S_i) \cup S_i = V_{i-1}$, we obtain

$$\pi(V_i) = \min\{\pi(C_i), \pi(V_{i-1})\}, \quad i = 2, \ldots, k.$$

This recursive formula can be reduced to

(45) $\pi(V) = \pi(V_n) = \min_{C \in \mathrm{Cliques}(G)} \pi(C),$

with $V_1 = C_1$.

We conclude that the independence relations of the form (44) are sufficient to verify the stated factorization.

"$\Leftarrow$": In the second part of the proof we have to show that G is a conditional independence graph of π if π factorizes as formalized in (45). More particularly, we have to verify that, for all disjoint subsets X, Y, and Z of vertices in V, if each vertex in X is separated from each vertex in Y by the subset Z of vertices, it is $I_\pi(X, Y|Z)$, which means that

(46) $\pi(X \cup Y|z) = \min\{\pi(X|z), \pi(Y|z)\}$ for all $z \in \Omega^Z$.

Our aim is to transform formula (45) into formula (46), by taking advantage of the separation property and its special consequences when dealing with hypertrees.

Note that separation by Z in an undirected graph means that there is no path from a vertex in X to a vertex in Y that does not contain a vertex from Z. In order to satisfy the separation property, it follows that X must not be in the local neighborhood of Y and vice versa, so that we do not find any hyperedge that contains vertices from X as well as vertices from Y.

Restricting ourselves at first glance to minimal separating subsets Z^*, we see that Z^* is a union of intersections of pairs of hyperedges in $H(G)$. This fact allows to divide the hyperedges of $H(G)$ into two classes $\mathcal{A}$ and $\mathcal{B}$ in the way that for any vertex $v \in V \setminus Z^*$ and all hyperedges $E \in \text{Cliques}(G)$ such that $v \in E$, it is either $E \in \mathcal{A}$ or $E \in \mathcal{B}$. On the other hand, if $v \in Z^*$, then there are two hyperedges E and E' such that $E \in \mathcal{A}$, $E' \in \mathcal{B}$ and $E \cap E' \subseteq Z^*$.

An algorithm that provides the mentioned classification reads as follows:

```
Initialize 𝒜 := {C₁} and ℬ:=∅;
set current class to 𝒜;
for i:=2 to k do
begin
    if Cᵢ ∩ Cₙ₍ᵢ₎ ⊆ Z*
    then change current class
         (i.e. from 𝒜 to ℬ or from ℬ to 𝒜);
    add Cᵢ to the current class of hyperedges
end
```

Using the above classification, it is

$$\begin{aligned} \pi(V) &= \min_{C \in \text{Cliques}(G)} \pi(C) \\ &= \min \left\{ \min_{C \in \mathcal{A}} \pi(C), \min_{C \in \mathcal{B}} \pi(C) \right\}. \end{aligned}$$

The major profit of this representation is that any instantiation of a variable in $V \setminus Z^*$ refers only to hyperedges in a single class and therefore modifies either the first or the second term, but not both of them. Further instantiations of the variables in Z^* do not appear, as all of these variables are presupposed to be instantiated for the conditioning process.

Generalizing our considerations to arbitrary separating subsets Z, we know that $Z = Z^* \cup W$, remembering that Z^* was chosen to be a minimal separating subset of vertices.

Let $z \in \Omega^Z$, $A \stackrel{\text{Df}}{=} \bigcup_{C \in \mathcal{A}} C$ and $B \stackrel{\text{Df}}{=} \bigcup_{C \in \mathcal{B}} C$. Since $\pi(X \cup Y|z)$ is trivial because of the properties of projection, it only remains to prove that $\pi(X \cup Y|z) \geq \min\{\pi(X|z), \pi(Y|z)\}$. We will do it by using the α-cut representation of possibility distributions: Let $\alpha \in [0,1]$, $\omega_A \in \Omega^{X \cup Z}$, $\omega_B \in \Omega^{Y \cup Z}$ such that $\omega_A(Z) = \omega_B(Z) = z$ and $\omega_A = [\pi(X|z)]_\alpha$, $\omega_B = [\pi(Y|z)]_\alpha$. Additionally, let $W_A \stackrel{\text{Df}}{=} W \cap A$ and $W_B \stackrel{\text{Df}}{=} W \cap B$. Then there exist

$\omega_A^* \in [\pi(A)]_\alpha$ such that $\omega_A^*(X \cup Z^* \cup W_A) = \omega_A(X \cup Z^* \cup W_A)$, and
$\omega_B^* \in [\pi(B)]_\alpha$ such that $\omega_B^*(Y \cup Z^* \cup W_B) = \omega_B(Y \cup Z^* \cup W_B)$,

because $X \cup Z^* \cup W_A \subseteq A$ and $Y \cup Z^* \cup W_B \subseteq B$.

We know that $A \cap B = Z^*$ and $\omega_A^*(Z^*) = \omega_B^*(Z^*)$, so that a combination and extension of the tuples ω_A^* and ω_B^* to a single tuple $\omega^* \in \Omega$ with $\omega^*(A) = \omega_A^*$, $\omega^*(B) = \omega_B^*$ is treatable, and all instantiations of the variables in $V \setminus Z^*$ are only referred either to $\omega^*(A)$ or to $\omega^*(B)$, but not to both tuples together.

From the chain

$$\pi(V) = \min_{C \in \mathrm{Cliques}(G)} \pi(C) \geq \min\{\pi(A), \pi(B)\} \geq \pi(V)$$

we see that $\omega_A^* \in [\pi(A)]_\alpha$ and $\omega_B^* \in [\pi(B)]_\alpha$ implies $\omega^* \in [\pi(V)]_\alpha$ and trivially $\omega^*(X \cup Y \cup Z) \in [\pi(X \cup Y \cup Z)]_\alpha$.

Hence, $\omega_A = \omega^*(X \cup Z) \in [\pi(X|z)]_\alpha$ and $\omega_B = \omega^*(Y \cup Z) \in [\pi(Y|z)]_\alpha$ implies $\omega^*(X \cup Y \cup Z) \in [\pi(X \cup Y \cup Z)]_\alpha$ and therefore

$$\pi(X \cup Y|z) \geq \min\{\pi(X|z), \pi(Y|z)\},$$

which completes the proof. ∎

We conclude this section with some remarks on the previous theorem. First, it might be an interesting observation that the hypergraphs $H(G_1)$ and $H(G_2)$ that we referred to in example 22 and 23, respectively, although satisfying the properties of a hypertree, nevertheless do not support a factorization of the type given in formula (43): It is

$$\pi_1 \neq \min\{\pi_1(\{A, B\}), \pi_1(\{B, D\}), \pi_1(\{C, D\})\} \quad \text{and}$$
$$\pi_2 \neq \min\{\pi_2(\{A, B, C\}), \pi_2(\{C, D, E\})\}.$$

Remember that $I_{\pi_1}(\cdot, \cdot|\cdot)$ and $I_{\pi_2}(\cdot, \cdot|\cdot)$ both do not agree with the global Markov property, but only satisfy the pairwise and the local Markov property, respectively.

This way we get an alternative confirmation for the need of the global Markov property in the definition of possibilistic conditional independence graphs: The global Markov property is adequate to provide a basis for the lossless join decomposability of possibility distributions, even if restricted to those distributions which have a most informative conditional independence graph (a minimal independency map) with hypertree structure.

Some additional remark concerning theorem 26 should be made with respect to more general aspects of decomposability: From a theoretical as well as from a practical point of view it is important to know whether a chosen factorization of a distribution function is **unique**, and how one can characterize graphs G for which a factorization is **full** in the sense that it applies every independence relation $I_G(\cdot, \cdot|\cdot)$ associated with G.

Following the ideas of probabilistic graphical modeling, a probability distribution p on Ω^V is called **reducible**, or, in other words, there exists a **decomposition** of p, if and only if there is a partitioning $V = X \cup Y \cup Z$ such that $X \neq \emptyset, Y \neq \emptyset$, $I_p(X, Y|Z)$ and the subgraph on Z in the conditional independence graph of p is complete. The marginal distributions $p(X \cup Z)$ and $p(Y \cup Z)$ w.r.t. $X \cup Z$ and

$Y \cup Z$, respectively, are called the **components** of this decomposition. In case that such a decomposition does not exist, p is said to be **irreducible**.

Related to probabilistic approaches, the concept of decomposition has been made explicit by [Lauritzen, 1982] and [Leimer, 1989]. The aim is to find a sequence of decompositions that reduces p to a system S of **maximal irreducible components**. Maximality avoids redundancy and means that for any distribution $p(A)$ of S there is no proper superset B of A such that $p(B)$ is also an element of S and thus referred to a higher-dimensional subspace Ω^B of Ω.

A probability distribution p on Ω^V is called **decomposable**, if and only is there exists a sequence of decompositions to complete maximal irreducable components.

Whenever one can refer to graph-isomorphs of uncertainty models, it is a pure graph theoretical task to prove that decomposability is given if and only is the associated conditional independence graph is triangulated. In this case, the set of complete maximal irreducible components is uniquely determined (cf. [Whittaker, 1990]).

That is one of the reasons why triangulated graphs have been studied very well in literature. They are also addressed as **rigid circuits** and **chordal graphs**, forming a subset of the **perfect graph**, see [Rose, 1970] and [Golumbic, 1980] for discussion. The characterization of probabilistic decomposable models as those that have triangulated independence graphs was first recognized in [Lauritzen *et al.*, 1984].

In our framework of possibilistic graphical models we again have to take into account that we are not dealing with graph-isomorphs, but only with independency maps. The relationship between the above mentioned decomposability and theorem 26 is as follows:

Let $G = (V, E)$ denote a conditional independence graph of a possibility distribution $\pi \in \text{POSS}(\Omega^V)$. Since G is an independency map of M_π, we know that $I_G(X, Y|Z)$ implies $I_\pi(X, Y|Z)$ for any X, Y, and Z. If G is triangulated, then $H(G)$ has hypertree structure, and π is decomposable with $\text{Cliques}(G)$ being the set of complete maximal irreducable components. Subsequent application of the factorization criterion for possibilistic conditional independence (theorem 21) leads to a sequence of decompositions of π, and finally it results in the factorization shown in formula (43). We thus have found an alternative proof of the first part of theorem 26.

In order to profit from factorization (43), we need a possibilistic conditional independence graph $G = (V, E)$ of π. If $H(G)$ has hypertree structure, then (43) holds. If $H(G)$ is not a hypertree, we transform G into a triangulated graph $G' = (V, E')$, $E \subseteq E'$ and use (43) with respect to G', tolerating a potential loss of information on independence relations when changing over from G to G'. The corresponding triangulation algorithms (so-called **fill-in-algorithms**) are well-known in graph theory, and they are efficient at least in the case that one accepts "good" triangulations instead of dealing with the NP-hard problem of getting triangulations that minimize the number of additional edges in E' compared to E. For more details on fill-in-algorithms, we refer to [Tarjan and Yannakakis, 1984].

4.2 *Tightest Hypertree Decompositions of Possibility Distributions*

Given a possibility distribution $\pi \in \mathrm{POSS}(\Omega^V)$ for an object state $\omega_0 \in \Omega^V$ under consideration, any hypergraph $H = (V, \mathcal{E})$ constitutes a family $(\pi(E))_{E\in\mathcal{E}}$ of marginal distributions that may also serve, accepting a certain loss of information, as a possibilistic specification of ω_0.

Calculating the join π_H of these distributions, defined by

(47) $\pi_H \overset{\mathrm{Df}}{=} \widehat{\mathrm{intersect}}_{E\in\mathcal{E}}(\mathrm{ext}_E^V(\pi(E)))$,

it is $\pi \leq \pi_H$, where equality holds in case of lossless join decomposability of π by H.

If we are given two hypergraphs $H_1 = (V, \mathcal{E}_1)$ and $H_2 = (V, \mathcal{E}_2)$, then we know that $\pi \leq \pi_{H_1}$ and π_{H_2}, but in general there is no specificity relation between π_{H_1} and π_{H_2}. For determining which of the two hypergraphs H_i establishes a "better" decomposition of π and thus a "tighter" approximation π_{H_i} of π, we need a **measure of nonspecificity** of possibility distributions that helps to quantify the loss of information that has to be tolerated when using π_{H_i} instead of π for a possibilistic specification of ω_0.

This section deals with defining such a measure of nonspecificity, and how to use it in order to find tightest hypertree decompositions of possibility distributions.

Reconsidering that possibility distributions represent two different kinds of imperfect information (uncertainty and set-valued imprecision), the measure of nonspecificity that we are searching for is supposed to have meaningful possibilistic counterparts of all properties of **Shannon entropy** [Shannon, 1948] as well as appropriate generalizations of the properties of **Hartley information** [Hartley, 1928]. The so called **U-Uncertainty** measure of nonspecificity [Higashi and Klir, 1983] can be verified as the only one that agrees with the corresponding set of required axioms, which are expansibility, subadditivity, restricted additivity, continuity, monotonicity, minimum and maximum property, branching, and normalization. For details, see [Klir and Mariano, 1987]. The unicity result is similar to that of Shannon entropy with respect to the probabilistic framework [Klir and Folger, 1988]. For practical reasons, we do not insist on normalization here, so that we use the following modified version of U-Uncertainty which satisfies all the mentioned axioms except of normalization when referred to $\sup(\pi)$ rather than 1 as the considered maximum degree of possibility.

DEFINITION 27. For any possibility distribution $\pi \in \mathrm{POSS}(\Omega)$ on a finite reference set Ω, the quantity

$$\mathrm{Nonspec}(\pi) \overset{\mathrm{Df}}{=} \int_0^{\sup(\pi)} \log_2 |[\pi]_\alpha| \, d\alpha$$

is called the **nonspecificity of** π.

$\mathrm{Nonspec}(\pi)$ is the expected amount of information (measured in bits) that has to be added in order to identify ω_0 within the set $[\pi]_\alpha$ of alternatives, assuming the uniform distribution on the set $[0,1]$ of possibilistic confidence levels α.

The loss of information (additional nonspecificity) that we have to accept when using π_H instead of π is quantified by

$$(48)\quad \mathrm{Loss}(\pi_H|\pi) \overset{\mathrm{Df}}{=} \mathrm{Nonspec}(\pi_H) - \mathrm{Nonspec}(\pi) = \int_0^{\sup(\pi)} \log_2 \frac{|[\pi_H]_\alpha|}{|[\pi]_\alpha|}\, d\alpha \geq 0.$$

The aim of decomposition of π is to find a hypergraph H_{opt} within a proper reference class $\mathfrak{H}$ of hypergraphs such that $\mathrm{Loss}(\pi_{H_{opt}}|\pi)$ is minimal relative to $\mathfrak{H}$.

In the following we confine to the reference class $\mathfrak{H}(\mathbf{V})$ of all hypertrees with V as their common set of nodes, and to the subclasses $\mathfrak{H}_l(\mathbf{V})$ of hypertrees whose maximum cardinality of hyperedges is $l \geq 2$.

The nice structural properties of hypertrees support the development of efficient propagation algorithms for inference in possibilistic graphical models, and they also provide a basis for an efficient approximation of tightest hypertree decompositions of possibility distributions.

The solution of our minimization problem is obtained in two steps: The first step is to find a tractable formula for calculating the nonspecificity of π_H, the second is to find (approximations of) tightest hypertree decompositions of π.

Calculating the Nonspecificity of π_H

Suppose that we are given a possibility distribution $\pi \in \mathrm{POSS}(\Omega^V)$ and a hypertree $H \in \mathfrak{H}(V)$ with a construction sequence $(E_1, \ldots, E_k)$. We will use this construction sequence for the successive calculation of $|[\pi_H]_\alpha|$ and thus for the calculation of $\mathrm{Nonspec}(\pi_H)$.

$(E_1, \ldots, E_k)$ induces the sequence $(H_1, \ldots, H_k)$ of hypertrees, defined by $H_i \overset{\mathrm{Df}}{=} (V_i, \mathcal{E}_i)$, $V_i \overset{\mathrm{Df}}{=} E_1 \cup \ldots \cup E_i$ and $\mathcal{E}_i \overset{\mathrm{Df}}{=} \{E_j \mid j \in (1, \ldots, l)\}$. For initialization, let $H_0 \overset{\mathrm{Df}}{=} (\emptyset, \emptyset)$. Analogously to the construction of $H = H_k$, based on the sequence $(H_1, \ldots, H_k)$, we can generate a construction sequence $(\pi_1, \ldots, \pi_k)$ of possibility distributions $\pi_i \in \mathrm{POSS}(\Omega^{V_i})$ such that $\pi_i \equiv \pi_{H_i}$, $i = 1, \ldots, k$, and thus $\pi_H \equiv \pi_k$.

In more formal detail, let

$$[\pi]_\alpha \overset{\mathrm{Df}}{=} \mathrm{ext}_{V_{i-1}}^{V_i}([\pi_{i-1}]_\alpha) \cap \mathrm{ext}_{E_i}^{V_i}([\pi(E_i)]_\alpha),$$

$i = 1, \ldots, k$, where $V_0 \overset{\mathrm{Df}}{=} \emptyset$, $[\pi_0]_\alpha \overset{\mathrm{Df}}{=} \Omega^V$ and therefore $[\pi_1]_\alpha = [\pi(E_1)]_\alpha$.

The cardinalities of these α-cuts can be computed by means of the following recursive scheme:

Let $S_i \overset{\mathrm{Df}}{=} E_i \cap V_{i-1}$ denote the set of all nodes in E_i that are already contained in the hypertree H_{i-1}, and let $R_i \overset{\mathrm{Df}}{=} E_i \setminus S_i$ be the set of new nodes that are added to H_{i-1} when the next hyperedge E_i is included.

Furthermore, for any arbitrarily chosen, but fixed $\alpha \in (0, 1]$, let

$$\begin{aligned} & \mathrm{proj}_{S_i}^{V_i}([\pi_{i-1}]_\alpha) \cap \mathrm{proj}_{S_i}^{E_i}([\pi(E_i)]_\alpha) \\ = & [\pi(S_i)]_\alpha \\ = & \{q_{i,1}, \ldots, q_{i,r_i}\} \end{aligned}$$

For $j = 1, \ldots, r_i$, let

$$\begin{aligned} a_{i,j} & \stackrel{\mathrm{Df}}{=} |\{\omega \in [\pi_{i-1}]_\alpha \mid \mathrm{proj}_{S_i}^{V_{i-1}}(\omega) = q_{i,j}\}|, i = 2, \ldots, k, \\ b_{i,j} & \stackrel{\mathrm{Df}}{=} |\{\omega \in [\pi(E_i)]_\alpha \mid \mathrm{proj}_{S_i}^{E_i}(\omega) = q_{i,j}\}|, i = 1, \ldots, k. \end{aligned}$$

Introducing $a_{1,j} \stackrel{\mathrm{Df}}{=} 1, j = 1, \ldots, r_1$, it follows that

$$(49) \quad |[\pi_i]_\alpha| = \sum_{j=1}^{r_i} a_{i,j} b_{i,j}, i = 1, \ldots, k.$$

According to the construction sequence $(E_1, \ldots, E_k)$, we simply calculate $|[\pi_1]_\alpha|, \ldots, |[\pi_k]_\alpha|$ and therefore $|[\pi_H]_\alpha| = |[\pi_k]_\alpha|$ in the k-th step.

A problem in the above calculations is to get the integers $a_{i,j}$, since the dimensionality and therefore the cardinality of $[\pi_{i-1}]_\alpha$ increases significantly during the construction process. Nevertheless, $|[\pi_H]_\alpha|$ is easy to compute when we choose an adequate organization of the necessary calculations. The following algorithm, written in a PASCAL-oriented notation, shows an efficient solution.

Algorithm 1. (calculates $|[\pi_H]_\alpha|$)
Let $\pi \in \mathrm{POSS}(\Omega^V)$ be a possibility distribution, and let $H \in \mathfrak{H}(V)$ be a hypertree with construction sequence $(E_1, \ldots, E_k)$. Furthermore, we declare and initialize the following integer variables:

```
for all i ∈ {1,...,k} do
    for all ω ∈ Ω^(E_i) do
        if ω ∈ [π_H(E_i)]_α
        then card[E_i, ω] := 1;
        else card[E_i, ω] := 0;
for all i ∈ {2,...,k} do
    for all ω ∈ Ω^(S_i) do
        card[S_i, ω] := 0;
```

Finally, let

$$\mathrm{successors}(i) := \{j \mid n(j) = i\}, \quad i = 1, \ldots, k,$$

denote the set of hyperedges E_j such that E_i is chosen as the hyperedge that screens E_j from the rest of hypertree H_j when E_j is added to H_{j-1} during the construction sequence of H.

Given any $\alpha \in (0, 1]$, the call of the following function delivers the cardinality of $[\pi_H]_\alpha$:

```
function cardinality :  integer;
    procedure collect_information_for_hyperegde(i:integer);
        if successors(i) ≠ ∅
        then for all j ∈ successors(i) do
                 collect_information_for_hyperedges(j)
        else begin
             for all ω ∈ Ω^{E_i} do
                 card[S_i, proj^{E_i}_{S_i}(ω)] := card[S_i, proj^{E_i}_{S_i}(ω)]
                                                 + card[E_i, ω];
             for all ω ∈ Ω^{E_n(i)} do
                 card[E_n(i), ω] := card[E_n(i), ω]
                                    * card[S_i, proj^{E_i}_{S_i}(ω)]
             end;

begin (* cardinality *)
    collect_information_for_hyperedge(1);
    cardinality := 0;
    for all ω ∈ Ω^{E_1} do
        cardinality := cardinality + card[E_1, ω];
end (* cardinality *)
```

□

The correctness of algorithm 1 is easy to see by recognizing that the then-part of the conditional statement in procedure `collect_information_for_hyperedge` results in

$$\mathrm{card}[E_i, \omega] = |\{\omega' \in [\pi_H(W_i)]_\alpha \,|\, \mathrm{proj}_{E_i}^{W_i}(\omega') = \omega\}|$$

for all $\omega \in \Omega^{E_i}$, where W_i has the following recursive definition:

$$W_i \overset{\mathrm{Df}}{=} \mathrm{nodes}(i),$$
$$\mathrm{nodes}(i) \overset{\mathrm{Df}}{=} E_i \cup \mathrm{nodes}(\mathrm{successors}(i)), \quad i = 1, \ldots, k.$$

Since $W_i = V$, the call of `collect_information_for_hyperedge(1)` yields

$$\mathrm{card}[E_i, \omega] = |\{\omega' \in [\pi_H]_\alpha \,|\, \mathrm{proj}_{E_i}^{V}(\omega') = \omega\}|$$

for all $\omega \in \Omega^{E_i}$ and therefore

$$\texttt{cardinality} = |[\pi_H]_\alpha|.$$

This result makes it simple to compute $\mathrm{Nonspec}(\pi_H)$ and $\mathrm{Loss}(\pi_H|\pi)$:

Let $\mathrm{degrees}(\pi) \overset{\mathrm{Df}}{=} \pi(\Omega^V) \cup \{0\}$ denote the set of all possibility degrees of π. By $\pi \in \mathrm{POSS}(\Omega^V)$, we assume that $|\pi(\Omega^V)| < \infty$, so that we can write degrees(π) $= (\alpha_0, \ldots, \alpha_q)$, where $0 = \alpha_0 < \alpha_1 < \ldots < \alpha_q$. With these presuppositions, we obtain:

$$\begin{aligned}\text{Loss}(\pi_H|\pi) &= \text{Nonspec}(\eta) - \text{Nonspec}(\pi) \\ &= \sum_{j=1}^{q}(\alpha_j - \alpha_{j-1}) \cdot \log_2 \frac{|[\pi_H]_{\alpha_j}|}{|[\pi]_{\alpha_j}|}.\end{aligned}$$

Tightest Hypertree Decomposition of π

Algorithm 1 is helpful for the calculation of $\text{Loss}(\pi_H|\pi)$ if a certain hypertree $H \in \mathfrak{H}_l(V)$ is given. However, in order to solve our minimization problem, we have to search for $H_{\text{opt}}\ in \mathfrak{H}(V)$ such that

(50) $\text{Nonspec}(\pi_{H_{\text{opt}}}) = \min_{H \in \mathfrak{H}(V)} \text{Nonspec}(\pi_H)$

Direct minimization is crucial because of the cardinality of $\mathfrak{H}(V)$, and its non-trivial subclasses $\mathfrak{H}_l(V)$. Furthermore, the complexity of the minimization problem itself does not support to explore a tractable procedure for an exact solution. Note that we have to deal with the problem of **strong identifiability**:

Let $\Pi = \text{POSS}(\Omega^V)$ denote the class of all possibility distributions on Ω^V, and let $\Pi(\mathfrak{H})$ be the subclass of all possibility distributions in Π that have a lossless join decomposition constituted by a hypergraph taken from a certain class $\mathfrak{H}$ of hypergraphs with node set V. The class $\Pi(\mathfrak{H})$ is called **identifiable** relative to Π, if there is an algorithm A, polynomial in the size of its input and output, such that for every possibility distribution $\pi \in \Pi$, the algorithm A decides whether $\pi \in \Pi(\mathfrak{H})$ (**recognition** of decomposability), and if the answer is positive, then A finds a hypergraph $H \in \mathfrak{H}$ with $\pi_H \equiv \pi$ (**description** of decomposability).

$\Pi(\mathfrak{H})$ is said to be **strongly identifiable** relative to Π, if it is identifiable relative to Π, and if A always finds a tightest approximation of π in $\Pi(\mathfrak{H})$, which means that the minimization problem (50) is solved.

Identifiability and strong identifiability have been considered for the more special problem of structure identification in relational data (decomposing a single α-cut $[\pi]_\alpha \subseteq \Omega^V$ of a given possibility distribution π is exactly this problem), so that we can profit from some results on complexity problems that have been found in the corresponding field of research:

If we are given a hypergraph H and a relation $R \subseteq \Omega^V$, then only in cases where H is tractable (for instance, if H is a hypertree) one can tractably decide the lossless join decomposability of R relative to H. Deciding lossless join decomposability relative to a **class** of hypergraphs seems to be a harder task which is presumably intractable even in cases where each individual member of the class is tractable [Dechter and Pearl, 1992].

For the general case of arbitrary hypergraphs, the NP-hardness of this problem is well-known from database theory.

At first glance, these results are not very encouraging, and there are in fact only few efficient structure identification algorithms for quite special classes of hypergraphs in the relational framework. As an example, it can be shown that the class of all tree-structured relations in Ω^V is identifiable relative to all relations in Ω^V. The description of the topology of a corresponding tree is obtained in $O((|R|+\log_2|V|)\cdot|V|^2)$ time [Dechter, 1990]. Assuming the uniform distribution on R, the underlying approach is based on a procedure for inducing a maximum weight spanning tree (MWST) of a tree-structured Bayesian network from data [Chow and Liu, 1968]. On the other hand, opposite to the probabilistic setting, it is not clear so far, whether the MWST algorithm leads to a tightest approximation in the sense of strong identifiability of tree-structured relations. A corresponding conjecture referred to experimental results has been formulated in [Dechter and Pearl, 1992].

As a consequence of the outlined complexity problems, heuristics are not avoidable when we search for an efficient algorithm that solves our possibilistic minimization problem (50). The idea is to substitute formula (49) for calculating $|[\pi_i]_\alpha|$ by a better treatable approximating formula which helps in finding a "good" decomposition of π relative to the class $\mathfrak{H}_l(V)$. The technique is similar to the MWST algorithm in the way that for all possible hyperedge $E \subseteq V$ such that $1 \leq |e| \leq l$, we calculate an ordering of weights $\omega_\pi(E)$ as a basis for constructing $H_{\text{opt}} \in \mathfrak{H}_l(V)$ or a good approximation of it.

A promising substitution of formula (49) reads as follows:

$$\text{(51)}\quad X_i \stackrel{\text{Df}}{=} \sum_{j=1}^{r_i} a_{i,j}\, b_{i,j} \stackrel{(49)}{=} |[\pi_i]_\alpha| \quad\approx\quad Y_i \stackrel{\text{Df}}{=} \frac{1}{r_i}\left(\sum_{j=1}^{r_i} a_{i,j}\right)\left(\sum_{j=1}^{r_i} b_{i,j}\right).$$

Evaluating the quality of this approximation is quite difficult, since considering $a_{i,j}$ and $b_{i,j}$ as random variables on sets $\{1,2,\ldots,s\}$ of natural numbers leads to non-trivial distributions of the random variables $|X_i - Y_i|$. Besides these structural problems, a further question is which distribution assumptions should be made for the variables $a_{i,j}$ and $b_{i,j}$ when applied to practical examples in order to calculate confidence intervals for $|X_i - Y_i|$. By investigating the performance in real world applications (see section 5), it turned out that the approximation $X_i \approx Y_i$ in many cases is sufficient to get the correct ordering of hyperedge weights.

Note that viewing the cardinalities $a_{i,j}$ and $b_{i,j}$ as uniformly distributed random variables on $\{1,2,\ldots,s\}$, it is easy to verify that, for $a \to \infty$ the random variables X_i and Y_i almost sure share the same limiting distribution.

Let us now have a more precise look at the calculations of the weights $\omega_\pi(E)$: First, we apply formula (51) to simplify the calculation of $|[\pi_i]_\alpha|$:

$$\begin{aligned}
|[\pi_i]_\alpha| \quad &\approx\quad \frac{1}{r_i}\left(\sum_{j=1}^{r_i} a_{i,j}\right)\left(\sum_{j=1}^{r_i} b_{i,j}\right)\\
&=\quad \frac{1}{r_i}|[\pi_{i-1}]_\alpha|\cdot|[\pi(E_i)]_\alpha|
\end{aligned}$$

$$= \quad |[\pi_{i-1}]_\alpha| \cdot \frac{|[\pi(E_i)]_\alpha|}{|[\pi(S_i)]_\alpha|},$$

and thus

$$|[\pi_H]_\alpha| = |[\pi_k]_\alpha| \approx \prod_{i=1}^{k} \frac{|[\pi(E_i)]_\alpha|}{|[\pi(E_i \cap V_{i-1})]_\alpha|}.$$

Further simplification is reached as follows: Let $\Omega_\pi^{(\alpha)} \overset{\text{Df}}{=} \times_{\nu \in V} [\pi(\{\pi\})]_\alpha$. Then

$$|[\pi_H]_\alpha| \approx |\Omega_\pi^{(\alpha)}| \cdot \prod_{i=1}^{k} \frac{|[\pi(E_i)]_\alpha|}{|[\pi(E_i \cap V_{i-1})]_\alpha| \cdot \prod_{\nu \in E_i \setminus V_{i-1}} |[\pi(\{\nu\})]_\alpha|}.$$

As an additional approximation, one may consider that

$$|[\pi(E_i \cap V_{i-1})]_\alpha| \leq \prod_{\nu \in E_i \setminus V_{i-1}} |[\pi(\{\nu\})]_\alpha|,$$

where equality holds if $|E_i \cap V_{i-1}| = 1$. In this case,

$$|[\pi_H]_\alpha| \approx |\Omega_\pi^{(\alpha)}| \cdot \prod_{i=1}^{k} \frac{|[\pi(E_i)]_\alpha|}{\prod_{\nu \in E_i} |[\pi(\{\nu\})]_\alpha|}$$

and therefore

$$\text{(52)} \quad \text{Nonspec}(\pi_H) \approx c + \sum_{i=1}^{k} \omega_\pi(E_i),$$

where

$$\text{(53)} \quad \omega_\pi(E_i) = \int_0^{\sup(\pi)} \log_2 \frac{|[\pi(E_i)]_\alpha|}{\prod_{\nu \in E_i} |[\pi(\{\nu\})]_\alpha|} \, d\alpha,$$

and

$$c = \int_0^{\sup(\pi)} \log_2(|\Omega_\pi^{(\alpha)}|) \, d\alpha$$

is a summand that does **not** depend on H.

An approximating candidate H_{appr} of an optimal hypertree $H_{\text{opt}} \in \mathfrak{H}_l(V)$ is easily found by calculating $\omega_\pi(E)$ for all $E \subseteq V$ such that $1 \leq |E| \leq l$, then ordering these hyperedges E with respect to their increasing weights $\omega_\pi(E)$, and finally constructing H_{appr} with respect to this ordering, where hyperedges of lower cardinality are preferred if decomposition in comparison to hyperedges with higher cardinality does not lose information.

Note that given three non-empty subsets E, E_1 and E_2 of nodes such that $E = E_1 \cup E_2$, it is

$$\begin{aligned} \mathrm{Nonspec}(\pi(E)) &= \int_0^{\sup(\pi)} \log_2 |[\pi(E)]_\alpha| \, d\alpha \\ &= \int_0^{\sup(\pi)} \log_2 \frac{|[\pi(E_1)]_\alpha| \cdot |[\pi(E_2)]_\alpha|}{|[\pi(E_1 \cap E_2)]_\alpha|} \, d\alpha \\ &= \mathrm{Nonspec}(\pi(E_1)) + \mathrm{Nonspec}(\pi(E_2)) \\ &\quad - \mathrm{Nonspec}(\pi(E_1 \cap E_2)). \end{aligned}$$

Nevertheless, we have to consider that approximation assumptions have been made in order to come to formula (52), so that we deal with the approximation

$$\omega_\pi(E) \approx \omega_\pi(E_1) + \omega_\pi(E_2) - \omega_\pi(E_1 \cap E_2),$$

where $\omega_\pi(\emptyset) \overset{\mathrm{Df}}{=} 0$.

For this reason, when constructing the hypertree, E_1 and E_2 are preferred to E, if and only if

$$|\omega_\pi(E) - \omega_\pi(E_1) - \omega_\pi(E_2) + \omega_\pi(E_1 \cap E_2)| < \epsilon(\pi, E),$$

where $\epsilon(\pi, E)$ is an adequate error bound.

In practical experiments (see section 5) it turned out that in most cases the decomposition results are not very sensitive to the particular choice of $\epsilon(\pi, E)$, which reflects one aspect of stability of the proposed structure induction techniques in real-world examples.

For instance,

$$\epsilon(\pi, E) := \left| \frac{\omega_{\pi(E)}}{50} \right|$$

occurs as a reasonable assignment.

Concerning the computational effort that is needed in order to find H_{appr}, we make the following short analysis: Provided that random access on the possibility degrees $\pi(\omega)$, $\omega \in \Omega$, is available, it is easy to see that a straight-forward implementation for the calculation of the weights $\omega_\pi(E)$ for all $E \subseteq V$ such that $1 \leq |E| \leq l$ runs in $O(T(n,l)mr^l)$ time, where $T(n,l) \overset{\mathrm{Df}}{=} \sum_{j=1}^{l} \binom{n}{j}$, and $r \overset{\mathrm{Df}}{=} \max\{|\Omega_i|, i = 1, \ldots, n\}$ is the maximal cardinality of the individual domains of the considered variables. The amount of calculations that are necessary to get a representation of all $\pi(\omega)$-values increases with the imprecision in the underlying database, indicated by the cardinality of the (set-valued) sample cases.

Performance can be improved by Greedy search methods in order to find a subset of hyperedges with promising weights that could be sufficient for generating an optimal hypertree or a good approximation of it.

General remarks on stability and approximation quality can only be made with the mentioned restrictions. The axioms that are satisfied by the chosen measure of nonspecificity guarantee that there is no counterintuitive behavior when decreasing or increasing the specificity of the possibility distribution that has to be decomposed. On the other hand, dealing with the NP-hardness of the decomposition problem itself, makes it sure that there are possibility distributions, for which the approximation in formula (52) leads to bad proposals of decompositions. However, the approximation quality can always be valuated by applying formula (48) in order to calculate the real loss of information when using the hypertree decomposition delivered by using the weights $\omega_\pi(E)$ instead of the original possibility distribution π.

Note that the calculation of $|[\pi]_\alpha|$ with respect to formula (48) is tractable, if π is not arbitrarily chosen, but representable as $\pi \equiv \pi_\Gamma$, where π_Γ is induced by an imperfect specification Γ that serves as an interpretation of a given database $\mathcal{D}$ of sample cases. The relatively low number of sample cases, when compared to the cardinality of Ω^V, limits the computational effort that is necessary in order to handle the possibility distribution π.

A final remark on formula (52) concerns its relationship to probabilistic decomposition techniques: Note that the term

$$\log_2 \frac{|[\pi(E_i)]_\alpha|}{\prod_{\nu \in E_i} |[\pi(\{\nu\})]_\alpha|}$$

can be viewed as the possibilistic counterpart of the **Kullback-Leibler cross entropy measure** [Kullback and Leibler, 1951; Shore, 1986] for the distance of nonspecificity between the two distributions $\pi(E_i)$ and

$$\begin{aligned} \pi'(E_i) &\stackrel{\text{Df}}{=} \widehat{\text{intersect}}_{\nu \in E_i}(\text{ext}_{(\nu)}^{E_i}(\pi(\{\nu\}))) \\ &= \min_{\nu \in E_i}(\pi(\{\nu\})), \end{aligned}$$

where $\pi'(E_i)$ is constructed by assuming the possibilistic independence of all variables in E_i.

5 APPLICATION

For a practical evaluation of possibilistic graphical models, we now present a non-trivial example that served as a referential application in the Esprit III BRA 6156 DRUMS (Defeasible Reasoning and Uncertainty Management Systems) with respect to knowledge representation and inference mechanisms. The example was provided by S. Lauritzen from Aalborg University and L.K. Rasmussen from Dina Foulum, Tjele, in Denmark. It concerns genotype determination and parentage verification of Danish Jersey cattle in the F-blood group system. In the original application, the modelling was referred to a non-singly connected Bayesian network, so that there is also a basis for a problem-oriented comparison of probabilistic graphical models with their possibilistic counterparts. From the viewpoint of this contribution, our special interest is devoted to a database of sample cases with lots of missing values that has been made available for this application. We use our learning algorithm to induce a possibilistic graphical model that best fits the database, with the goal of obtaining a conditional independence graph that corresponds to the DAG structure in the Bayesian network, which in the original application was designed by an expert for cattle breeding.

Calculations are supported by POSSINFER [Gebhardt and Kruse, 1996d], which is a software tool for reasoning with possibilistic graphical models, mainly developed at the Mathematics and Computer Science Department of the University of Braunschweig, but within cooperations also made available to Deutsche Aerospace in Friedrichshafen, Germany, for a data fusion demonstrator [Beckmann *et al.*, 1994], and the Daimler-Benz Research Center in Ulm, Germany, for data mining problems [Borgelt *et al.*, 1996]. In its current version, POSSINFER fully integrates all concepts, theorems, and algorithms of this contribution, referred to evidence propagation algorithms as well as learning techniques.

5.1 Example: Blood Typing in Cattle Breeding Programs

In many laboratories throughout the world, bovine blood typing is carried out as a regular part of modern cattle breeding programs. Due to the introduction of embryo transplantation technology and increasing international trade with semen and embryos, the primary purpose is to establish sophisticated methods for genotype determination and parentage verification for pedigree registration. [Rasmussen, 1992; Rasmussen, 1995] contain a description of using Bayesian networks and the software tool HUGIN [Andersen *et al.*, 1989] as an appropriate modelling framework for the associated reasoning problems. The application domain refers to 52 blood group factors for blood typing of Danish Jersey cattle, controlled by 10 blood group systems. Two of these systems are rather complicated, controlling 26 and 10 of the 52 blood group factors, respectively. For presentational issues, the F-blood group system is a convenient one, since it is of sufficient complexity, although restricted to only 4 different blood group factors. This is the reason why we will use it for the following discussion.

The F-blood group system

Blood group determination of cattle is supported by photometer measurements. For each blood group factor, the reaction between an antibody to this blood group factor and the blood group factor itself is quantified by lysis values, which are integer values varying from 0 (no reaction, the blood group factor is absent) through 7 (very high reaction, the blood group factor is present). The F-blood group system distinguishes 4 blood group factors (F1, F2, V1, and V2). The presence or absence of a blood group factor depends on the underlying genotype, which is specified by one of the six values F1/F1, F1/V1, F1/V2, V1/V1, V1/V2, and V2/V2, respectively. The chosen notation reflects that a genotype is composed of two alleles.

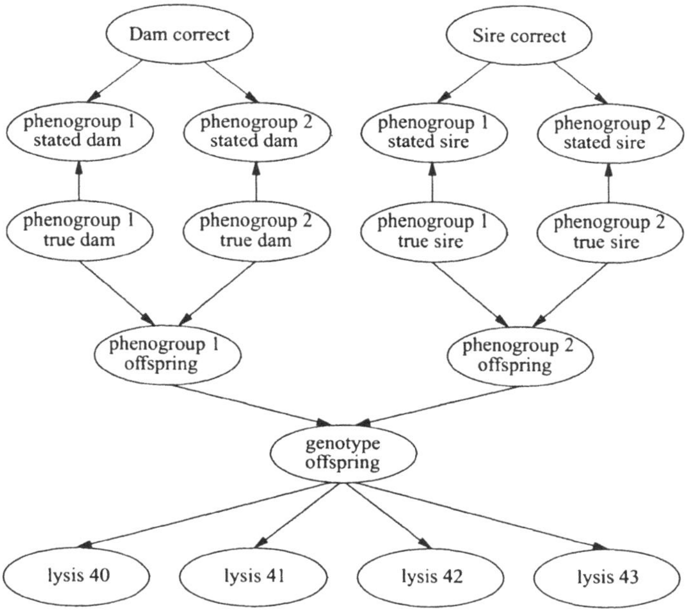

Figure 4. Graphical model for genotype determination and parentage verification of Danish Jersey cattle in the F-blood group system

Each allele contains a subset of the observed blood group factors, called its associated phenotype.

The blood group factors occur, however, only in a small number of combinations, the so-called phenogroups. In the F-blood group system, three phenogroups are known, namely F1F2, V1V2, and V2F2, abbreviated by F1, V1, and V2, where F1 implies the occurrence of F2, V1 implies the occurrence of V2, and V2 implies the occurrence of F2. The two phenogroups associated with the genotype are influenced by the four phenogroups associated with the genotypes of the considered animal's parents. These phenogroups are therefore denoted by phenogroup 1 offspring, phenogroup 2 offspring, phenogroup 1 true dam, phenogroup 2 true dam, phenogroup 1 true sire, and phenogroup 2 true sire, respectively. In case that the stated parentage is not true, the phenogroups of the real parents are not observable, but they might be deducable with uncertainty from evidential information. Inference is based on the dependency structure between the true phenogroups, the stated phenogroups, and a potential parental error, referred to one or both of the parents.

The qualitative part of the general knowledge about the dependency structure of all mentioned variables is shown in the directed acyclic (probabilistic) independence graph of figure 4.

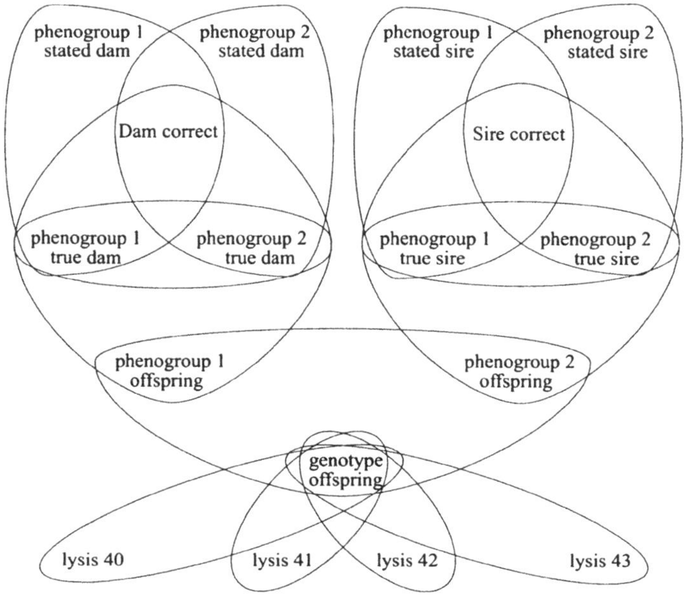

Figure 5. Hypertree representation of the DAG in figure 4

As already indicated, knowledge representation in the original application of this real-world example is referred to a Bayesian network provided by an expert for cattle breeding [Rasmussen, 1992].

The conditional independence graph reflects decomposability of the 17-dimensional probability distribution (as the quantitative part of the network) in that the expert only needs to specify 336 conditional probabilities on subspaces of at most three dimensions, whereas the whole universe of discourse Ω has 5.804.752.896 elements.

5.2 *Solution with Possibilistic Graphical Models*

We use the blood group determination example as a practical basis for the discussion of the theoretical results obtained in the previous sections of this chapter. Since we deal with hypertree-structured possibilistic graphical models, we do not refer to the directed independence graph in figure 4, but transfer it, by moralization and triangulation, to its corresponding hypertree representation. The result is shown in figure 5.

The aim is to learn this hypertree (i.e., its associated conditional independence graph) from a database of sample cases made available by [Rasmussen, 1992]. Unfortunately, the provided database is restricted to samples that refer only to the lysis factors, the phenogroups of the stated parents, and the genotype.

```
ORDER OF ATTRIBUTES:
lysis 40

phenogroup 2 true dame
phenogroup 1 true sire
phenogroup 2 true sire
dam correct
sire correct

SAMPLE CASES:
123:     0 6 0 6 f1 v2 f1 v2 v2v2 v2 v2 f1 v2 f1 v2 y y
124:     7 6 0 7 f1 v2 ** ** f1v2 ** ** f1 v2 ** ** y y
125:     7 7 0 0 f1 v2 f1 f1 f1f1 f1 f1 f1 v2 f1 f1 y y
126:     7 7 0 0 f1 v2 f1 f1 f1f1 f1 f1 f1 v2 f1 f1 y y
127:     7 7 0 7 f1 v2 f1 v1 f1v2 v2 f1 f1 v2 f1 v1 y y
128:     6 6 0 0 f1 f1 ** ** f1f1 f1 f1 f1 f1 ** ** y y
129:     0 5 4 5 f1 v1 ** ** v1v2 v1 v2 f1 v1 ** ** y y
130:     7 7 6 7 f1 v2 f1 v1 f1v1 f1 v1 f1 v2 f1 v1 y y
131:     7 7 7 7 f1 v2 ** ** f1v1 f1 v1 f1 v2 ** ** y y
132:     7 7 0 0 f1 v1 ** ** f1f1 f1 f1 f1 v1 ** ** y y
133:     7 7 0 7 f1 v2 f1 f1 f1v2 v2 f1 f1 v2 f1 f1 y y
```

Figure 6. Part of the database for the blood group determination example

We extend this database to an artificial database by sampling the 8 remaining variables with respect to the given probabilistic expert knowledge that was used to design the Bayesian network in figure 4. The resulting database $\mathcal{D}^*$ consists of 500 sample cases with lots of missing values, denoted by "**".

To give an impression of this database, part of it is shown in figure 6.

Missing values are interpreted differently from total ignorance of a variable's instantiation in the sense that sample cases with missing values are substituted by the set of all tuples that can be derived when instantiating the variables with missing values, but excluding those tuples that do not occur in any other sample case. These tuples are not supported by the rest of the database and are thus supposed to be non-occurring, irrelevant as possible candidates for any current object state $\omega_0 \in \Omega$.

The next step of our consideration is to learn from the modified database a possibilistic graphical model $\mathcal{N} = (G, \Pi)$ such that $H(G)$ constitutes (an approximation of) the tightest hypertree decomposition of the possibility distribution π_Γ that is induced by the database's representation in form of an imperfect specification $\Gamma = (\Omega, C, \gamma, P_C)$, with denoting the probability measure for the uniform distribution on the set C of specification contexts for the sample cases.

Figure 7 shows the initial window of a POSSINFER learning session presenting the whole set V of variables of interest. Each variable has to be specified by fixing

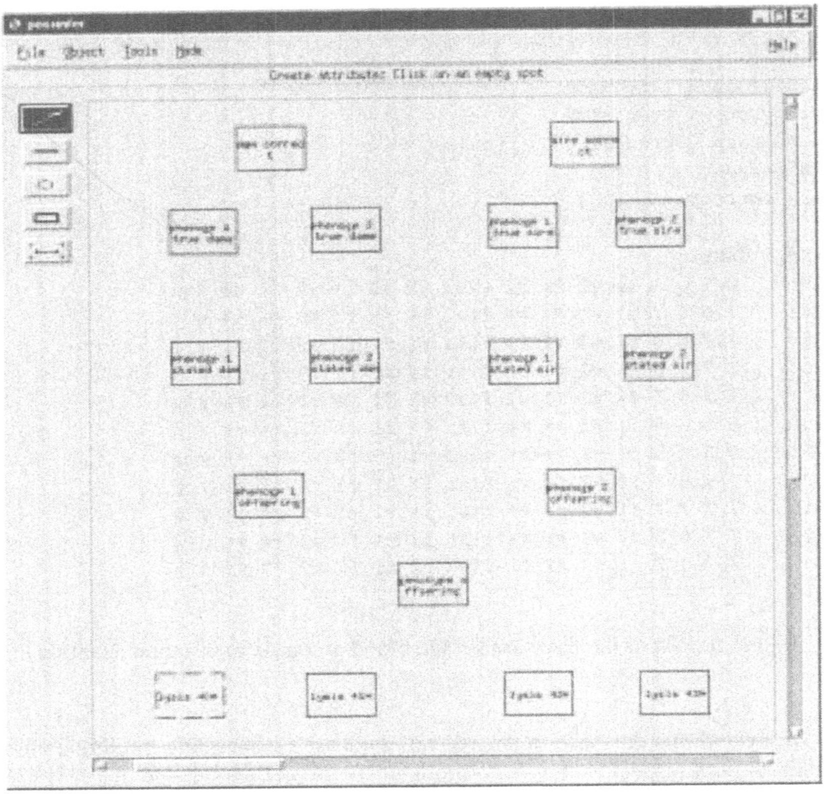

Figure 7. POSSINFER - Initial state of the learning process with all variables of the blood group determination example

its domain (real, integer, char, enum), where the choice of ranges and partitions is to ensure the finiteness of the subset of those representative values out of the domain that are referred to in the quantitative part of the graphical model. Furthermore, parameterized classes of possibility distributions may be specified and named in order to support the input of (possibilistic) evidence for inference carried out after the graphical model has been generated (see figure 8 and figure 9).

A learning editor (figure 10) supports the selection of those variables for which a database of sample cases is available, so that the linking edges to additional variables can be included afterwards based on expert knowledge.

As a first example, we want to induce the possibilistic graphical model that best fits our database, but excluding the two variables *dam correct* and *sire correct*. This means, looking at the original Bayesian network in figure 4, that we deal with a

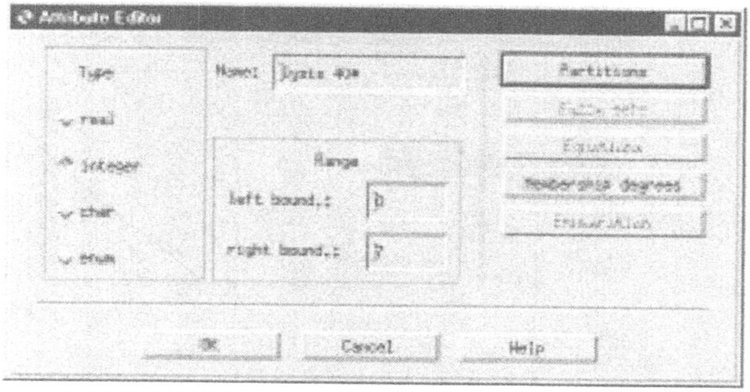

Figure 8. Editor for the variable lysis 40

singly-connected subgraph of its directed acyclic conditional independence graph. The learning algorithm (based on formula (52)) is applied to the class $\mathfrak{H}_3(V)$ of hypertrees with 3 as their maximum cardinality of hyperedges, fixing an upper error bound $\epsilon(\pi, E) := 3\%$.

Starting with 500 sample cases in $\mathcal{D}^*$, the described treatment of missing values leads to a modified database $\mathcal{D}$ with 384 supporting (imprecise) sample cases (figure 11). The induced hypertree H (figure 12) totally coincides with the structure that the expert provided in the original application (figure 4).

The approximation quality can be quantified as follows: The nonspecificity of the possibility distribution π as our information-compressed representation and interpretation of $\mathcal{D}$ is $\text{Nonspec}_{\mathcal{D}}(\pi) = 3.1618$, which means that the expected amount of information to be added in order to identify any current object state $\omega_0 \in \Omega$ with the aid of π is 3.1618 bits. Using the proposed approximating hypertree decomposition π_H of π, the nonspecificity increases to $\text{Nonspec}_{\mathcal{D}}(\pi_H) = 4.3338$ (see figure 11). This is the price that we have to pay when reducing considerations from the 15-dimensional common domain of all variables, except for *dam correct* and *sire correct*, to the family of 3-dimensional domains associated with the hypertree decomposition.

To give an example for evidence propagation through the induced possibilistic graphical model, suppose that we examine a particular calf, for which we get the instantiations *lysis40* = 4, *lysis41* = 5, *lysis42* = 0, and *lysis43* = 3. Figure 13 shows the posterior possibility distribution for the genotype of this calf. It says that F1/F1 and F1/V2 remain as the only plausible values for the genotype, where F1/F1 is preferred to F1/V2. From a decision support point of view, these propagation results correspond to those that are obtained when using the Bayesian network in figure 4, where the posterior probabilities of the genotype are calculated as P(*genotype* = F1/F1) = 0.61 and P(*genotype* = F1/V2) = 0.34 (see [Rasmussen, 1992]).

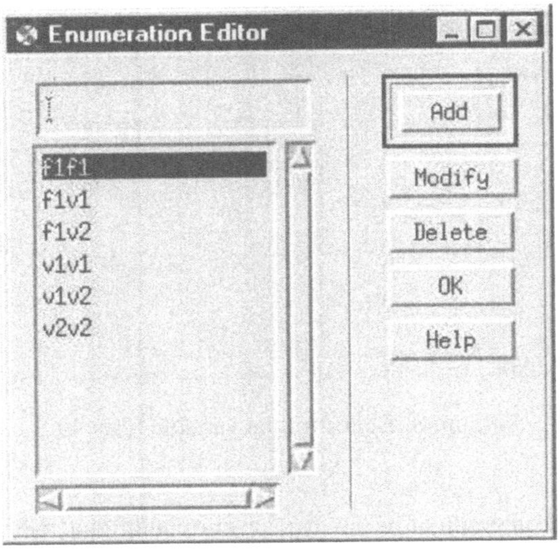

Figure 9. The six possible values of the variable genotype

The four other possible values of this variable have small positive posterior probabilities, more expressive than the possibilistic result, but also not taken into account for crisp decisions with respect to the genotype.

Involving the variables *dam correct* and *sire correct* into our considerations, we refer to the more complicated structure of a non-singly connected directed acyclic conditional independence graph (see figure 4). In opposite to relational dependencies between genotype and phenogroups, and sufficiently supported uncertain dependencies between genotype and measurable lysis factors, there is only little support in the database for parentage verification, since parental errors are exceptional cases with a relative frequency of less than 2%.

The information compression that occurs when interpreting the database $\mathcal{D}$ by π leads to the problem that rarely supported dependencies partly lose their extractability from π. The learning algorithm is thus not able to identify them and delivers the network structure shown in figure 14, even when enlarging the database $\mathcal{D}^*$ to an artificial database of 2000 sample cases.

The provided hypertree is an extension of the hypertree shown in figure 12, by linking the nodes *dam correct* and *sire correct* to lysis factors. An improved selection technique for hyperedges when generating the hypertree decomposition, leads to the graph in figure 15.

This hypertree coincides with that of figure 5, except of linking the variable *dam correct* to the sire's phenogroups. Since the relative frequency of parental errors purely caused by stating the wrong dam is only 0.18%, the information compression in the possibilistic approach identifies such cases as erroneous samples rather

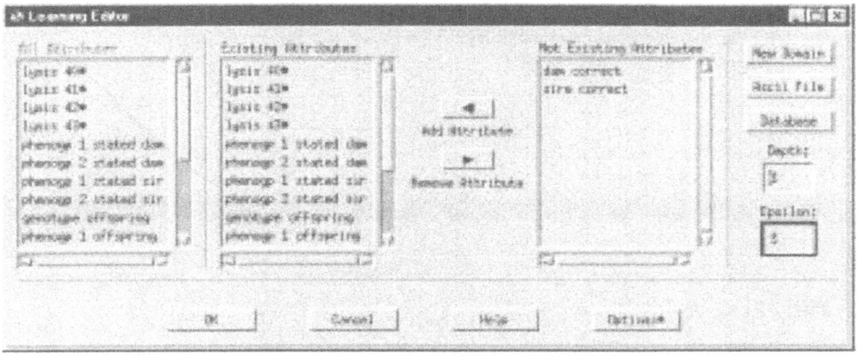

Figure 10. Selection of variables for the learning algorithm

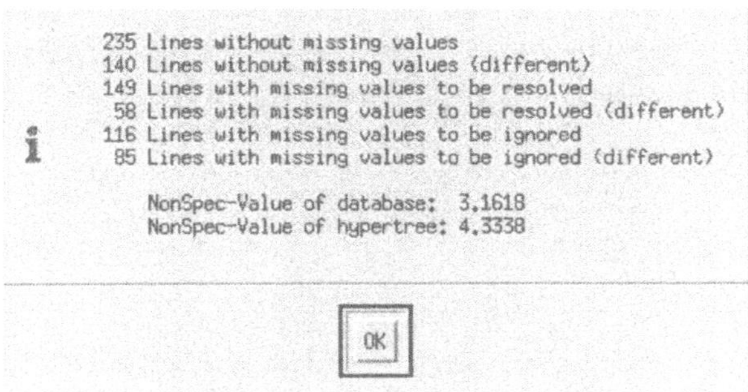

Figure 11. Information provided by the learning algorithm

than important exceptions to be recognized. From a nonspecificity point of view it turns out that the variable *dam correct* has very low influence to the other variables: The nonspecificity of the hypertree decomposition in figure 14 is 4.3929. With respect to figure 15 it decreases to 4.3865, whereas the nonspecificity of the decomposition relative to the hypertree in figure 5 is 4.3734. Additional tests referred to the quality of the decomposition delivered by the learning algorithm confirmed its robustness against reasonable variations of the error bound $\epsilon(\pi, E)$ and reasonable variations of the nonspecificity of the sample cases. More remarks on evaluation aspects, concerning possibilistic graphical models to their own, but also compared to probabilistic graphical models, are left to the concluding discussion in the next section.

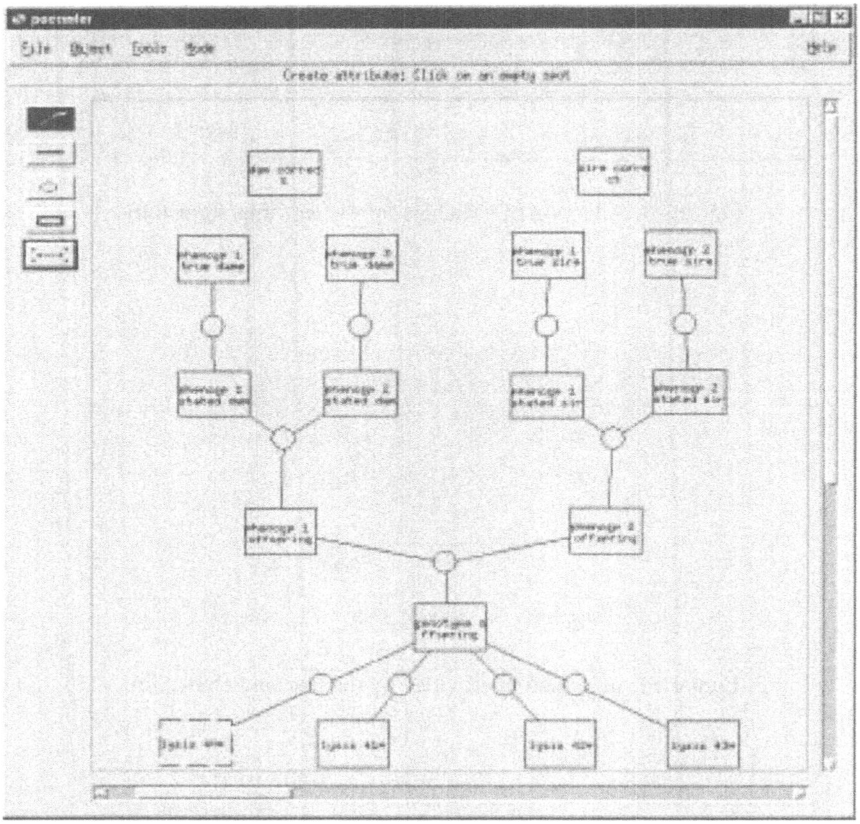

Figure 12. Hypertree Structure of the possibilistic graphical model of database D delivered by the learning algorithm, excluding the variables *dam correct* and *sire correct*

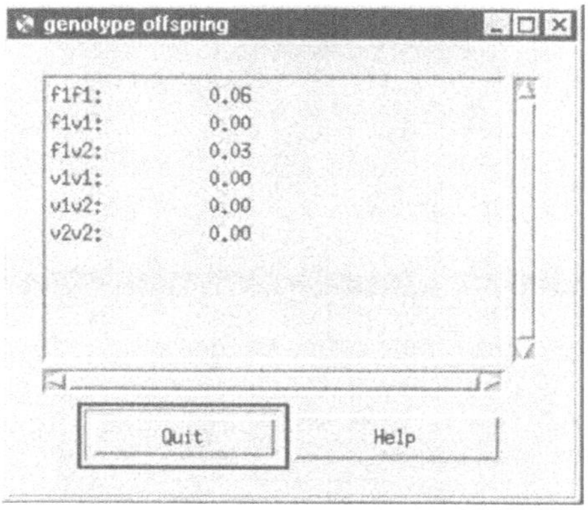

Figure 13. Possibility distribution of the genotype after evidence propagation

6 CONCLUSIONS

This contribution introduces a non-standard background semantics and the corresponding theoretical underpinnings of possibilistic graphical models as a framework for reasoning and learning under uncertainty and imprecision. It shows how to operate on possibility distributions in a consistent way, studies the concepts of possibilistic independence and possibilistic independence graphs, as well as the factorization and decomposition of possibility distributions, and deals with learning possibilistic graphical models from data, supported by the development of a software tool that has successfully been applied to a practical example. The theoretical appropriateness and practical usefulness of the approach has been justified, so that we now want to focus our evaluation of possibilistic graphical models on the discussion of their benefits and limits compared to probabilistic graphical models.

Our investigation of semantic aspects has clarified that any attempt to defend one framework against the other is rather destructive than constructive, at least in case that one is more interested in supporting the various facets of dealing with imperfect information in knowledge-based systems than trying to push particular scientific communities.

Probability theory and possibility theory both provide concepts for the treatment of uncertain data. The following characteristic properties are relevant for working out the advantages of each of the two frameworks:

Probability Theory

- serves for the exact modeling of uncertain, but precise data. Imprecise data that provide sets of alternatives without any preference structure among these alternatives are not representable by single probability distributions. Appropriate additional assumptions, for example applying the insufficient

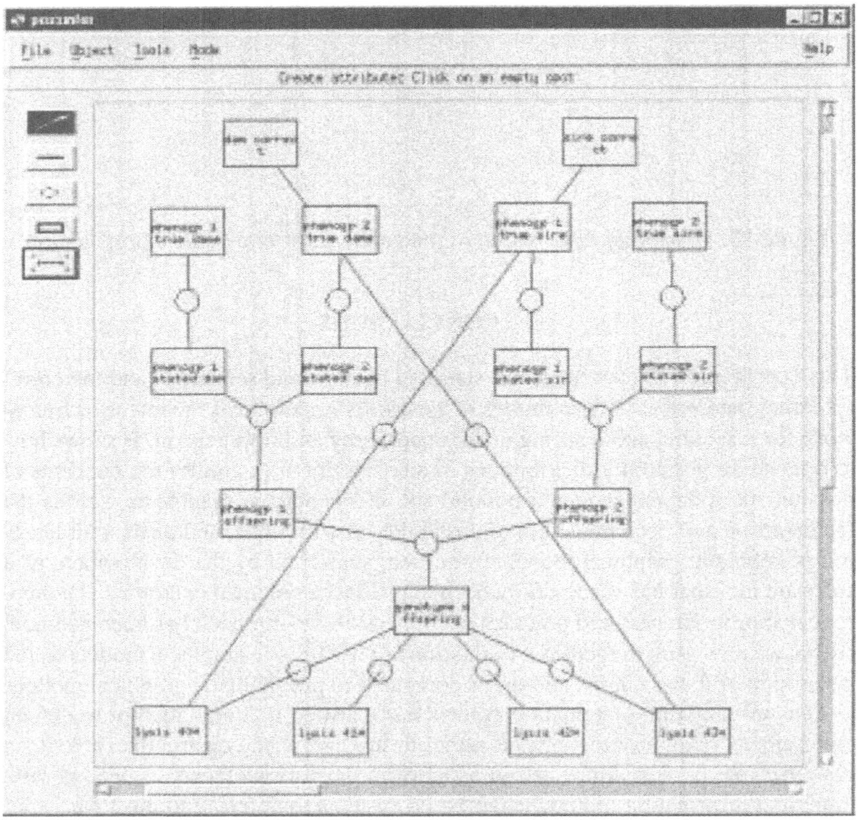

Figure 14. Hypertree structure of the possibilistic graphical model of D delivered by the learning algorithm, including the variables *dam correct* and *sire correct*

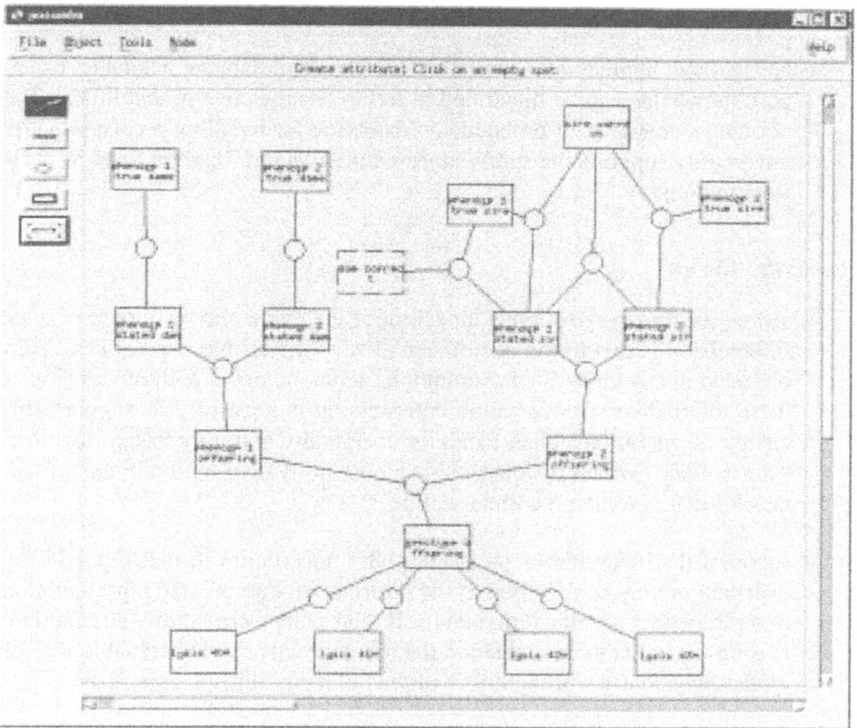

Figure 15. Hypertree structure on the base of an improved hyperedge selection algorithm when generating the hypertree decomposition

reason principle, have to be made in order to transform imprecise data into sets of precise data.

- supports the modeling of imprecision by using families of probability distributions. Related approaches refer, for instance, to probability intervals, upper and lower probabilities, and random sets, respectively. On the other hand, these approaches often give rise to crucial complexity problems, so that they are of limited power for efficient reasoning in knowledge-based systems.

- has its main application domain, where statistical data are available, or expert knowledge can be quantified in terms of subjective probabilities. The normative character of probabilistic modeling for handling precision under uncertainty supports the surely most justified and transparent form of decision making.

Possibility Theory

- serves for the approximative modeling of uncertain and/or imprecise data. Imprecision is due to the occurrence of sets of possible alternatives, without making any unjustified assumptions about unknown preferences among these alternatives. Information compression is necessary to support efficiency of operations when handling uncertainty and imprecision together. This is made explicit by interpreting a possibility distribution in the semantic setting that we dealt with in section 3.

- supports the modeling of precision under uncertainty in that a possibility distribution may be regarded as the information-compressed representation of an imperfect specification that itself represents a probability distribution. It is obvious that in this situation the more effective and reasonable way of modeling is to use a probabilistic instead of a possibilistic approach.

- has its main application domain in those fields of uncertainty modeling, where a relevant part of the available information is imprecise, and should be considered as being imprecise. The proposed semantic background of possibility theory provides a basis for well-founded techniques of possibilistic inference and possibilistic decision support, although possibility theory primarily is viewed as a framework for knowledge representation.

Probabilistic as well as possibilistic graphical models are equipped with efficient evidence propagation and learning algorithms of comparable complexity. Nevertheless, considering the special aspects of graphical modeling and the individual properties of the two addressed calculi, the following differences can be found:

Probabilistic graphical models

- support exact evidence propagation under uncertainty, since dealing with precise data does not make it necessary to tolerate any loss of information in order to be be efficient. The exactness of the reasoning process has the advantage that probabilistic graphical models are able to provide significant decision support, even when operating on exceptional events with a very low probability of occurrence. Avoiding the treatment of imprecision allows to focus on a sensitive consideration of uncertainty aspects. Nevertheless, sometimes this tends to be an over-modeling. A typical example is medical diagnosis, where evidence propagation in probabilistic graphical models often results in the necessity of choosing sets of alternative instantiations of decision variables in order to reach a probability that an expert accepts as his comfort level for decision making. This means that uncertain reasoning with precise data does not always prevent us from obtaining imprecise decisions.

- can be induced by databases of sample cases, where the respective learning algorithms make use of approximation techniques or greedy search methods to circumvent the inherent complexity of the learning problem. In most approaches, it is required that the cases in the database are precise and complete (without missing values). A problem that is still pending concerns the availability of adequate treatable performance measures, for example the evaluation of the representational quality of the induced graphical model relative to the database, and the minimum number of sample cases needed in order to reach this quality.

Possibilistic graphical models

- support approximate evidence propagation under uncertainty and imprecision. The necessity of information compression may lead to an under-modeling of the available knowledge, so that the quality of decision support is affected by correct, but nonspecific results of the reasoning process. Referred to the possibilistic interpretation of a database of sample cases, a typical situation for the potential inability of decision making occurs when reasoning with events that are only rarely supported by the database. This behavior is welcome in systems, where exceptions indicate noisy information rather than interesting dependencies. A benefit of considering, but not overestimating exceptions consists in its support on the stability of approximate reasoning techniques.

- can be induced by databases of sample cases that may consist of imprecise as well as missing values. Due to the NP-hardness of the learning problem, possibilistic structure identification algorithms need to use approximation or greedy search methods in order to be efficient. Performance evaluation refers to measuring the nonspecificity of the possibility distribution associated with the delivered graphical model, compared to the nonspecificity of the possibility distribution that is associated with the referential database. Note that due to the information compression step, those uncertain dependencies between variables that are only weakly supported by the database, may no longer be extractable from the interpreting possibility distribution. In this case, the learning algorithm will provide a good decomposition, but it has no chance to identify structural properties that have been lost during the information compression step.

The above discussion has indicated that probabilistic and possibilistic graphical models are useful in quite different domains of knowledge representation, which makes them cooperative rather than competitive. It is therefore one topic of future work in this field to study, in which way probabilistic and possibilistic data that have been obtained from expert knowledge and/or databases of sample cases, can be combined and then represented as the quantitative part of a unified type of graphical model.

REFERENCES

[Andersen *et al.*, 1989] S.K. Andersen, K.G. Olesen, F.V. Jensen, and F. Jensen. HUGIN — A shell for building Bayesian belief universes for expert systems. *Proceedings 11th IJCAI*, 1080–1085. Morgan Kaufmann, San Mateo, CA, 1989.

[Beckmann *et al.*, 1994] J. Beckmann, J. Gebhardt, and R. Kruse. Possibilistic inference and data fusion. *Proceedings 2nd European Congress on Fuzzy and Intelligent Technologies (EUFIT'94)*, 46–47. Verlag Mainz, Aachen, 1994.

[Benferhat *et al.*, 1994] S. Benferhat, D. Dubois, and H. Prade. Expressing independence in a possibilistic framework and its application to default reasoning. *Proc. 11th ECAI*, 150–154, 1994.

[Besag *et al.*, 1991] J. Besag, J. York, and A. Mollie. Bayesian image restoration with two applications in spatial statistics. *Ann. Inst. Statist. Math.*, 43:1–59, 1991.

[Bonissone *et al.*, 1987] P.R. Bonissone, S.S. Gans, and K.S. Decker. RUM: A layered architecture for reasoning with uncertainty. *Proc. 10th IJCAI, Milan*, 891–898. Morgan Kaufmann, San Mateo, CA, 1987.

[Borgelt *et al.*, 1996] C. Borgelt, J. Gebhardt, and R. Kruse. Concepts for probabilistic and possibilistic induction of decision trees on real world data. *Proc. 4th European Congress on Fuzzy and Intelligent Technologies (EUFIT'96)*, Vol. 3, 1556–1560. Verlag Mainz, Aachen, 1996.

[Buchanan and Shortliffe, 1984] B.G. Buchanan and E.H. Shortliffe. *Rules-based expert systems.* Addison-Wesley, Reading, MA, 1984.

[Buckley *et al.*, 1986] J.J. Buckley, W. Siler, and D.M. Tucker. A fuzzy expert system. *Fuzzy Sets and Systems*, 20:1–16, 1986.

[Buntine, 1994] W. Buntine. Operations for learning graphical models. *J. of Artificial Intelligence Research*, 2:159–224, 1994.

[de Campos *et al.*, 1995] L.M. de Campos, J. Gebhardt, and R. Kruse. Axiomatic treatment of possibilistic independence. In: C. Froidevaux and J. Kohlas, eds. *Symbolic and Quantitative Approaches to Reasoning and Uncertainty*, LNCS 946, 77–88. Springer, Berlin, 1995.

[Castillo *et al.*, 1997] E. Castillo, J.M. Gutierrez, and A.S. Hadi. *Expert Systems and Probabilistic Network Models.* Springer, New York, NY, 1997.

[Farinas del Cerro and Herzig, 1994] L. Farinas del Cerro and A. Herzig. Possibility theory and independence. *Proc. 5th IPMU Conference*, 820–825, Paris 1994.

[Cheeseman, 1986] P. Cheeseman. Probability versus fuzzy reasoning. In: L.N. Kanal and J.F. Lemmer, eds. *Uncertainty in Artificial Intelligence*, 85–102. North-Holland, Amsterdam, 1986.

[Chow and Liu, 1968] C.K. Chow and C.N. Liu. Approximating discrete probability distributions with dependence trees. *IEEE Trans. on Information Theory*, 14:462–467, 1968.

[Cohen, 1973] L.J. Cohen. A note on inductive logic. *J. of Philosophy*, 70:27–40, 1973.

[Cooman and Kerre, 1994] G. de Cooman and E.E. Kerre. A new approach to possibilistic independence. *Proc 3rd IEEE International Conference on Fuzzy Systems*, 1446–1451, 1994.

[Cooper and Herskovits, 1992] G. Cooper and E. Herskovits. A bayesian method for the induction of probabilistic networks from data. *Machine Learning*, 9:309–347, 1992.

[Cowell, 1992] R. Cowell. BAIES — a probabilistic expert system shell with qualitative and quantitative learning. In: J. Bernardo, J. Berger, A. Dawid, and A.Smith, eds. *Bayesian Statistics 4*, 595–600. Oxford University Press, New York, NY, 1992.

[Cox, 1946] R. Cox. Probability, frequency, and reasonable expectation. *American J. of Physics*, 14:1–13, 1946.

[Dawid, 1979] A.P. Dawid. Conditional independence in statistical theory. *J. Royal Stat. Soc. Series B*, 41:1–31, 1979.

[Dechter, 1990] R. Dechter. Decomposing a relation into a tree of binary relations. *J. Comput. Syst. Sci.*, 41:2–24, 1990.

[Dechter and Pearl, 1992] R. Dechter and J. Pearl. Structure identification in relational data. *Artificial Intelligence*, 58:237–270, 1992.

[Dempster, 1967] A.P. Dempster. Upper and lower probabilities induced by a multivalue mapping. *Ann. Math. Stat.*, 38:325–339, 1967.

[Dempster, 1968] A.P. Dempster. Upper and lower probabilities generated by a random closed interval. *Ann. Math. Stat.*, 39:957–966, 1968.

[Domotor, 1980] Z. Domotor. Probability kinematics and representation of belief change. *Philosophy of Science*, 47:384–403, 1980.

[Dubois *et al.*, 1993] D. Dubois, S. Moral, and H. Prade. A semantics for possibility theory based on likelihoods. Technical Report ESPRIT III BRA 6156 DRUMS II, CEC, 1993.

[Dubois and Prade, 1988a] D. Dubois and H. Prade. *Possibility Theory*. Plenum Press, New York, NY, 1988.

[Dubois and Prade, 1988b] D. Dubois and H. Prade. Modeling uncertainty and inductive inference. *Acta Psychologica*, 68:53–78, 1988.

[Dubois and Prade, 1990] D. Dubois and H. Prade. The logical view of conditioning and its application to possibility and evidence theories. *Int. J. of Approximate Reasoning*, 4:23–46, 1990.

[Dubois and Prade, 1991] D. Dubois and H. Prade. Fuzzy sets in approximate reasoning, part 1: Inference with possibility distributions. *Fuzzy Sets and Systems*, 40:143–202, 1991.

[Dubois and Prade, 1992] D. Dubois and H. Prade. Belief change and possibility theory. In: P. Gardenförs, ed. *Belief Revision*, 142–182. Cambridge University Press, Cambridge, MA, 1992.

[Elkan, 1994] C. Elkan. The paradoxical success of fuzzy logic. *Proc. AAAI '93*, 698–703. MIT Press, Cambridge, MA, 1994.

[Fagin, 1977] R. Fagin. Multi-valued dependencies and a new normal form for relational databases. *ACM Trans. on Database Systems 2*, 3:262–278, 1977.

[Fayyad *et al.*, 1996] U. Fayyad, G. Piatetsky-Shapiro, P. Smyth, and R. Uthurusamy. *Advances in Knowledge Discovery and Data Mining*. AAAI Press, Menlo Park, CA, 1996.

[Kampé de Fériet, 1982] J. Kampé de Fériet. Interpretation of membership functions of fuzzy sets in terms of plausibility and belief. In: M. M. Gupta and E. Sanchez, eds. *Fuzzy Information and Decision Processes*, 13–98. North-Holland, Amsterdam, 1982.

[Fine, 1973] T.L. Fine. *Theories of Probability: An Examination of Foundations*. Academic Press, New York, NY, 1973.

[de Finetti, 1974] B. de Finetti. *Theory of Probability*. J. Wiley and Sons, London, 1974.

[Fishburn, 1996] P.C. Fishburn. The axioms of subjective probability. *Statistical Science*, 1:335–358, 1996.

[Fisher and Lenz, 1996] D. Fisher and H.-J. Lenz. Learning from data. *Artificial Intelligence and Statistics 5*. Springer, New York, NY, 1996.

[Fonck, 1994] P. Fonck. Conditional independence in possibility theory. *Proc. 10th UAI conference*, 221–226. Morgan Kaufmann, San Mateo, CA, 1994.

[Gärdenfors, 1988] P. Gärdenfors. *Knowledge in Flux, Modeling the Dynamics of Epistemic States*. MIT Press, Cambridge, MA, 1988.

[Gebhardt, 1992] J. Gebhardt. *Ein integrierender Ansatz zur possibilistischen Inferenz in wissensbasierten Systemen*. PhD thesis, University of Braunschweig, 1992.

[Gebhardt and Kruse, 1993a] J. Gebhardt and R. Kruse. The context model — an integrating view of vagueness and ncertainty. *Int. J. of Approximate Reasoning*, 9:283–314, 1993.

[Gebhardt and Kruse, 1993b] J. Gebhardt and R. Kruse. A new approach to semantic aspects of possibilistic reasoning. In: M. Clarke, S. Moral, and R. Kruse, eds. *Symbolic and Quantitative Approaches to Reasoning and Uncertainty*, 151–159. Springer, Berlin, 1993.

[Gebhardt and Kruse, 1996a] J. Gebhardt and R. Kruse. Learning possibilistic networks from data. In: D.H. Fisher and H. Lenz, eds. *Learning from Data, Artificial Intelligence and Statistics 5*, 143–153. Springer, New York, NY, 1996.

[Gebhardt and Kruse, 1996b] J. Gebhardt and R. Kruse. Automated construction of possibilistic networks from data. *J. of Applied Mathematics and Computer Science*, 6:101–136, 1996.

[Gebhardt and Kruse, 1996c] J. Gebhardt and R. Kruse. Tightest hypertree decompositions of multivariate possibility distributions. *Proc. 6th IPMU Conference*, 923–927, Granada, 1996.

[Gebhardt and Kruse, 1996d] J. Gebhardt and R. Kruse. POSSINFER — a software tool for possibilistic inference. In: D. Dubois, H. Prade, and R. Yager, eds. *Fuzzy Set Methods in Information Engineering — A Guided Tour of Applications*. J. Wiley and Sons, New York, NY, 1996.

[Gebhardt and Kruse, 1997a] J. Gebhardt and R. Kruse. Parallel combination of information sources. In: D. Gabbay and P. Smets, eds. *Handbook of Defeasible Reasoning and Uncertainty Management Systems, Vol.1: Updating Uncertain Information*. Kluwer, Dordrecht, 1997.

[Gebhardt and Kruse, 1997b] J. Gebhardt and R. Kruse. Background and perspectives of possibilistic graphical models. In: D. Gabbay and R. Kruse, eds. *Qualitative and Quantitative Practical Reasoning*. Springer, Berlin, 1997.

[Geiger and Heckerman, 1995] D. Geiger and D. Heckerman. A characterization of the Dirichlet distribution with application to learning bayesian networks. *Proc. 11th UAI Conference*, 196–207. Morgan Kaufmann, San Mateo, CA, 1995.

[Godo *et al.*, 1989] L. Godo, R. Lopez de Mantaras, C. Sierra, and A. Verdegay. Milord: The architecture and the management of linguistically expressed uncertainty. *Int. J. of Intelligent Systems*, 4:471–501, 1989.

[Golumbic, 1980] M.C. Golumbic. *Algorithmic Graph Theory and Perfect Graphs*. Academic Press, London, 1980.

[Goodman, 1982] I.R. Goodman. Fuzzy sets as equivalence classes of random sets. In: R. R. Yager, ed. *Recent Developments in Fuzzy Sets and Possibility Theory*, 327–343. Pergamon Press, New York, NY, 1982.

[Gyssens *et al.*, 1994] M. Gyssens, P.G. Jeavons, and D.A. Cohen. Decomposing constraint satisfaction problems using database techniques. *Artificial Intelligence*, 66:57–89, 1994.

[Hammersley and Clifford, 1971] J.M. Hammersley and P. Clifford. Markov fields on finite graphs and lattices. manuscript cited in [Isham, 1981], 1971.

[Hartley, 1928] R.V.L. Hartley. Transmission of information. *The Bell Systems Technical J.*, 7:535–563, 1928.

[Heckerman, 1986] D.E. Heckerman. Probabilistic interpretation for MYCIN's certainty factors. In: J.F. Lemmer and L.N. Kanal, eds. *Uncertainty in Artificial Intelligence*, 167–196. North-Holland, Amsterdam, 1986.

[Heckerman, 1988] D.E. Heckerman. An axiomatic framework for belief updates. In: J.F. Lemmer and L.N. Kanal, eds. *Uncertainty in Artificial Intelligence*, Vol. 2, 11–22. North-Holland, Amsterdam, 1988.

[Heckerman, 1991] D.E. Heckerman. *Probabilistic Similarity Networks*. MIT Press, Cambridge, MA, 1991.

[Hestir *et al.*, 1991] K. Hestir, H.T. Nguyen, and G.S. Rogers. A random set formalism for evidential reasoning. In: I.R. Goodman, M.M. Gupta, H.T. Nguyen, and G.S. Rogers, eds. *Conditional Logic in Expert Systems*, 209–344. North-Holland, Amsterdam, 1991.

[Higashi and Klir, 1983] M. Higashi and G.J. Klir. Measures of uncertainty and information based on possibility distributions. *Int. J. of General Systems*, 43–58, 1983.

[Hisdal, 1978] E. Hisdal. Conditional possibilities, independence, and non-interaction. *Fuzzy Sets and Systems*, 1:283–297, 1978.

[Howard and Matheson, 1981] R.A. Howard and J.E. Matheson. Influence diagrams. In: R.A. Howard and J.E. Matheson, eds. *Readings on the Principles and Applications of Decision Analysis*, Vol. 2, 721–762. Strategic Decisions Group, Menlo Park, CA, 1981.

[Hunter, 1991] D. Hunter. Graphoids and natural condition functions. *Int. Journal of Approximate Reasoning*, 5:489–504, 1991.

[Isham, 1981] V. Isham. An introduction to spatial point processes and Markov random fields. *Int. Statist. Rev.*, 49:21–43, 1981.

[Klawonn *et al.*, 1995] F. Klawonn, J. Gebhardt, and R. Kruse. Fuzzy control on the basis of equality relations with an example from idle speed control. *IEEE Transactions on Fuzzy Systems*, 3:336–350, 1995.

[Klir and Folger, 1988] G.J. Klir and T.A. Folger. *Fuzzy Sets, Uncertainty and Information*. Prentice Hall, Englewood Cliffs, NJ, 1988.

[Klir and Mariano, 1987] G.J. Klir and M. Mariano. On the uniqueness of possibilistic measure of uncertainty and information. *Fuzzy Sets and Systems*, 197–219, 1987.

[Kohlas and Monney, 1995] J. Kohlas and P.A. Monney. *A Mathematical Theory of Hints: An Approach to Dempster-Shafer Theory of Evidence*. Springer, Berlin, 1995.

[Kruse *et al.*, 1994] R. Kruse, J. Gebhardt, and F. Klawonn. *Foundations of Fuzzy Systems*. J. Wiley and Sons, Chichester, 1994.

[Kruse *et al.*, 1991] R. Kruse, E. Schwecke, and J. Heinsohn. *Uncertainty and Vagueness in Knowledge Based Systems: Numerical Methods*. Springer, Berlin, 1991.

[Kullback and Leibler, 1951] S. Kullback and R.A. Leibler. Information and sufficiency. *Ann. Math. Statistics*, 22:79–86, 1951.

[Kyburg, 1987] H.E. Kyburg. Bayesian and non-Bayesian evidential updating. *Artificial Intelligence*, 31:271–293, 1987.

[Lauritzen, 1982] S.L. Lauritzen. *Lectures on Contingency Tables*. University of Aalborg Press, Denmark, 1982.

[Lauritzen, 1997] S.L. Lauritzen. *Graphical Models*. Oxford University Press, Oxford, 1997.

[Lauritzen *et al.*, 1990] S.L. Lauritzen, A.P. Dawid, B.N. Larsen, and H.G. Leimer. Independence properties of directed Markov fields. *Networks*, 20:491–505, 1990.

[Lauritzen *et al.*, 1984] S.L. Lauritzen, T.P. Speed, and K. Vijayan. Decomposable graphs and hypergraphs. *J. Austral. Math. Soc.*, 12–29, 1984.

[Lauritzen and Spiegelhalter, 1988] S.L. Lauritzen and D.J. Spiegelhalter. Local computations with probabilities on graphical structures and their application to expert systems. *Journal of the Royal Stat. Soc., Series B*, 50:157–224, 1988.

[Lauritzen *et al.*, 1993] S.L. Lauritzen, B. Thiesson, and D.J. Spiegelhalter. Diagnostic systems created by model selection methods — a case study. *Proc. 4th Int. Workshop on AI and Statistics, Ft. Lauderdale, FL*, 93–105, 1993.

[Leimer, 1989] H.G. Leimer. Triangulated graphs with marked vertices. *Ann. Discrete Math.*, 311–324, 1989.

[Lindley, 1982] D.V. Lindley. Scoring rules and the inevitability of probability. *International Statistical Review*, 50:1–26, 1982.

[Lindley, 1987] D.V. Lindley. The probability approach to the treatment of uncertainty in artificial intelligence and expert systems. *Stat. Sciences*, 2:17–24, 1987.

[Luo and Kay, 1995] R.C. Luo and M.G. Kay. *Multisensor Integration and Fusion for Intelligent Machines and Systems*. Ablex Publishing Corporation, Norwood, NJ, 1995.

[Madigan and Raftery, 1994] D. Madigan and A. Raftery. Model selection and accounting for uncertainty in graphical models using Occam's window. *J. of the American Statistical Association*, 89:1535–1545, 1994.

[Maier, 1983] D. Maier. *The Theory of Relational Databases*. Computer Science Press, Rockville, MD, 1983.

[Matheron, 1975] G. Matheron. *Random Sets and Integral Geometry*. J. Wiley and Sons, New York, NY, 1975.

[Nguyen, 1978] H.T. Nguyen. On random sets and belief functions. *J. of Mathematical Analysis and Applications*, 65:531–542, 1978.

[Pearl, 1986] J. Pearl. Fusion, propagation, and structuring in belief networks. *Artificial Intelligence*, 29:241–288, 1986.

[Pearl, 1990] J. Pearl. Reasoning with belief functions: An analysis of compatibility. *Int. J. of Approximate Reasoning*, 5:363–390, 1990.

[Pearl, 1992a] J. Pearl. *Probabilistic Reasoning in Intelligent Systems: Networks of Plausible Inference* (2nd edition). Morgan Kaufmann, New York, NY, 1992.

[Pearl, 1992b] J. Pearl. Rejoinder to comments on "Reasoning with belief functions: An analysis of compatibility.". *Int. J. of Approximate Reasoning*, 6:425–443, 1992.

[Pearl and Paz, 1985] J. Pearl and A. Paz. Graphoids — a graph based logic for reasoning about relevance relations. Technical Report 850038 (R-53-L), Cognitive Systems Laboratory, University of California, Los Angeles, CA, 1985.

[Pearl and Wermuth, 1993] J. Pearl and N. Wermuth. When can association graphs admit a causal interpretation. *4th Int. Workshop on AI and Statistics, Ft. Lauderdale, FL*, 141–150, 1993.

[Rasmussen, 1992] L.K. Rasmussen. Blood group determination of Danish jersey cattle in the f-blood group system. Technical Report Dina research Rep. No. 8, Dina Foulum, Tjele, 1992.

[Rasmussen, 1995] L.K. Rasmussen. Bayesian network for blood typing and parentage verification of cattle. Technical Report Dina research Report No. 38, Aalborg University, Department of Mathematics and Computer Science, 1995.

[Rescher, 1976] N. Rescher. *Plausible Reasoning: An Introduction to the Theory and Practice of Plausible Inference*. Van Gorcom, Assen, Amsterdam, 1976.

[Rose, 1970] D.J. Rose. Triangulated graphs and elimination processes. *J. Math. Anal. Appl.*, 597–609, 1970.

[Savage, 1954] L.J. Savage. *The Foundations of Statistics*. J. Wiley and Sons, New York, NY, 1954.

[Shackle, 1961] G.L.S. Shackle. *Decision, Order, and Time in Human Affairs*. Cambridge University Press, Cambridge, MA, 1961.

[Shafer, 1976] G. Shafer. *A Mathematical Theory of Evidence*. Princeton University Press, Princeton, NJ, 1976.

[Shafer and Pearl, 1990] G. Shafer and J. Pearl. *Readings in Uncertain Reasoning*. Morgan Kaufmann, San Mateo, CA, 1990.

[Shafer and Shenoy, 1988] G. Shafer and P.P. Shenoy. Local computation in hypertrees. Technical Report Working paper 201, School of Business, University of Kansas, Lawrence, KS, 1988.

[Shafer *et al.*, 1987] G. Shafer, P.P. Shenoy, and K. Mellouli. Propagating belief functions in qualitative Markov trees. *Int. J. of Approximate Reasoning*, 1:349–400, 1987.

[Shannon, 1948] C.E. Shannon. The mathematical theory of communication. *The Bell System Technical Journal*, 27:379–423, 623–656, 1948.

[Shenoy, 1989] P.P. Shenoy. A valuation-based language for expert systems. *Int. J. of Approximate Reasoning*, 3:383–411, 1989.

[Shenoy, 1992] P.P. Shenoy. Conditional independence in uncertainty theories. *Proc. 8th UAI Conference*, 284–291. Morgan Kaufmann, San Mateo, CA, 1992.

[Shenoy, 1993] P.P. Shenoy. Valuation networks and conditional independence. *Proc. 9th UAI Conference*, 191–199. Morgan Kaufmann, San Mateo, CA, 1993.

[Shenoy and Shafer, 1990] P.P. Shenoy and G.R. Shafer. Axioms for probability and belief-function propagation. In: R.D. Shachter, T.S. Levitt, L.N. Kanal, and J.F. Lemmer, eds. *Uncertainty in AI*, Vol. 4, 169–198. North-Holland, Amsterdam, 1990.

[Shore, 1986] J.E. Shore. Relative entropy, probabilistic inference, and AI. In: L.N. Kanal and J.F. Lemmer, eds. *Uncertainty in Artificial Intelligence*. North-Holland, Amsterdam, 1986.

[Shortliffe, 1976] E.H. Shortliffe. *Computer-based medical consultations: MYCIN*. Elsevier, New York, NY, 1976.

[Shortliffe and Buchanan, 1975] E.H. Shortliffe and B.G. Buchanan. A model for inexact reasoning in medicine. *Mathematical Biosciences*, 23:351–379, 1975.

[Singh and Valtorta, 1993] M. Singh and M. Valtorta. An algorithm for the construction of Bayesian network structures from data. *Proc. 9th. UAI Conference*, 259–265. Morgan Kaufmann, San Mateo, CA, 1993.

[Smets, 1988] P. Smets. Belief functions. In: P. Smets, A. Mamdani, D. Dubois, and H. Prade, eds. *Nonstandard Logics for Automated Reasoning*, 253–286. Academic Press, London, 1988.

[Smets, 1992] P. Smets. Resolving misunderstandings about belief functions: A response to many criticisms raised by J. Pearl. *Int. J. of Approximate Reasoning*, 6:321–344, 1992.

[Smets, 1993] P. Smets. Probability of deductibility and belief functions. In: M. Clarke, R. Kruse, and S. Moral, eds. *Symbolic and Quantitative Approaches to Reasoning and Uncertainty*, LNCS 747, 332–340. Springer, Berlin, 1993.

[Smets and Kennes, 1994] P. Smets and R. Kennes. The transferable belief model. *Artificial Intelligence*, 66:191–234, 1994.

[Smith, 1989] J.Q. Smith. Influence diagrams for statistical modeling. *The Annals of Statistics*, 17:654–672, 1989.

[Smithson, 1989] M. Smithson. *Ignorance and Uncertainty*. Springer, New York, NY, 1989.

[Spirtes and Glymour, 1991] P. Spirtes and C. Glymour. An algorithm for fast recovery of sparse causal graphs. *Social Science Computing Review*, 9:62–72, 1991.

[Spohn, 1988] W. Spohn. Ordinal conditional functions: A dynamic theory of epistemic states. In: W. Harper and B. Skyrms, eds. *Causation in Decision, Belief Change, and Statistics*, 105–134, 1988.

[Spohn, 1990] W. Spohn. A general non-probabilistic theory of inductive reasoning. In: R.D. Shachter, T.S. Levitt, L.N. Kanal, and J.F. Lemmer, eds. *Uncertainty in Artificial Intelligence*, 149–158. North-Holland, Amsterdam, 1990.

[Strassen, 1964] V. Strassen. Meßfehler und Information. *Zeitschrift Wahrscheinlichkeitstheorie und verwandte Gebiete*, 2:273–305, 1964.

[Studeny, 1993] M. Studeny. Formal properties of conditional independence in different calculi of AI. In: M. Clarke, R. Kruse, and S. Moral, eds. *Symbolic and Quantitative Approaches to Reasoning and Uncertainty*, LNCS 747, 341–348. Springer, Berlin, 1993.

[Tarjan and Yannakakis, 1984] R.E. Tarjan and M. Yannakakis. Simple linear-time algorithms to test chordality of graphs, test acyclicity of hypergraphs, and selectively reduce acyclic hypergraphs. *SIAM Journal of Computing*, 13:566–579, 1984.

[Ullman, 1988] J.D. Ullman. *Principles of Database and Knowledge-Based Systems, Vol. 1*. Computer Science Press, Rockville, MD, 1988.

[Ullman, 1989] J.D. Ullman. *Principles of Database and Knowledge-Based Systems, Vol. 2*. Computer Science Press, Rockville, MD, 1989.

[Verma and Pearl, 1992] T. Verma and J. Pearl. An algorithm for deciding whether a set of observed independencies has a causal explanation. *Proc. 8th UAI Conference*, 323–330, 1992.

[Walley, 1991] P. Walley. *Statistical Reasoning with Imprecise Probabilities*. Chapman and Hall, New York, NY, 1991.

[Walley and Fine, 1982] P. Walley and T. Fine. Towards a frequentist theory of upper and lower probability. *The Annals of Statistics*, 10:741–761, 1982.

[Wang, 1983] P.Z. Wang. From the fuzzy statistics to the falling random subsets. In P.P. Wang, ed. *Advances in Fuzzy Sets, Possibility and Applications*, 81–96. Plenum Press, New York, NY, 1983.

[Whittaker, 1990] J. Whittaker. *Graphical Models in Applied Multivariate Statistics*. J. Wiley and Sons, New York, NY, 1990.

[Williams, 1980] P. Williams. Bayesian conditionalization and the principle of minimum information. *British J. of the Philosophy of Science*, 31:131–144, 1980.

[Zadeh, 1975] L.A. Zadeh. The concept of a linguistic variable and its application to approximate reasoning. *Information Sciences*, 9:43–80, 1975.

[Zadeh, 1978] L.A. Zadeh. Fuzzy sets as a basis for a theory of possibility. *Fuzzy Sets and Systems*, 1:3–28, 1978.

LUIS M. DE CAMPOS, JUAN F. HUETE, SERAFÍN MORAL

INDEPENDENCE IN UNCERTAINTY THEORIES AND ITS APPLICATIONS TO LEARNING BELIEF NETWORKS

1 INTRODUCTION

The concept of independence is essential to perform efficiently reasoning tasks, because of its capacity to modularise the available knowledge. Using independence relationships for the variables in a given domain of knowledge, we can construct graphical representations for this domain, called belief networks, where inference can be done by means of local computations. When the independence relationships between the variables are not given directly but we are using a data set, containing instances of the variables in the domain, to estimate them, the task of learning the graphical structure can be considered as a kind of abductive reasoning, because we are trying to find the most adequate structure given the data, i.e., the structure that best explains the observations.

So far, most of the attention paid to belief networks has been focused in Bayesian belief networks, i.e., networks where the underlying formalism for managing uncertainty is probability theory. However, many properties of stochastic independence can be abstracted from their probabilistic roots, thus making possible to consider belief networks for uncertainty theories different from probability. In principle, we only need an appropriate concept of independence for each theory. So, two are the objectives of this chapter: first, to give an overview of several of the existing techniques to build, learn or estimate belief networks from qualitative, non numerical, information about independencies; second, to study how to define independence relationships in some (non purely probabilistic) theories of uncertainty.

The chapter is organized as follows: In Section 2, after introducing the axioms that capture in an abstract way the intuitive notion of conditional independence, we show how a dependency model can be represented by means of graphical structures (undirected graphs and directed acyclic graphs). Next, in Section 3, we study methods for estimating (or learning) these belief networks, although we restrict our research to deal only with directed graphs. There are several methodologies, based on different principles, to carry out this task, and we focus on methods which are based on the identification of the independence relationships between the variables in the domain. We start from algorithms for restricted graphs and incrementally increase the complexity of the structures. So, first we study learning algorithms for trees and polytrees. These graphs have the property of being simply connected (there is no more than one undirected path between any two variables), which makes possible the use of purely local propagation methods. Next we consider the

D. M. Gabbay and R. Kruse (eds.), Abductive Reasoning and Learning, 391–434.

so called simple graphs (where every pair of variables with a common direct child have no common ancestors nor is one an ancestor of the other) and finally we study the general case. Having studied algorithms for learning belief networks based on conditional independence, if we want to use them in connection with any particular theory of uncertainty, then it is necessary to define a concept of independence within this theory. Section 4 addresses this topic for two different uncertainty formalisms, namely possibility measures and probability intervals. Several alternatives are considered for each theory, and their properties are analized. The study also includes concepts related with that of independence, such as marginalization and conditioning. Finally, section 5 contains the concluding remarks.

2 DEPENDENCY MODELS

The words dependence and relevance are used everyday in the human discourse, and the concept they represent is perceived as one of the basic components of human reasoning: dependence is a relationship that establishes a potential change in our current belief about an event or variable, as a consequence of a specific change of our knowledge about another event or variable. Therefore, if a variable X is considered as independent of another variable Y, given a state of knowledge Z, then our belief about X will not change if we obtain additional information about Y. Independence allows us to modularise the knowledge in such a way that we only need to consult the information that is relevant to the specific question which we are interested in, instead of having to explore a complete knowledge base; thus, using independence is important to perform efficiently reasoning tasks in extensive and/or complex domains of knowledge. Moreover, people are able to easily identify qualitative dependencies, without any reference to numerical values.

So, given the importance of the notion of dependence, the following two topics need to be considered: first, to identify the essence of the qualitative notion of dependence, which should be common to most of the different formalisms of knowledge representation; this is carried out in the next subsection by means of an axiomatic study. Second, to find ways to explicitly and efficiently encode knowledge about dependencies in a reasoning system. It has been pointed out [Pearl, 1988] that graphs constitute the most usual models of human memory, so that it is natural to use them to encode conceptual dependencies; this topic is addressed in subsection 2.2.

2.1 *Axioms for Independence*

In this subsection, first we are going to describe the notation and some basic concepts used throughout the paper, although we shall omit the description of basic terminology for graphs. Next we shall introduce an axiomatic view of the concept of conditional independence.

A *Dependency Model* [Pearl, 1988] is a pair $M = (U, I)$, where U is a finite set of elements or variables, and $I(.,. \mid .)$ is a rule that assigns truth values to a three place predicate whose arguments are disjoint subsets of U. Single elements of U will be denoted by standard lowercase letters, such as x, y, $z \ldots$, whereas subsets of U will be represented by capital letters, such as X, Y, $Z \ldots$ The intended interpretation of $I(X, Y \mid Z)$ (read X is independent of Y given Z) is that having observed Z, no additional information about X could be obtained by also observing Y. For example, in a probabilistic model [Dawid, 1979; Geiger *et al.*, 1991; Lauritzen *et al.*, 1990; Spohn, 1980; Studený, 1989], $I(X, Y \mid Z)$ holds if and only if

(1) $P(\mathtt{x} \mid \mathtt{z}, \mathtt{y}) = P(\mathtt{x} \mid \mathtt{z})$ whenever $P(\mathtt{z}, \mathtt{y}) > 0$,

for every instantiation $\mathtt{x}$, $\mathtt{y}$ and $\mathtt{z}$ of the subsets of the variables X, Y and Z respectively. However, dependency models are applicable to many situations far beyond probabilistic models [de Campos, 1995; de Campos and Huete, 1993; de Campos and Huete, 1993; de Campos and Moral, 1995; Pearl, 1988; Shenoy, 1992; Studený, 1993; Wilson, 1994]. In any case, the study of the concept of conditional independence in probability theory and that of embedded multivalued dependence in database theory [Fagin, 1977] has resulted in the identification of several properties that may be reasonable to demand of any relationship which attempts to capture the intuitive notion of independence. These properties are the following:

(A1) *Symmetry*:
$(I(X, Y \mid Z) \Rightarrow I(Y, X \mid Z))\ \forall X, Y, Z \subseteq U.$

(A2) *Decomposition*:
$(I(X, Y \cup W \mid Z) \Rightarrow I(X, Y \mid Z))\ \forall X, Y, W, Z \subseteq U.$

(A3) *Weak Union*:
$(I(X, Y \cup W \mid Z) \Rightarrow I(X, W \mid Z \cup Y))\ \forall X, Y, W, Z \subseteq U.$

(A4) *Contraction*:
$(I(X, Y \mid Z)$ and $I(X, W \mid Z \cup Y) \Rightarrow I(X, Y \cup W \mid Z))\ \forall X, Y, W, Z \subseteq U.$

(A5) *Intersection*:
$(I(X, Y \mid Z \cup W)$ and $I(X, W \mid Z \cup Y) \Rightarrow I(X, Y \cup W \mid Z))\ \forall X, Y, W, Z \subseteq U.$

The intuitive interpretation of these axioms is as follows: Symmetry asserts that in any state of knowledge Z, if Y tells us nothing new about X, then X tells us nothing new about Y. Decomposition establishes that if two combined pieces of information Y and W are considered irrelevant to X, then each separate piece of information is also irrelevant. Weak Union asserts that learning the irrelevant information Y cannot help the irrelevant information W become relevant to X. Contraction states that if two pieces of information, X and W, are irrelevant to each other after knowing irrelevant information Y, then they were irrelevant before knowing Y too. Together, Weak Union and Contraction mean that irrelevant information should not modify the nature of being relevant or irrelevant of other propositions in the system. Finally, Intersection asserts that if two combined items of information, Y and W, are relevant to X, then at least one of them is also relevant to X, when the other is added to our previous state of knowledge, Z. Dependency models are called *semi-graphoids* if they verify the axioms A1–A4, and *graphoids* if they satisfy the axioms A1–A5. It is well-known that the relation of probabilistic conditional independence defined by eq. (1) satisfies all the properties A1–A4, and it also verifies A5 for probability distributions which are strictly positive.

The main reasons that make an axiomatic view of independence interesting are that the axioms can be considered as general inference rules able to derive new independencies from some initial set (without any reference to numerical quantities) and even new inference rules. Moreover, an axiomatic system represents the essential of the concept of independence, and provides an appropriate way for comparing the properties of different formalisms of independence [Pearl, 1988].

2.2 *Dependency Models and Graphs*

We have already commented that making use of information about dependencies and independencies is important to design efficient reasoning systems. Therefore, the question is how to represent this kind of information. Graphs constitute an intuitive way of structuring the knowledge: the links in a graph (and the absence of links) allow us to represent explicitly dependence assertions (and independence ones, respectively). So, it makes sense to use graphs to incorporate structural information about dependencies and independencies in a reasoning system.

A *graphical representation* of a dependency model M is a direct correspondence between the elements in U and the set of nodes or vertices in a given graph, G, such that the topology of G reflects some properties of M. For simplicity in the notation, a node in the graph G will be referred to by the element in M associated with it. The way we relate independence assertions in M with some topological property of a graph depends on the kind of graph we use; this property is *separation* for the case of undirected graphs [Lauritzen, 1982; Pearl, 1988] and *d-separation* for directed acyclic ones (dags) [Pearl, 1988; Verma and Pearl, 1990a]:

- *Separation*: Given an undirected graph G, two subsets of nodes, X and Y, are said to be separated by the set of nodes Z, and this is denoted by $< X, Y \mid Z >_G$, if Z intercepts all chains between the nodes in X and those in Y, or, in other words, if the removal of the set of nodes Z from the graph together with their associated edges, disconnects the nodes in X from those in Y.

- *d-separation*: Given a dag G, a trail C (a trail in a directed graph is a sequence of adjacent nodes, the direction of the arrows does not matter) from node x to node y is said to be blocked by the set of nodes Z, if there is a vertex $z \in C$ such that, either
  - $z \in Z$ and arrows of C do not meet head-to-head at z, or
  - $z \notin Z$, nor has z any descendants in Z, and the arrows of C do meet head-to-head at z.

  A trail that is not blocked by Z is said to be active. Two subsets of nodes, X and Y, are said to be d-separated by Z, and this is denoted by $< X,Y \mid Z >_G$, if all trails between the nodes in X and the nodes in Y are blocked by Z (separation and d-separation are denoted in the same way, because the context will avoid confusing them). In [Lauritzen *et al.*, 1990], a criterion equivalent to d-separation was proposed (in terms of the separation of X from Y by Z in the moral graph of the smallest ancestral set containing $X \cup Y \cup Z$).

Of course, every directed acyclic graph (undirected graph, respectively) together with the criterion of d-separation (separation, respectively), constitutes a dependency model which is a graphoid.

For a given dependency model it is not always possible to find a graph displaying all of its dependencies and independencies, and we shall need to reduce our requirements:

Given a dependency model, M, we say that an undirected graph (a dag, respectively), G, is an *Independency map* or *I-map* [Pearl, 1988] if every separation (d-separation respectively) in G implies an independence in M:

$$< X,Y \mid Z >_G \Rightarrow I(X,Y \mid Z).$$

On the other hand, an undirected graph (a dag, respectively), G, is called a *Dependency map* or *D-map* [Pearl, 1988] if every independency relation in the model implies a separation (d-separation respectively) in the graph:

$$I(X,Y \mid Z) \Rightarrow < X,Y \mid Z >_G .$$

A graph, G, is a *Perfect map* of M [Pearl, 1988] if it is both an I-map and a D-map. M is said to be *graph-isomorphic* if a graph exists which is a perfect map of M.

We can say that the graph structure represents the dependency model qualitatively. However, if we want to use it for decision or inference purposes, we must also give quantitative information about the strength of the dependencies displayed by the graph. In principle, we shall consider probabilistic models, so that, quantitatively, the model is represented by means of a joint probability distribution for the variables. The assignment of numerical parameters to a graphical model is different for undirected and directed graphs.

In the case of directed acyclic graphs, this is a simple matter: we only have to assign to each variable x_i in the dag a conditional probability distribution for every instantiation of the variables that form the parent set of x_i in the graph, $\Pi(x_i)$. The product of these local distributions constitutes a complete and consistent specification, i.e., a joint probability distribution (which also preserves the independence relationships displayed by the dag). In other words, the model is represented by means of a joint probability distribution for the variables, which can be factorized in the following way:

$$(2) \quad P(\mathtt{x_1}, \mathtt{x_2}, \ldots, \mathtt{x_n}) = \prod_{i=1}^{n} P(\mathtt{x_i} \mid \Pi(\mathtt{x_i}))$$

where $\mathtt{x_i}$ represents any assignment of values to the variable x_i, and $\Pi(\mathtt{x_i})$ denotes any assignment of values to the variables in the set $\Pi(x_i)$. In that case, the concept of probabilistic conditional independence between variables allows us to represent a joint probability distribution efficiently by decomposing it as a product of psychologically meaningful conditional probability distributions, which quantify the strengths of the influences. Moreover, when we look for an explanation for a given set of observations, independence allows us to identify which information is relevant and which is superfluous.

However, the case of undirected graphs is different: constructing a complete and consistent quantitative specification while preserving the dependence structure of an arbitrary undirected graph can be done using the method of Gibb's potential [Lauritzen, 1982] (which assigns compatibility functions to the cliques of the graph), but it is considerably more complicated, in terms of both computational effort and meaningfulness of the parameters, than the simple method used for dags.

In that follows, we are going to focus in directed representations of dependency models, that we shall call *belief networks*.

3 ALGORITHMS FOR LEARNING BELIEF NETWORKS BASED ON CONDITIONAL INDEPENDENCE

Once a complete belief network has been built, it constitutes an efficient device to perform inferences [Lauritzen and Spiegelhalter, 1988; Pearl, 1988]. However, there remains the previous problem of building such a network, i.e., to provide the graph structure and the numerical parameters necessary for characterising the network. So, an interesting task is then to develop methods able to learn the network directly from raw data, as an alternative or a complement to the (difficult and time-consuming) method of eliciting opinions from the experts. An increasing number of methods of automatic learning of networks have been proposed (there are also a number of algorithms for learning undirected graphs [Andersen *et al.*, 1991; Buntine, 1994; de Campos and Huete, 1997a; Wermuth and Lauritzen, 1983; Wong and Xiang, 1994; Xiang *et al.*, 1997], although we shall not study them).

A possible classification could divide these methods into two main groups (which are neither completely exhaustive nor strictly exclusive):

- Methods using some type of heuristic measure or scoring metric to fit a 'good' network to a data base of examples.
- Methods based on using conditional independence tests between variables.

The problem is how to construct a directed acyclic graph from a data base (from which we can obtain an approximation of the joint probability distribution). In the first type of methods, the idea is to find a graph having the minimum number of links that 'adequately' represents the data. The adequacy of the graph is measured by means of some kind of heuristic measure or scoring metric. This metric, together with a search procedure (implicit or explicit), are the two basic components of these methods. All the algorithms use some kind of heuristic search (mainly greedy search techniques, due to the super-exponential size of the search space). However, the type of metric is more diverse: we have methods using Bayesian techniques [Buntine, 1994; Chickering, 1996; Chickering, 1996a; Cooper and Herskovits, 1992; Cooper, 1995; Friedman and Goldszmidt, 1996; Friedman and Goldszmidt, 1996a; Friedman *et al.*, 1997; Geiger and Heckerman, 1995; Heckerman, 1996; Heckerman *et al.*, 1995; Larrañaga *et al.*, 1996a; Larrañaga *et al.*, 1996b; Madigan and Raftery, 1994; Ramoni and Sebastiani, 1997; Spiegelhalter *et al.*, 1993], methods based on the Minimum Description Length principle [Bouckaert, 1993; Friedman and Goldszmidt, 1996; Lam and Bacchus, 1994; Suzuki, 1993; Wedelin, 1993], and methods based on entropy and cross-entropy [Acid and de Campos, 1994; Acid *et al.*, 1991; Acid *et al.*, 1991a; de Campos, 1998; Cheng *et al.*, 1997; Chow and Liu, 1968; Geiger, 1992; Herskovits *et al.*, 1990; Molina *et al.*, 1993; Rebane and Pearl, 1989; Sarkar, 1993].

The methods based on conditional independence tests are more abstract than the previous ones, in the sense than they are more independent of the quantitative values represented in the network: their inmediate objective is not to find a network where the joint probability distribution that it represents is very similar to the original one (fits the original data), but they make a qualitative study of the independence properties of the model, and then try to find a network representing these properties. So, the inputs to the algorithms are conditional independence relationships between sets of variables in the model (which are usually estimated from a data base), and the outputs are graphs displaying as much as possible these relationships. Next, they estimate the different conditional probability distributions for each variable from either a data set or an expert. The different learning algorithms based on independence criteria mainly differ in:

- The kind of graph they recover.
- The efficiency of the method, in the sense of:
  - the number of conditional independence tests.

  - the order of these tests[1]. Note that testing conditional independence statements from data requires a time which is exponential with the order of the test. Moreover, the reliability of a test drastically decreases as the order of the test increases.

- Their guarantee of finding a correct solution (e.g., a minimal I-map or a perfect map of the model).

- Their robustness (for example with respect to a small number of errors in the estimation of the test results[2]).

There are a number of learning algorithms based on independence criteria. Some of them recover trees or polytrees [de Campos, 1998; Geiger *et al.*, 1990; Huete and de Campos, 1993], or simple graphs [de Campos and Huete, 1997; Geiger *et al.*, 1993]; other algorithms recover general (unrestricted) dags [Pearl and Verma, 1991; Spirtes *et al.*, 1991; Spirtes *et al.*, 1993; Verma and Pearl, 1990; Verma and Pearl, 1990a; Verma and Pearl, 1992].

Finally, there also exist hybrid approaches for learning belief networks, which simultaneously use a search technique directed by some metric and conditional independence tests [Acid and de Campos, 1996; Acid and de Campos, 1996a; Acid and de Campos, 1997; Singh and Valtorta, 1993; Singh and Valtorta, 1995].

In this section, we shall study only methods based on conditional independence. The main reason is that we can make the algorithms for learning the graph structure independent from the uncertainty formalism used to codify the numerical parameters, and therefore we could also use other uncertainty models different from probability theory, as possibility theory of probability intervals.

We start from algorithms for restricted graphs and incrementally increase the complexity of the structures. First, we study learning algorithms for polytrees, i.e., simply connected graphs, next we consider simple graphs (where every pair of variables with a common direct child have no common ancestors nor is one an ancestor of the other) and finally we study the general case.

3.1 Algorithms for Polytrees

In this subsection we shall study algorithms for learning a special but important type of belief networks: *polytrees* (which include trees as particular case). These are networks where no more than one (undirected) path connects every two nodes, or in other words, graphs having no undirected cycle. Only for polytrees the purely local propagation procedures [Pearl, 1988] are valid. In fact, several propagation methods for general networks [Becker and Geiger, 1994; Lauritzen and Spiegelhalter, 1988; Pearl, 1986] are based on the idea of transforming the graph and turning

[1] The order of a test refers to the number of variables involved.

[2] This is related to the asymptotic behaviour of the algorithm, i.e., the reliability of the test results, depending on the size of the data base used to estimate them.

it singly connected. Moreover, their simplified structure allows us learning polytrees in a much more efficient way than general belief networks. The price we have to pay is a less expressive power, because the kind of independence relationships that may be representent is more restricted for polytrees than for general (multiply connected) networks (see [de Campos, 1998] for an axiomatic characterization of polytree-isomorphic dependency models).

We are going to describe two algorithms for learning polytrees, under the assumption that the dependency models we are considering are polytree-isomorphic (in [Geiger *et al.*, 1990] an algorithm for finding polytrees which are I-maps of dependency models, which requires less restrictive assumptions, is described; however, it needs conditional independence tests of high order, which are unreliable and computationally expensive when performing them from a data set).

The first algorithm is based on the following result [de Campos, 1998]: Let M be a polytree-isomorphic dependency model and let G_M be its directed graphical representation. Then, for every pair $x, y \in U$, the edge x—y is in the skeleton of G_M (the undirected graph obtained from G_M if the directionality of the links is ignored) if and only if $\neg I(x, y \mid z)\ \forall z \in U \setminus \{x, y\}$, and $\neg I(x, y \mid \emptyset)$. The EP (Exact Polytree) algorithm is shown in Figure 1.

The algorithm first finds the skeleton by using only marginal and first order conditional independence tests, and next directs it by making marginal independence tests between variables having a common neighbour. It can be shown [de Campos, 1998] that whenever the dependency model M is polytree-isomorphic, the EP algorithm gives a polytree which is isomorphic to M as output. Moreover, for dag-isomorphic dependency models which are not polytree-isomorphic, the EP algorithm either returns 'fail' (meaning that the model is not polytree-isomorphic) or produces a dag (non necessarily singly connected), which is at least an I-map of the model, as output. The algorithm is efficient: it takes at most $O(n^3)$, even if we have to estimate the truth values of the independence statements from a data set.

If, even in the case that the model is not polytree-isomorphic, we want to obtain a polytree approximation of the model, then the EP algorithm can be modified to cope with this task: we need to replace the idea of an independence relationship being true or false by a *graded dependence relationship* which measures the degree of dependence between two variables: we can use a dependence function $Dep(X, Y \mid Z)$, whose value is zero if $I(X, Y \mid Z)$ holds, and such that the more dependent X and Y are, given Z, the greater $Dep(X, Y \mid Z)$ is. For example, in a probabilistic dependency model we could use the well-known Kullback-Leibler cross-entropy measure [Kullback, 1968] (or any other of the Dep functions used in [Acid *et al.*, 1991a]). The basic idea is to preserve those edges representing the strongest dependence relationships, but with the restriction that the resultant structure must be singly connected. For each pair of nodes $x, y \in U$, we can calculate several degrees of dependence, one for each other node in the model, $Dep(x, y \mid z)$, $\forall z \in U \setminus \{x, y\}$, and $Dep(x, y \mid \emptyset)$. In order to have a single measure of dependence, we could aggregate them in some way; any triangular norm $\otimes$

EP Algorithm

1. Start with a complete undirected graph G
2. for every pair of nodes $x, y \in U$ do
   2.1. $removed$ ←false
   2.2. $S \leftarrow U \setminus \{x, y\}$
   2.3. if $I(x, y \mid \emptyset)$ =true
      then
      2.3.1. Remove the edge x—y from G
      2.3.2. $removed$ ←true
   2.4. while $S \neq \emptyset$ and $removed$ =false do
      2.4.1. Select a node $z \in S$
      2.4.2. $S \leftarrow S \setminus \{z\}$
      2.4.3. if $I(x, y \mid z)$ =true
         then
         2.4.3.1. Remove the edge x—y from G
         2.4.3.2. $removed$ ←true
3. for every triplet of nodes $x, y, z \in U$ such that x—z, z—$y \in G$ do
   3.1. if $I(x, y \mid \emptyset)$ =true
      then direct the subgraph x—z—y as $x \rightarrow z \leftarrow y$
4. Direct the remaining edges without introducing directed cycles or
   new head-to-head connections
   if this is possible then return G, else return 'fail'

Figure 1. Algorithm for learning polytrees.

[Schweizer and Sklar, 1983], such as, for example, the minimum or the product, could be an appropriate conjunctive operator. So, we define the global degree of dependence $Dep_p(x, y)$ by means of

$$Dep_p(x, y) = Dep(x, y \mid \emptyset) \otimes \left(\bigotimes_{z \in U \setminus \{x, y\}} Dep(x, y \mid z) \right)$$

To preserve the strongest dependencies compatible with a singly connected structure, we can use any Maximum Weight Spanning Tree (MWST) algorithm. Next the algorithm tries to direct the edges of the skeleton by using the following idea: in a head-to-head pattern $x \rightarrow z \leftarrow y$, the instantiation of the head-to-head node z should normally increase the degree of dependence between the variables x and y, whereas in a non head-to-head pattern such as $x \leftarrow z \rightarrow y$, the instantiation of the middle node z should produce the opposite effect, decreasing the degree of dependence between x and y. So, the idea is to compare the degree of dependence between x and y after the instantiation of z, $Dep(x, y \mid z)$, with the degree of dependence between x and y before the instantiation of z, $Dep(x, y \mid \emptyset)$, and directing the edges towards z if the former is greater than the latter. The proposed algorithm, called PA (Polytree Approximation) algorithm, is depicted in Figure 2. In this algorithm $\otimes$ is a triangular norm and ϵ is a small non-negative real number (that may be zero) representing a threshold used to detect independence.

PA Algorithm

1. Start with an empty graph G
2. for every pair of nodes $x, y \in U$ do
   2.1. Compute $Dep(x, y \mid \emptyset)$
   2.2. $Dep_p(x, y) \leftarrow Dep(x, y \mid \emptyset)$
   2.3. for every node $z \in U \setminus \{x, y\}$ do
      2.3.1. Compute $Dep(x, y \mid z)$
      2.3.2. $Dep_p(x, y) \leftarrow Dep_p(x, y) \otimes Dep(x, y \mid z)$
3. Rank the $n(n-1)/2$ values $Dep_p(x, y)$ in decreasing order
4. repeat
   Select the next largest value $Dep_p(x, y)$ and add the edge x—y to G
   unless it creates a cycle in G or $Dep_p(x, y) \leq \epsilon$
   until $n-1$ edges have been added to G or
   the last value which has been considered verifies $Dep_p(x, y) \leq \epsilon$
5. for every triplet of nodes $x, y, z \in U$ such that x—z, z—$y \in G$ do
   5.1. if $Dep(x, y \mid \emptyset) < Dep(x, y \mid z)$
   then direct the subgraph x—z—y as $x \rightarrow z \leftarrow y$
6. Direct the remaining edges without introducing new head-to-head connections
7. return G

Figure 2. Algorithm building a polytree approximation of a dependency model.

There is another algorithm, developed in [Rebane and Pearl, 1989], that constructs polytree approximations for probabilistic dependency models and also guarantees finding a polytree isomorphic to the model if such a one exists. It also uses a MWST algorithm for building the skeleton, and next directs it by using tests of marginal independence. A comparison of both algorithms can be found in [de Campos, 1998]

The second (exact) algorithm for learning polytrees is based on the same idea that the first one: two non adjacent nodes in a polytree must be either marginally independent or conditionally independent given a third node. However, the way of organising the computations is more sophisticated, aiming to obtain a greater efficiency. The result is an $O(n^2)$ algorithm.

For each variable $x \in U$, we consider the set $\Lambda_x = \{y \in U \mid \neg I(x, y \mid \emptyset)\}$, i.e., the set of variables marginally dependent with x. The set of nodes which are connected directly with x, Ψ_x, called the Sheaf structure for node x, is obviously a subset of Λ_x. The idea is to iteratively check the inclusion of every node of Λ_x into the set Ψ_x: we start from an empty Sheaf structure, $\Psi_x^0 = \emptyset$, and in each step i, given the current Sheaf structure Ψ_x^{i-1}, we select a node $x_i \in \Lambda_x$ and test whether to include it in the new Sheaf structure Ψ_x^i. This test can be carried out in the following way [Huete and de Campos, 1993]:

1. If $\exists y \in \Psi_x^{i-1}$ such that $I(x, x_i \mid y)$ then $\Psi_x^i = \Psi_x^{i-1}$.

2. Otherwise (we have in this case $\forall y \in \Psi_x^{i-1}\ \neg I(x, x_i \mid y)$), let $\mathcal{J}^i = \{y \in \Psi_x^{i-1} \mid I(x, y \mid x_i)\}$. Then $\Psi_x^i = \Psi_x^{i-1} \cup \{x_i\} \setminus \mathcal{J}^i$.

In the first case, if we find a variable y that renders x and x_i conditionally independent, then x_i cannot be adjacent to x, and the Sheaf structure is not modified (see Figure 3.a). In the second case, we cannot separate (at this moment) x from x_i, so that x_i will be included in the new Sheaf structure; moreover, as a consequence of the inclusion of x_i in Ψ_x^i, some other variables have to be removed from Ψ_x^i: those that are independent of x, given x_i (see Figure 3.b).

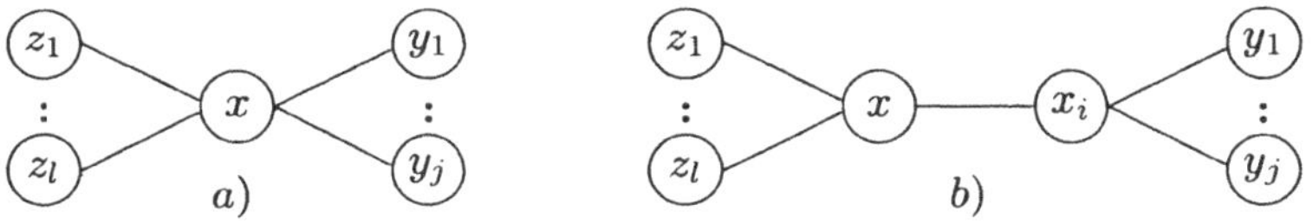

Figure 3. a) Case 1: $\Psi_x^i = \Psi_x^{i-1} = \{z_1, \ldots, z_l, y_1, \ldots, y_j\}$. b) Case 2: $\Psi_x^i = \Psi_x^{i-1} \cup \{x_i\} \setminus \mathcal{J}^i = \{z_1, \ldots, z_l, x_i\}$

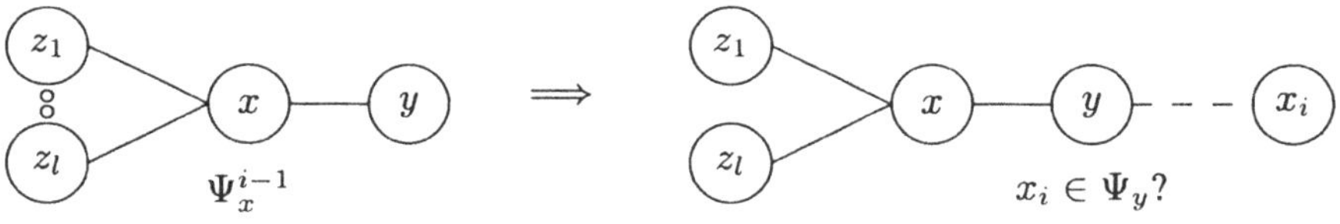

Figure 4. Case 1: $x_i \notin \Psi_x^i$, and $\exists y \in \Psi_x^{i-1}$ such that $I(x, x_i \mid y)$

At the end of the process, once we have checked all the variables in Λ_x for inclusion in the Sheaf structure, we obtain the correct Sheaf structure $\Psi_x^{|\Lambda_x|} = \Psi_x$. This procedure can be carried out in an independent way for each variable x, thus making possible to easily use distributed computation. If, however, we have a single processor, we can modify the algorithm to reduce the number of independence tests needed: in the case 1 above, we know that x_i is not in Ψ_x and that y is in the path connecting x and x_i; therefore, in addition to simply not including x_i in Ψ_x, we can direct the search by also studying at this moment whether x_i belongs to the Sheaf structure for node y, Ψ_y; if x_i does not belong to Ψ_y, this is due to the fact that we find another variable, say z, such that $I(y, x_i \mid z)$, and then we study whether x_i belongs to Ψ_z. We continue in this way until we find the correct place for x_i (see Figure 4). By using this technique of intermingling the calculus of different Sheaf structures, we can avoid some unnecessary independence tests. The resultant algorithm is shown in Figure 5.

Iterative Algorithm

1. For each variable x in U do
   1.1. $\Lambda_x = \emptyset$; $\Psi_x = \emptyset$; Visited[x]=False ; Expanded[x]=False
   1.2. For each variable y in $U \setminus \{x\}$ do
      1.2.1. If $I(x, y \mid \emptyset)$ =False Then $\Lambda_x = \Lambda_x \cup \{y\}$
2. Select a node x from U, and include it in G ; Visited[x]=True
3. While there are non expanded nodes in G do
   3.1. Select a non expanded node x from G ; Expanded[x]= True
   3.2. While there are non visited nodes in Λ_x do
      3.2.1. Select a non visited node z from Λ_x
      3.2.2. Visited[z]=True ; Forward=True ; Inserted=False
      3.2.3. While Forward=True do
         3.2.3.1. If every $y \in \Psi_x$ has been tested
            then Forward=False
            else
            3.2.3.1.1. Select a non tested y in Ψ_x
            3.2.3.1.2. If $I(x, z \mid y)$ =True then $x = y$
      3.2.4. For every y in Ψ_x do
         3.2.4.1. If $(x, y \mid z)$ =True
            then
            3.2.4.1.1. Include z in Ψ_x and Ψ_y; include x and y in Ψ_z
            3.2.4.1.2. Delete y from Ψ_x and x from Ψ_y
            3.2.4.1.3. Inserted=True
      3.2.5. If Inserted=False then include z in Ψ_x and x in Ψ_z
      3.2.6. Reset x to its original value
4. Fuse every Ψ_x to obtain G

Figure 5. Algorithm building the skeleton of a polytree.

The algorithm in Figure 5 only builds the skeleton of the polytree, the directionality of the edges may be detected using marginal independence tests like those in steps 3 and 4 of the EP algorithm. The overall process takes $O(n^2)$.

3.2 Algorithms for Simple Networks

In this subsection, we study algorithms for learning another special type of belief networks, although more general than polytrees: *simple graphs*. They are directed acyclic graphs where every pair of nodes with a common direct child have no common ancestor nor is one an ancestor of the other [Geiger *et al.*, 1993]. This means that in a simple graph only a special kind of undirected cycles are permitted: simple cycles, those containing at least two head-to-head nodes. Simple graphs allow us to represent a richer set of independence relationships, to use fast updating procedures [Heckerman, 1990], as well as efficient learning algorithms; therefore, simple graphs are worthy of pursuit.

We shall describe two learning algorithms for simple graphs. Both of them [Geiger *et al.*, 1993; de Campos and Huete, 1997] use specific independence properties, which are true in simple graphs, to develop the algorithms. The first one [Geiger *et al.*, 1993] is more general, in the sense that it does not assume the dependency model to be dag-isomorphic, and needs a polynomial number, $O(n^2)$, of conditional independence tests, each one of order $n-2$, where n is the number of variables in the model.

In the second algorithm [de Campos and Huete, 1997], more restrictive, a dag-isomorphic dependency model is assumed. This assumption permits us to reduce to the minimum (which may be zero) the number of conditional independence tests of order greater than one. The algorithm gives as the output a simple graph isomorphic to the model, or a 'fail' signal if such graph does not exist. The assumption of dag-isomorphy avoids some of the practical problems of the algorithm in [Geiger *et al.*, 1993]: when we have to learn the graph from statistical data, testing conditional independence statements of order $n-2$ requires an amount of time which is exponential with n. Therefore, although the algorithm is polynomial (order $O(n^2)$) in the number of independence tests, the overall complexity is still exponential. Moreover, to reliably test high order conditional independence statements we need a enormous amount of data. When the model is isomorphic to a simple graph, our algorithm will recover the model using only zero and first order conditional independence tests (in this case, each test can be carried out in polynomial time).

First, let us briefly describe the algorithm proposed in [Geiger *et al.*, 1993]. It assumes that the dependency model M is a graphoid. The algorithm outputs a simple graph which is a minimal I-map that represents M well, or acknowledgement that no such graph exists. A graph G is said to represent M well if whenever two nodes x and y are connected with a simple trail, then x and y are marginally dependent, i.e., $\neg I(x, y \mid \emptyset)$ (this concept could be called *marginal D-mapness*). A trail between two nodes x and y is said to be simple if every node in the trail is not head-to-head. The algorithm is depicted in Figure 6. Proofs of its correctness are found in [Geiger *et al.*, 1993].

Geiger, Paz and Pearl's Algorithm

1. Start with a complete undirected graph G
2. Remove every edge x—y for which $I(x, y \mid U \setminus \{x, y\})$ holds
3. Remove every edge x—y for which $I(x, y \mid \emptyset)$ holds
4. Orient every pair of edges x—y and y—z towards y whenever x—y—z is in the graph, and $I(x, z \mid \emptyset)$ holds
5. Orient the remaining links without introducing new head-to-head connections and such that the resultant dag is simple. If the resultant orientation is not feasible, then return 'fail'
6. If the resultant simple dag does not represent M well then return 'fail'. Otherwise, output the resultant network

Figure 6. Algorithm for learning simple graphs.

Now, we are going to introduce another algorithm for learning simple graphs, that assumes that the dependency model is dag-isomorphic. The development of this algorithm is motivated by the following question: does an independence relationship between the nodes connected directly to a node x exist, such that it identifies whether a dag is simple or not?. The following result [de Campos and Huete, 1997] gives a positive answer to this question: Let G be any dag isomorph to a dependency model M. Then, G is simple if and only if for every node $x \in U$ the relationship $I(p_x, c_x \mid x)$ holds, $\forall p_x \in \text{Parents}_x$, $c_x \in \text{Children}_x$.

The property above raises the question of the possibility of recovering a simple graph using only conditional independence relationships of order zero and one. That is, given a dependency model which is isomorphic to a simple graph G, could we recover this simple graph using as the input a list of marginal and first-order conditional independence relationships obtained from the model?. Again, the answer is positive. The algorithm displayed in Figure 7 achieves this objective. The idea is to search, for each variable x in the model, the set of nodes connected directly with x, and them fuse these elements to build the graph. The sets of nodes Λ_x, Ψ_x, $K_x(y)$, $\Omega^*_x(y)$ and $\Omega_x(y)$ that appear in the algorithm can be calculated as indicate the following equations:

Λ_x: The set of variables marginally dependent with x, i.e.

$$\Lambda_x = \{y \in U \mid \neg I(x, y \mid \emptyset)\}$$

Ψ_x: The set of variables having neither zero nor first order conditional independence relationships with x, i.e.

$$\Psi_x = \{y \in \Lambda_x \mid \neg I(x, y \mid z) \forall z \in U \setminus \{x, y\}\}$$

$K_x(y)$: The initial set of candidate nodes:

$$K_x(y) = \{w \in \Psi_x \setminus \{y\} \mid \neg I(y, w \mid x)\}$$

$\Omega^*_x(y)$: The set of cadidate nodes:

$$\Omega^*_x(y) = \{w_i \in K_x(y) \text{ such that } \exists w_j \in K_x(y) \text{ satisfying } \quad 1.)\ I(w_i, w_j \mid \emptyset) \quad 2.)\ \neg I(w_i, w_j \mid y)\} \tag{3}$$

$\Omega_x(y)$: It can be constructed by refining the set of candidate nodes $\Omega^*_x(y)$ through:

$$\Omega_x(y) = \Omega^*_x(y) \setminus \{w_i \text{ such that either a) } \exists z \in \Psi_y, z \in \Lambda_{w_i} \text{ such that } I(z, x \mid \emptyset) \text{ and } \neg I(z, x \mid y), \text{ or b) } \exists z \in \Psi_y, z \in \Lambda_{w_i} \text{ such that } \neg I(z, x \mid \emptyset) \text{ and } I(z, x \mid y) \text{ and } \neg I(z, w_i \mid y)\} \tag{4}$$

CH Algorithm

1. For each x in U do
   1.1. Calculate Λ_x
   1.2. Calculate Ψ_x
   1.3. For each $y \in \Psi_x$ do
      1.3.1. Calculate $K_x(y)$. If $K_x(y) = \emptyset$ go to step 1.3.
      1.3.2. Calculate $\Omega^*_x(y)$. If $\Omega^*_x(y) = \emptyset$ go to step 1.3.
      1.3.3. Calculate $\Omega_x(y)$. If $\Omega_x(y) \neq \emptyset$ then exclude y from Ψ_x
   1.4. For each pair $y, z \in \Psi_x$, If $I(y, z \mid \emptyset)$ holds, put the nodes y, z as parents of x
2. Fuse every Ψ_x, to obtain G
3. Direct the remaining links without introducing new head-to-head connections

Figure 7. Algorithm for recovering simple graphs.

Now, let us explain how and why the algorithm works: first, whenever two nodes x and y are marginally dependent ($x \in \Lambda_y$ and $y \in \Lambda_x$), there exists a simple trail linking them in the dag that represents the model. Moreover, for the simple trails in a dag, we can establish the following classification:

$HT(x, y)$: Those simple trails linking x and y with a head connection at x and a tail connection at y, i.e., directed paths from y to x.

$TH(x, y)$: Those simple trails linking x and y with a tail connection at x and a head connection at y, i.e., directed paths from x to y.

$HH(x, y)$: Those simple trails linking x and y with a head connection at x and a head connection at y. In these trails, we can find a node z, such that we have directed subpaths from z to x and from z to y.

It can be seen [de Campos and Huete, 1997] that in a simple graph G, whenever there are more than one simple trail linking two nodes x and y, then all these trails necessarily are of type $HH(x, y)$. Note that these trails form a simple cycle, that will be called active simple cycle. The variables which are head-to-head in this cycle are called closing variables. When we are considering a simple-graph-isomorphic model, if there exists an active simple cycle with closing variables x and y, then there are neither zero nor first order conditional independence relationships between these variables. Obviously, these relationships neither hold if a direct connection between x and y exists. This fact permits us to ensure that, after executing step 1.2 of the algorithm, the set Ψ_x will include, in addition to the direct parents and children of x, only those nodes closing an active simple cycle with x.

Moreover, given two variables x and y in a simple graph G such that there are at least two $HH(x, y)$ simple trails, then we have $I(x, y \mid \rho_x(y))$, where $\rho_x(y)$ is the set of nodes formed by the parents of x in any trail $HH(x, y)$. In this case, the node y could be excluded from the set of nodes connected directly to x. So, it is necessary to have a method to find those variables $y \in \Psi_x$ such that there are at least two simple trails of type $HH(x, y)$. We shall successively refine an initial set of candidate nodes (obtained from Ψ_x) to separate x and y (step 1.3 in the algorithm).

The set of candidate nodes, $\Omega_x^*(y)$, is formed by the variables that may belong to an active simple cycle with closing variables x and y. Those variables must verify the independence relationships given in eq. (3). But, as ilustrated in the graph of Figure 8, it is possible to obtain a non-empty candidate set $\Omega_x^*(y)$ even for a direct link x—y in the model (with $x = 12$ and $y = 9$ in Figure 8). So, before excluding a node y from Ψ_x, we must assure that the link x—y does not belong to the model. To do that, we use the refined set $\Omega_x(y)$, obtained by eq. (4). After this process, we may decide whether remove the link x—y or not, in the following way [de Campos and Huete, 1997]: If there is an arc from x to y or an arc from y to x in G, then $\Omega_x(y) = \emptyset$; otherwise $\Omega_x(y) \neq \emptyset$. This explains step 1.3 in the algorithm.

Finally, in step 1.4, the nodes in Ψ_x which are parents of x can be distinguished from the children by testing marginal independence relationships. Under the assumption that the dependency model is isomorphic to a simple graph, the algorithm gives as the output a simple graph with the same skeleton and the same head-to-head connections than the original, i.e., a simple graph isomorphic to the model.

Now, let us discuss how efficient the recovery algorithm is:

- The recovery algorithm can be implemented locally in part, the calculations are independent for each variable in the model: only the transition from

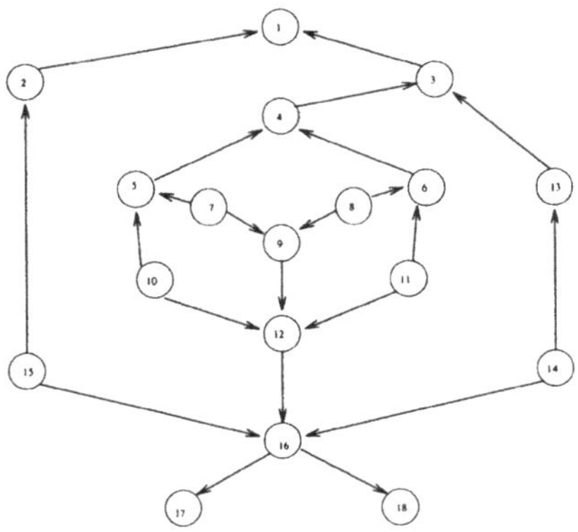

Figure 8. Simple Graph.

$\Omega_x^*(y)$ to $\Omega_x(y)$ cannot be made in a purely local way. This would make it possible to use distributed computation.

- The algorithm needs a polynomial number of independence tests $O(n^3)$.
- The independence tests needed are marginal independence tests and conditional independence tests given only one variable. Therefore, we can calculate the results of the tests in polynomial time.
- Knowing the result of the tests, the algorithm obtains a simple graph in polynomial time, $O(n^4)$.

The following example shows how the recovery algorithm works. We use the simple graph in Figure 8 as a hidden model, where the set of variables is $U = \{1, \ldots, 18\}$ and we study how the algorithm constructs the set of neighbours for nodes 12 and 16. For node 16 we have the fact that $\Lambda_{16} = U \setminus \{16\}$, and $\Psi_{16} = \{1, 3, 12, 14, 15, 17, 18\}$. We can construct $K_{16}(1) = \{3, 12, 14, 15\}$ and then, by considering that $3 \in K_{16}(1)$, $I(3, 15 \mid \emptyset)$ and $\neg I(3, 15 \mid 1)$, we can conclude that $3 \in \Omega_{16}^*(1)$. Using analogous reasoning we obtain the set $\Omega_{16}^*(1) = \{3, 12, 14, 15\}$. Similarly, $\Omega_{16}^*(3) = \{12, 14\}$. The other sets $\Omega_{16}^*(.)$ are equal to the empty set. The next step is to construct the sets $\Omega_{16}(.)$, i.e., $\Omega_{16}(1)$ and $\Omega_{16}(3)$; in this case, we find that there is no element that may be excluded from $\Omega_{16}(1)$ and $\Omega_{16}(3)$ and then, nodes $1, 3$ are eliminated from Ψ_{16}, so Neighbour(16)= $\{12, 14, 15, 17, 18\}$.

To conclude the example, we now consider node 12. We know that $\Lambda_{12} = U \setminus \{2, 13, 14, 15\}$, and $\Psi_{12} = \{4, 5, 6, 9, 10, 11, 16\}$. We obtain the sets $\Omega_{12}^*(4) = \{5, 6, 9, 10, 11\}$, $\Omega_{12}^*(5) = \{9, 10\}$, $\Omega_{12}^*(6) = \{9, 11\}$ and $\Omega_{12}^*(9) = \{5, 6\}$, with the other sets $\Omega_{12}^*(.)$ being empty. Again we find that the sets $\Omega_{12}(.)$ are all non-empty, except for $\Omega_{12}(9)$, in which nodes 7 and 8 verify the conditions that exclude the nodes 5 and 6, respectively. Finally, we find that Neighbour(12)=

$\{9, 10, 11, 16\}$.

When we use the algorithm given in Figure 7 to learning simple graphs, it is assumed that the dependency model is isomorphic to a simple graph. In this case, the algorithm gives a simple graph as the output (and does it efficiently). The independence tests can be obtained either from empirical data or from expert judgments, or a combination thereof. In any case, it avoids two main problems in learning algorithms: the need for a great number of independence tests and, when we use empirical data, the exponential cost necessary to calculate these tests.

The problem arises when we have a dag-isomorphic dependency model, but we do not know whether or not it may be represented by a simple graph. In this case, the algorithm can be modified in such a way that it gives an error code as the output, if the model cannot be represented by a simple graph. There are two possibilities that must be checked:

A *The output of the algorithm is not a simple graph.* If the model can be represented by a simple graph, then all the head-to-head connections must be detected in step 1.4. So, if when directing the remaining links in step 3, a new head-to-head connection is obtained, then we have a marginal independence relationship that is not in the model, so we can give an error code as the output. But it may be possible to direct the remaining links without including new head-to-head connections. In that case, we must test whether the resultant graph is simple, i.e., if there is no simple trail connecting x with x.

B *The output of the algorithm is a simple graph.* If the model is not simple, we find that a link has been be incorrectly eliminated by the algorithm. By looking at the algorithm, we see that links are eliminated in steps 1.2 or 1.3. In the first case, the independence relationships are tested, and so the arcs are eliminated correctly. In step 1.3, a link is eliminated by the assumption that we are considering a simple graph. The problem arises from this assumption. So, in step 1.3.3, when there is a set $\Omega_x(y) \neq \emptyset$, we must test the relationship $I(x, y \mid \rho_x(y))$ before eliminating the arc; if this relationship does not hold we can give an error code as the output. In this process we need to use higher order independence tests, but we can delay these tests until the end of the algorithm. The modified algorithm is shown in Figure 9.

3.3 Algorithms for General Networks

In this subsection we consider algorithms for learning belief networks which make no restriction about the topology of the graphs.

The first method we study is based on the concept of *boundary strata* or *recursive basis* [Verma and Pearl, 1990a]: Let M be a dependency model defined on a set of variables U, and let d be a complete ordering on U, so that with respect to this ordering U can be written as $U = \{x_1, \ldots, x_n\}$. Let us denote the set of variables preceding x_i in the ordering d as $\text{Pred}_d(x_i)$, i.e., $\text{Pred}_d(x_i) = \{x_1, \ldots, x_{i-1}\}$. The

MCH Algorithm

1. For each x in U do
   1.1. Calculate Λ_x
   1.2. Calculate Ψ_x
   1.3. For each $y \in \Psi_x$ do
      1.3.1. Calculate $K_x(y)$. If $K_x(y) = \emptyset$ go to step 1.3.
      1.3.2. Calculate $\Omega_x^*(y)$. If $\Omega_x^*(y) = \emptyset$ go to step 1.3.
      1.3.3. Calculate $\Omega_x(y)$. If $\Omega_x(y) \neq \emptyset$ exclude y from Ψ_x
   1.4. For each pair $y, z \in \Psi_x$, If $I(y, z \mid \emptyset)$ holds, put the nodes y, z as parents of x
2. Fuse every Ψ_x, to obtain G
3. Direct the remaining links without introducing new head-to-head connections.
   If this orientation is not feasible, return 'fail'
4. Test whether the output graph is simple. If it is not simple, return 'fail'
5. For each $\Omega_x(y) \neq \emptyset$, if $I(x, y \mid \Omega_x(y) \cap \text{Parents}_x)$ does not hold, return 'fail'

Figure 9. Modified algorithm.

boundary strata of x_i relative to *d*, B_i, is a minimal set satisfying $B_i \subseteq \text{Pred}_d(x_i)$ and $I(x_i, \text{Pred}_d(x_i) \setminus B_i \mid B_i)$. The important result is the following [Pearl, 1988; Verma and Pearl, 1990a]: Let M be a semi-graphoid and, given an ordering *d*, let G be the graph (called *Boundary graph* of M) obtained by designating each B_i as the parent set of the variable x_i. Then G is a minimal I-map of M. So, we can construct a belief network (given any ordering) by searching, for each variable x_i, a minimal subset of $\text{Pred}_d(x_i)$ rendering x_i conditionally independent from the rest of $\text{Pred}_d(x_i)$. The number of conditional independence tests necessary may be exponential (as well as the cost of performing some of these tests from a database). Different orderings may produce different graphs (although all of them are I-maps), and some graphs may be more sparse than other. As sparsity is a desirable property, it would be convenient to choose an ordering producing the most sparse graph. In [Srinivas *et al.*, 1990] an algorithm based on the boundary strata method, which uses a greedy heuristic for generating sparse graphs, is proposed. Another alternative is to estimate a good ordering before constructing the network (algorithms carrying out this task can be found in [Bouckaert, 1994; de Campos and Huete, 1998; Larrañaga *et al.*, 1996]).

The definition of a *Recursive diagram* [Wermuth and Lauritzen, 1983] can be considered as another method for building belief networks: given an ordering *d* on U, for each pair of variables x_i and x_j, x_j is a parent of x_i in the graph if and only if $x_j < x_i$ (for the ordering *d*) and $\neg I(x_i, x_j \mid \text{Pred}_d(x_i) \setminus \{x_j\})$. If the dependency model M is a graphoid, then the Recursive diagram is a minimal I-map of

M (and coincides with the Boundary graph of M for the same ordering).

The algorithms in the rest of this subsection work under the assumption that the dependency model is dag-isomorphic, and eliminate the necessity of providing an ordering on U. The basis for the algorithms are the following properties [Spirtes *et al.*, 1993; Verma and Pearl, 1990]:

- If M is dag-isomorphic, then M is isomorphic to graph G if and only if

  (i) for all nodes x, y of G, x and y are adyacent if and only if x and y are dependent conditional on every set of nodes that does not include x or y; and

  (ii) for all nodes x, y, z such that x is adjacent to y and y is adjacent to z, and x and z are not adjacent, $x \rightarrow y \leftarrow z$ is a subgraph of G if and only if x and z are dependent conditional on every set containing y but not x or z.

- In a dag G, if x is not a descendant of y, and x and y are not adjacent, then x and y are d-separated by Parents_y.

The SGS algorithm [Spirtes *et al.*, 1993] is shown in Figure 10. This algorithm is very inefficient because for each pair of variables, it must test conditional independence relationships given all possible subsets of the remaining variables, in the worst case, and the worst case is also the expected case for edges in the true graph. Alternative versions of the SGS algorithm have been proposed in [Pearl and Verma, 1991; Verma and Pearl, 1990; Verma and Pearl, 1992].

SGS Algorithm

1. Form the complete undirected graph H on the vertex set U
2. For each pair of nodes x and y, if there exists a subset S of $U \setminus \{x, y\}$ such that x and y are conditionally independent given S, remove the edge x—y from H
3. For each triple of nodes x, y, z such that x—y—z is a subgraph of H but x and z are not adjacent in H, orient x—y—z as $x \rightarrow y \leftarrow z$ if and only if there is no subset S of $U \setminus \{x, z\}$ such that $I(x, z \mid S \cup \{y\})$ holds
4. repeat
   4.1. if $x \rightarrow y$ is in H, y and z are adjacent, x and z are not adjacent, and there is no arrowhead at y, then orient y—z as $y \rightarrow z$
   4.2. if there is a directed path from x to y, and x—y is in H, then orient x—y as $x \rightarrow y$
   until no more edges can be oriented

Figure 10. SGS algorithm for learning dag-isomorphic models.

The SGS algorithm needs an exponential number of conditional independence tests, which in addition may be of higher order. The PC algorithm [Spirtes *et al.*, 1991; Spirtes *et al.*, 1993] tries to minimize these two problems: it starts by forming the complete undirected graph, then thins that graph by removing edges with zero order conditional independence relationships, thins again with first order conditional independence relationships, and so on. The set of variables conditioned on needs only be a subset of the set of variables adjacent to one or the other of the variables conditioned. The PC algorithm is displayed in Figure 11 (there, Adj(G, x) stands for the set of nodes adjacent to x in the graph G, which is constantly changing as the algorithm progresses).

PC Algorithm

1. Form the complete undirected graph G on the vertex set U
2. $n = 0$
3. repeat
   3.1. repeat
      select an ordered pair of variables x and y that are adjacent in G such that Adj$(G, x) \setminus \{y\}$ has cardinality greater than or equal to n, and a subset S of Adj$(G, x) \setminus \{y\}$ of cardinality n, and if x and y are conditionally independent given S delete edge x—y from G and record S in Sepset(x, y) and Sepset(y, x)
   until all ordered pairs of adjacent variables x and y such that
      Adj$(G, x) \setminus \{y\}$ has cardinality greater than or equal to n and all subsets S of Adj$(G, x) \setminus \{y\}$ of cardinality n have been tested for conditional independence
   3.2. $n = n + 1$
until for each ordered pair of adjacent nodes x, y, Adj$(G, x) \setminus \{y\}$ is of cardinality less than n
4. For each triple of nodes x, y, z such that x—y—z is a subgraph of G but x and z are not adjacent in G, orient x—y—z as $x \rightarrow y \leftarrow z$ if and only if y is not in Sepset(x, z)
5. repeat
   5.1. if $x \rightarrow y$ is in G, y and z are adjacent, x and z are not adjacent, and there is no arrowhead at y, then orient y—z as $y \rightarrow z$
   5.2. if there is a directed path from x to y, and x—y is in G, then orient x–y as $x \rightarrow y$
   until no more edges can be oriented

Figure 11. PC algorithm for learning dag-isomorphic models.

The complexity of PC is bounded by the largest degree in G. If k is the maximal degree of any node and n is the number of nodes, the number of conditional independence tests required by PC is bounded by $n^2(n-1)^{k-1}/(k-1)!$. A slighty different version of PC, called PC* [Spirtes *et al.*, 1993], tests for conditional independence of any two variables x, y given subsets of variables adjacent to x and subsets of variables adjacent to y that are also on undirected paths between x and y. The idea is to avoid unnecessary tests, although as the number of undirected paths may be very large, PC* is only feasible for a relatively small number of variables.

Another algorithm which also assumes an underlying dag-isomorphic dependency model [de Campos and Huete, 1999b] is based on an intensive use of low order conditional independence tests: it first uses zero and first order independence statements to obtain a prior skeleton of the network, and also to fix and remove arrows from this skeleton. Then, a refinement procedure, based on minimum cardinality d-separating sets [Acid and de Campos, 1996], which uses a small number of conditional independence tests of higher order, is carried out to produce the final graph.

4 INDEPENDENCE IN NON PROBABILISTIC THEORIES OF UNCERTAINTY

In the previous section we have studied algorithms for learning belief networks based on conditional independence. If we want to use them in connection with any specific uncertainty formalism, it is necessary to define a concept of conditional independence within this formalism.

There are several formalisms to deal with numerical uncertainty in artificial intelligence. Some of them, at least in a formal sense, are hierarchycally ordered, going from the general to the specific. Among others, we have general fuzzy measures [Sugeno, 1974], lower and upper probabilities, Choquet capacities of order two [Choquet, 1953], and belief and plausibility functions [Dempster, 1967; Shafer, 1976], which include both possibility measures [Zadeh, 1978; Dubois and Prade, 1988] and probabilities. In the framework of systems which handle uncertain knowledge, the concepts of independence and conditional independence have been studied in-depth only for probability measures (see, for example [Dawid, 1979; Lauritzen *et al.*, 1990; Spohn, 1980]), although there is an increasing interest in studying independence for other theories of uncertain information ([de Campos and Huete, 1993; de Campos and Huete, 1993; de Campos and Moral, 1995; Chrisman, 1996; Shenoy, 1992; Studený, 1993]). In this section we address this topic for different uncertainty theories. More precisely, we shall focus in two cases: possibility measures and probability intervals (although some of the ideas presented here could be applied to other theories). Several alternatives are considered for each theory, and their properties are analized.

4.1 Possibility Measures

We are going to investigate several ways to define independence relationships in the framework of possibility theory [Zadeh, 1978; Dubois and Prade, 1988]. Some other works about the same topic have recently appeared [Benferhat *et al.*, 1994; de Campos, 1995; de Campos *et al.*, 1995; de Campos and Huete, 1999; de Campos and Huete, 1999a; de Campos *et al.*, 1995a; de Cooman and Kerre, 1994; Dubois *et al.*, 1994; Fariñas del Cerro and Herzig, 1994; Fonck, 1994; Fonck, 1997; Huete, 1995; Sangüesa *et al.*, 1998; Studený, 1993]. First, we describe the concepts of marginalization and conditioning for possibility measures that we are going to use. Next, we study different definitions of possibilistic independence, based on a view of possibility measures as particular cases of plausibility functions. Finally, other definitions of independence are considered, based on a different view of possibility measures, closer to fuzzy sets.

Possibility Measures. Marginalization and Conditioning

Consider a finite referential $D_x = \{\mathtt{x}_1, \mathtt{x}_2, \ldots, \mathtt{x}_n\}$ (where a variable x takes its values). A possibility measure defined on D_x is a set function

$$\Pi : \mathcal{P}(D_x) \to [0,1],$$

verifying

1. $\Pi(D_x) = 1$,
2. $\Pi(\mathtt{A} \cup \mathtt{B}) = \max(\Pi(\mathtt{A}), \Pi(\mathtt{B})),\ \forall \mathtt{A}, \mathtt{B} \subseteq D_x$.

Associated with the possibility measure Π, the possibility distribution

$$\pi : D_x \to [0,1],$$

is defined by means of

$$\pi(\mathtt{x}) = \Pi(\{\mathtt{x}\}),\ \forall \mathtt{x} \in D_x.$$

The information contained in the possibility distribution is sufficient to recover the possibility measure, since $\forall \mathtt{A} \subseteq D_x,\ \Pi(\mathtt{A}) = \max_{\mathtt{x} \in \mathtt{A}} \pi(\mathtt{x})$. So, we can restrict our study to deal only with possibility distributions.

Given a bidimensional possibility distribution π for two variables x and y, defined on the cartesian product $D_x \mathrm{x} D_y$, the marginal possibility distribution on one variable, x (analogously for y), π_x, is formalized by means of the following definition:

(5) $\pi_x(\mathtt{x}) = \max_{\mathtt{y} \in D_y} \pi(\mathtt{x}, \mathtt{y}),\ \forall \mathtt{x} \in D_x.$

From now on, to simplify the notation of a marginal possibility distribution, we shall drop the subindex, thus writing $\pi(\mathtt{x})$ instead of $\pi_x(\mathtt{x})$.

However, the concept of conditional possibility distribution is not so universal, there exist different alternatives to define it. We shall mainly consider two alternatives:

1.- The possibility distribution on x conditioned to the event $[y = \mathtt{y}]$, denoted by $\pi_d(.|\mathtt{y})$, is defined as

$$(6)\quad \pi_d(\mathrm{x}|\mathrm{y}) = \frac{\pi(\mathrm{x},\mathrm{y})}{\pi(\mathrm{y})} = \frac{\pi(\mathrm{x},\mathrm{y})}{\max_{\mathrm{x}'\in D_x}\pi(\mathrm{x}',\mathrm{y})} \quad \forall \mathrm{x} \in D_x.$$

This is Dempster rule of conditioning, specialized to possibility measures, which are consonant plausibility functions [Shafer, 1976].

2.- The possibility distribution on x conditioned to the event $[y = \mathrm{y}]$, denoted by $\pi_h(.|\mathrm{y})$, is defined as

$$(7)\quad \pi_h(\mathrm{x}|\mathrm{y}) = \begin{cases} \pi(\mathrm{x},\mathrm{y}) & \text{if } \pi(\mathrm{x},\mathrm{y}) < \pi(\mathrm{y}) \\ 1 & \text{if } \pi(\mathrm{x},\mathrm{y}) = \pi(\mathrm{y}) \end{cases} \quad \forall \mathrm{x} \in D_x.$$

A slightly different version of this definition was considered in [Hisdal, 1978], as the solution of the equation $\pi(\mathrm{x},\mathrm{y}) = \pi(\mathrm{x}|\mathrm{y}) \wedge \pi(\mathrm{y})$, $\forall \mathrm{x}$ (this definition is based on Bayes' rule for probabilities, where the minimum operator replaces the product). The definition above is the least specific solution of this equation [Dubois and Prade, 1993].

Obviously these definitions of marginal and conditional possibility distributions for two variables can be inmediately extended to the n-dimensional case.

Observe that the difference between these two definitions of conditioning is the different kind of normalization used. In both cases we first focus on the values compatible with the available information $[y = \mathrm{y}]$, i.e., $\{(\mathrm{x}',\mathrm{y}'), \mid \mathrm{x}' \in D_x, \mathrm{y}' = \mathrm{y}\}$, and next we must normalize the values $\pi(\mathrm{x}',\mathrm{y})$ to get a possibility distribution. In the first case the normalization is carried out by dividing by the greatest value $\max_{\mathrm{x}'\in D_x}\pi(\mathrm{x}',\mathrm{y})$ $(=\pi(\mathrm{y}))$, whereas in the second case the normalization is achieved by increasing the greatest values of $\pi(\mathrm{x}',\mathrm{y})$ up to 1.

Definitions of Independence Based on $\pi_d(.)$

Consider a finite set of n variables, U, and a n-dimensional possibility distribution π defined on U. Given three disjoint subsets of variables in U, X,Y and Z, we shall represent by means of $I(X,Y \mid Z)$ the sentence X is conditionally independent of Y given Z, for the model associated to the distribution π. We shall denote by $\mathrm{x},\mathrm{y},\mathrm{z}$ the values that the respective sets of variables can take on. The values of, for example, $Y \cup Z$, will be denoted simply by yz.

The most obvious way to define the conditional independence is to proceed in a way similar to the probabilistic case (eq. (1)). This is the idea considered in [Shenoy, 1992] for the more general framework of valuation-based systems. In our case this idea gives rise to:

$$(8)\quad I(X,Y \mid Z) \Longleftrightarrow \pi_d(\mathrm{x}|\mathrm{yz}) = \pi_d(\mathrm{x}|\mathrm{z}),$$

that is to say, the knowledge about the value of the variable Y does not modify our belief about the values of X, when the value of variable Z is known; $\mathrm{x},\mathrm{y},\mathrm{z}$ are arbitrary values of the variables X,Y,Z, with the only restriction that the conditional distributions involved must be defined, i.e., $\pi(\mathrm{yz}) \neq 0$. This definition has also been considered in [Studený, 1993], and in [Wilson, 1994] for generalized conditional probabilities.

The definition above verifies the properties A1–A4, and it also satisfies A5 if the possibility distribution π is strictly positive [de Campos and Huete, 1999a].

One problem with this definition could be, in our opinion, that it is too strict, because it requires the equality of the distributions, and these distributions represent an imprecise knowledge. The problem becomes worse when the distributions must be estimated from data or human judgments.

The previous definition, eq. (8), represents that our information about X remains unchanged after conditioning to Y. A different idea could be to assert the independence when we do not gain additional information about the values of X after conditioning to Y (but we could lose some information). The idea of a possibility distribution being more or less informative than another one is adequately captured by the well-known definition of inclusion [Dubois and Prade, 1986]: Given two possibility distributions π and π', π' is said to be included in (or gives less information than) π if and only if $\pi(\mathrm{x}) \leq \pi'(\mathrm{x})\ \forall \mathrm{x}$.

Using the relationship of inclusion between possibility distributions, the previous idea can be expressed more formally by means of

(9) $I(X,Y \mid Z) \Longleftrightarrow \pi_d(\mathrm{x}|\mathrm{yz}) \geq \pi_d(\mathrm{x}|\mathrm{z})\ \ \forall \mathrm{x},\mathrm{y},\mathrm{z}$ such that $\pi(\mathrm{yz}) \neq 0$.

Unfortunately this definition does not verify the property A3 (Weak Union), although it verifies A1–A2 and A4 [de Campos and Huete, 1999a]. We think that the problem is that the idea of independence as a non-gain of information has not been carried out till the finish: if after conditioning we lose information, it seems more convenient to keep the initial information. This can be debatable, but it represents a sort of default rule: if in a very specific context we do not have much information, then use the information available in a less specific context. In practical terms, this idea implies a change in the definition of conditioning:

(10) $$\pi_{d_c}(\mathrm{x}|\mathrm{y}) = \begin{cases} \pi(\mathrm{x}) & \text{if } \pi(\mathrm{xy}) \geq \pi(\mathrm{x})\pi(\mathrm{y})\ \forall \mathrm{x} \\ \pi_d(\mathrm{x}|\mathrm{y}) & \text{if } \exists \mathrm{x}' \text{ such that } \pi(\mathrm{x}'\mathrm{y}) < \pi(\mathrm{x}')\pi(\mathrm{y}) \end{cases}$$

that is, if $\pi_d(\mathrm{x}|\mathrm{y})$ is always greater than or equal to $\pi(\mathrm{x})$, that is, it gives less information than $\pi(\mathrm{x})$, then maintain $\pi(\mathrm{x})$; otherwise use the previous conditioning.

Using this conditioning, the new definition of independence is

(11) $I(X,Y \mid Z) \Longleftrightarrow \pi_{d_c}(\mathrm{x}|\mathrm{yz}) = \pi_{d_c}(\mathrm{x}|\mathrm{z})\ \ \forall \mathrm{x},\mathrm{y},\mathrm{z}$.

This definition satisfies the properties A2–A5 (even the distributions which are not strictly positive verify Intersection). It does not satisfy Symmetry (we could obtain Symmetry by defining a new relationship $I'(.,. \mid .)$ by means of $I'(X,Y \mid Z) \Leftrightarrow I(X,Y \mid Z)$ and $I(Y,X \mid Z)$, but then it is not clear so far that all the other axioms A2–A5 will still be satisfied).

Another approach to define independence consists in considering that the possibility distributions $\pi_d(.|\mathrm{yz})$ and $\pi_d(.|\mathrm{z})$ are similar in some sense. So, if $\simeq$ is a relationship in the set of possibility distributions defined on X, we could define the independence by means of

(12) $I(X, Y \mid Z) \Longleftrightarrow \pi_d(.|\mathbf{yz}) \simeq \pi_d(.|\mathbf{z})\ \ \forall \mathbf{y}, \mathbf{z}$ such that $\pi(\mathbf{yz}) \neq 0$.

In order to define the relationship $\simeq$ we can consider different options. Let us see some of them:

If we think that a possibility distribution essentially establishes an ordering among the values that the variable we are considering can take on, and the numbers (the possibility degrees) only have a secondary importance, then we can say that two possibility distributions are similar if they establish the same ordering. More formally, we can define the relation $\simeq$ by means of

(13) $\pi \simeq \pi' \Leftrightarrow \forall \mathbf{x}, \mathbf{x}' \, [\pi(\mathbf{x}) < \pi(\mathbf{x}') \Leftrightarrow \pi'(\mathbf{x}) < \pi'(\mathbf{x}')]$.

The independence obtained through this relationship (we could call it 'isoordering') satisfies the properties A2–A4, and also A5 for strictly positive distributions; however it does not verify A1.

Another option is to speak about similarity between distributions when the possibility degrees of each distribution for each value are alike. More concretely, let m be any positive integer, and $\{\alpha_k\}_{k=0,\ldots,m}$ such that $\alpha_0 < \alpha_1 < \ldots < \alpha_{m-1} < \alpha_m$, with $\alpha_0 = 0$ and $\alpha_m = 1$. If we denote $I_k = [\alpha_{k-1}, \alpha_k)$, $k = 1, \ldots, m-1$, and $I_m = [\alpha_{m-1}, \alpha_m]$, then we define the relationship $\simeq$ by means of

(14) $\pi \simeq \pi' \Leftrightarrow \forall \mathbf{x}\ \exists k \in \{1, \ldots, m\}$ such that $\pi(\mathbf{x}), \pi'(\mathbf{x}) \in I_k$.

This definition is equivalent to the following, which is established in terms of the α-cuts of the distribution:

(15) $\pi \simeq \pi' \Leftrightarrow C(\pi, \alpha_k) = C(\pi', \alpha_k)\ \forall k = 1, \ldots, m-1$,

where $C(\pi, \alpha) = \{\mathbf{x} \mid \pi(\mathbf{x}) \geq \alpha\}$. This idea is also equivalent to discretize the interval $[0, 1]$, and to say that two distributions are similar if their respective discrete versions coincide. The independence defined in these terms satisfies the properties A2–A5 (although A5 only for strictly positive distributions), and again Symmetry fails.

Finally, a third option to define $\simeq$ is based on the following idea: consider a threshold α_0, and suppose that only from values greater than α_0 it is considered interesting to differentiate between the possibility degrees of two distributions; the values whose possibility degrees are below the threshold are not considered relevant. In terms of the α-cuts, this relationship $\simeq$ can be expressed as follows:

(16) $\pi \simeq \pi' \Leftrightarrow C(\pi, \alpha) = C(\pi', \alpha),\ \forall \alpha \geq \alpha_0$,

which is equivalent to

(17) $\pi \simeq \pi' \Leftrightarrow C(\pi, \alpha_0) = C(\pi', \alpha_0)$ and $\pi(\mathbf{x}) = \pi'(\mathbf{x})\ \forall \mathbf{x} \in C(\pi, \alpha_0)$.

Once again, the independence defined in terms of this relationship $\simeq$ verifies the properties A2–A4 (and A5 for distributions strictly positive), and it does not satisfy A1.

Now, let us study in an abstract way what kind of properties of $\simeq$ are sufficient to guarantee that the associated independence relationship verifies some of the properties A1–A5:

First, it is evident that transitivity of $\simeq$ guarantees A4 (Contraction). If in addition $\simeq$ is symmetric, then it can be deduced that A2 (Decomposition) is verified if and only if A3 (Weak Union) is also verified. Moreover, in order to verify the basic coherence property $I(X, \emptyset \mid Z)$ (called Trivial Independence in [Wilson, 1994]), it is obvious that $\simeq$ should be reflexive. Therefore, it seems that the equivalence relations $\simeq$ are good candidates for defining independence.

A sufficient condition for the fulfilment of A2 is that $\simeq$ verify the following property [de Campos and Huete, 1999a]:

Let $\{\pi_s\}$ a family of possibility distributions such that

$$\pi_s(\mathbf{x}) = \frac{f_s(\mathbf{x})}{\lambda_s} \quad \forall \mathbf{x}$$

where λ_s are positive real numbers less than or equal to 1 (therefore $\max_{\mathbf{x}} f_s(\mathbf{x}) = \lambda_s \ \forall s$). Let π_m be the possibility distribution obtained by means of

$$\pi_m(\mathbf{x}) = \frac{\max_s f_s(\mathbf{x})}{\max_s \lambda_s} \quad \forall \mathbf{x}$$

Then

(18) $\pi_s \simeq \pi \ \ \forall s \Rightarrow \pi_m \simeq \pi$

Moreover, for the case of strictly positive distributions, the fulfilment of the property above also guarantees A5 (Intersection).

So, every possibilistic independence relationship defined as in eq. (12), in terms of an equivalence relationship $\simeq$ verifying (18), satisfies the properties A2–A5 (although the last one only for strictly positive distributions). The only property excluded of this context is Symmetry; this is somewhat surprising, because this is one of the properties of independence seemingly more intuitive. Obviously, the three relationships $\simeq$ defined before are equivalence relationships and they verify (18).

Definitions of Independence Based on $\pi_h(.)$

Now, we are going to propose some definitions of independence based on $\pi_h(.)$, the other concept of possibilistic conditioning being considered. If we follow a development parallel to that one carried out for $\pi_d(.)$, the first definition of conditional independence is

(19) $I(X,Y \mid Z) \Longleftrightarrow \pi_h(\mathrm{x}|\mathrm{yz}) = \pi_h(\mathrm{x}|\mathrm{z}) \ \forall \mathrm{x}, \mathrm{y}, \mathrm{z}.$

This definition verifies A2–A5 but does not verify A1 [de Campos and Huete, 1999]. The new relationship, $I'(X,Y \mid Z) \Leftrightarrow I(X,Y \mid Z)$ and $I(Y,X \mid Z)$, satisfies all the properties A1–A5 [Fonck, 1994] (although it leads to a somewhat restrictive concept of independence; for example, it can be seen that $I'(X,Y \mid Z)$ implies $\pi(\mathrm{x}) = 1 \, \forall \mathrm{x}$ or $\pi(\mathrm{y}) = 1 \, \forall \mathrm{y}$ [de Campos and Huete, 1999]).

Using again the idea of independence as a non-gain of information after conditioning, we can define an independence relationship as in eq. (9), by means of

(20) $I(X,Y \mid Z) \Longleftrightarrow \pi_h(\mathrm{x}|\mathrm{yz}) \geq \pi_h(\mathrm{x}|\mathrm{z}) \ \forall \mathrm{x}, \mathrm{y}, \mathrm{z}.$

In this case the definition above verifies the properties A1–A4 and does not verify A5 in general [de Campos and Huete, 1999] (so there is a difference between this definition using π_d and using π_h). Moreover this definition turns out to be equivalent to the following:

(21) $I(X,Y \mid Z) \Longleftrightarrow \pi(\mathrm{xyz}) = \pi(\mathrm{xz}) \wedge \pi(\mathrm{yz}) \ \forall \mathrm{x}, \mathrm{y}, \mathrm{z}.$

If we consider the case of marginal independence ($Z = \emptyset$), from eq. (21) we obtain

(22) $I(X,Y \mid \emptyset) \Longleftrightarrow \pi(\mathrm{xy}) = \pi(\mathrm{x}) \wedge \pi(\mathrm{y}) \ \forall \mathrm{x}, \mathrm{y},$

and we have the well-known concept of non-interactivity for possibility measures or fuzzy sets [Zadeh, 1978]. So, this concept of independence is that we could call conditional non-interactivity.

Observe that for this kind of conditioning, π_h, the only necessary operation is comparison. So, we could easily consider generalized possibility distributions taking values in sets different from the $[0, 1]$ interval: it suffices to use a set $(\mathcal{L}, \preceq)$, where

$$\mathcal{L} = \{L_0, L_1, \ldots, L_n\}$$

and

$$L_0 \preceq L_1 \preceq \ldots \preceq L_n,$$

that is to say, a totally ordered set (for example, a set of linguistic labels), and then define possibility measures by means of

$$\Pi : \mathcal{P}(D_x) \to \mathcal{L}$$

verifying:

1. $\Pi(D_x) = L_n,$

2. $\Pi(\mathtt{A} \cup \mathtt{B}) = \vee_{\preceq}(\Pi(\mathtt{A}), \Pi(\mathtt{B})), \ \forall \mathtt{A}, \mathtt{B} \subseteq D_x,$

where $\vee_{\preceq}$ is the maximum (supremum) operator associated to the ordering $\preceq$. We could speak about these generalized possibility measures as qualitative possibility measures. In these conditions we can define the conditioning and the independence in exactly the same way as before (eqs.(7) and (20)), obtaining the same properties.

A different way of extension (analogous to that one previously considered for the other conditioning, eq. (12)) would consist in considering a relationship $\approx$ in the set of possibility distributions, and then define the independence by means of

(23) $I(X, Y \mid Z) \Longleftrightarrow \pi(\mathrm{xyz}) \approx \pi(\mathrm{xz}) \wedge \pi(\mathrm{yz})$.

In this case, a sufficient condition for this relationship satisfying the axioms A1–A4, is that $\approx$ is an equivalence relationship compatible with the marginalization and combination of possibility distributions (using the minimum as the combination operator) [de Campos and Huete, 1999], i.e.:

- $\approx$ is an equivalence relationship for possibility distributions.
- If $\pi(\mathrm{xy}) \approx \pi'(\mathrm{xy})$ then $\max_{\mathrm{x}} \pi(\mathrm{xy}) \approx \max_{\mathrm{x}} \pi'(\mathrm{xy})$.
- If $\pi_1(\mathrm{x}) \approx \pi_1'(\mathrm{x})$ and $\pi_2(\mathrm{x}) \approx \pi_2'(\mathrm{x})$ then $\pi_1(\mathrm{x}) \wedge \pi_2(\mathrm{x}) \approx \pi_1'(\mathrm{x}) \wedge \pi_2'(\mathrm{x})$.

Finally, we could also define a new concept of conditioning, analogous to that of eq. (10), by means of

(24) $$\pi_{h_c}(\mathrm{x}|\mathrm{y}) = \begin{cases} \pi(\mathrm{x}) & \text{if } \pi_h(\mathrm{x}|\mathrm{y}) \geq \pi(\mathrm{x})\ \forall \mathrm{x} \\ \pi_h(\mathrm{x}|\mathrm{y}) & \text{if } \exists \mathrm{x}' \text{ such that } \pi_h(\mathrm{x}'|\mathrm{y}) < \pi(\mathrm{x}') \end{cases}$$

Using this conditioning, the new definition of independence is

(25) $I(X, Y \mid Z) \Longleftrightarrow \pi_{h_c}(\mathrm{x}|\mathrm{yz}) = \pi_{h_c}(\mathrm{x}|\mathrm{z})\ \ \forall \mathrm{x}, \mathrm{y}, \mathrm{z}$.

This definition is similar to that of eq. (11), but replacing π_{d_c} by π_{h_c}, and we get the same properties: A2–A5, and Symmetry fails. This relationship is more restrictive than the conditional non-interactivity, since every independence statement with definition (25) implies the corresponding statement of conditional non-interactivity, eqs. (20) and (21), but the converse is not true.

4.2 Probability Intervals

In this subsection we are going to study the concept of independence for other formalism to deal with numerical uncertainty, very different from possibility theory, and more similar to probability theory: *probability intervals*. This formalism generalizes probability theory and almost maintains its characteristics with respect to efficient computation. First, we shall briefly introduce the main concepts and tools concerning probability intervals (a detailed study of probability intervals can be found in [de Campos *et al.*, 1994; de Campos *et al.*, 1995]). Next, we shall discuss several alternative definitions of independence for probability intervals.

Probability Intervals. Marginalization, Conditioning and Product

Let us consider a variable x taking its values in a finite set $D_x = \{\mathtt{x}_1, \mathtt{x}_2, \ldots, \mathtt{x}_n\}$ and a family of intervals $L = \{[l(\mathtt{x}_i), u(\mathtt{x}_i)],\ i = 1, \ldots, n\}, 0 \leq l(\mathtt{x}_i) \leq u(\mathtt{x}_i) \leq 1, \forall i$. These intervals are interpreted as a set of bounds of probability by defining the set $\mathcal{P}_L$ of probability distributions on D_x as

$$(26)\quad \mathcal{P}_L = \{P \in \mathcal{P}(D_x) \mid l(\mathtt{x}_i) \leq p(\mathtt{x}_i) \leq u(\mathtt{x}_i)\ \forall i\},$$

where $\mathcal{P}(D_x)$ denotes the set of probabilities defined on D_x. As $\mathcal{P}_L$ is obviously a convex set, we can consider probability intervals as a particular case of convex sets of probabilities [Kyburg, 1987].

In order to avoid for the set $\mathcal{P}_L$ to be empty, it is necessary to impose a condition to the intervals $[l(\mathtt{x}_i), u(\mathtt{x}_i)]$, which is:

$$(27)\quad \sum_i l(\mathtt{x}_i) \leq 1 \leq \sum_i u(\mathtt{x}_i).$$

A set of probability intervals verifying condition (27) is called *proper*. We shall always use proper probability intervals, because non proper intervals are useless.

In addition to a convex set $\mathcal{P}_L$, we can also associate with the proper set L a pair (l, u) of lower and upper probabilities ([de Campos *et al.*, 1988; Huber, 1981; Walley, 1991]) (also called a probability envelope or a pair of representable measures) through $\mathcal{P}_L$ as follows:

$$(28)\quad l(\mathtt{A}) = \inf_{P \in \mathcal{P}_L} P(\mathtt{A}),\ u(\mathtt{A}) = \sup_{P \in \mathcal{P}_L} P(\mathtt{A}),\ \forall \mathtt{A} \subseteq D_x.$$

So, probability intervals can also be considered as particular cases of lower and upper measures, where the set of associated probabilities is defined by restrictions affecting only the individual probabilities $p(\mathtt{x}_i)$.

In order to maintain consistency between both views of probability intervals, it would be interesting that the restriction of $l(.)$ and $u(.)$ to the singletons (sets with only one element) be equal to the original bounds, that is to say, that

$$(29)\quad l(\{\mathtt{x}_i\}) = l(\mathtt{x}_i),\ u(\{\mathtt{x}_i\}) = u(\mathtt{x}_i),\ \forall i = 1, \ldots, n.$$

These equalities are not always true: in general, we only have the inequalities $l(\{\mathtt{x}_i\}) \geq l(\mathtt{x}_i), u(\{\mathtt{x}_i\}) \leq u(\mathtt{x}_i)\ \forall i$. But we can get the equality by modifying the bounds $l(\mathtt{x}_i)$ and $u(\mathtt{x}_i)$ without altering the set $\mathcal{P}_L$, i.e., without changing the set of possible probabilities. The conditions that the intervals $[l(\mathtt{x}_i), u(\mathtt{x}_i)]$ must verify in order to get the equalities (29) are:

$$(30)\quad \sum_{j \neq i} l(\mathtt{x}_j) + u(\mathtt{x}_i) \leq 1 \text{ and } \sum_{j \neq i} u(\mathtt{x}_j) + l(\mathtt{x}_i) \geq 1,\ \forall i.$$

A set of probability intervals verifying the conditions in eq. (30) is called *reachable*. Given the proper set of probability intervals $L = \{[l(\mathtt{x_i}), u(\mathtt{x_i})], i = 1, \ldots, n\}$, and its associated convex set of probabilities $\mathcal{P}_L$, we define a new set of probability intervals $L' = \{[l'(\mathtt{x_i}), u'(\mathtt{x_i})],\ i = 1, \ldots, n\}$ by means of

$$(31)\quad l'(\mathtt{x_i}) = l(\mathtt{x_i}) \vee (1 - \sum_{j \neq i} u(\mathtt{x_j})),\ u'(\mathtt{x_i}) = u(\mathtt{x_i}) \wedge (1 - \sum_{j \neq i} l(\mathtt{x_j})),\ \forall i.$$

It can be proven that the set of probabilities associated to L' is also $\mathcal{P}_L$. In this way we can replace the original probability intervals by those defined in eq. (31) without affecting the set $\mathcal{P}_L$. This replacement permits us to refine the probability bounds that define $\mathcal{P}_L$ in such a way that these bounds can be always reached (because the probability intervals defined in eq. (31) verify conditions (30)). So, as the narrower intervals $[l'(\mathtt{x_i}), u'(\mathtt{x_i})]$ constitute a more accurate representation of the probabilities in $\mathcal{P}_L$ than $[l(\mathtt{x_i}), u(\mathtt{x_i})]$, we shall perform the substitution in those cases where L does not satisfy eq. (30), and thus we shall always use reachable probability intervals.

Reachable sets of probability intervals guarantee that the values of the lower and upper associated measures (l, u) for singletons, $l(\{\mathtt{x_i}\})$ and $u(\{\mathtt{x_i}\})$, coincide with the initial probability bounds $l(\mathtt{x_i})$ and $u(\mathtt{x_i})$, as eq. (29) asserts. The values of $l(.)$ and $u(.)$ for the other subsets of D_x which are not singletons can be calculated directly from the values $l(\mathtt{x_i})$ and $u(\mathtt{x_i})$ by means of the following expressions:

$$l(\mathtt{A}) = (\sum_{\mathtt{x_i} \in \mathtt{A}} l(\mathtt{x_i})) \vee (1 - \sum_{\mathtt{x_i} \notin \mathtt{A}} u(\mathtt{x_i})),\ \forall \mathtt{A} \subseteq D_x$$

$$(32)\quad u(\mathtt{A}) = (\sum_{\mathtt{x_i} \in \mathtt{A}} u(\mathtt{x_i})) \wedge (1 - \sum_{\mathtt{x_i} \notin \mathtt{A}} l(\mathtt{x_i})),\ \forall \mathtt{A} \subseteq D_x.$$

Now let us consider two variables x and y taking values in the sets $D_x = \{\mathtt{x_1}, \mathtt{x_2}, \ldots, \mathtt{x_n}\}$ and $D_y = \{\mathtt{y_1}, \mathtt{y_2}, \ldots, \mathtt{y_m}\}$ respectively, and a set of reachable bidimensional probability intervals $L = \{[l(\mathtt{x_i}, \mathtt{y_j}), u(\mathtt{x_i}, \mathtt{y_j})], i = 1, \ldots, n, j = 1, \ldots, m\}$, defined on the cartesian product $D_x \mathrm{x} D_y$, representing the joint available information on these two variables. To define the marginals of L we can use the interpretation of L as a pair of lower and upper measures (l, u). Given (l, u), the marginal measures (l_x, u_x) on x (for the marginals on y is analogous) are defined as: $l_x(\mathtt{A}) = l(\mathtt{A} \mathrm{x} D_y)$, $u_x(\mathtt{A}) = u(\mathtt{A} \mathrm{x} D_y)$, $\forall \mathtt{A} \subseteq D_x$. Alternatively, we could use the interpretation of probability intervals as convex sets of probabilities, and define the marginal of L on x as the set $\mathcal{P}_x$ of marginal probabilities of the probabilities in the convex set $\mathcal{P}_L$, being $\mathcal{P}_L$ the set of probabilities associated to L, i.e., $\mathcal{P}_x = \{P \in \mathcal{P}(D_x) \mid \exists Q \in \mathcal{P}_L \text{ such that } p(\mathtt{x_i}) = \sum_j q(\mathtt{x_i}, \mathtt{y_j}),\ \forall i\}$. Both definitions are equivalent, in the sense that $\mathcal{P}_x$ is just the set of probabilities associated to (l_x, u_x), i.e., $l_x(\mathtt{A}) = \min_{P \in \mathcal{P}_x} P(\mathtt{A})$, $u_x(\mathtt{A}) = \max_{P \in \mathcal{P}_x} P(\mathtt{A})$, $\forall \mathtt{A} \subseteq D_x$.

Moreover, it can be proven that these marginal lower and upper measures are in fact the probability intervals $L_x = \{\left[\sum_j l(\mathrm{x_i}, \mathrm{y_j}), \sum_j u(\mathrm{x_i}, \mathrm{y_j})\right], i = 1, \ldots, n\}$, and using eq. (31) the equivalent but reachable expression for L_x is $L_x = \{[l(\mathrm{x_i}), u(\mathrm{x_i})]\}$, where

$$l(\mathrm{x_i}) = \sum_j l(\mathrm{x_i}, \mathrm{y_j}) \vee (1 - \sum_j \sum_{k \neq i} u(\mathrm{x_k}, \mathrm{y_j})), \quad \forall i$$

$$u(\mathrm{x_i}) = \sum_j u(\mathrm{x_i}, \mathrm{y_j}) \wedge (1 - \sum_j \sum_{k \neq i} l(\mathrm{x_k}, \mathrm{y_j})), \quad \forall i. \tag{33}$$

To define the conditioning of probability intervals we shall again use their interpretation as lower and upper measures, because several definitions of conditioning in this framework are available (see [Moral and de Campos, 1991] for a review). Two of them will be considered here:

Given a pair of lower and upper measures (l, u) defined on a domain D, and a subset $\mathrm{B} \subseteq D$, the conditional lower and upper measures given that we know B, $(l(.|\mathrm{B}), u(.|\mathrm{B}))$, are defined as

– Campos/Moral's conditioning [de Campos *et al.*, 1990] (valid for any fuzzy measure):

$$l(\mathrm{A}|\mathrm{B}) = \frac{l(\mathrm{A} \cap \mathrm{B})}{l(\mathrm{A} \cap \mathrm{B}) + u(\bar{\mathrm{A}} \cap \mathrm{B})}, \quad u(\mathrm{A}|\mathrm{B}) = \frac{u(\mathrm{A} \cap \mathrm{B})}{u(\mathrm{A} \cap \mathrm{B}) + l(\bar{\mathrm{A}} \cap \mathrm{B})} \quad \forall \mathrm{A} \subseteq D. \tag{34}$$

– Shafer's conditioning [Shafer, 1976] (defined for evidence measures, although it may also be used in general):

$$l(\mathrm{A}|\mathrm{B}) = \frac{u(\mathrm{B}) - u(\bar{\mathrm{A}} \cap \mathrm{B})}{u(\mathrm{B})}, \quad u(\mathrm{A}|\mathrm{B}) = \frac{u(\mathrm{A} \cap \mathrm{B})}{u(\mathrm{B})} \quad \forall \mathrm{A} \subseteq D. \tag{35}$$

In our case, we have a set of bidimensional probability intervals L, and we want to calculate the conditional probability intervals given that we know the value of one variable, say $[y = \mathrm{y_j}]$, $L(x|\mathrm{y_j})$. Taking into account the expressions for the lower and upper measures associated to probability intervals given in eq. (32), and the previous expressions (34) and (35), the conditional set of probability intervals $L(x|\mathrm{y_j}) = \{[l(\mathrm{x_i}|\mathrm{y_j}), u(\mathrm{x_i}|\mathrm{y_j}]\}$ can be written as

$$l(\mathrm{x_i}|\mathrm{y_j}) = \frac{l(\mathrm{x_i}, \mathrm{y_j})}{l(\mathrm{x_i}, \mathrm{y_j}) + (\sum_{k \neq i} u(\mathrm{x_k}, \mathrm{y_j}) \wedge (1 - \sum_k \sum_{h \neq j} l(\mathrm{x_k}, \mathrm{y_h}) - l(\mathrm{x_i}, \mathrm{y_j})))}$$

$$u(\mathrm{x_i}|\mathrm{y_j}) = \frac{u(\mathrm{x_i}, \mathrm{y_j})}{u(\mathrm{x_i}, \mathrm{y_j}) + (\sum_{k \neq i} l(\mathrm{x_k}, \mathrm{y_j}) \vee (1 - \sum_k \sum_{h \neq j} u(\mathrm{x_k}, \mathrm{y_h}) - u(\mathrm{x_i}, \mathrm{y_j})))} \tag{36}$$

using Campos/Moral's conditioning, and

$$l(\mathrm{x_i}|\mathrm{y_j}) = 1 - \frac{\sum_{k\neq i} u(\mathrm{x_k},\mathrm{y_j}) \wedge (1 - \sum_k \sum_{h\neq j} l(\mathrm{x_k},\mathrm{y_h}) - l(\mathrm{x_i},\mathrm{y_j}))}{\sum_i u(\mathrm{x_i},\mathrm{y_j}) \wedge (1 - \sum_{h\neq j} \sum_i l(\mathrm{x_i},\mathrm{y_h})}$$

$$(37)\quad u(\mathrm{x_i}|\mathrm{y_j}) = \frac{u(\mathrm{x_i},\mathrm{y_j})}{\sum_i u(\mathrm{x_i},\mathrm{y_j}) \wedge (1 - \sum_{h\neq j} \sum_i l(\mathrm{x_i},\mathrm{y_h})}$$

using Shafer's conditioning, respectively.

Note that the calculus of the conditional probability intervals is very simple in both cases. Moreover, these intervals are always reachable, and therefore it is not necessary to transform them in reachable intervals by using eq. (31).

Finally, we also want to define a concept of product of probability intervals. Although other definitions could be used [de Campos and Huete, 1993], we shall adopt the following: Given the sets of probability intervals $L_x = \{[l(\mathrm{x_i}), u(\mathrm{x_i})],\ i = 1,\ldots,n\}$ and $L_y = \{[l(\mathrm{y_j}), u(\mathrm{y_j})],\ j = 1,\ldots,m\}$ defined for the two variables x and y respectively, the product of L_x and L_y is $L_x \times L_y = \{[l(\mathrm{x_i},\mathrm{y_j}), u(\mathrm{x_i},\mathrm{y_j})], i = 1,\ldots,n, j = 1,\ldots,m\}$, defined as follows:

$$(38)\quad l(\mathrm{x_i},\mathrm{y_j}) = l(\mathrm{x_i})l(\mathrm{y_j}),\ u(\mathrm{x_i},\mathrm{y_j}) = u(\mathrm{x_i})u(\mathrm{y_j}),\ \forall i,j.$$

It can be seen that $L_x \times L_y$ is a set of reachable bidimensional probability intervals.

Definitions of Independence for Probability Intervals

In probability theory, independence between two variables holds when the bidimensional probability distribution coincides with the product obtained by multiplication of the marginal distributions. This condition is equivalent to the equality between the marginal and each conditional probability distribution. So, one way to define an independence relationship for probability intervals is to compare the original set of bidimensional probability intervals L and the product of its marginal sets of probability intervals, $L_x \times L_y$. This can be done in several ways. As the claim of equality between L and $L_x \times L_y$ seems us unrealistic, we can impose a weaker condition:

Given a set L of bidimensional probability intervals defined on two variables x and y, whose associated convex set of probabilities is $\mathcal{P}_L$, let L_x and L_y be its marginal sets of probability intervals, let $L_x \times L_y$ be their product, and let $\mathcal{P}_p$ be the convex set of probabilities associated to the product. Then we say that x and y are p-weakly independent (with respect to L) if $\mathcal{P}_L$ and $\mathcal{P}_p$ have non-empty intersection, $\mathcal{P}_L \cap \mathcal{P}_p \neq \emptyset$ (i.e., L and $L_x \times L_y$ are compatible, in the terminology used in [de Campos *et al.*, 1994]).

This is a very weak definition of independence: it rather points out the possibility of independence. As soon as there exists one probability in the convex set $\mathcal{P}_L$ that coincides with the product of its marginals, the previous definition asserts the independence of x and y. Anyway, taking into account the characterization of compatibility between sets of probability intervals given in [de Campos *et al.*, 1994], we can also characterize this definition of independence in the following terms:

x and y are p-weakly independent (with respect to L) if and only if

1. $l(\mathtt{x_i}, \mathtt{y_j}) \leq u(\mathtt{x_i})u(\mathtt{y_j}) \ \forall i, j.$
2. $l(\mathtt{x_i})l(\mathtt{y_j}) \leq u(\mathtt{x_i}, \mathtt{y_j}) \ \forall i, j.$
3. $\sum_{i,j}(l(\mathtt{x_i}, \mathtt{y_j}) \vee l(\mathtt{x_i})l(\mathtt{y_j})) \leq 1 \leq \sum_{i,j}(u(\mathtt{x_i}, \mathtt{y_j}) \wedge u(\mathtt{x_i})u(\mathtt{y_j})).$

A stronger definition of independence is the following:

The variables x and y are said p-strongly independent (with respect to L) if $\mathcal{P}_L$ is included in $\mathcal{P}_p$, $\mathcal{P}_L \subseteq \mathcal{P}_p$ (i.e., in the terminology used in [de Campos *et al.*, 1994], L is included in $L_x \times L_y$). Obviously, if x and y are p-strongly independent, then they are also p-weakly independent.

A characterization of p-strong independence (based on a characterization of the inclusion of probability intervals, [de Campos *et al.*, 1994]) is the following:

x and y are p-strongly independent (with respect to L) if and only if

(39) $l(\mathtt{x_i})l(\mathtt{y_j}) \leq l(\mathtt{x_i}, \mathtt{y_j}) \leq u(\mathtt{x_i}, \mathtt{y_j}) \leq u(\mathtt{x_i})u(\mathtt{y_j}) \ \forall i, j.$

The motivation for the definition of p-strong independence is the following: let us suppose that we start from two sets of probability intervals, L_x^1 for variable x and L_y^2 for variable y, and construct their product $L^p = L_x^1 \times L_y^2$ (we are assuming the independence of x and y). Now, we calculate the marginals of L^p, L_x^p and L_y^p, and their product, $L_x^p \times L_y^p$. Then it can be shown that L^p is included in $L_x^p \times L_y^p$, i.e., x and y are p-strongly independent with respect to L^p.

A different approach would be obtained if we compare the conditional and marginal probability intervals instead of the bidimensional and the product. This can also be done in a number of ways. We may impose the equality of the marginal and all the conditionals. This is a very strong condition, difficult to satisfy.

Another option is to use the same idea we already employed for possibility distributions: to assert the independence of x from y when we do not gain additional information about the values of x after conditioning to y, altough we could lose some information. To precise the meaning of 'not to get additional information', we use the concept of inclusion of probability intervals [de Campos *et al.*, 1994], already mentioned. Then, we say that the variable x is c-independent from variable y (with respect to L) if the marginal set of probability intervals L_x is included in all the conditional sets of probability intervals $L(x|\mathtt{y_j}), \mathtt{y_j} \in D_y$, i.e.,

(40) $l(\mathtt{x_i}|\mathtt{y_j}) \leq l(\mathtt{x_i}) \leq u(\mathtt{x_i}) \leq u(\mathtt{x_i}|\mathtt{y_j}) \ \forall i, j.$

Observe that we obtain different definitions, depending on the kind of conditioning we use. It can be proven (due to the fact that probability intervals always are

Choquet capacities of order two [de Campos *et al.*, 1994]) that if x is c-independent from y using Shafer's conditioning, then x is also c-independent from y using Campos/Moral's conditioning.

It can be also proven that if x is c-independent from y using either Campos/moral's or Shafer's conditioning, then x and y are p-weakly independent. However, there is no clear connection between c-independence using any of the two conditionings and p-strong independence.

With the previous definition, an independence relationship is established if we maintain or lose (but never gain) precision in our information after conditioning. A different option is to replace the requirement of equality between the marginal and the conditional probability intervals by the weaker condition of compatibility (as we did with the product and the bidimensional probability intervals): assert independence if the marginal and all the conditional probability intervals are compatible. In that case, let $\mathcal{P}_{x|\mathrm{y_j}}$ be the convex set of probabilities associated to the set of conditional probability intervals $L(x|\mathrm{y_j})$, and let $\mathcal{P}_x$ be the convex set associated to the marginal set of probability intervals L_x. Then, we say that x is c-weakly independent from y if $\mathcal{P}_{x|\mathrm{y_j}}$ and $\mathcal{P}_x$ have non-empty intersection, for all $\mathrm{y_j} \in D_y$. This definition of independence, once again associated with the concept of compatibility of convex sets of probabilities, may be characterized as follows:

x is c-weakly independent from y (with respect to L) if and only if

1. $\max_j l(\mathrm{x_i}|\mathrm{y_j}) \leq u(\mathrm{x_i}) \;\; \forall i$.
2. $l(\mathrm{x_i}) \leq \min_j u(\mathrm{x_i}|\mathrm{y_j}) \;\; \forall i$.
3. $\sum_i (l(\mathrm{x_i}|\mathrm{y_j}) \vee l(\mathrm{x_i})) \leq 1 \leq \sum_i (u(\mathrm{x_i}|\mathrm{y_j}) \wedge u(\mathrm{x_i})) \;\; \forall j$.

Semantically, we assert the independence whenever the information obtained after conditioning is compatible with the information that we had before conditioning. In this case we can lose or gain precision in our information, but in any case there must exist some information in common before and after conditioning.

A slighty different alternative is to call for compatibility only among the conditional probability intervals. That is, x is said cc-weakly independent from y if every pair of conditional sets of probability intervals have non-empty intersection, i.e., $\mathcal{P}_{x|\mathrm{y_j}} \cap \mathcal{P}_{x|\mathrm{y_k}} \neq \emptyset, \; \forall \mathrm{y_j}, \mathrm{y_k} \in D_y$. This condition becomes equivalent to $\bigcap_j \mathcal{P}_{x|\mathrm{y_j}} \neq \emptyset$. Once again, this definition can be reformulated in the following way:

x is cc-weakly independent from y (with respect to L) if and only if

1. $\max_j l(\mathrm{x_i}|\mathrm{y_j}) \leq \min_j u(\mathrm{x_i}|\mathrm{y_j}) \;\; \forall i$.
2. $\sum_i \max_j l(\mathrm{x_i}|\mathrm{y_j}) \leq 1 \leq \sum_i \min_j u(\mathrm{x_i}|\mathrm{y_j})$.

As it happened with c-independence, if x is c-weakly (cc-weakly, respectively) independent from y using Shafer's conditioning, then x is also c-weakly (cc-weakly, respectively) independent from y using Campos/Moral's conditioning. Moreover, it is clear that c-independence of x from y implies both c-weak and cc-weak independence of x from y (for the same conditioning). However, the relationship between c-weak and cc-weak independence is not clear so far: we know that c-weak independence does not imply cc-weak independence, but we do not know whether cc-weak independence implies c-weak independence, although we conjecture it.

Finally, a different approach can be obtained in we relax the requirement of equality between the marginal and the conditional probability intervals (i.e., 'not to modify the information after conditioning') by a less restrictive one, i.e., to 'obtain similar information after conditioning'. So, we can use a similarity relationship, $\simeq$, between sets of probability intervals. In that case, we assert the independence if

(41) $L_x \times L_y \simeq L$

if we base our definition on the product, or

(42) $L(x|\mathrm{y_j}) \simeq L_x, \quad \forall \mathrm{y_j} \in D_y$

if we use conditioning instead.

For example, we can use the following similarity relationships. The first one, based on the same idea that we already used for possibility distributions, consists on discretizing the $[0,1]$ interval, and to establish the similarity between two sets of probability intervals whenever their respective discrete versions coincide. Formally, let $I_1, \ldots, I_m$ be m subintervals in $[0,1]$, such that $0 = \alpha_0 < \alpha_1 < \ldots < \alpha_m - 1 < \alpha_m = 1$, and $I_k = [\alpha_{k-1}, \alpha_k)$ for each $k = 1, \ldots, m-1$, and $I_m = [\alpha_m - 1, \alpha_m]$. Then, a similarity relationship between two sets of probability intervals, L and L', can be defined as follows:

$$L \simeq L' \Leftrightarrow \quad \forall \mathrm{x} \in D\ \exists k, k' \in \{1, \ldots, m\} \text{ such that}$$

(43) $$l(\mathrm{x}), l'(\mathrm{x}) \in I_k \text{ and } u(\mathrm{x}), u'(\mathrm{x}) \in I_{k'}.$$

A second way of defining a similarity relationship is based on a distance criterion between the values that probability intervals take on. In that case, we need a threshold ϵ to make the test. So, $\simeq$ can be defined, for example, as:

$$L \simeq L' \Leftrightarrow \forall \mathrm{x} \in D, \quad |l(x) - l'(x)| \leq \epsilon \text{ and } |u(x) - u'(x)| \leq \epsilon,$$

where $|. - .|$ represents the absolute difference.

All the previous definitions of (marginal) independence between two variables can easily be generalized to deal with conditional independence of two variables given a third variable (or a set of variables). It is also very easy to see that for probability distributions (a particular case of probability intervals, where the lower and upper bounds coincide), all the definitions (except those based on similarity) reproduce the usual concept of probabilistic independence.

5 CONCLUDING REMARKS

In the first part of this work we have given an overview of the existing techniques for learning belief networks based on the identification of conditional independence statements, including also new proposals for the specific cases of polytrees and simple graphs. Anyway, we think that the topic of learning belief networks still is not exhausted, there is room for new algorithms, that we plan to study in the future.

On the other hand, the concept of independence, as in the case of probability theory, is also important for other formalisms of uncertainty management. In the second part of this work, several alternative definitions of independence for possibility theory and probability intervals have been proposed. In the possibilistic case, most of them are based on the concept of conditioning, although using different formulations of this concept. We have also studied their properties with respect to a well-known set of axioms, which try to capture the intuitive idea of independence. Further research should include, for example, the study of the consequences of a non-symmetric definition of independence with respect to its graphical representation, and the use of other points of view for defining independence, which are not necessarily based on the conditioning. For probability intervals, we should study the properties that the considered definitions of independence verify. The study of ways to extract both possibility distributions and probability intervals from raw data, and merge this information with subjective judgments from experts is also interesting from a practical point of view [de Campos and Huete, 2000]. Finally, methods for learning possibility-based [Gebhardt and Kruse, 1996] and probability interval-based belief networks that do not rely on the identification of independences but on heuristic measures, are also worthy of pursuit.

ACKNOWLEDGEMENTS

This work has been supported by the European Union under Project Esprit III b.r.a. 6156 (DRUMS II).

Departamento de Ciencias de la Computación e I.A., Universidad de Granada, 18071-Granada, Spain.

REFERENCES

[Acid and de Campos, 1994] S. Acid and L. M. de Campos. Approximations of causal networks by polytrees: An empirical study. In *Advances in Intelligent Computing*, B. Bouchon-Meunier, R. R. Yager and L. A. Zadeh, eds. pp. 149–158. Lecture Notes in Computer Science 945, Springer Verlag, Berlin, 1995.

[Acid and de Campos, 1996] S. Acid and L. M. de Campos. An algorithm for finding minimum d-separating sets in belief networks. In *Proceedings of the Twelfth Conference on Uncertainty in Artificial Intelligence*, E. Horvitz and F. Jensen, eds. pp. 3–10. Morgan Kaufmann, San Mateo, 1996).

[Acid and de Campos, 1996a] S. Acid and L. M. de Campos. BENEDICT: An algorithm for learning probabilistic belief networks. In *Proceedings of the Sixth International Conference of Information Processing and Management of Uncertainty in Knowledge-Based Systems (IPMU)*, pp. 979–984, 1996.

[Acid and de Campos, 1997] S. Acid and L. M. de Campos. Algoritmos híbridos para el aprendizaje de redes de creencia. In *Proceedings of the VII Conferencia de la Asociación Española para la Inteligencia Artificial*, pp. 499–508, 1997. In Spanish.

[Acid *et al.*, 1991] S. Acid, L. M. de Campos, A. González, R. Molina and N. Pérez de la Blanca. CASTLE: A tool for bayesian learning. In *Proceeding of the ESPRIT'91 Conference*, pp. 363–377. Commission of the European Communities, 1991.

[Acid *et al.*, 1991a] S. Acid, L. M. de Campos, A. González, R. Molina and N. Pérez de la Blanca. Learning with CASTLE. In *Symbolic and Quantitative Approaches to Uncertainty*, R. Kruse and P. Siegel, eds. pp. 99–106. Lecture Notes in Computer Science 548, Springer Verlag, Berlin, 1991.

[Andersen *et al.*, 1991] L. R. Andersen, J. H. Krebs and J. Damgaard. STENO: an expert system for medical diagnosis based on graphical models and model search, *Journal of Applied Statistics*, **18**, 139–153, 1991.

[Becker and Geiger, 1994] A. Becker and D. Geiger. Approximation algorithms for the loop cutset problem. In *Uncertainty in Artificial Intelligence, Proceedings of the Tenth Conference*, D. Poole and R. López de Mántaras, eds. pp. 60–68. Morgan Kaufmann, San Francisco, 1994.

[Benferhat *et al.*, 1994] S. Benferhat, D. Dubois and H. Prade. Expressing independence in a possibilistic framework and its application to default reasoning. In *Proceedings of the Eleventh ECAI Conference*, pp. 150–154, 1994.

[Bouckaert, 1993] R. R. Bouckaert. Belief network construction using the minimum description length principle. In *Symbolic and Quantitative Approaches to Reasoning and Uncertainty*, M. Clarke, R. Kruse and S. Moral, eds. pp. 41–48. Lecture Notes in Computer Science 747, Springer Verlag, Berlin, 1993.

[Bouckaert, 1994] R. R. Bouckaert. Optimizing causal orderings for generating DAGs from data. In *Uncertainty in Artificial Intelligence: Proceedings of the Eighth Conference*, D. Dubois, M. P. Wellman, B.D. D'Ambrosio and P. Smets, eds. pp. 102–109. Morgan Kaufmann, San Mateo, 1994.

[Buntine, 1994] W. Buntine. Operations for learning with graphical models, *Journal of Artificial Intelligence Research*, **2**, 159–225, 1994.

[de Campos, 1995] L. M. de Campos. Independence relationships in possibility theory and their application to learning belief networks. In *Mathematical and Statistical Methods in Artificial Intelligence*, G. Della Riccia, R. Kruse and R. Viertl, eds. pp. 119–130. CISM Courses and Lectures 363, Springer Verlag, Wien, 1995.

[de Campos, 1998] L. M. de Campos. Independency relationships and learning algorithms for singly connected networks, *Journal of Experimental and Theoretical Artificial Intelligence*, **10**, 511–549, 1998.

[de Campos *et al.*, 1995] L. M. de Campos, J. Gebhardt and R. Kruse. Axiomatic treatment of possibilistic independence. In *Symbolic and Quantitative Approaches to Reasoning and Uncertainty*, C. Froidevaux and J. Kohlas, eds. pp. 77–88. Lecture Notes in Artificial Intelligence 946, Springer Verlag, Berlin, 1995.

[de Campos and Huete, 1993] L. M. de Campos, J. F. Huete. Independence concepts in upper and lower probabilities. In *Uncertainty in Intelligent Systems*, B. Bouchon-Meunier, L. Valverde and R. R. Yager, eds. pp. 49–59. North-Holland, Amsterdam, 1993.

[de Campos and Huete, 1993] L. M. de Campos and J. F. Huete. Learning non probabilistic belief networks. In *Symbolic and Quantitative Approaches to Reasoning and Uncertainty*, M. Clarke, R. Kruse and S. Moral, eds. pp. 57–64. Lecture Notes in Computer Science 747, Springer Verlag, Berlin, 1993.

[de Campos and Huete, 1997] L. M. de Campos and J. F. Huete. On the use of independence relationships for learning simplified belief networks, International Journal of Intelligent Systems 12, 495–522, 1997.

[de Campos and Huete, 1997a] L. M. de Campos and J. F. Huete. Algorithms for learning decomposable models and chordal graphs, in Proceedings of the Thirteenth Conference on Uncertainty in Artificial Intelligence, D. Geiger, P.P. Shenoy (Eds.) (Morgan and Kaufmann, San Mateo) 46–53, 1997.

[de Campos and Huete, 1998] L. M. de Campos and J. F. Huete. Aproximación de una ordenación de variables en redes causales mediante algoritmos genéticos, Inteligencia Artificial 4, 30–39. 1998, (in Spanish).

[de Campos and Huete, 1999] L. M. de Campos and J. F. Huete. Independence concepts in possibility theory: Part I, *Fuzzy Sets and Systems*, **103**, 127–152, 1999.

[de Campos and Huete, 1999a] L. M. de Campos and J. F. Huete. Independence concepts in possibility theory: Part II, *Fuzzy Sets and Systems*, **103**, 487–505, 1999.

[de Campos and Huete, 1999b] L. M. de Campos and J. F. Huete. A new approach for learning belief networks using independence criteria, *International Journal of Approximate Reasoning*, to appear.

[de Campos and Huete, 2000] L. M. de Campos and J. F. Huete. Deriving possibility distributions from data sets using confidence intervals of probabilities. In *Information, Uncertainty and Fusion*, B. Bouchon-Meunier, R. R. Yager and L. A. Zadeh, eds. Kluwer Scientific Publishers, to appear.

[de Campos *et al.*, 1994] L. M. de Campos and J. F. Huete, S. Moral, Probability intervals: A tool for uncertain reasoning, International Journal of Uncertainty, Fuzziness and Knowledge-Based Systems 2, 167–196, 1994.

[de Campos *et al.*, 1995] L. M. de Campos, J. F. Huete and S. Moral. Uncertainty management using probability intervals. In *Advances in Intelligent Computing*, B. Bouchon-Meunier, R. R. Yager and L. A. Zadeh, eds. pp. 190–199. Lecture Notes in Computer Science 945, Springer Verlag, Berlin, 1995.

[de Campos *et al.*, 1995a] L. M. de Campos, J. F. Huete and S. Moral. Possibilistic independence. In *Proceedings of the Third European Congress on Intelligent Techniques and Soft Computing (EUFIT)*, pp. 69–74, 1995.

[de Campos *et al.*, 1988] L .M. de Campos, M. T. Lamata and S. Moral. Logical connectives for combining fuzzy measures. In *Methodologies for Intelligent Systems 3*, Z. W. Ras and L. Saitta, eds. pp. 11–18. North-Holland, New York, 1988.

[de Campos *et al.*, 1990] L. M. de Campos, M. T. Lamata and S. Moral. The concept of conditional fuzzy measure, *International Journal of Intelligent Systems*, **5**, 237–246, 1990.

[de Campos and Moral, 1995] L. M. de Campos and S. Moral. Independence concepts for convex sets of probabilities. In *Proceedings of the Eleventh Conference on Uncertainty in Artificial Intelligence*, P. Besnard and S. Hanks, eds. pp. 108–115. Morgan Kaufmann, San Mateo, 1995.

[Cano *et al.*, 1993] J. E. Cano, S. Moral and J. F. Verdegay. Propagation of convex sets of probabilities in directed acyclic networks. In *Uncertainty in Intelligent Systems*, B. Bouchon-Meunier, L. Valverde and R. R. Yager, eds. pp. 15–26. North-Holland, Amsterdam, 1993.

[Cheng *et al.*, 1997] J. Cheng, D. A. Bell and W. Liu. An algorithm for Bayesian belief network construction from data. In *Proceedings of the Seventh International Workshop on Artificial Intelligence and Statistics*, pp. 83–90. 1997.

[Chickering, 1996] D. M. Chickering. Learning equivalence classes of Bayesian network structures. In *Proceedings of the Twelfth Conference on Uncertainty in Artificial Intelligence*, E. Horvitz and F. Jensen, eds. pp. 150–157. Morgan Kaufmann, San Mateo, 1996.

[Chickering, 1996a] D. M. Chickering. Learning Bayesian networks is NP-Complete. In *Learning from Data*, D. Fisher and H. J. Lenz, eds. pp. 121–130. Lecture Notes in Statistics 112, Springer, New York, 1996.

[Choquet, 1953] G. Choquet. Theory of capacities, *Ann. Inst. Fourier*, **5**, 131–292, 1953.

[Chow and Liu, 1968] C. K. Chow and C. N. Liu. Approximating discrete probability distribution with dependence trees, *IEEE Transactions on Information Theory*, **14**, 462-467, 1968.

[de Cooman and Kerre, 1994] G. de Cooman and E. E. Kerre. A new approach to possibilistic independence. In *Proceedings of the Third IEEE International Conference on Fuzzy Systems*, pp. 1446–1451, 1994.

[Cooper and Herskovits, 1992] G. F. Cooper and E. Herskovits. A Bayesian method for the induction of probabilistic networks from data, *Machine Learning*, **9**, 309–347, 1992.

[Cooper, 1995] G. F. Cooper. A method for learning belief networks that contain hidden variables, *Journal of Intelligent Information Systems*, **4**, 71–88, 1995.

[Chrisman, 1996] L. Chrisman. Independence with lower and upper probabilities. In *Uncertainty in Artificial Intelligence: Proceedings of the Twelfth Conference*, E. Horvitz and F. Jensen, eds. pp. 169–177. Morgan Kaufmann, San Mateo, 1996.

[Dawid, 1979] A. P. Dawid. Conditional independence in statistical theory, *J.R. Statist. Soc. Ser. B*, **41**, 1–31, 1979.

[Dempster, 1967] A. P. Dempster. Upper and lower probabilities induced by a multivalued mapping, *Ann. Math. Stat.*, **38**, 325–339, 1967.

[Dubois and Prade, 1986] D. Dubois and H. Prade. A set-theoretic view of belief functions, *International Journal of General Systems*, **12**, 193–226, 1986.

[Dubois and Prade, 1988] D. Dubois and H. Prade. *Possibility Theory: An approach to computerized processing of uncertainty*, Plenum Press, New York, 1988.

[Dubois and Prade, 1993] D. Dubois and H. Prade. Belief revision and updates in numerical formalisms—An overview, with new results for the possibilistic framework. In *Proceedings of the 13th IJCAI Conference*, pp. 620–625. Morgan Kaufmann, 1993.

[Dubois *et al.*, 1994] D. Dubois, L. Fariñas del Cerro, A. Herzig and H. Prade. An ordinal view of independence with applications to plausible reasoning. In *Uncertainty in Artificial Intelligence: Proceedings of the Tenth Conference*, R. López de Mántaras and D. Poole, eds. pp. 195–203. Morgan Kaufmann, San Francisco, 1994.

[Fagin, 1977] R. Fagin. Multivalued dependencies and a new form for relational databases, *ACM Transactions on Database Systems*, **2**, 262–278, 1977.

[Fariñas del Cerro and Herzig, 1994] L. Fariñas del Cerro and A. Herzig. Possibility theory and independence. In *Proceedings of the Fifth IPMU Conference*, pp. 820–825, 1994.

[Fertig and Breese, 1990] K. W. Fertig and J. S. Breese. Interval influence diagrams. In *Uncertainty in Artificial Intelligence 5*, M. Henrion, R. D. Shachter, L. N. Kanal and J. F. Lemmer, eds. pp. 149–161. North-Holland, Amsterdam, 1990.

[Fonck, 1994] P. Fonck. Conditional independence in possibility theory. In *Uncertainty in Artificial Intelligence: Proceedings of the Tenth Conference*, R. López de Mántaras and D. Poole, eds. pp. 221–226. Morgan Kaufmann, San Francisco, 1994.

[Fonck, 1997] P. Fonck. A comparative study of possibilistic conditional independence and lack of interaction, *International Journal of Approximate Reasoning*, **16**, 149–171, 1997.

[Friedman and Goldszmidt, 1996] N. Friedman and M. Goldszmidt. Learning Bayesian networks with local structure. In *Proceedings of the Twelfth Conference on Uncertainty in Artificial Intelligence*, E. Horvitz and F. Jensen, eds. pp. 252–262. Morgan Kaufmann, San Mateo, 1996.

[Friedman and Goldszmidt, 1996a] N. Friedman and M. Goldszmidt. Building classifiers using Bayesian networks. In *Proceedings of the National Conference on Artificial Intelligence*, pp. 1277–1284. AAAI Press, Menlo Park, 1996.

[Friedman *et al.*, 1997] N. Friedman, D. Geiger and M. Goldszmidt. Bayesian network classifiers, *Machine Learning*, **29**, 131–163, 1997.

[Gebhardt and Kruse, 1996] J. Gebhardt and R. Kruse. Learning possibilistic networks from data. In *Learning from Data*, D. Fisher and H. J. Lenz, eds. pp. 143–153. Lecture Notes in Statistics 112, Springer, New York, 1996.

[Geiger, 1992] D. Geiger. An entropy-based learning algorithm of bayesian conditional trees. In *Uncertainty in Artificial Intelligence: Proceedings of the Eighth Conference*, D. Dubois, M. P. Wellman, B. D. D'Ambrosio and P. Smets, eds. pp 92–97. Morgan Kaufmann, San Mateo, 1992.

[Geiger and Heckerman, 1995] D. Geiger and D. Heckerman. A characterisation of the Dirichlet distribution with application to learning Bayesian networks. In *Uncertainty in Artificial Intelligence: Proceedings of the Eleventh Conference*, P. Besnard and S. Hanks, eds. pp. 196–207. Morgan Kaufmann, San Francisco, 1995.

[Geiger *et al.*, 1990] D. Geiger, A. Paz and J. Pearl. Learning causal trees from dependence information. In *Proceedings of the Eighth National Conference on Artificial Intelligence*, pp. 770–776, 1990.

[Geiger *et al.*, 1991] D. Geiger, A. Paz and J. Pearl. Axioms and algorithms for inferences involving probabilistic independence, *Information and Computation*, **91**, 128–141, 1991.

[Geiger *et al.*, 1993] D. Geiger, A. Paz and J. Pearl. Learning simple causal structures, *International Journal of Intelligent Systems*, **8**, 231–247, 1993.

[Heckerman, 1990] D. Heckerman. A tractable inference algorithm for diagnosing multiple diseases. In *Uncertainty in Artificial Intelligence 5*, R. D . Shachter, T. S. Levitt, L. N. Kanal and J. F. Lemmer, eds.. pp. 163–171. North-Holland, Amsterdam, 1990.

[Heckerman, 1996] D. Heckerman. Bayesian networks for knowledge discovery. In *in Advances in Knowledge Discovery and Data Mining*, U. M. Fayyad, G. Piatetsky-Shapiro, P. Smyth and R. Uthurusamy, pp. 273–305. MIT Press, Cambridge, 1996.

[Heckerman *et al.*, 1995] D. Heckerman, D. Geiger and D. M. Chickering. Learning bayesian networks: The combination of knowledge and statistical data, *Machine Learning*, **20**, 197–243, 1995.

[Herskovits *et al.*, 1990] E. Herskovits, G. Cooper and Kutato. An entropy-driven system for construction of probabilistic expert systems from databases. In *Proceedings of the Sixth Conference on Uncertainty in Artificial Intelligence*, P. Bonissone, ed. pp. 54–62. Cambridge, 1990.

[Hisdal, 1978] E. Hisdal. Conditional possibilities, independence and noninteraction, *Fuzzy Sets and Systems*, **1**, 283–297, 1978.

[Huber, 1981] P. J. Huber. *Robust Statistics*, Wiley, New York, 1981.

[Huete, 1995] J. F. Huete. *Aprendizaje de Redes de Creencia mediante la Detección de Independencias: Modelos no Probabilísticos*, Ph.D. Thesis, Universidad de Granada, 1995. (In Spanish.)

[Huete and de Campos, 1993] J. F. Huete and L.M. de Campos. Learning causal polytrees. In *Symbolic and Quantitative Approaches to Reasoning and Uncertainty*, M. Clarke, R. Kruse and S. Moral, eds pp. 180–185. Lecture Notes in Computer Science 747, Springer Verlag, Berlin, 1993.

[Kullback, 1968] S. Kullback. *Information Theory and Statistics*, Dover Publications, 1968.

[Kyburg, 1987] H. E. Kyburg. Bayesian and non-Bayesian evidential updating, *Artificial Intelligence*, **31**, 271–293, 1987.

[Lam and Bacchus, 1994] W. Lam and F. Bacchus. Learning Bayesian networks, an approach based on the MDL principle, *Computational Intelligence*, **10**, 269–293, 1994.

[Larrañaga *et al.*, 1996] P. Larrañaga, C. M. Kuijpers, R. H. Murga and Y. Yurramendi. Learning bayesian networks structure by searching for the best ordering with genetic algorithms, *IEEE Transactions on Systems Man and Cybernetics- Part A: Systems and Humans*, **26**, 487–493, 1996.

[Larrañaga *et al.*, 1996a] P. Larrañaga, M. Poza, Y. Yurramendi, R. H. Murga and C. M. H. Kuijpers. Structure learning of Bayesian networks by genetic algorithms: A performance analysis and control parameters, *IEEE Transactions on Pattern Analysis and Machine Intelligence*, **18**, 912–926, 1996.

[Larrañaga *et al.*, 1996b] P. Larrañaga, R. H. Murga, M. Poza and C. M. H. Kuijpers. Structure learning of Bayesian networks by hybrid genetic algorithms. In *Learning from Data: AI and Statistics*, V. D. Fisher and H. J. Lenz, eds. pp. 165–174. Springer, New York, 1996.

[Lauritzen, 1982] S. L. Lauritzen. *Lectures on contingency tables 2nd ed.* University of Aalborg Press, Aalborg, Denmark, 1982.

[Lauritzen *et al.*, 1990] S. L. Lauritzen, A. P. Dawid, B. N. Larsen and H.-G. Leimer. Independence properties of directed Markov fields, *Networks*, **20**, 491–505, 1990.

[Lauritzen and Spiegelhalter, 1988] S. L. Lauritzen and D. J. Spiegelhalter. Local computations with probabilities on graphical structures and their application to expert systems, *J.R. Statist. Soc. Ser. B*, **50**, 157–224, 1988.

[Madigan and Raftery, 1994] D. Madigan and A. Raftery. Model selection and accounting for model uncertainty in graphical models using Occam's window, *Journal of the American Statistics Association*, **89**, 1535–1546, 1994.

[Molina *et al.*, 1993] R. Molina, L. M. de Campos and J. Mateos. Using Bayesian Algorithms for Learning Causal Networks in Classification Problems. In *Uncertainty in Intelligent Systems*, B. Bouchon-Meunier, L. Valverde, R. R. Yager, eds. pp. 49–59. North-Holland, Amsterdam, 1993.

[Moral and de Campos, 1991] S. Moral and L. M. de Campos. Updating uncertain information. In *Uncertainty in Knowledge Bases*, B. Bouchon-Meunier, R. R. Yager and L. A. Zadeh, eds. pp. 58–67. Lecture Notes in Computer Science 521, Springer Verlag, Berlin, 1991.

[Pearl, 1986] J. Pearl. Fusion, propagation and structuring in belief networks, *Artificial Intelligence*, **29**, 241–288, 1986.

[Pearl, 1988] J. Pearl. *Probabilistic Reasoning in Intelligent Systems: Networks of Plausible Inference*, Morgan and Kaufmann, San Mateo, 1988.

[Pearl and Verma, 1991] J. Pearl and T. S. Verma. A theory of inferred causation. In *Principles of Knowledge Representation and Reasoning: Proceedings of the Second International Conference*, J. A. Allen, R. Fikes and E. Sandewall, eds. pp. 441–452. Morgan Kaufmann, San Mateo, 1991.

[Ramoni and Sebastiani, 1997] M. Ramoni, P. Sebastiani. Discovering Bayesian networks in incomplete databases, KMI-TR-46 Technical Report, The Open University, 1997.

[Rebane and Pearl, 1989] G. Rebane and J. Pearl. The recovery of causal polytrees from statistical data, *Uncertainty in Artificial Intelligence 3*, L. N. Kanal, T. S. Levitt and J. F. Lemmer, eds. pp. 175–182. North-Holland, Amsterdam, 1989.

[Sangüesa *et al.*, 1998] R. Sangüesa, J. Cabós and U. Cortés. Possibilistic conditional independence: A similarity-based measure and its application to causal network learning, *International Journal of Approximate Reasoning*, **18**, 145–167, 1998.

[Sarkar, 1993] S. Sarkar. Using tree-decomposable structures to approximate belief networks. In *Uncertainty in Artificial Intelligence, Proceedings of the Ninth Conference*, D. Heckerman and A. Mamdani, eds. pp. 376–382. Morgan Kaufmann, San Mateo, 1993.

[Schweizer and Sklar, 1983] B. Schweizer and A. Sklar. *Probabilistic Metric Spaces*, North-Holland, New York, 1983.

[Shafer, 1976] G. Shafer. *A Mathematical Theory of Evidence*, Princeton University Press, Princeton, 1976.

[Shenoy, 1992] P. P. Shenoy. Conditional independence in uncertainty theories. In *Uncertainty in Artificial Intelligence: Proceedings of the Eighth Conference*, D. Dubois, M. P. Wellman, B. D'Ambrosio and P. Smets, eds. pp. 284–291. Morgan Kaufmann, San Mateo, 1992.

[Singh and Valtorta, 1993] M. Singh and M. Valtorta. An algorithm for the construction of Bayesian network structures from data. In *Uncertainty in Artificial Intelligence: Proceedings of the Ninth Conference*, D. Heckerman and A. Mamdani, eds. pp. 259–265. Morgan Kaufmann, San Francisco, 1993.

[Singh and Valtorta, 1995] M. Singh and M. Valtorta. Construction of Bayesian network structures from data: A brief survey and an efficient algorithm, *International Journal of Approximate Reasoning*, **12**, 111–131, 1995.

[Spiegelhalter *et al.*, 1993] D. Spiegelhalter, A. Dawid, S. Lauritzen and R. Cowell. Bayesian analysis in expert systems, *Statistical Science*, **8**, 219–283, 1993.

[Spirtes *et al.*, 1991] P. Spirtes, C. Glymour and R. Scheines. An algorithm for fast recovery of sparse causal graphs, *Social Science Computer Review*, **9**, 62–72, 1991.

[Spirtes *et al.*, 1993] P. Spirtes, C. Glymour and R. Scheines. *Causation, Prediction and Search*, Lecture Notes in Statistics 81, Springer Verlag, New York, 1993.

[Spohn, 1980] W. Spohn. Stochastic independence, causal independence and shieldability, *Journal of Philosophical Logic*, **9**, 73–99, 1980.

[Srinivas *et al.*, 1990] S. Srinivas, S. Russell and A. Agogino. Automated construction of sparse bayesian networks from unstructured probabilistic models and domain information. In *Uncertainty in Artificial Intelligence 5*, M. Henrion, R. D. Shachter, L. N. Kanal and J. F. Lemmer, eds. pp. 295–308. North-Holland, Amsterdam, 1990.

[Studený, 1989] M. Studený. Attempts at axiomatic description of conditional independence, *Kybernetika*, **25**, Supplement to n. 3, 72–79, 1989.

[Studený, 1993] M. Studený. Formal properties of conditional independence in different calculi of AI. In *Symbolic and Quantitative Approaches to Reasoning and Uncertainty*, M. Clarke, R. Kruse and S. Moral, eds. pp. 341–348. Lecture Notes in Computer Science 747, Springer Verlag, Berlin, 1993.

[Sugeno, 1974] M. Sugeno. *Theory of fuzzy integrals and its applications*, Ph.D. Thesis, Tokyo Institute of Technology, Tokyo, 1974.

[Suzuki, 1993] J. Suzuki. A construction of bayesian networks from databases based on the MDL principle. In *Uncertainty in Artificial Intelligence: Proceedings of the Ninth Conference*, D. Heckerman and A. Mamdani, eds. pp. 266–273. Morgan Kaufmann, San Francisco, 1993.

[Verma and Pearl, 1990] T. Verma and J. Pearl. Equivalence and synthesis of causal models. In *Uncertainty in Artificial Intelligence: Proceedings of the Sixth Conference (Mass)*, pp. 220–227, 1990.

[Verma and Pearl, 1990a] T. Verma and J. Pearl. Causal networks: Semantics and expressiveness. In *Uncertainty in Artificial Intelligence 4*, R. D. Shachter, T. S. Levitt, L. N. Kanal and J. F. Lemmer, eds. pp. 69–76. North-Holland, Amsterdam, 1990.

[Verma and Pearl, 1992] T. Verma, J. Pearl. An algorithm for deciding if a set of observed independencies has a causal explanation. In *Uncertainty in Artificial Intelligence: Proceedings of the Eighth Conference*, D. Dubois, M. P. Wellman, B. D'Ambrosio and P. Smets, eds. pp. 323–330. Morgan Kaufmann, San Mateo, 1992.

[Walley, 1991] P. Walley. *Statistical Reasoning with Imprecise Probabilities*, Chapman and Hall, London, 1991.

[Wedelin, 1993] D. Wedelin. *Efficient algorithms for probabilistic inference, combinatorial optimization and the discovery of causal structure from data*, Ph.D. Thesis, Chalmers University of Thechnology, Goteborg, 1993.

[Wermuth and Lauritzen, 1983] N. Wermuth and S. L. Lauritzen. Graphical and recursive models for contingency tables, *Biometrika*, **70**, 537–552, 1983.

[Wilson, 1994] N. Wilson. Generating graphoids from generalized conditional probability. In *Uncertainty in Artificial Intelligence: Proceedings of the Tenth Conference*, D. Pooleand R. López de Mántaras, eds. pp. 583–590. Morgan Kaufmann, San Francisco, 1994.

[Wong and Xiang, 1994] S. K. M Wong and Y. Xiang. Construction of a Markov network from data for probabilistic inference. In *Proceedings of the Third International Workshop on Rough Sets and Soft Computing*, pp. 562–569, 1994.

[Xiang *et al.*, 1997] Y. Xiang, S. K. M. Wong and N. Cercone. A 'microscopic' study of minimum entropy search in learning decomposable Markov models, *Machine Learning*, **26**, 65–92, 1997.

[Zadeh, 1978] L. A. Zadeh. Fuzzy sets as a basis for a theory of possibility, *Fuzzy Sets and Systems*, **1**, 3–28, 1978.

INDEX